ORGANIC SYNTHESES

Collective Volume 2

A REVISED EDITION OF
ANNUAL VOLUMES X–XIX

Edited by A. H. BLATT, *Secretary to the Board*
Queens College, Flushing, N. Y.

JOHN WILEY & SONS, INC.

NEW YORK   LONDON

PREFACE

Collective Volume 2 contains in revised form the material which appeared in annual volumes X–XIX of *Organic Syntheses*. It is a companion to Collective Volume 1 and in its general plan follows the pattern of that volume. In the revision of the material for Collective Volume 2, errors found in the original printings have been corrected; calculations and references have been checked; modifications and improvements in procedures, which were noted in appendices in the annual volumes, have been incorporated in the text; and eleven new and improved checked procedures have been added. These new procedures are for the preparation of 2-carbethoxycyclopentanone, 1,2-dibromocyclohexane, ethyl adipate, ethyl methylmalonate, ethyl α-naphthoate, 4-nitrophthalimide, nitrosomethylurea (two procedures), pimelic acid (two procedures), and triphenylmethylsodium.

The section on Methods of Preparation under each preparation has been revised to include those methods of preparative value found in the literature covered by *Chemical Abstracts* through Volume 35, for 1941. There are a number of references to articles published in 1942, but the coverage of the 1942 literature is not complete. For the convenience of those who may have to make a complete survey of the literature on any preparation, the *Chemical Abstracts* indexing name for each preparation is given as a subtitle when that name differs at all from the title of the preparation.

Whenever a compound whose preparation is described in this volume can be purchased for about five dollars or less per kilogram, the directions for its preparation have been marked with an asterisk. In many laboratories the preparation of these compounds, other than for the experience gained, may be an extravagance.

The references to certain pieces of equipment, such as the modified Claisen flask and the gas absorption trap, whose use has become almost standard practice in *Organic Syntheses*, have become so frequent in recent annual volumes that their inclusion in Collective Volume 2 would unduly clutter up the text. Consequently these references have been omitted from this volume. For the convenience of those who are not familiar with *Organic Syntheses* practice, it may be noted that modified Claisen flasks are described on p. 18 of this volume and in Collective Volume 1, **1941**, on p. 130, and that gas absorption traps are described on p. 4 of this volume and in Collective Volume 1, **1941**, on p. 97. A comparable policy of omission has been adopted for special reagents. If a procedure calls for absolute alcohol, for example, no reference is made to the several methods described in the volume for preparing this reagent; the methods are listed in the general index. If, however, a procedure calls for absolute alcohol prepared in a particular way, a reference is given to the directions for its preparation.

Attention is called to the following facts: cotton may generally be substituted for such chemical drying agents as calcium chloride in drying tubes (Org. Syn. **20, 9**); the enameled-steel utensils known as bain-marie ware can often be used advantageously in place of the larger sizes of beakers; and a mercury-sealed stirrer can often be replaced by the simpler rubber-tube sealed stirrer shaft (Org. Syn. **21**, 40). These items of information come through the courtesy of F. P. Pingert, C. F. H. Allen, and L. P. Kyrides, respectively.

The editors are grateful for the many suggestions, corrections, and improvements which have been called to their attention. They welcome additional suggestions, corrections, and directions for inclusion in future volumes of *Organic Syntheses*. This material should be sent to the secretary, in duplicate.

It is a pleasure to acknowledge the assistance of James R. Kreuzer and Therese Herman in the preparation of this volume.

CONTENTS

vii

CONTENTS

α-ACETAMINOCINNAMIC ACID

(Cinnamic acid, α-acetamido-)

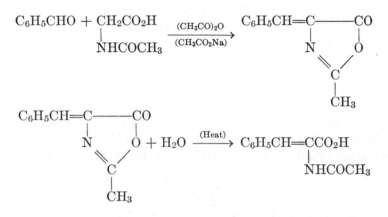

Submitted by R. M. Herbst and D. Shemin.
Checked by Reynold C. Fuson and E. A. Cleveland.

1. Procedure

A mixture of 58.5 g. (0.5 mole) of acetylglycine (p. 11) (Note 1), 30 g. (0.37 mole) of anhydrous sodium acetate, 79 g. (0.74 mole) of freshly distilled benzaldehyde, and 134 g. (1.25 moles) of 95 per cent acetic anhydride in a loosely corked 1-l. Erlenmeyer flask is warmed on the steam bath with occasional stirring until solution is complete (ten to twenty minutes). The resulting solution is boiled for one hour under reflux, cooled, and placed in a refrigerator overnight. The solid mass of yellow crystals is treated with 125 cc. of cold water and broken up with a stirring rod. The crystals are then transferred to a Büchner funnel and washed thoroughly with cold water (Note 2). After being dried in a vacuum desiccator over phosphorus pentoxide and potassium hydroxide, the crude azlactone weighs 69–72 g. (74–77 per cent of the theoretical amount). The product melts at 148–150°, and is sufficiently pure for preparative purposes (Note 3).

In a 1-l. round-bottomed, short-necked flask 47 g. (0.25 mole) of the crude azlactone of α-acetaminocinnamic acid is dissolved by boiling with a mixture of 450 cc. of acetone and 175 cc. of water. Hydrolysis is completed by boiling under reflux for four hours. Most of the acetone is then removed by distillation at ordinary pressure on a steam bath. The residual solution is diluted with 400 cc. of water, heated to boiling for five minutes to ensure complete solution of the acetamino acid, and filtered (Notes 4 and 5). A small amount of undissolved material (0.2–0.5 g.) which remains on the filter is washed with 50–75 cc. of boiling water. Any crystals which separate from the filtrate are redissolved by heating, after which the solution is boiled for five minutes with 10 g. of Norite and filtered with the aid of gentle suction while still almost at the boiling point (Note 5). The Norite is washed thoroughly on the funnel with two to four 50-cc. portions of boiling water to remove the crystals which separate during the filtration, and the washings are added to the main filtrate. After standing in a refrigerator overnight the colorless, crystalline needles are collected on a Büchner funnel (Note 6), washed with 150–200 cc. of ice-cold water, and dried for several hours at 90–100°. The yield is 41–46 g. (80–90 per cent of the theoretical amount) of practically pure material, m.p. 191–192° (Note 7).

2. Notes

1. The azlactone of α-acetaminocinnamic acid may also be prepared by substituting the equivalent amount of glycine for acetylglycine and increasing the amount of acetic anhydride to three molecular proportions, but the yield is only about 45–50 per cent of the theoretical amount.

2. If the excess benzaldehyde is not almost completely removed by repeated washing with water, a final wash with 50–75 cc. of ether may be advantageous, although this causes some loss of azlactone owing to its solubility in ether.

3. The azlactone can be recrystallized from alcohol, from carbon tetrachloride, or from ethyl acetate with addition of petroleum ether. Aqueous solvents should be avoided, since the azlactone ring is easily opened by water. When alcohol is used for recrystallization, there is some danger of opening the azlactone ring with the formation of an ester, particularly on prolonged heating of the solution.

4. The solution may be filtered by gravity through a large folded filter (preferably in a steam-jacketed funnel), or through a Büchner funnel with gentle suction.

5. The solubility of α-acetaminocinnamic acid in water decreases very rapidly on cooling below the boiling point of the solution. Since the solution is very nearly saturated with the product, a large share of the acid will crystallize in the funnel during filtration if the solution is allowed

to cool too much. This property of the product makes it inadvisable to work with larger quantities.

6. Occasionally after treatment with Norite the solution is green owing to traces of iron and phenylpyruvic acid. If the crystals are still yellow at this point, the treatment with Norite should be repeated before the product is collected on a filter.

7. If further purification is desired, the product may be recrystallized from 600 cc. of boiling water, with a loss of about 5 per cent. The loss is due in part to hydrolysis of the product with the formation of phenyl-pyruvic acid.

3. Methods of Preparation

The azlactone of α-acetaminocinnamic acid has been prepared by heating a mixture of glycine, benzaldehyde, acetic anhydride, and anhydrous sodium acetate; [1,2] and from *N*-chloroacetylphenylalanine by treatment with acetic anhydride.[2]

α-Acetaminocinnamic acid has been prepared from the corresponding azlactone by hydrolysis with either aqueous sodium hydroxide [1] or with boiling water alone.[2]

ACETO-*p*-CYMENE

(Acetophenone, 5-isopropyl-2-methyl-)

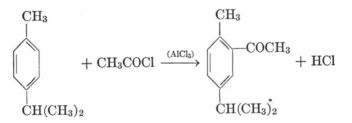

Submitted by C. F. H. Allen.
Checked by Reynold C. Fuson and Charles F. Woodward.

1. Procedure

A 1-l. three-necked flask is fitted with a dropping funnel, a stirrer, a thermometer for reading low temperatures (Note 1), and a condenser, to the upper end of which is attached a tube for disposing of the hydrogen

[1] Erlenmeyer, Jr., and Früstück, Ann. **284**, 48 (1895).
[2] Bergmann and Stern, ibid. **448**, 26 (1926).

chloride evolved (Note 2). A mixture of 200 cc. of carbon disulfide and 180 g. (1.35 moles) of anhydrous aluminum chloride is placed in the flask which is then immersed in an ice-salt freezing mixture and stirred very vigorously until the temperature of the mixture is $-5°$ or below. A mixture of 175 g. (1.3 moles) of p-cymene and 110 g. (100 cc., 1.4

Fig. 1.

moles) of acetyl chloride is added from the dropping funnel at such a rate that the temperature never rises above 5°. This addition requires about three and one-third hours (Note 3). The mixture is allowed to stand overnight and is then poured upon 1 kg. of cracked ice to which 200 cc. of concentrated hydrochloric acid has been added. The mixture is extracted with three 700-cc. portions of ether; the ether solution is dried over anhydrous calcium chloride and distilled at ordinary pressure from a Claisen flask provided with an indented column, until the temperature reaches 190°. The material that remains in the flask is fractionally distilled twice under diminished pressure. The principal fraction is aceto-p-cymene, a pale yellow oil boiling at 124–125°/12 mm. (155–157°/30 mm.). It weighs 115–125 g. (50–55 per cent of the theoretical amount) (Note 4). About 50 g. of cymene is recovered (Note 5), and there is a small amount (10–12 g.) of residual oil left in the flask (Note 6).

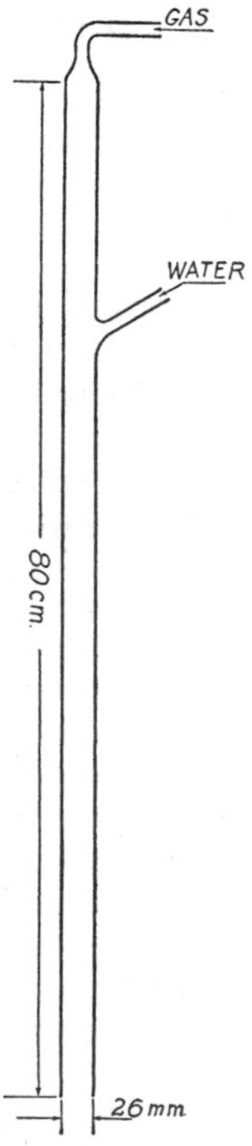

2. Notes

1. Since it is impossible to read that part of the thermometer scale which extends into the reaction flask, a thermometer should be used which when in position has the zero point above the stopper of the flask. A thermometer reading from $-50°$ to $+50°$ is recommended.

2. A gas trap of the type shown in Fig. 1 is suitable for this purpose. Another gas trap is shown in Org. Syn. Coll. Vol. I, 1941, 97.

3. After about two-thirds of the mixture has been added the rate of addition may be increased somewhat. The time required for the addition depends on the efficiency of the cooling and stirring; the stirring must be vigorous. With one-half of these amounts in a 500-cc. flask, the time required is only about one and one-third hours since, under these conditions, it is easier to control the temperature.

4. From the first fractionation a fraction boiling over a 20° range is

taken as crude ketone; e.g., at 28–30 mm. the fraction is taken which boils at 145–165°. Much trouble is caused by the tendency of the ketone to become superheated.

5. Acetyl chloride gives a better yield and less high-boiling residue than acetic anhydride.

6. This procedure has also been used successfully in the acetylation of cumene and *tert.*-butylbenzene. At the low temperatures employed there is very little decomposition, as is shown by the small amount of high-boiling residue.

3. Method of Preparation

Aceto-*p*-cymene can be prepared by the action of acetyl chloride on *p*-cymene in the presence of anhydrous aluminum chloride [1] or ferric chloride.[2]

ACETOL

(2-Propanone, 1-hydroxy-)

$$HCO_2C_2H_5 + KOH \rightarrow HCO_2K + C_2H_5OH$$

$$CH_3COCH_2Br + HCO_2K \rightarrow CH_3COCH_2OCHO + KBr$$

$$CH_3COCH_2OCHO + CH_3OH \rightarrow CH_3COCH_2OH + HCO_2CH_3$$

Submitted by P. A. LEVENE and A. WALTI.
Checked by FRANK C. WHITMORE and J. PAULINE HOLLINGSHEAD

1. Procedure

IN a 3-l. round-bottomed flask fitted with a 75-cm. Liebig condenser is placed 210 g. of potassium hydroxide (purified with alcohol) dissolved in 1.5 l. of anhydrous methyl alcohol. The solution is cooled to below 50° (Note 1), 300 g. of purified ethyl formate is added, and the mixture is refluxed for two hours (Notes 2 and 3).

Then 410 g. (251 cc., 3 moles) of bromoacetone (p. 88) is added, and the mixture is refluxed for sixteen hours on a water bath at 95–97°. At the end of the operation the solution is cooled to 0° in an ice-salt bath. The potassium bromide which settles is filtered on a cooled suction filter, and the filtrate is fractionated.

[1] Lacourt, Bull. soc. chim. Belg. **38**, 17 (1929); Claus, Ber. **19**, 232 (1886); Klages and Lickroth, ibid. **32**, 1563 (1899); Verley, Bull. soc. chim. (3) **17**, 910 (1897).

[2] Meissel, Ber. **32**, 2421 (1899).

The fraction boiling at 23–35°/12 mm. is discarded, as it contains very little acetol. The main fraction distils at 35–47°/12 mm. and weighs 160 g. This material is refractionated, and the portion boiling at 40–43°/12 mm. is collected. The yield is 120–130 g. (54–58 per cent of the theoretical amount) (Note 4).

2. Notes

1. It is necessary to cool the mixture below 50° to prevent loss of the volatile ethyl formate.

2. Technical ethyl formate was purified by washing with 3 per cent sodium carbonate solution, then with cold water, drying over anhydrous sodium sulfate, filtering, and fractionating. Compare p. 180. It is very important that all the materials used in the synthesis of acetol be anhydrous, as otherwise condensation products are formed.

3. If commercial potassium formate is used it should be dried under reduced pressure at 80°. One and one-half to two moles should be used per mole of the bromo compound.

4. Acetol polymerizes very readily on standing but remains unchanged when dissolved in an equal volume of methyl alcohol.

3. Methods of Preparation

Acetol has usually been prepared by the reaction between bromoacetone and sodium or potassium formate or acetate, followed by hydrolysis of the ester with methyl alcohol.[1, 2] Treatment of glycerol[3] or propylene glycol[4] at 200–300° with a dehydrogenating catalyst leads to the formation of acetol, while the direct oxidation of acetone with Baeyer and Villiger's acetone-peroxide reagent furnishes acetol together with pyruvic acid.[5]

[1] Nef, Ann. **335**, 247, 260 (1904).

[2] Urion, Ann. chim. (11) **1**, 78 (1934).

[3] Holmes, Brit. pat. 428,462 [C. A. **29**, 6908 (1935)].

[4] Carbide and Carbon Chemicals Corporation, U. S. pat. 2,143,383 [C. A. **33**, 2914 (1939)].

[5] Pastureau, Bull. soc. chim. (4) **5**, 227 (1909).

ACETONE CYANOHYDRIN

(Isobutyronitrile, α-hydroxy-)

$$(CH_3)_2CO + NaCN + H_2SO_4 \rightarrow (CH_3)_2C(OH)CN + NaHSO_4$$

Submitted by R. F. B. Cox and R. T. Stormont.
Checked by Reynold C. Fuson and Madison Hunt.

1. Procedure

In a 5-l. three-necked, round-bottomed flask fitted with an efficient stirrer (Note 1), a separatory funnel, and a thermometer in a well is placed a solution of 500 g. (9.7 moles) of powdered 95 per cent sodium cyanide in 1.2 l. of water and 900 cc. (713 g., 12.3 moles) of acetone. The flask is surrounded by an ice bath, and the solution is stirred vigorously. When the temperature falls to 15°, 2.1 l. (8.5 moles) of 40 per cent sulfuric acid (Note 2) is added over a period of three hours, the temperature being kept between 10° and 20°. After all the acid has been added the stirring is continued for fifteen minutes and then the flask is set aside for the salt to settle. Usually a layer of acetone cyanohydrin forms and is decanted and separated from the aqueous layer. The sodium bisulfate is removed by filtration and washed with three 50-cc. portions of acetone. The combined filtrate and acetone washings are added to the aqueous solution, which is then extracted three times with 250-cc. portions of ether (Note 3). The extracts are combined with the cyanohydrin previously separated and dried with anhydrous sodium sulfate. The ether and acetone are removed by distillation from a water bath, and the residue is distilled under reduced pressure. The low-boiling portion is discarded, and acetone cyanohydrin is collected at 78–82°/15 mm. The yield is 640–650 g. (77–78 per cent of the theoretical amount) (Note 4).

2. Notes

1. It is advantageous to use a heavy metal stirrer because of the increased viscosity of the mixture toward the end of the reaction. Since some hydrogen cyanide may escape from the reaction mixture, the stopper carrying the stirrer should be fitted with a tube for leading off the gas or the reaction should be carried out under a hood.

2. The 40 per cent sulfuric acid is prepared by adding 650 cc. of concentrated sulfuric acid (sp. gr. 1.84) to 1.8 l. of water.

3. Extraction and distillation should be started as soon as possible

after the completion of the reaction, and the distillation should be done as rapidly as possible to avoid decomposition.

4. The preparation of acetone cyanohydrin from potassium cyanide and the bisulfite addition product of acetone is described in Org. Syn. **20,** 43. The procedure given there furnishes a less pure cyanohydrin which, however, is suitable for some synthetic uses.

3. Methods of Preparation

Acetone cyanohydrin has been prepared from acetone and anhydrous hydrogen cyanide in the presence of a basic catalyst such as potassium carbonate, potassium hydroxide, or potassium cyanide;[1] by the reaction of potassium cyanide on the sodium bisulfite addition product of acetone;[2] and by the action of hydrogen cyanide, prepared directly in the reaction mixture, on an aqueous solution of acetone.[3] More modern industrial procedures employ acetone, liquid hydrogen cyanide, and a basic catalyst.[4]

2-ACETOTHIENONE

(Ketone, methyl 2-thienyl)

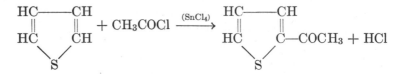

Submitted by JOHN R. JOHNSON and G. E. MAY.
Checked by REYNOLD C. FUSON and E. A. CLEVELAND.

1. Procedure

IN a 500-cc. round-bottomed, three-necked flask provided with a thermometer, dropping funnel, a liquid-sealed stirrer, and calcium

[1] Urech, Ann. **164,** 256 (1872); Ultée, Ber. **39,** 1857 (1906); Rec. trav. chim. **28,** 7 (1909).

[2] Bucherer and Grolée, Ber. **39,** 1225 (1906); Ger. pat. 141,509 (Chem. Zentr. **1903,** I, 1244); Org. Syn. **20,** 43.

[3] Welch and Clemo, J. Chem. Soc. **1928,** 2629.

[4] Triplex Safety Glass Company Ltd., Brit. pat. 416,007 [C. A. **29,** 814 (1935)] and 452,285 [C. A. **31,** 417 (1937)]; I. G. Farbenind. A.-G., Fr. pat. 804,124 [C. A. **31,** 2615 (1937)]; Deutsche Gold- und Silber-Scheideanstalt vorm. Roessler, Fr. pat. 812,366 [C. A. **32,** 958 (1938)].

chloride tube are placed 16.8 g. (0.2 mole) of thiophene (p. 578), 15.6 g. (14 cc., 0.2 mole) of acetyl chloride, and 200 cc. of dry benzene. The solution is cooled to 0°, and 52 g. (23 cc., 0.2 mole) of freshly distilled stannic chloride is added dropwise, with efficient stirring, during the course of about forty minutes. The reaction mixture assumes a purple color when the first drops of stannic chloride are added, and soon a purple solid precipitates.

After all the stannic chloride has been added, the cooling bath is removed and the mixture stirred for one hour longer. The addition product is hydrolyzed by the slow addition of a mixture of 90 cc. of water and 10 cc. of concentrated hydrochloric acid. The yellow benzene layer is separated, washed with 25 cc. of water, and dried over 5–10 g. of anhydrous calcium chloride. Benzene and unchanged thiophene (Note 1) are distilled through a short fractionating column (using an oil bath), and the residual liquid is distilled under reduced pressure. The yield of 2-acetothienone, b.p. 89–91°/9 mm., is 20–21 g. (79–83 per cent of the theoretical amount) (Note 2).

2. Notes

1. By shaking the recovered benzene-thiophene mixture with a solution of 5.5 g. of mercuric chloride, 10 g. of sodium acetate, and 10 cc. of alcohol in 80 cc. of water, the unchanged thiophene is converted to the 2-chloromercurithiophene (containing a small amount of the dimercurichloride); from this the free thiophene can be obtained by treatment with hydrochloric acid. The recovered thiophene amounts to 2–2.5 g.

2. 2-Acetothienone has the following physical constants: d_4^{20} 1.168, n_D^{20} 1.566. Its semicarbazone melts at 186–187° (corr.).

3. Methods of Preparation

2-Acetothienone has been prepared by treating thiophene with acetyl chloride in the presence of aluminum chloride [1] or stannic chloride,[2] and by treating 2-chloromercurithiophene with acetyl chloride.[3] The present method is essentially that of Stadnikoff and Goldfarb.[2] Stannic chloride is superior to aluminum chloride as a catalyst for this reaction as aluminum chloride induces polymerization of the thiophene.

[1] Peter, Ber. **17**, 2643 (1884); Biedermann, ibid. **19**, 636 (1886).

[2] Stadnikoff and Goldfarb, ibid. **61**, 2341 (1928).

[3] Volhard, Ann. **267**, 178 (1892); Steinkopf and Baumeister, ibid. **403**, 69 (1914).

ACETYLENEDICARBOXYLIC ACID

$$HO_2CCHBrCHBrCO_2H + 4KOH \rightarrow$$
$$KO_2CC{\equiv}CCO_2K + 2KBr + 4H_2O$$
$$KO_2CC{\equiv}CCO_2K + H_2SO_4 \rightarrow KO_2CC{\equiv}CCO_2H + KHSO_4$$
$$KO_2CC{\equiv}CCO_2H + H_2SO_4 \text{ (excess)} \rightarrow HO_2CC{\equiv}CCO_2H + KHSO_4$$

Submitted by T. W. Abbott, Richard T. Arnold, and Ralph B. Thompson.
Checked by Reynold C. Fuson and W. E. Holland.

1. Procedure

A solution of potassium hydroxide is prepared by dissolving 122 g. (2.2 moles; 1.5 times the theoretical amount) of potassium hydroxide in 700 cc. of 95 per cent methyl alcohol (Note 1) contained in a 2-l. round-bottomed flask provided with a reflux condenser. To this alkaline solution is added 100 g. (0.36 mole) of α,β-dibromosuccinic acid (p. 177), and the mixture is refluxed for one hour and fifteen minutes on a steam bath. The reaction mixture is cooled and filtered with suction. The mixed salts are washed with 200 cc. of methyl alcohol (Note 2) and dried by pressing between filter papers; when dry the product weighs 144–150 g.

This salt mixture is dissolved in 270 cc. of water, and the acid potassium salt is precipitated by adding 8 cc. of concentrated sulfuric acid in 30 cc. of water. After standing for three hours, or overnight, the mixture is filtered with suction (Note 3). The acid salt is then dissolved in 240 cc. of water to which 60 cc. of concentrated sulfuric acid has been added, and the solution is extracted with five 100-cc. portions of ether. The combined ether solutions are evaporated to dryness on a steam bath, leaving pure hydrated crystals of acetylenedicarboxylic acid. After drying for two days over concentrated sulfuric acid in a vacuum desiccator the crystals decompose sharply at 175–176°. The yield is 30–36 g. (73–88 per cent of the theoretical amount).

2. Notes

1. The yield is slightly lower when 95 per cent ethyl alcohol is used.
2. This salt mixture is composed of potassium bromide and potassium acetylenedicarboxylate.
3. This acid salt is practically bromine-free and does not require additional washing.

3. Method of Preparation

The procedure described is essentially that of Bandrowski [1] and Baeyer [2] as modified by Ruggli.[3] The same general method has also been used by Backer and van der Zanden,[4] and by Moureu and Bongrand.[5]

ACETYLGLYCINE

(Aceturic acid)

$$H_2NCH_2CO_2H + (CH_3CO)_2O \rightarrow CH_3CONHCH_2CO_2H + CH_3CO_2H$$

Submitted by R. M. HERBST and D. SHEMIN.
Checked by REYNOLD C. FUSON and E. A. CLEVELAND.

1. Procedure

IN a 1-l. Erlenmeyer flask provided with a mechanical stirrer are placed 75 g. (1 mole) of glycine (Org. Syn. Coll. Vol. I, **1941**, 298) and 300 cc. of water. The mixture is stirred vigorously until the glycine is almost completely dissolved, when 215 g. (2 moles) of 95 per cent acetic anhydride (Note 1) is added in one portion. Vigorous stirring is continued for fifteen to twenty minutes, during which time the solution becomes hot and acetylglycine may begin to crystallize. The solution is placed in the refrigerator (Note 2) overnight to effect complete crystallization. The precipitate is collected on a Büchner funnel, washed with ice-cold water, and dried at 100–110°. This product weighs 75–85 g. and melts at 207–208°. The combined filtrate and washings are evaporated to dryness under reduced pressure on a water bath at 50–60°. The residue on recrystallization from 75 cc. of boiling water yields a second fraction, of 20–30 g., which melts at 207–208° after being washed with ice-cold water and dried at 100–110°. An additional 4–6 g. of only slightly less pure product may be obtained from the mother liquor by concentration. The total yield is 104–108 g. (89–92 per cent of the theoretical amount) (Note 3).

[1] Bandrowski, Ber. **10,** 838 (1877).
[2] Baeyer, ibid. **18,** 677 (1885).
[3] Ruggli, Helv. Chim. Acta **3,** 564 (1920).
[4] Backer and van der Zanden, Rec. trav. chim. **47,** 776 (1928).
[5] Moureu and Bongrand, Ann. chim. (9) **14,** 9 (1920).

2. Notes

1. The equivalent quantity of 90 per cent acetic anhydride may be used.

2. The refrigerator used by the checkers maintained a temperature of 5–7°.

3. The method may be employed to acetylate most α-amino acids with only slight modifications depending upon the solubility of the particular amino acid. When optically active amino acids are acetylated, there is little or no racemization.[1]

3. Methods of Preparation

Acetylglycine has been prepared by the interaction of acetyl chloride and the silver salt of glycine in dry ether or benzene;[2,3] by the action of acetic anhydride on glycine suspended in warm benzene;[3] by heating glycine with acetic anhydride;[4] by treating an aqueous solution of glycine or its sodium salt with ketene;[5] and by treating an aqueous alkaline solution of glycine with acetic anhydride.[6]

ACONITIC ACID

$$\begin{array}{c}\text{CH}_2\text{CO}_2\text{H}\\|\\\text{C(OH)CO}_2\text{H}\\|\\\text{CH}_2\text{CO}_2\text{H}\end{array}\xrightarrow{\text{(H}_2\text{SO}_4)}\begin{array}{c}\text{CHCO}_2\text{H}\\||\\\text{CCO}_2\text{H}\\|\\\text{CH}_2\text{CO}_2\text{H}\end{array}+\text{H}_2\text{O}$$

Submitted by WILLIAM F. BRUCE.
Checked by LOUIS F. FIESER and C. H. FISHER.

1. Procedure

IN a 1-l. round-bottomed flask equipped with a reflux condenser (Note 1) are placed 210 g. (1 mole) of powdered citric acid monohydrate and a solution of 210 g. (115 cc., 2 moles) of concentrated sulfuric acid in 105

[1] Behr and Clarke, J. Am. Chem. Soc. **54**, 1631 (1932).
[2] Kraut and Hartmann, Ann. **133**, 105 (1865).
[3] Curtius, Ber. **17**, 1665 (1884).
[4] Radenhausen, J. prakt. Chem. (2) **52**, 437 (1895).
[5] Bergmann and Stern, Ber. **63**, 437 (1930).
[6] Chattaway, J. Chem. Soc. **1931**, 2495.

cc. of water. The mixture is heated in an oil bath kept at a temperature of 140–145° for seven hours. The light brown solution is poured into a shallow dish, and the flask is rinsed with 10 cc. of hot glacial acetic acid. The liquid is allowed to cool slowly to 41–42° (Note 2), with occasional stirring to break up the solid mass of aconitic acid which separates, and the solid is collected on a suction funnel (Note 3). The material is pressed and drained thoroughly until practically dry, when it is removed and stirred to a homogeneous paste with 70 cc. of concentrated hydrochloric acid, cooled in an ice bath. The solid is collected on a suction funnel (Note 3), washed with two 10-cc. portions of cold glacial acetic acid, sucked thoroughly, and spread out in a thin layer on porous plate or paper for final drying (Note 4). This product contains practically no sulfate and is pure enough for most purposes. It is colorless, and when dry weighs 71–77 g. (41–44 per cent of the theoretical amount) (Note 5). The point of decomposition determined under controlled conditions (Note 6) varies from 180° to 200°.

For purification the acid is crystallized from about 150 cc. of glacial acetic acid, using an acid-resistant filter for the hot solution (Note 7). Aconitic acid separates as small, colorless needles weighing 50–60 g., and about 10 g. more can be secured by concentrating the mother liquor under reduced pressure to one-third of its volume. The material is dried in the air and then in a desiccator containing sodium hydroxide in order to remove all traces of acetic acid. One crystallization usually is sufficient to bring the point of decomposition to 198–199° (Note 6).

2. Notes

1. A ground-glass connection is highly desirable.

2. By filtering at this point rather than at a lower temperature, a separation from a small amount of low-melting material is accomplished without much loss of aconitic acid.

3. This material may be filtered conveniently by means of a sintered glass funnel, or by using a pad of pure wool flannel in an 8-cm. Büchner funnel.

4. In humid weather the solid often deliquesces, and this necessitates drying in a desiccator. The material retains acetic acid very tenaciously, and drying should be continued until the odor of the solvent no longer can be noticed.

5. A determination by the method of Pucher, Vickery, and Leavenworth [1] showed that 26 g. of citric acid remained in the sulfuric acid solution. It is inadvisable to use this solution for another run; the

[1] Pucher, Vickery, and Leavenworth, Ind. Eng. Chem., Anal. Ed. **6**, 190 (1934).

accumulation of water and by-products reduces the yield and the quality of the product considerably.

6. When heated in a capillary tube aconitic acid decomposes rather suddenly with vigorous gas evolution at a temperature which is closely dependent upon the rate of heating and the temperature at which the sample is introduced. In the literature [2] "melting points" ranging from 182.5° to 194.5° are recorded. The uncrystallized aconitic acid, when introduced at 180° into a small bath provided with mechanical agitation and heated at the rate of 2–3° per minute, usually decomposed at 189–190°. The once recrystallized material, introduced at 190°, decomposed at 198–199°; introduced at 195°, it decomposed at 204–205°. A determination on the Dennis bar,[3] the most reliable method for this type of compound, showed a decomposition point of 209°. The sample must be thoroughly dry to obtain the highest figures.

7. The hot solution is very destructive to filter paper. A convenient filter is made by preparing a 1–2 mm. mat of asbestos in a 6-cm. Büchner funnel, dusting onto this a 2–3 mm. layer of Norite, washing by suction, and heating the unit, together with a suction flask, in an oven at 120°. When dry and hot, the apparatus is ready for use.

3. Methods of Preparation

Aconitic acid has been prepared from citric acid by the action of sulfuric acid [4] or hydrogen chloride,[5] or by heating.[6] It has been prepared also from methyl acetylcitrate [7] and from acetylcitric anyhdride.[8] The method described is essentially that of Hentschel.[4] Phosphoric acid (85 per cent) can be used in place of sulfuric acid, but much closer regulation of the conditions seems necessary and the yield is not greatly improved.

The effects of acid strength and temperature on the reaction between sulfuric acid and citric acid have been reported by Quartaroli and Belfiori: the use of pyrosulfuric acid, cold, or sulfuric acid of less than 94 per cent strength, hot, leads to the formation of aconitic acid.[9]

[2] Malachowski and Maslowski, Ber. **61,** 2521 (1928).

[3] Dennis and Shelton, J. Am. Chem. Soc. **52,** 3128 (1930).

[4] Hentschel, J. prakt. Chem. (2) **35,** 205 (1887).

[5] Hunäus, Ber. **9,** 1751 (1876).

[6] Pawolleck, Ann. **178,** 153 (1875).

[7] Anschütz and Klingemann, Ber. **18,** 1953 (1885).

[8] Easterfield and Sell, J. Chem. Soc. **61,** 1007 (1892).

[9] Quartaroli and Belfiori, Ann. chim. applicata **28,** 297 (1938) [C. A. **33,** 1669 (1939)].

ACRIDONE

$$o\text{-}ClC_6H_4CO_2H + C_6H_5NH_2 \xrightarrow[\text{(K}_2\text{CO}_3\text{)}]{\text{(CuO)}} o\text{-}C_6H_5NHC_6H_4CO_2H$$

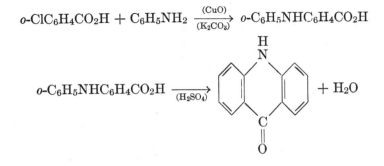

$$o\text{-}C_6H_5NHC_6H_4CO_2H \xrightarrow[\text{(H}_2\text{SO}_4\text{)}]{} \quad\quad + H_2O$$

Submitted by C. F. H. ALLEN and G. H. W. McKEE.
Checked by W. W. HARTMAN and A. WEISSBERGER.

1. Procedure

(A) *N-Phenylanthranilic Acid.*—In a 1-l. round-bottomed flask fitted with an air-cooled condenser, a mixture of 155 g. (1.66 moles) of aniline, 41 g. (0.26 mole) of o-chlorobenzoic acid (Note 1), 41 g. (0.3 mole) of technical anhydrous potassium carbonate, and 1 g. of copper oxide is refluxed for two hours, using an oil bath. The excess aniline is removed by distillation with steam (about three hours is required), and 20 g. of decolorizing carbon (Note 2) is added to the brown residual solution. The mixture is boiled for fifteen minutes and filtered by suction. The filtrate is added, with stirring, to a mixture of 30 cc. of concentrated hydrochloric acid and 60 cc. of water. The precipitated acid is filtered with suction when cold. After drying to constant weight in the air, the yield is 46–52 g. (82–93 per cent of the theoretical amount) of a nearly white product; m.p. 179–181° with preliminary shrinking (Notes 3, 4, and 5).

(B) *Acridone.*—In a 500-cc. flask a solution of 42.7 g. (0.2 mole) of N-phenylanthranilic acid (Note 6) in 100 cc. of concentrated sulfuric acid (sp. gr. 1.84) is heated on a boiling water bath for four hours and then poured into 1 l. of boiling water. Spattering is minimized by allowing the solution to run down the wall of the container. The yellow precipitate is filtered after boiling for five minutes, and the filtrate is saved (Note 7). The moist solid is boiled for five minutes with a solution of 30 g. (0.28 mole) of sodium carbonate in 400 cc. of water, collected with suction (Note 8), and washed well with water. After drying in the air the crude acridone weighs 35.5–37.5 g. (91–96 per cent of the theoretical amount) and melts at 344–346° (Note 9). This material is pure enough for many purposes; it may be recrystallized from a mixture of aniline and acetic acid, using 10 cc. of aniline and 25 cc. of acetic acid for every 2 g. of solid. The recovery is about 90 per cent, and the recrystallized product melts at 348–352° (Note 10).

2. Notes

1. Sixty grams of technical o-chlorobenzoic acid is dissolved in 200 cc. of hot water containing 20 g. of sodium carbonate, 10 g. of decolorizing carbon is added, and, after boiling for ten minutes, the mixture is filtered by suction. The filtrate is added to hydrochloric acid prepared by diluting 31 cc. of the concentrated acid with an equal volume of water. The air-dried product weighs 41 g. and is used directly. This purification is essential to obtain a good yield and a product of good quality. If it is omitted, a blue to black acid results, from which the color is removed only with difficulty.

Directions for preparing o-chlorobenzoic acid by the oxidation of o-chlorotoluene are given on p. 135.

2. Ordinary animal charcoal and Darco gave equally good results.

3. N-Phenylanthranilic acid decomposes slowly at elevated temperatures. Before the melting point is reached, there is considerable preliminary shrinkage. If the "dip method" is used, the melting point is 182–183°. The literature values vary from 181° to 184° for the pure acid.

4. This acid is pure enough for all ordinary purposes. The melting point is only slightly raised by dissolving 5 g. in 100 cc. of water containing 2.5 g. of sodium carbonate, adding 2.5 g. of decolorizing carbon, boiling for five minutes, filtering, and acidifying. The recovery is 4.6 g. If the product is colored this procedure must be followed to get acridone of light color.

5. For recrystallization, 5 g. of the acid is dissolved in 25 cc. of boiling alcohol and 5 cc. of water added. The recovery is 4.8 g., and the melting point is 182–183°. Acetic acid (2 cc. per gram) may be substituted for alcohol; it is more convenient for recrystallizing large quantities.

6. The N-phenylanthranilic acid may be used without recrystallization if it has been prepared from purified o-chlorobenzoic acid. If not, the crude N-phenylanthranilic acid must be decolorized as described in Note 4; otherwise a greenish acridone is obtained which, however, has the proper melting point.

7. On standing overnight, the filtrate deposits 1.6 to 2 g. of a very impure product which melts at about 315°.

8. On acidification of the filtrate, a little (1.5–2 g.) N-phenylanthranilic acid is always precipitated.

9. The crude acridone shrinks in a capillary tube at 330–335° and melts to a dark-colored liquid at 344-346°.

10. Isoamyl alcohol may also be used for recrystallization; 1 g. of acridone requires 120 cc. of solvent and yields 0.75 g. of material melting at 354°, using a Berl-Kullman copper block.[1]

3. Methods of Preparation

The practical methods of preparation of N-phenylanthranilic acid are the action of aniline on o-chloro- or o-bromobenzoic acid,[2, 3] or the action of bromobenzene on anthranilic acid,[4, 5, 6] copper or its salts being used in both instances.

The only method of preparative value for acridone is by ring closure of N-phenylanthranilic acid.[7]

ACROLEIN ACETAL

(Acrolein, diethyl acetal)

$$ClCH_2CH_2CH(OC_2H_5)_2 + KOH \rightarrow$$
$$CH_2{=}CHCH(OC_2H_5)_2 + KCl + H_2O$$

Submitted by E. J. Witzemann, Wm. Lloyd Evans, Henry Hass, and E. F. Schroeder.
Checked by Frank C. Whitmore and Harry T. Neher.

1. Procedure

To 340 g. (6 moles) of dry, powdered potassium hydroxide (Note 1) in a 500-cc. short-necked round-bottomed flask (Note 2) is added 167 g. (1 mole) of β-chloropropionaldehyde acetal (p. 137). The mixture is

[1] Berl and Kullman, Ber. **60**, 811 (1927); for an elaborate modification, see Walsh, Ind. Eng. Chem., Anal. Ed. **6**, 468 (1934).

[2] Ullmann, Ber. **36**, 2383 (1903); Ann. **355**, 322 (1907).

[3] Meister, Lucius, and Brüning, Ger. pat. 145,189 (Chem. Zentr. **1903**, II, 1097).

[4] Goldberg, Ber. **39**, 1691 (1906).

[5] Houben and Brassert, ibid. **39**, 3238 (1906).

[6] Goldberg and Ullmann, Ger. pat. 173,523 (Chem. Zentr. **1906**, II, 931); Goldberg, Ger. pat. 187,870 (Chem. Zentr. **1907**, II, 1465).

[7] Graebe and Lagodzinski, Ber. **25**, 1734 (1892); Ann. **276**, 35 (1893); Matsumura, J. Am. Chem. Soc. **57**, 1533 (1935).

shaken vigorously and attached at once to a three-bulbed Glinsky or other suitable column, connected to a water condenser set for distillation (Note 3). The flask is heated in an oil bath at 210–220° until nothing more distils (Note 4). The distillate is transferred to a separatory funnel and the lower aqueous layer is removed. The acrolein acetal is dried over 10 g. of potassium carbonate, filtered, and distilled from a modified Claisen flask (Fig. 2). The yield is 98 g. (75 per cent of the theoretical amount) of a product which boils at 122–126°.

Fig. 2.

2. Notes

1. The powdered potassium hydroxide should pass a 60-mesh sieve (24 per cm.). The dryness of the powder is of the utmost importance. Water must be avoided as much as possible. Therefore, the potassium hydroxide should be fused at 350° for two hours and then pulverized as rapidly as possible. A 24-cm. (10-in.) disk pulverizer having a capacity of about 200 g. per minute is recommended. If the potassium hydroxide is not fused before pulverizing, the yield of acetal drops to about 60 per cent.

2. The concentrated potassium hydroxide left at the completion of the reaction attacks the glass rapidly. For this reason iron retorts made from 10-cm. (4-in.) pipe should be used when many runs are to be made.

3. The reaction between the acetal and the alkali is very vigorous.

4. A large low-boiling fraction indicates too much moisture in the potassium hydroxide used. See Note 1.

3. Methods of Preparation

Acrolein acetal has been prepared by treatment of β-chloropropionaldehyde acetal with dry, powdered potassium hydroxide,[1] and from acrolein, ethyl orthoformate, and ammonium nitrate in boiling alcohol.[2]

[1] Wohl, Ber. **31,** 1798 (1898); Witzemann, J. Am. Chem. Soc. **36,** 1911 (1914); Spoehr and Young, Carnegie Inst. Washington Yearbook, **25,** 176 (1925–1926); Expt. Sta. Record, **57,** 817 (1927) [C. A. **22,** 2368 (1928)].

[2] Fischer and Baer, Helv. Chim. Acta **18,** 516 (1935).

β-ALANINE

$$\begin{array}{c} CH_2CO \\ | \\ CH_2CO \end{array}\!\!\!\!> NH + KOBr + 2KOH \rightarrow \begin{array}{c} CH_2NH_2 \\ | \\ CH_2CO_2H \end{array} + KBr + K_2CO_3$$

Submitted by H. T. Clarke and Letha Davies Behr.
Checked by W. H. Carothers and W. L. McEwen.

1. Procedure

To a cold (0–5°) solution of 302 g. of potassium hydroxide sticks in 2720 cc. of distilled water is added slowly, with stirring, 96.6 g. (30.8 cc., 0.6 mole) of bromine. This solution is chilled to 0°, and 59.4 g. (0.6 mole) of succinimide (p. 562) is added with hand stirring. The mixture is warmed in a water bath to 55–60°, when it becomes colorless, and is held at that temperature for two hours (Note 1). After standing overnight at room temperature, it is acidified to Congo red with concentrated hydrochloric acid (about 380 cc., sp. gr. 1.18) (Note 2) and evaporated to dryness on a steam bath under reduced pressure. The residue is treated with 1 l. of warm 95 per cent alcohol; the undissolved potassium bromide is filtered and washed with 150–200 cc. of cold alcohol in small portions. The filtrate and washings are combined and evaporated to dryness under reduced pressure, and the residue is extracted with 100 cc. of 95 per cent alcohol. The resulting solution is again evaporated to dryness and the residue finally extracted with 140 cc. of hot absolute alcohol (Note 3). After distilling the bulk of the alcohol, this extract is diluted with about 200 cc. of distilled water and shaken out twice with 80-cc. portions of ether. The ether extracts are discarded (Note 4).

The aqueous solution is freed of ether and alcohol and then boiled under reflux for one to one and a half hours in order to hydrolyze any β-alanine ester. After evaporating under reduced pressure to remove as much as possible of the excess hydrochloric acid, the residue is dissolved in water and diluted to exactly 1 l. A 5-cc. portion of this solution is withdrawn for determination of total halides. A suspension of silver oxide prepared from 10 per cent more than the equivalent quantity of silver nitrate (Note 5) is added to the remaining portion of the solution, and the mixture is stirred well in order to bring about complete precipitation of the halides. After standing overnight the precipitate is filtered and washed with water. The filtrate and washings are concentrated under reduced pressure to about 400 cc., saturated with hydrogen sulfide, and filtered through a thin layer of decolorizing carbon. The colorless filtrate is evaporated to a volume of about 100 cc., treated with decolorizing carbon if necessary, concentrated on the steam bath until crystallization begins, and chilled. The crystals are filtered with suction, washed with a little cold alcohol, and dried. A further crop is obtained by concentrating the mother liquor and again chilling (Note 6). The combined crops (28–30 g., m.p. 189–192°) are recrystallized from water, employing the same procedure, and yield 22–24 g. (41–45 per cent of the theoretical amount) of pure β-alanine, which melts at 197–198° (corr.) with decomposition. About 2 g. of less pure product can be secured from the final mother liquors.

2. Notes

1. The odor of ammonia is perceptible, indicating some hydrolysis.

2. On acidification a small amount of bromine may be liberated; this is removed rapidly during the subsequent evaporation.

3. In the last extraction the alcohol-insoluble material may be removed advantageously with a centrifuge.

4. This ether extraction removes small quantities of succinic acid and its esters.

5. The silver oxide is prepared by dissolving the silver nitrate in about five parts of cold water and adding a slight excess of pure sodium hydroxide in 10 per cent solution. The precipitate is well stirred, collected by filtration or centrifuging, and washed free of sodium salts. It should not be dried before use.

6. The final mother liquor consists of a rather viscous solution containing uncrystallizable by-products.

3. Methods of Preparation

The above directions are based upon the methods of Hoogewerff and Van Dorp,[1] as modified by Holm [2] and by Hale and Honan.[3] β-Alanine has also been prepared by the action of hypobromite upon succinamide and hydrolysis of the resulting β-ureidopropionic acid; [4] by the action of ammonia upon β-iodopropionic acid; [5] by the hydrolysis of methyl carbomethoxy-β-aminopropionate, obtained by the action of sodium

[1] Hoogewerff and Van Dorp, Rec. trav. chim. 10, 5 (1891).

[2] Holm, Arch. Pharm. 242, 597 (1904).

[3] Hale and Honan, J. Am. Chem. Soc. 41, 774 (1919).

[4] Weidel and Roithner, Monatsh. 17, 172 (1896).

[5] Heintz, Ann. 156, 25 (1870); Mulder, Ber. 9, 1902 (1876); Abderhalden and Fodor, Z. physiol. Chem. 85, 114 (1913).

methoxide on succinbromimide; [6] by the reduction of β-nitrosopropionic acid; [7] by heating ethyl acrylate with alcoholic ammonia; [8] from succinylglycine ester by the azide synthesis; [9] by the action of liquid ammonia upon methyl acrylate; [10] and by the reduction of cyanoacetic acid,[11] or its ethyl ester followed by hydrolysis.[12]

ALLANTOIN

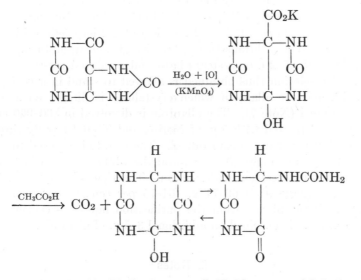

Submitted by W. W Hartman, E. W. Moffett, and J. B. Dickey.
Checked by W. H. Carothers and W. L. McEwen.

1. Procedure

ONE HUNDRED grams of uric acid (0.595 mole) and 4.5 l. of hot (70–85°) water are placed in a 12-l. round-bottomed flask equipped with a mechanical stirrer. The stirrer is started, and a solution of 80 g. (2

[6] Lengfeld and Stieglitz, Am. Chem. J. **15**, 215, 504 (1893).

[7] v. Pechmann, Ann. **264**, 288 (1891).

[8] Wender, Gazz. chim. ital. **19**, 437 (1889).

[9] Curtius and Hechtenberg, J. prakt. Chem. (2) **105**, 289 (1923).

[10] Morsch, Monatsh. **63**, 220 (1933).

[11] E. Merck, Ger. pat. 597,305 [C. A. **28**, 5078 (1934)]; Ruggli and Businger, Helv. Chim. Acta **25**, 35 (1942).

[12] Weygand, Ber. **74**, 256 (1941).

moles) (Note 1) of commercial sodium hydroxide in 120 cc. of water is added. Stirring is continued until the uric acid is in solution (Note 2), after which the solution is cooled by means of a stream of water directed against the flask. When the temperature has fallen to 25–30°, 50 g. (0.32 mole) (Note 3) of potassium permanganate is added all at once (Note 4) to the vigorously stirred solution. Stirring is continued for fifteen to twenty minutes (Note 5), and the mixture is filtered (Note 6) at once through a 19-cm. Büchner funnel. The first fraction of the filtrate contains a small amount of manganese dioxide. This fraction must be collected separately and returned to the funnel. As soon as the filtrate becomes clear it is collected in a 12-l. round-bottomed flask which contains 130 cc. (137 g., 2.2 moles) of glacial acetic acid. The filtrate is tested with litmus to be sure that it is acid, and evaporated to a volume of 1.5–2 l. on a steam bath under reduced pressure (20–30 mm.). The solution thus obtained is allowed to stand in a cool place overnight, and the allantoin which crystallizes is filtered on a 9-cm. Büchner funnel (Note 7). The allantoin is dissolved in 800–900 cc. of boiling water, treated with 5 g. of Norite, and filtered rapidly through a fluted filter paper in a steam funnel. The filtrate is allowed to stand in a cool place overnight (Note 8), and the white crystals of allantoin are separated by filtration with suction. The yield of product melting at 230–231° (Note 9) is 60–71 g. (64–75 per cent of the theoretical amount). If the filtrate from the purification liquors is concentrated to 100 cc., there is obtained an additional 3–5 g. of allantoin.

2. Notes

1. The use of more than 80 g. of sodium hydroxide does not increase the yield, but, if not neutralized immediately upon completion of the reaction, it causes decomposition of some allantoin.

2. It is essential that the uric acid be completely in solution; otherwise not all of it will be oxidized. When the solution is cooled a small amount of white precipitate sometimes separates, but this does not affect the yield.

3. The amount of potassium permanganate can be varied between 50 and 62 g. (0.32–0.39 mole) without changing the yield.

4. The potassium permanganate must be added rapidly (one to five minutes).

5. If the period of stirring is reduced to ten minutes, some unchanged uric acid is recovered. The period can be extended slightly beyond twenty minutes without decreasing the yield of allantoin, but if it is extended beyond one hour the yield is appreciably decreased.

6. Filtration must be as rapid as possible; this necessitates the use of a large Büchner funnel.

7. The filtrate is discarded since the amount of allantoin is not sufficient to repay attempts to separate it from the various other compounds present.

8. Crystallization can be hastened by stirring in an ice bath.

9. The melting point appears to depend somewhat on the rate of heating. The melting point 228–230° is observed in a capillary tube in a bath heated slowly from room temperature. If the capillary is placed in a bath already heated to 228°, the specimen melts at 233–234°. On a copper block still higher melting points are obtained.

3. Methods of Preparation

Allantoin has been prepared by the oxidation of uric acid with potassium permanganate,[1] lead dioxide,[2] potassium ferricyanide,[3] oxygen,[4] manganese dioxide,[5] ozone,[6] or hydrogen peroxide,[7] and by the electrolytic oxidation of lithium urate.[8] It is also formed by heating urea with glyoxylic acid[9] or with any one of a number of disubstituted acetic acids such as, for instance, dichloroacetic acid.[10]

[1] Claus, Ber. **7**, 226 (1874); Sundwik, Z. physiol. Chem. **41**, 343 (1904); Behrend, Ann. **333**, 141 (1904); Biltz, Ber. **43**, 1999 (1910); Biltz and Giesler, ibid. **46**, 3410 (1913); Biltz and Max, ibid. **54**, 2451 (1921); Neubauer, Ann. **99**, 206 (1856).

[2] Wöhler and Liebig, ibid. **26**, 241 (1838); Mulder, ibid. **159**, 349 (1871).

[3] Schlieper, ibid. **67**, 214 (1848).

[4] Biltz and Max, Ber. **54**, 2451 (1921).

[5] Wheeler, Zeit. für Chem. **1866**, 746.

[6] Gorup-Besanez, Ann. **110**, 94 (1859).

[7] Venable, J. Am. Chem. Soc. **40**, 1099 (1918).

[8] Fichter and Kern, Helv. Chim. Acta **9**, 429 (1926).

[9] Grimaux, Ann. chim. phys. (5) **11**, 389 (1877).

[10] Merck and Company, Inc., U. S. pat. 2,158,098 [C. A. **33**, 6350 (1939)].

ALLYLAMINE

$$CH_2{=}CHCH_2NCS + H_2O + HCl \rightarrow CH_2{=}CHCH_2NH_2 \cdot HCl + COS$$
$$CH_2{=}CHCH_2NH_2 \cdot HCl + KOH \rightarrow CH_2{=}CHCH_2NH_2 + KCl + H_2O$$

Submitted by M. T. Leffler.
Checked by W. W. Hartman and E. J. Rahrs.

1. Procedure

In a 5-l. round-bottomed flask, equipped with a reflux condenser connected to a gas trap (Note 1), are placed 2 l. (12.1 moles) of 20 per cent hydrochloric acid and 500 g. (5.05 moles) of allyl isothiocyanate (Note 2). The mixture is refluxed over a free flame until the upper layer of allyl isothiocyanate has completely disappeared, about fifteen hours being required for the hydrolysis. When the reaction is complete, the solution is poured into a 3-l. beaker and concentrated on the steam bath until crystals begin to form in the hot solution. This occurs when the volume is approximately 400 cc. (Note 3).

The warm residue is then diluted with water to a volume of 500–550 cc. and placed in a 2-l. three-necked, round-bottomed flask equipped with a 500-cc. dropping funnel, a mercury-sealed mechanical stirrer, and a condenser arranged for distillation. The lower end of the condenser is fitted to a receiver consisting of a 500-cc. suction flask, whose side arm is connected to a reflux condenser (Note 4). The receiver is placed in an ice-salt bath and the three-necked flask in a water bath. The temperature of the water bath is raised to 95–98°, the stirrer is started, and a solution of 400 g. (7.1 moles) of potassium hydroxide in 250 cc. of water is added dropwise from the funnel. As soon as the free hydrochloric acid is neutralized, the amine begins to distil. The rate of the addition of the alkali is regulated so as to maintain a dropwise distillation of allylamine and, after the addition is complete, heating and stirring are continued until all the amine has distilled.

The distillate is then dried over solid potassium hydroxide for twenty-four hours and finally over metallic sodium (Note 5). The allylamine is distilled from a water bath held at 70–78°, through a 12-in. fractionating column, into a receiver immersed in an ice bath. Two fractions are collected: up to 54°/746 mm., and 54–57°/746 mm. The lower fraction amounts to 14–16 g. and on redistillation yields 6–8 g. of pure material. The total yield of pure allylamine boiling at 54–57°/746 mm. is 200–210 g. (70–73 per cent of the theoretical amount).

2. Notes

1. It is desirable to use a gas trap in order to prevent vapors of allyl isothiocyanate from escaping into the room. The gas trap described on p. 4 is suitable.

2. Eastman's "practical" grade (b.p. 150–152°) of allyl isothiocyanate was used in this preparation.

3. The rate of evaporation is greatly accelerated by allowing a stream of air to blow across the surface of the hot liquid. It is advantageous for the evaporation to proceed as far as possible in order to remove most of the free hydrochloric acid, and no harm is done if it continues until the solution turns to a semi-solid mass of crystals.

4. Care must be taken throughout to prevent loss of the product by volatilization. Furthermore, the vapors should not be allowed to come into contact with the nasal passages, as violent sneezing is produced.

5. The distillate should be kept cold (5–10°) during the drying process and should be separated from the potassium hydroxide before drying with sodium.

3. Method of Preparation

Allylamine has been prepared by the hydrolysis of allyl isothiocyanate with dilute sulfuric [1] or hydrochloric [2] acid.

γ-AMINOBUTYRIC ACID

(Butyric acid, γ-amino-)

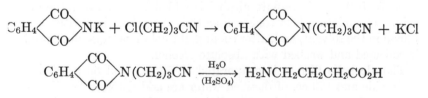

Submitted by C. C. DeWitt.
Checked by W. W. Hartman and A. J. Schwaderer.

1. Procedure

In a 1-l. round-bottomed flask fitted with a tightly fitting cork stopper carrying an air condenser are placed 100 g. (0.54 mole) of finely

[1] Hofmann, Ber. **1**, 183 (1868).
[2] Gabriel and Eschenbach, ibid. **30**, 1124 (1897).

powdered potassium phthalimide (Org. Syn. Coll. Vol. I, **1941**, 119) and 52 g. (0.5 mole) of γ-chlorobutyronitrile (Org. Syn. Coll. Vol. I, **1941**, 156). The flask is heated in an oil bath maintained at 150–180° for one and one-half hours (Note 1) and then allowed to cool. The excess potassium phthalimide and the potassium chloride formed are removed by extraction with several portions of boiling distilled water until the wash water gives no test for chloride ion. The flask is then cooled and the product caused to solidify, and the remaining water is decanted as completely as possible. The solid is treated with 140 cc. of concentrated sulfuric acid, and the mixture is warmed gently in an oil bath under a reflux condenser until all the γ-phthalimidobutyronitrile is brought into solution. Through the reflux condenser 200 cc. of distilled water is added carefully and the solution is refluxed vigorously for three hours. The mixture is cooled and allowed to stand overnight, and the phthalic acid is filtered. The filtrate is transferred to a large evaporating dish, 1 l. of distilled water is added, and then an excess of barium carbonate (about 550 g.) is added in small portions (Note 2). The mixture is evaporated nearly to dryness on the steam bath, and the residue is stirred throroughly with 1 l. of distilled water and again evaporated (Note 3). Finally 1 l. of distilled water is stirred with the solid and the mixture is filtered on a large Büchner funnel. The precipitate is washed with three 200-cc. portions of hot distilled water, and the filtrate and washings are concentrated to a volume of 200 cc. on the steam bath. After the addition of 2 g. of activated carbon the solution is filtered by suction, using a No. 42 Whatman paper, and the charcoal is washed with several small portions of hot distilled water. The filtrate is concentrated on the steam bath to the point of crystallization (about 75 cc.), and 375–500 cc. of absolute alcohol is added to precipitate the amino acid. The mixture is stirred well so that the yellow impurities are retained in the solvent, and, after cooling, the colorless, crystalline product is collected and washed with absolute alcohol.

The alcoholic filtrate is evaporated to 50 cc., and 50 g. of barium hydroxide and 150 cc. of distilled water are added (Note 4). The mixture is refluxed for two hours and the excess barium hydroxide is precipitated with carbon dioxide. The barium carbonate is removed by filtration and washed with hot distilled water. A slight excess of sulfuric acid is added to the filtrate to liberate the amino acid from its barium salt, and an excess of barium carbonate is added to remove sulfate ion. The mixture is digested on the steam bath until effervescence ceases, and then it is filtered and the precipitate is washed with hot distilled water. The filtrate and washings are concentrated on the steam bath to a volume of 100 cc., decolorized with 1 g. of active carbon, filtered, and concentrated to the point of crystallization (about 25 cc.). The amino acid is precipitated by the addition of 150 cc. of absolute alcohol, and the product is collected and washed with absolute alcohol.

The combined yield is 24–32 g. (47–62 per cent on the basis of the γ-chlorobutyronitrile used). The amino acid may be recrystallized by dissolving it in the least possible amount of distilled water and adding 5 to 7 volumes of absolute alcohol.

2. Notes

1. It is advisable to interrupt the heating after about forty-five minutes and thoroughly mix the pasty material by means of a glass rod. Longer heating of the reaction mass, although unnecessary, does no harm.

2. The reagent neutralizes the sulfuric acid and decomposes the ammonium sulfate, but it does not react with the amino acid.

3. Usually the ammonia is removed completely by these two treatments, but if it is not the thorough mixing of the solid with water and the subsequent evaporation must be repeated.

4. The alcoholic filtrate contains appreciable amounts of pyrrolidone. The treatment with excess barium hydroxide converts this into the barium salt of the amino acid.[1, 2]

3. Methods of Preparation

γ-Aminobutyric acid has been prepared by the electrolytic reduction of succinimide to pyrrolidone and hydrolysis of the pyrrolidone by means of barium hydroxide;[1] by the oxidation of piperylurethan with fuming nitric acid and treatment of the resulting product with concentrated hydrochloric acid in sealed tubes at 100°;[3] by the hydrolysis of the condensation product of N-(β-bromoethyl)-phthalimide and sodiomalonic ester;[4] and by the method described above which is a slight modification of that of Gabriel.[2]

[1] Tafel and Stern, Ber. **33**, 2224 (1900).
[2] Gabriel, ibid. **22**, 3335 (1889); **23**, 1771 (1890).
[3] Schotten, ibid. **16**, 643 (1883).
[4] Aschan, ibid. **24**, 2450 (1891).

ε-AMINOCAPROIC ACID

(Caproic acid, ε-amino-)

$$(CH_2)_5 \diagdown \genfrac{}{}{0pt}{}{NH}{CO} + H_2O \rightarrow NH_2(CH_2)_5CO_2H$$

Submitted by J. C. Eck.
Checked by Louis F. Fieser and C. H. Fisher.

1. Procedure

In a 500-cc. round-bottomed flask, 50 g. (0.44 mole) of 2-ketohexa-methylenimine (p. 371) is added to a solution of 45 cc. of concentrated hydrochloric acid (sp. gr. 1.19) in 150 cc. of water. The solution is boiled for about one hour until it becomes clear (Note 1) and evaporated to dryness under reduced pressure on a steam bath.

The resulting ε-aminocaproic acid hydrochloride is converted to the free acid by a procedure similar to that used in the preparation of dl-alanine (Org. Syn. Coll. Vol. I, 1941, 21). The hydrochloride is dissolved in 1 l. of water in a 1.5-l. beaker and treated successively with 50 g. of powdered litharge, 25 g. of powdered litharge, 5 g. of freshly precipitated lead hydroxide, 25 g. of powdered silver oxide (Note 2), and, finally, hydrogen sulfide. During this procedure, the original volume is maintained by the addition of small amounts of water.

After the complete removal of halogen and metallic ions, the solution is concentrated to a volume of about 100 cc., and 300 cc. of absolute alcohol is added. Then the amino acid is precipitated by slowly adding 500 cc. of ether with stirring and cooling.

The resulting ε-aminocaproic acid is collected on a suction filter and dried in a desiccator. The yield of ε-aminocaproic acid melting at 201–203° is 52.5–53.5 g. (90–92 per cent of the theoretical amount).

2. Notes

1. This indicates that the hydrolysis is complete.
2. The exact amount of silver oxide required may be determined by titrating a sample of the solution with silver nitrate by the Volhard method.

3. Methods of Preparation

ε-Aminocaproic acid has been prepared by the hydrolysis of ε-benzoylaminocapronitrile,[1] by the hydrolysis of ethyl δ-phthalimidobutylmalonate,[2] and from cyclohexanone oxime by rearrangement and hydrolysis.[3]

α-AMINOISOBUTYRIC ACID

(Isobutyric acid, α-amino-)

$$CH_3COCH_3 + NaCN + NH_4Cl \rightarrow (CH_3)_2C(OH)CN + NaCl + NH_3$$

$$(CH_3)_2C(OH)CN + NH_3 \rightarrow (CH_3)_2C(NH_2)CN + H_2O$$

$$(CH_3)_2C(NH_2)CN + 2H_2O + 2HBr \rightarrow$$
$$(CH_3)_2C(NH_2 \cdot HBr)CO_2H + NH_4Br$$

$$(CH_3)_2C(NH_2 \cdot HBr)CO_2H + C_5H_5N \rightarrow$$
$$(CH_3)_2C(NH_2)CO_2H + C_5H_5N \cdot HBr$$

Submitted by H. T. CLARKE and H. J. BEAN.
Checked by C. S. MARVEL and C. F. BAILEY.

1. Procedure

A FILTERED solution of 200 g. (3.7 moles) of ammonium chloride in 500 cc. of water is placed in a 3-l. round-bottomed flask. The flask is surrounded by an ice bath and cooled to 5–10°. A solution of 175 g. (3 moles) of acetone in 500 cc. of ether is added with stirring (Note 1). Then a solution of 160 g. (3.2 moles) of sodium cyanide in 350 cc. of water is added, with stirring, at such a rate that the temperature never exceeds 10° (Note 2).

The reaction mixture is stirred for one hour after all the cyanide has been added and then is allowed to stand overnight. The ether layer is separated and the aqueous liquor is extracted with six 300-cc. portions of ether. The ether extracts are combined and the ether is distilled. The residue, which consists mainly of acetone cyanohydrin, is diluted with 800 cc. of methyl alcohol. The solution is cooled and saturated

---

[1] von Braun and Steindorff, Ber. **38**, 176 (1905); von Braun, ibid. **40**, 1839 (1907); Ruzicka and Hugoson, Helv. Chim. Acta **4**, 479 (1921); Marvel, MacCorquodale, Kendall, and Lazier, J. Am. Chem. Soc. **46**, 2838 (1924).

[2] Gabriel and Maass, Ber. **32**, 1266 (1899).

[3] Wallach, Ann. **312**, 188 (1900); Eck and Marvel, J. Biol. Chem. **106**, 387 (1934).

with ammonia gas (Note 3). The reaction mixture is allowed to stand for two or three days (Note 4), and the excess ammonia is expelled by a current of air. The methyl alcohol is removed by distillation as completely as possible, and 600 cc. of water is added to the residue. Then 1 kg. of 48 per cent hydrobromic acid is added and the mixture is refluxed for two hours.

The hydrobromic acid is distilled under reduced pressure on a steam bath. The residue is treated with 400–500 cc. of water, and the solution is again concentrated under reduced pressure to remove as much hydrobromic acid as possible (Note 5).

The residue is dissolved in fifteen to twenty times its weight of methyl alcohol (Note 6) and filtered, and an excess of pyridine (Note 7) is added. The free amino acid separates on standing overnight. It is collected on a Büchner funnel, washed thoroughly with methyl alcohol, and dried. The yield is 92–102 g. (30–33 per cent of the theoretical amount). If a pyridine-free product is desired, it is dissolved in 200 cc. of warm water and filtered, and the filtrate is poured into 2 l. of methyl alcohol (Note 8). There is less than 10 g. of product in the mother liquors. It may be isolated by evaporating to dryness, washing with methyl alcohol, and purifying by reprecipitation in the same way.

2. Notes

1. Vigorous stirring is necessary to obtain the best results.

2. The reaction temperature may rise to 15° without lowering the yield. If the temperature falls to 0°, the reaction does not take place readily.

3. The excess of ammonia is necessary to cause the formation of the aminonitrile from the acetone cyanohydrin formed in the first stage of the process.

4. In some runs this time was only twenty-four hours and no serious diminution of the yield was noted.

5. After addition of the water and subsequent evaporation almost to dryness it is well to add another small portion of water (25–75 cc.) and again evaporate to dryness to ensure the complete removal of hydrobromic acid. This should be done several times if necessary.

6. The amount of methyl alcohol should not exceed 3 l.; otherwise the amino acid will be precipitated incompletely. Long stirring in the cold may be necessary to effect complete solution, though apparently no difficulty is encountered if the residue does not dissolve completely.

7. The minimum amount of pyridine necessary is determined by the amount of hydrobromic acid remaining in the residue. An excess of

pyridine does no harm. If it is desired to use the minimum amount, pyridine is added in small portions until the solution is neutral to Congo red, and then an additional 250 g. is added.

8. A further small quantity may be obtained by evaporating the mother liquor to a small volume on a steam bath, allowing it to crystallize, and washing the crystals with methyl alcohol.

3. Methods of Preparation

The only satisfactory method of preparing α-aminoisobutyric acid is the Strecker synthesis[1] in one or another of its modifications.[2] The process of isolating the product by treating an alcoholic solution of the hydrobromide with pyridine is essentially the same as that developed for glycine;[3] alternatively, aniline may be used.[4]

2-AMINO-4-METHYLTHIAZOLE

(Thiazole, 2-amino-4-methyl-)

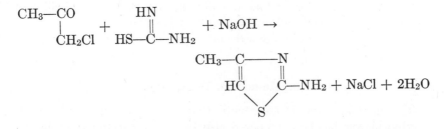

Submitted by J. R. Byers and J. B. Dickey.
Checked by Louis F. Fieser and Byron Riegel.

1. Procedure

Seventy-six grams (1 mole) of thiourea is suspended in 200 cc. of water (Note 1) in a 500-cc. flask equipped with a reflux condenser, dropping funnel, and mechanical stirrer. The stirrer is started, and

[1] Strecker, Ann. **75**, 28 (1850).

[2] Tiemann and Friedländer, Ber. **14**, 1970 (1881); Marckwald, Neumark, and Stelzner, ibid. **24**, 3283 (1891); Gulewitsch, ibid. **33**, 1900 (1900); Hellsing, ibid. **37**, 1923 (1904); Gulewitsch and Wasmus, ibid. **39**, 1184 (1906); Zelinsky and Stadnikoff, ibid. **39**, 1726 (1906); Cocker and Lapworth, J. Chem. Soc. **1931**, 1391.

[3] Org. Syn. **4**, 31 (1925).

[4] Benedict, J. Am. Chem. Soc. **51**, 2277 (1929).

92.5 g. (80 cc., 1 mole) of chloroacetone (Note 2) is run in during thirty minutes. As the reaction proceeds the thiourea dissolves and the temperature rises. The yellow solution is refluxed for two hours and cooled, and, while the mixture is stirred continuously but not so rapidly as to produce a troublesome emulsion, 200 g. of solid sodium hydroxide is added with cooling. The upper, oily layer is separated in a separatory funnel and the aqueous layer is extracted three times with ether, using a total of 300 cc. (Note 3). The dark red oil is combined with the ethereal extract, and the solution is dried over 30 g. of solid sodium hydroxide and filtered by gravity to remove small amounts of tar. The ether is removed by distillation from a steam bath, and the oil is distilled at reduced pressure. After a very small fore-run, 2-amino-4-methyl-thiazole is collected at 117–120°/8 mm., or 130–133°/18 mm. The yield of material melting at 44–45° is 80–85.5 g. (70–75 per cent of the theoretical amount).

2. Notes

1. The reaction may be conducted without this diluent, but it is then likely to become violent.

2. Commercial chloroacetone was distilled and the fraction boiling at 116–122° taken; nearly all of this boiled at 118–120°.

3. If a precipitate is produced and causes an emulsion, add ice and water until it dissolves.

3. Methods of Preparation

The method given is essentially that of Traumann.[1] 2-Amino-4-methylthiazole has been prepared also from chloroacetone and ammonium thiocyanate;[2] from chloroacetone and ammonium thiocyanate in ammonia water;[3] by the action of ammonium thiocyanate on thiocyano-acetone;[3] by saponifying and decarboxylating the cyclic ester from ethyl γ-bromoacetoacetate and thiourea;[4] and from thiocyanoacetone and ammonia in absolute ether.[5]

[1] Traumann, Ann. **249**, 37 (1888).

[2] Hantzsch and Traumann, Ber. **21**, 938 (1888); Hantzsch and Weber, ibid. **20**, 3118 (1887).

[3] Tscherniac and Norton, ibid. **16**, 345 (1883); Tcherniac, J. Chem. Soc. **115**, 1071 (1919).

[4] Steude, Ann. **261**, 33 (1891).

[5] Hantzsch, Ber. **61**, 1785 (1928).

1,2-AMINONAPHTHOL HYDROCHLORIDE

(2-Naphthol, 1-amino-, hydrochloride)

I

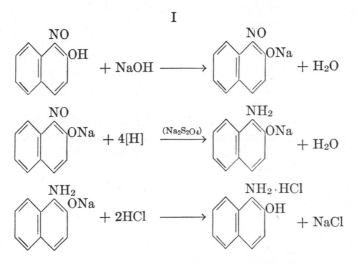

Submitted by J. B. Conant and B. B. Corson.
Checked by Frank C. Whitmore and A. L. Osterhof.

1. Procedure

In an 8-l. (2-gal.) earthenware crock, equipped with a mechanical stirrer and a tube for introducing steam, are placed 240 g. (1.39 moles) of nitroso-β-naphthol (from 200 g. of β-naphthol, Org. Syn. Coll. Vol. I, **1941**, 411) (Note 1), 1.5 l. of water, and 300 cc. of 5 N sodium hydroxide. The lumps are broken up with a rod, and the mixture is stirred for about thirty minutes. At the end of this time, practically all the nitroso compound is dissolved and 1.2 l. of 5 N sodium hydroxide is added. Steam is passed in until the temperature of the mixture is 35°, and then the steam is shut off and 600 g. of technical sodium hydrosulfite (at least 85 per cent pure) is added while the solution is being stirred. The solution is stirred continuously for five minutes while the temperature rises to 60–65°, and then for one minute at five-minute intervals during one-half hour. At the end of about fifteen minutes the solution becomes clear and light yellow in color. A small amount of black scum floats on the surface.

The solution is now cooled to 20° by the addition of about 1 kg. of ice and 500 cc. of technical concentrated hydrochloric acid (sp. gr. 1.16) is added with stirring. This precipitates the aminonaphthol as a voluminous, almost white precipitate (Note 2) which is collected on two 20-cm. Büchner funnels. It is rapidly pressed as free from mother liquor as possible and transferred quickly (Note 3) to an 8-l. (2-gal.) crock containing 2.5 l. of water and 250 cc. of technical concentrated hydrochloric acid. The large lumps are broken up with a rod, steam is passed in, and the stirrer is started. Steam is introduced at such a rate that a temperature of 85–90° is reached within ten minutes; the mixture is stirred at this temperature for one hour longer. The hot mixture is then filtered through a 15-cm. Büchner funnel, and the filtrate is cooled to 35–40° in an ice bath.

The claret-colored solution (5.2–5.8 l.) is filtered through a fluted filter paper in a 15-cm. funnel into 1.2 l. of concentrated hydrochloric acid in an 8-l. bottle. The aminonaphthol hydrochloride starts to precipitate immediately. The mixture is allowed to stand for at least two hours with occasional agitation to ensure complete precipitation. The hydrochloride is collected on a 15-cm. Büchner funnel and washed successively with three small portions of 20 per cent hydrochloric acid and three 50-cc. portions of ether (Note 4). It is then dried in the air in thin layers on filter paper. The yield is 180–200 g. of anhydrous material (Note 5) (66–74 per cent of the theoretical amount based on 200 g. of β-naphthol). The hydrochloride is unstable; in solution it decomposes rapidly, but this decomposition can largely be prevented by the addition of sodium bisulfite. The dry solid slowly changes and should be used within a few weeks of its preparation (Note 6).

2. Notes

1. It is usually not convenient to dry the nitroso-β-naphthol. The amount given in these directions corresponds to 200 g. of β-naphthol, and the yield is calculated on this basis. The nitroso-β-naphthol dissolves in one mole of sodium hydroxide, forming a green solution. There is left in suspension a small amount of amorphous brown material which need not be removed.

2. The amounts of sodium hydroxide, sodium hydrosulfite, and hydrochloric acid used are such that complete precipitation of the aminonaphthol results at this point. It is well to test for complete precipitation, however, by adding a few drops of alkali to one portion of the filtrate and a little acid to another.

3. The aminonaphthol is very sensitive to atmospheric oxidation. When first precipitated it is white, but it becomes purple in the air and therefore should be handled rapidly. A small amount of sodium bisulfite

may be added at this point to decrease the oxidation, but this is usually unnecessary since some bisulfite from the reduction adheres to the precipitate and is carried through to the final precipitation.

4. Washing with ether greatly facilitates the drying of the product and also lightens its color somewhat. If the material is to be used without being dried, the washing with ether may be omitted.

5. Air-dried material, when apparently completely dry, contains 10–15 per cent of moisture. In order to obtain the true weight of the material an aliquot sample should be dried over sodium hydroxide under reduced pressure.

6. The product, as first formed, is a very light purple but darkens on long standing. It may be purified by dissolving in hot water containing sodium bisulfite, filtering, and reprecipitating with hydrochloric acid.

II

$$+ \; ^+N_2C_6H_4SO_3^- + NaOH \rightarrow$$

$$+ H_2O$$

$$+ \; 2Na_2S_2O_4 + 4H_2O \rightarrow$$

$$+ \; 4NaHSO_3 + H_2NC_6H_4SO_3Na$$

Submitted by Louis F. Fieser.
Checked by C. R. Noller and W. R. White.

1. Procedure

(A) *Diazotization of Sulfanilic Acid.*—A mixture of 105 g. (0.5 mole) of sulfanilic acid dihydrate, 26.5 g. (0.25 mole) of anhydrous sodium carbonate, and 500 cc. of water is heated and stirred until all the sulfanilic acid has dissolved, and the solution is then cooled in an ice bath to 15° (sodium sulfanilate begins to crystallize at this temperature). A solution of 37 g. (0.54 mole) of sodium nitrite in 100 cc. of water is added and the resulting solution is poured at once onto a mixture of 106 cc. (1.25 moles) (Note 1) of concentrated hydrochloric acid (sp. gr. 1.18) and 600 g. of ice contained in a 2-l. beaker. The solution, from which p-benzenediazonium sulfonate separates on stirring, is allowed to stand in an ice bath for fifteen to twenty-five minutes, during which time the naphthoxide solution is prepared.

(B) *Coupling: Orange II.*—Seventy-two grams of β-naphthol (0.5 mole) is dissolved in the warm solution obtained by dissolving 110 g. (2.75 moles) (Note 2) of sodium hydroxide in 600 cc. of water in a 5-l. flask, and the solution is cooled to about 5° by the addition of 400 g. of ice. The suspension of the diazonium salt then is added and the mixture is stirred well and allowed to stand without external cooling for one hour (Note 3). The azo compound soon separates from the red solution and eventually forms a stiff paste.

(C) *Reduction: 1,2-Aminonaphthol Hydrochloride.*—The suspension of Orange II is heated to 45–50°, when all the material dissolves with slight evolution of gas. About one-tenth of 230 g. (about 1.1 moles) of technical sodium hydrosulfite (Note 4) is added cautiously and the mixture is stirred until the froth subsides; the remainder is then added without delay. The yellow material which first separates (probably the hydrazo compound) soon is converted into the nearly colorless amino-naphthol. In order to complete the reduction and to give an easily filterable product the mixture is heated strongly until it begins to froth; it is then cooled to 25° by stirring in an ice bath, and the pink or cream-colored product is collected and washed free from the slightly yellow mother liquor with water.

The crude aminonaphthol is washed into a beaker containing a solution at 30° of 2 g. of stannous chloride dihydrate and 53 cc. (0.63 mole) of concentrated hydrochloric acid in 1 l. of water. When the mixture is stirred the amine soon dissolves, leaving in suspension a small amount of fluffy material which is easily distinguishable from the original lumps (Note 5). The solution is clarified by stirring (without heating) for five minutes with 10 g. of decolorizing carbon, and it is then filtered by suction. The pale yellow solution is treated with 50 cc. of concentrated hydrochloric acid and heated to the boiling point, a second 50 cc. of the acid being added as the heating progresses. The color becomes somewhat fainter during this process. The vessel containing the hot solution is transferred to an ice bath and allowed to cool undisturbed, and the 1-amino-2-naphthol hydrochloride soon separates in the form of large, perfectly colorless needles. When fairly cold, 100 cc. of concentrated hydrochloric acid is added and the solution is cooled to 0° before collecting the product (Note 6). The hydrochloride is washed with a cold solution of 50 cc. of concentrated hydrochloric acid in 200 cc. of water and dried on a filter paper at a temperature not above 30–35°. The yield is 70–83 g. (72–85 per cent of the theoretical amount). The material will remain colorless, or very nearly so, if protected from the light in storage. The fresh solution in water is only faintly colored and leaves but a trace of residue on filtration.

Although this material is suitable for most purposes, it may be purified further in the following manner. It is dissolved by heating in a solution of 2 g. of stannous chloride and 2 cc. of concentrated hydrochloric acid in 1 l. of water, and the hot solution is clarified by filtration through a 5-mm. mat of decolorizing carbon (Note 7). The yellow or red color which may develop disappears on reheating to the boiling point. After the addition of 100 cc. of concentrated hydrochloric acid the solution is allowed to cool in an ice bath, treated with a second 100 cc. of acid, cooled to 0°, and collected and washed as before. The crystalline product is colorless, ash-free, and of analytical purity. The loss in the crystallization of an 80-g. lot amounts to 5–10 g. (6–12 per cent).

2. Notes

1. Diazotization can be accomplished by the use of just one equivalent of acid (0.5 mole), but the solution of the diazoic acid, $NaO_3SC_6H_4N{=}NOH$, so formed is much less stable than the suspension of the inner salt which results from the use of more acid.

2. The excess alkali is not required for the process of coupling, but rather to provide conditions suitable for the reduction.

3. The yield is not improved by allowing a longer period for the reaction. Under the conditions specified, the mixture during the coupling remains at a temperature of 5–10°.

4. If the hydrosulfite is of poor quality more will be needed, and an additional amount of sodium hydroxide should also be added.

5. The solution is highly supersaturated, but it will remain so unless allowed to stand for an undue amount of time. It is also a mistake to add the quantity of concentrated hydrochloric acid specified to a suspension of the aminonaphthol, for this may initiate crystallization.

6. An alternative method of crystallization is to add all the hydrochloric acid (200 cc.) to the boiling solution and to allow this to cool slowly; very large, thick needles result. In the presence of stannous chloride there is no danger of a darkening of the solution as the result of oxidation.

7. This method is preferable to the usual one when dealing with a substance sensitive to air oxidation.

3. Methods of Preparation

1-Amino-2-naphthol has been obtained from β-naphthylamine [1] and, more practically, from β-naphthol through the nitroso compound or an azo compound. Nitroso-β-naphthol has been reduced in alkaline solution with hydrogen sulfide [2, 3] or sodium hydrosulfite,[4] but early workers encountered difficulty in converting the amine into its hydrochloride without undue oxidation. Sulfur dioxide was employed as an antioxidant, but it is wholly inadequate. Reduction in an acidic medium, usually with stannous chloride, has been more satisfactory. The isolation of the amine stannochloride and its tedious decomposition with hydrogen sulfide [5] are unnecessary, for the amine hydrochloride can be caused to crystallize essentially free from tin by avoiding an excess of the reducing agent.[2, 6, 7] Nitroso-β-naphthol has been reduced also with zinc dust and sulfuric acid,[8] but the quality of the material, used for conversion to the quinone, is in some doubt.

The nitroso derivative has disadvantages as an intermediate in that it involves handling either a voluminous precipitate or a large volume of solution, and in that some tar is likely to form; hence an azo compound is preferable. Technical Orange II has been reduced in a neutral or alkaline medium with sodium sulfide [2] or sodium hydrosulfite,[9] the sulfanilic acid being eliminated as the soluble sodium salt. With stannous chloride, the necessity of isolating the amine stannochloride [1] can be avoided by using just the calculated amount of reagent, the resulting mixture of amine hydrochloride and sulfanilic acid being separated with an alkaline buffer.[2] Witt [10] found that the sulfanilic acid can be kept in solution if this is sufficiently acidic, and with this improvement Russig [11] worked out a procedure for the preparation and reduction of

[1] Liebermann and Jacobson, Ann. **211**, 49 (1882).

[2] Groves, J. Chem. Soc. **45**, 294 (1884); Stenhouse and Groves, Ann. **189**, 153 (1877).

[3] Lagodzinski and Hardine, Ber. **27**, 3075 (1894); Böeseken, Rec. trav. chim. **34**, 272 (1915); Porai-Koschitz, Ger. pat. 463,519 (Chem. Zentr. **1928**, II, 1384).

[4] See p. 33.

[5] Grandmougin and Michel, Ber. **25**, 974 (1892).

[6] Zincke, Ann. **268**, 274 (1892).

[7] Paul, Z. angew. Chem. **10**, 48 (1897).

[8] Skita and Rohrmann, Ber. **63**, 1482 (1930).

[9] Grandmougin, ibid. **39**, 3561 (1906).

[10] Witt, ibid. **21**, 3472 (1888).

[11] Russig, J. prakt. Chem. (2) **62**, 56 (1900); Böeseken, Rec. trav. chim. **41**, 780 (1922).

Orange II which was reported to give excellent yields but which, judging from the results of conversion to the quinone, affords a poor product.

Orange II also has been reduced with zinc dust and hydrochloric acid,[12] by electrolysis,[13] and by catalytic hydrogenation,[14] while a number of other azo dyes derived from β-naphthol have been reduced to 1,2-aminonaphthol with hydrogen and Raney nickel.[15]

The method described above is novel chiefly in that it makes use of stannous chloride as an antioxidant in preparing and crystallizing the amine hydrochloride. The method is applicable with slight modifications to the preparation of many other aminophenols.[16]

1,4-AMINONAPHTHOL HYDROCHLORIDE

(1-Naphthol, 4-amino-, hydrochloride)

Submitted by Louis F. Fieser.
Checked by C. R. Noller and W. R. White.

1. Procedure

ONE HUNDRED AND FIVE grams (0.5 mole) of sulfanilic acid dihydrate is diazotized exactly as described in the last paragraph of p. 35. While

[12] Zincke, Ann. **278**, 188 (1894).

[13] Boehringer and Sons, Ger. pat. 121,835 (Chem. Zentr. **1901**, II, 152): Hubbuch and Lowy, C. A. **23**, 343 (1929).

[14] Tetralin G.M.B.H., Ger. pat. 406,064 [Frdl. **14**, 395 (1921–25)].

[15] Whitmore and Revukas, J. Am. Chem. Soc. **62**, 1687 (1940).

[16] Fieser and Fieser, ibid. **57**, 491 (1935).

the suspension of p-benzenediazonium sulfonate is kept in an ice bath, 72 g. (0.5 mole) of α-naphthol (Note 1) is dissolved in the warm solution obtained by dissolving 110 g. (2.75 moles) of sodium hydroxide in 600 cc. of water in a 5-l. flask. The naphthoxide solution is cooled to 25°, 400 g. of ice is added, and the suspension of the diazonium salt is then introduced (Note 2). The mixture is stirred well and allowed to stand for one hour, during which time the formation of the dye, Orange I, goes to completion.

The deep purple-red solution of Orange I is warmed to 45–50°, about one-tenth of 230 g. (about 1.1 moles) of technical sodium hydrosulfite (Note 3) is added cautiously, and the mixture is stirred until the froth subsides; the remainder of the hydrosulfite is then added without delay. The tan suspension of the aminonaphthol is heated to about 70° to effect sufficient coagulation to permit filtering. It is then cooled quickly to 25° by stirring in an ice bath, and the precipitate is filtered and washed with fresh 1 per cent sodium hydrosulfite solution.

The crude aminonaphthol is transferred quickly to a beaker containing a solution of 2 g. of stannous chloride dihydrate and 63 cc. of hydrochloric acid in 800 cc. of water at 30°. On stirring and warming, the amine dissolves. The solution, which is usually a deep red, is filtered with suction; treatment with charcoal is not necessary. One hundred cubic centimeters of concentrated hydrochloric acid is added, the solution is heated to boiling for five to ten minutes, and a second 100-cc. portion of the acid is added. During this heating the color fades to a light yellow, and on cooling to 0° a mass of small, nearly colorless crystals is obtained.

This material (dry weight, 77–80 g.) (Note 4), without being dried, is dissolved by heating in a solution of 2 g. of stannous chloride and 2 cc. of concentrated hydrochloric acid in 700 cc. of water. The hot solution is clarified by filtration through a 5-mm. mat of decolorizing carbon. One hundred cubic centimeters of concentrated hydrochloric acid is added to the hot solution, which is then cooled in an ice bath, treated with a second 100-cc. portion of acid, and cooled to 0°. The precipitate is filtered and washed with a cold solution of 50 cc. of concentrated hydrochloric acid in 200 cc. of water. The 1,4-aminonaphthol hydrochloride forms small, nearly colorless needles of a high degree of purity. The solution in water is faintly pink, and the crystals may acquire a slight pink color after a few weeks. The yield is 70–73 g. (72–75 per cent of the theoretical amount) (Note 5).

2. Notes

1. The α-naphthol should be free from the β-isomer; if the material is very highly colored it is advisable to purify it by distillation at atmospheric pressure. Material melting at 95–96° is satisfactory.

2. A low temperature is required during the coupling in order to avoid the formation of the disazo compound.

3. If the hydrosulfite is of poor quality more will be needed, and an additional amount of sodium hydroxide should also be added.

4. The material is slightly yellow and may redden on drying; it probably contains a trace of 2,4-diamino-1-naphthol.

5. These directions are very similar to the directions on p. 35 for preparing 1,2-aminonaphthol and should be compared with those directions. The differences between the two procedures are the result of the greater solubility and the greater sensitivity to air oxidation of the 1,4-aminonaphthol.

3. Methods of Preparation

The usual method of preparing 1,4-aminonaphthol has been from α-naphthol through an azo dye. The majority of investigators have reduced technical Orange I with stannous chloride;[1, 2, 3, 4, 5, 6] benzeneazo-α-naphthol has been reduced by the same reagent.[7, 8] In order to make possible the use of crude technical α-naphthol a method has been developed[9] for the preparation of the benzeneazo compound, its separation from the isomeric dye coming from the β-naphthol present as well as from any disazo compound by extraction with alkali, and the reduction of the azo compound in alkaline solution with sodium hydrosulfite. The process, however, is tedious and yields an impure product.

1,4-Aminonaphthol can be prepared by reduction of 1,4-nitrosonaphthol, but this is not practical because the starting material is not readily available.[7] The aminonaphthol can also be prepared with a 60 per cent overall yield by the reduction of α-nitronaphthalene to the naphthyl hydroxylamine and rearrangement of the hydroxylamine.[10]

[1] Liebermann and Jacobson, Ann. 211, 49 (1882).
[2] Russig, J. prakt. Chem. (2) 62, 56 (1900); Böeseken, Rec. trav. chim. 41, 780 (1922).
[3] Liebermann, Ann. 183, 247 (1876).
[4] Liebermann, Ber. 14, 1796 (1881).
[5] Seidel, ibid. 25, 423 (1892).
[6] Zincke and Wiegand, Ann. 286, 70 (1895).
[7] Grandmougin and Michel, Ber. 25, 974 (1892).
[8] Fuchs and Pirak, ibid. 59, 2456 (1926).
[9] Conant, Lutz, and Corson, Org. Syn. Coll. Vol. I, 1941, 49.
[10] Neunhoeffer and Liebich, Ber. 71, 2247 (1938).

The method described above is applicable with slight modifications to the preparation of both the ortho and para aminonaphthols and to many homologs, benzologs, and heterocyclic isologs of these substances. The chief feature of novelty is in the use of stannous chloride as an anti-oxidant in preparing and crystallizing the amine hydrochlorides.[11]

1-AMINO-2-NAPHTHOL-4-SULFONIC ACID

(2-Naphthol-4-sulfonic acid, 1-amino-)

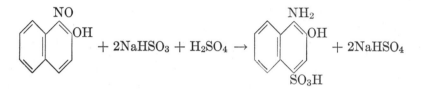

$$+ 2NaHSO_3 + H_2SO_4 \rightarrow \qquad + 2NaHSO_4$$

Submitted by Louis F. Fieser.
Checked by Frank C. Whitmore and D. J. Loder.

1. Procedure

THREE HUNDRED grams of β-naphthol (2.1 moles) is converted into nitroso-β-naphthol (Org. Syn. Coll. Vol. I, **1941**, 411) (Note 1), and the product is transferred to a 6-l. (1.5 gal.) crock which is wide enough to admit the Büchner funnel (30-cm.) employed. A cold solution of 600 g. (5.8 moles) of sodium bisulfite and 100 cc. of 6 N sodium hydroxide solution in 2 l. of water (Note 2) is used to rinse the material adhering to the funnel into the crock. The mixture is diluted with water to 4–4.5 l. and stirred until solution of the nitroso-β-naphthol is complete (about fifteen minutes). The dark solution is siphoned onto a large Büchner funnel and filtered by suction, thus removing a small amount of tarry material which is always present. The clear, yellowish brown filtrate is transferred to an 8- to 10-l. wide-mouthed bottle and diluted with water to 7 l. While the solution is vigorously stirred, 400 cc. of concentrated sulfuric acid is poured slowly down the walls of the bottle; the mixture is then placed in the hood and protected from the light (Note 3). The temperature rises from 20–25° to 35–40° at once and to about 50° in the course of two hours, when the reaction is nearly complete. After standing for a total of five hours or more (Note 4), the precipitate, which sets to a stiff paste in the bottle, is collected on a

[11] Fieser and Fieser, J. Am. Chem. Soc. **57**, 491 (1935).

filter. The residue is transferred to a 1-l. beaker and washed with 200 cc. of water. The mixture is filtered, and the residue is washed with 300 cc. of water on the filter. The moist material weighs 700–800 g. No appreciable decomposition takes place on drying the product to constant weight at 120°. A light powder of fine, gray needles is thus obtained. The yield is 410–420 g. (82–84 per cent of the theoretical amount based on the β-naphthol used) (Note 5).

2. Notes

1. In preparing nitroso-β-naphthol in the quantity here required it is convenient to use a 7- to 8-l. bottle, with an 8- to 10-cm. opening, equipped with a stirrer of heavy glass rod having four or five right-angle bends which extend to the top of the bottle and which are just small enough to fit the mouth. The bottle is placed in a bucket containing a salt-ice mixture which is stirred occasionally by hand. With this arrangement a temperature of 0° may be maintained without internal cooling.

2. Sodium hydroxide is added to the bisulfite solution in order to neutralize any acid which has not been removed by washing, and which would liberate sulfur dioxide and thus cause some reduction of the nitroso compound before the addition product is formed. An excess of alkali, as employed above, aids in the solution of the material.

3. The aminonaphtholsulfonic acid becomes rose colored on long exposure to the light, especially when moist.

4. The time required for the process may be shortened somewhat by adding the sodium bisulfite-sodium hydroxide solution to the suspension of the crude nitroso compound, thus avoiding a long filtration. The amounts of water employed in the various operations should be reduced to a minimum, and enough additional sodium hydroxide solution should be used to neutralize the excess acid present. The product is of a slightly inferior quality, and the yield is 4–5 per cent lower.

5. This gray material is not quite pure and contains water of crystallization so that the percentage yield reported is in error. A better product can be obtained by stirring the mixture of nitroso-β-naphthol and sodium bisulfite solution vigorously by hand with a wooden paddle thus causing all the soluble product to dissolve in three to four minutes. The suspension is then filtered as rapidly as possible using two 15-cm. Büchner funnels and changing filter papers frequently. The clear, golden-yellow filtrate is acidified immediately on completion of the filtration. The product is then light gray, whereas, if much time elapses before the bisulfite solution is acidified, the solution turns red and the

aminonaphtholsulfonic acid may be deep purple-gray in color. After the product has been collected and washed with water, it is washed with warm alcohol until the filtrate is colorless, 1.5–2 l. being required. The product is washed with two 100-cc. portions of ether and dried to constant weight at 60–80° in the absence of light. A pure white, dust-dry product is thus obtained, weighing 370–380 g. (75–78 per cent of the theoretical amount based on the β-naphthol). The wash alcohol does not dissolve an appreciable amount of the aminonaphtholsulfonic acid since evaporation of the deep red wash liquor gives a dark residue weighing only 3–4 g. (E. L. MARTIN and LOUIS F. FIESER, private communication.)

3. Methods of Preparation

1-Amino-2-naphthol-4-sulfonic acid has been prepared by warming 2-naphthoquinone-1-chloroimide with sodium bisulfite solution;[1] by reduction of 1-benzeneazo-2-naphthol-4-sulfonic acid with stannous chloride and hydrochloric acid;[2] by treatment of 1-amino-2-naphthol hydrochloride with sodium sulfite;[3] and by treatment of nitroso-β-naphthol with sodium bisulfite and hydrochloric acid.[4]

4-AMINOVERATROLE

(Aniline, 3,4-dimethoxy-)

$$2(CH_3O)_2C_6H_3CN + 2H_2O_2 \xrightarrow{(KOH)} 2(CH_3O)_2C_6H_3CONH_2 + O_2$$
$$(CH_3O)_2C_6H_3CONH_2 + NaOCl + 2NaOH \rightarrow$$
$$(CH_3O)_2C_6H_3NH_2 + NaCl + Na_2CO_3 + H_2O$$

Submitted by J. S. BUCK and W. S. IDE.
Checked by JOHN R. JOHNSON and H. B. STEVENSON.

1. Procedure

IN a 5-l. flask fitted with a mechanical stirrer are placed 1.8 kg. (1.6 moles) of fresh 3 per cent hydrogen peroxide solution, 100 g. of 25 per cent potassium hydroxide solution, and 57 g. (0.35 mole) of veratronitrile (p. 622). The mixture is warmed slowly to 45°, with stirring, and

[1] Friedländer and Reinhardt, Ber. 27, 241 (1894).
[2] Marschalk, Bull. soc. chim. (4) 45, 660 (1929).
[3] Marschalk, ibid. (4) 45, 662 (1929).
[4] Schmidt, J. prakt. Chem. (2) 44, 522 (1891); Böniger, Ber. 27, 23 (1894).

the source of heat is then withdrawn. The reaction proceeds with evolution of oxygen, and the temperature continues to rise (Note 1). The amide soon begins to separate; in about fifty minutes the reaction is complete and the temperature begins to fall. The mixture is cooled to 3–5° and allowed to remain in the cooling bath for one and one-half to two hours. The white crystalline product is filtered with suction and dried in the air. The veratric amide melts at 162.5–163.5° and weighs 55–58 g. (87–92 per cent of the theoretical amount).

An alkaline solution of sodium hypochlorite is prepared by passing chlorine (0.412 gram for each gram of the amide) (Note 2) into a mixture of 300 g. of cracked ice and a cold solution of 80 g. of sodium hydroxide in 500 cc. of water, contained in a 2-l. round-bottomed flask. The whole of the veratric amide (55–58 g.) is added in one portion, and the mixture is warmed slowly in a water bath, with mechanical stirring. The material soon darkens in color, and at 50–55° (internal temperature) oily droplets begin to separate. The temperature is raised gradually to 70° and maintained at this point for one hour. A solution of 120 g. of sodium hydroxide in 120 cc. of water is added slowly, and the temperature is increased to 80° for an additional hour.

Upon cooling the mixture, the oily layer of amine solidifies to a red crystalline mass. The crude amine is filtered with suction, washed with two 60-cc. portions of ice-cold water, pressed thoroughly, and transferred to an ordinary 125-cc. Claisen flask. The filtrate is extracted with three 60-cc. portions of benzene, the extracts are transferred separately to the Claisen flask, and the benzene is distilled at atmospheric pressure (Note 3). The residual amine is distilled under reduced pressure and is collected at 172–174°/24 mm. (Note 4). The distillate solidifies quickly to a mass of colorless crystals which melt sharply at 87.5–88° (Note 5). From 58 g. (0.32 mole) of veratric amide there is obtained 39–40 g. (80–82 per cent of the theoretical amount) of 4-aminoveratrole.

2. Notes

1. A large amount of frothing occurs, and the temperature rises to 52–55° during the reaction. The flask should be removed at intervals and shaken by hand to bring the material carried up by the froth into contact with the solution.

2. The quantity of chlorine indicated is 5 per cent more than the theoretical amount. The chlorine may be obtained from a cylinder, but for small amounts it is more accurate to generate chlorine by treating a weighed amount of potassium permanganate with an excess of concentrated hydrochloric acid (0.367 g. $KMnO_4 \approx 0.412$ g. Cl_2). For

55–58 g. of veratric amide the weight of permanganate to be used is 20.2–21.3 g.

For generating the chlorine, the requisite weight of potassium permanganate is placed in a 500-cc. distilling flask fitted with a separatory funnel and suspended so that the flask can be agitated. Concentrated hydrochloric acid (about 130 cc. is required) is allowed to drop slowly upon the permanganate crystals, and as the reaction progresses the flask is heated. After all the acid has been added the mixture is boiled gently for a few minutes to expel the last portions of chlorine, and the delivery tube is removed rapidly from the alkaline hypochlorite solution to prevent the solution from being sucked back into the generator. A safety bottle should be inserted between the side tube of the generator and the tube which delivers the chlorine into the alkali.

3. Distillation of the benzene extracts in this way serves to remove the water, and in the subsequent distillation of the amine no watery fore-run is obtained.

4. Since the aminoveratrole tends to solidify in the side tube of the distilling flask, it is advisable to distil the material rapidly and to employ a bath temperature about 60° higher than that of the distilling vapor. Care must be taken to avoid contaminating the distillate with a small amount of colored material which comes over if the distillation is carried too far.

5. The melting point of 4-aminoveratrole obtained by reduction of 4-nitroveratrole is reported as 85–86°.[1] The amine discolors on exposure to air and light; it should be stored in a sealed, dark container.

3. Methods of Preparation

4-Aminoveratrole has been prepared by the reduction of 4-nitroveratrole with tin,[1,2] or stannous chloride,[3] and hydrochloric acid. The present procedure is based upon a method used for the preparation of aminopiperole (3,4-methylenedioxyaniline).[4]

[1] Moureu, Bull. soc. chim. (3) **15**, 647 (1896).

[2] Simonsen and Rau, J. Chem. Soc. **113**, 28 (1918); Pollecoff and Robinson, ibid. **113**, 645 (footnote) (1918).

[3] Heinisch, Monatsh. **15**, 232 (1894).

[4] Rupe and Majewski, Ber. **33**, 3401 (1900).

n-AMYLBENZENE

(Benzene, amyl-)

$$C_6H_5CH_2Cl + Mg \rightarrow C_6H_5CH_2MgCl$$

$$C_6H_5CH_2MgCl + 2p\text{-}CH_3C_6H_4SO_3C_4H_9 \rightarrow$$
$$C_6H_5(CH_2)_4CH_3 + C_4H_9Cl + (p\text{-}CH_3C_6H_4SO_3)_2Mg$$

Submitted by HENRY GILMAN and J. ROBINSON.
Checked by C. S. MARVEL and S. S. ROSSANDER.

1. Procedure

ONE mole of benzylmagnesium chloride is prepared in a 2-l. three-necked, round-bottomed flask from 24.3 g. (1 gram atom) of magnesium turnings, 126.5 g. (115 cc., 1 mole) of benzyl chloride, and 500 cc. of anhydrous ether, according to the directions given in Org. Syn. Coll. Vol. I, **1941**, 471.

The solution of benzylmagnesium chloride is cooled with running water, and 456 g. (2 moles) of *n*-butyl *p*-toluenesulfonate (Note 1) dissolved in about twice its volume of anhydrous ether is then added slowly with stirring through the separatory funnel at such a rate that the ether just boils. The time required for the addition is about two hours. A white solid soon forms and the mixture assumes the consistency of a thick cream. Stirring is continued, without cooling, for about two hours, and the mixture is hydrolyzed by pouring onto crushed ice to which is then added about 125 cc. of concentrated hydrochloric acid (Note 2).

The ether layer is separated and combined with a 200-cc. ether extract of the aqueous layer. The combined ether solution is washed once with about 100 cc. of water and then dried by shaking for a few minutes with about 10 g. of anhydrous potassium carbonate. After filtration, the ether is distilled on a water bath. When practically all the ether has been removed, about 5 g. of sodium, freshly cut and in thin slices, is added and the mixture is boiled for about two hours (Note 3). The solution is decanted and then distilled, using an efficient fractionating column. The fraction boiling at 190–210° is collected. This on redistillation yields 74–88 g. (50–59 per cent of the theoretical amount) of *n*-amylbenzene boiling at 198–202° (Note 4).

2. Notes

1. Directions for the preparation of n-butyl p-toluenesulfonate are given in Org. Syn. Coll. Vol. I, **1941**, 145, and in Org. Syn. **20**, 51.

2. The hydrolysis is preferably carried out in a 5-l. Erlenmeyer flask. The magnesium p-toluenesulfonate is sparingly soluble in hydrochloric acid, and complete solution is brought about by the subsequent addition of about 2 l. of water.

3. Refluxing with sodium helps to remove the small amount of benzyl alcohol formed by the atmospheric oxidation of benzylmagnesium chloride.

4. The major part of the n-amylbenzene distils at 199–201°. A careful fractionation of the distillate that comes over around 75° yields 24 g. (26 per cent of the theoretical amount) of n-butyl chloride boiling at 76–80°.

3. Methods of Preparation

n-Amylbenzene has been prepared by the action of sodium on a mixture of benzyl and butyl bromides;[1] by the reaction between benzyl sodium and butyl chloride;[2] by the reduction of n-valerophenone with formic acid over copper at 300°,[3] or with zinc and hydrochloric acid;[4] by the action of sodium ethoxide on the hydrazone[5] and the semicarbazone[6] of n-valerophenone; and by the procedure described, which is an adaptation of the directions of Gilman and Heck[7] and Rossander and Marvel.[8]

[1] Schramm, Ann. **218**, 388 (1883).

[2] Morton and Fallwell, Jr., J. Am. Chem. Soc. **60**, 1429 (1938).

[3] Mailhe and de Godon, Bull. soc. chim. (4) **21**, 62 (1917).

[4] Stenzl and Fichter, Helv. Chim. Acta **17**, 677 (1934).

[5] Schmidt, Hopp, and Schoeller, Ber. **72**, 1893 (1939).

[6] Ziegler, Dersch, and Wollthan, Ann. **511**, 38 (1934).

[7] Gilman and Heck, J. Am. Chem. Soc. **50**, 2223 (1928).

[8] Rossander and Marvel, ibid. **50**, 1491 (1928).

d-ARGININE HYDROCHLORIDE

Gelatine $\xrightarrow{\text{Hydrolysis}}$ Amino acids

$$NH_2C(NH)NHCH_2CH_2CH_2CH(NH_2)CO_2H + C_6H_5CHO \rightarrow$$
$$NH_2C(NH)NHCH_2CH_2CH_2CH(NCHC_6H_5)CO_2H \cdot H_2O$$

$$C_{13}H_{18}O_2N_4 \cdot H_2O + HCl \rightarrow$$
$$NH_2C(NH)NHCH_2CH_2CH_2CH(NH_2HCl)CO_2H + C_6H_5CHO$$

Submitted by E. BRAND and M. SANDBERG.
Checked by H. T. CLARKE and S. GRAFF.

1. Procedure

Benzylidenearginine, Method 1.—To 500 g. of gelatine (Note 1) is added 1.5 l. of concentrated hydrochloric acid (sp. gr. 1.19); the mixture is warmed on the steam bath for thirty minutes and boiled over a free flame for eight to ten hours (Note 2) under a reflux condenser provided with a trap for hydrogen chloride gas. The solution is concentrated to 400 cc. on the steam bath under reduced pressure, employing the apparatus shown in Org. Syn. Coll. Vol. I, **1941**, 427. The syrupy residue is then diluted with 500 cc. of distilled water and again concentrated to 400 cc. This process of dilution and evaporation is repeated twice more (Note 3). The final residue is dissolved in 500 cc. of hot distilled water and decolorized by adding 15 g. of decolorizing carbon and heating for ten minutes on the steam bath. The filtrate is chilled in an ice-salt bath and treated with 250–350 cc. of a 40 per cent solution of sodium hydroxide until slightly alkaline to litmus, keeping the temperature below 10°. An additional 70-cc. portion of 40 per cent sodium hydroxide solution is added, keeping the temperature below 5°; this is followed by the addition, in four portions, of 225 cc. of benzaldehyde, with vigorous shaking after each addition, the temperature being held below 5° throughout (Note 4). The addition of the benzaldehyde occupies about ten minutes.

The resulting emulsion is allowed to stand overnight in the refrigerator at 0–5°; the crystalline precipitate is filtered by suction and washed first with 80 cc. of ice-cold water in four portions, then with 50 cc. of a mixture of two volumes of ether and one volume of methyl alcohol, and, finally, with ether (Note 5) until the washings are colorless and free of benzaldehyde. After drying in a vacuum desiccator the product weighs 35–40 g.; it melts with decomposition at 206–207° (corr.).

Benzylidenearginine, Method 2.—The hydrolysis of 500 g. of gelatine and the removal of excess hydrochloric acid are conducted as described above. After treatment with 15 g. of decolorizing carbon the filtrate is diluted to 2.5 l., heated almost to boiling, and treated with 110 g. of 2,4-dinitro-1-naphthol-7-sulfonic ("flavianic") acid (Note 6) dissolved in 400 cc. of hot water. The mixture is boiled (Note 7) for about three minutes and diluted with boiling water to a total volume of 4 l. The mixture is cooled rapidly to 45° and then allowed to cool slowly to room temperature, with occasional stirring and vigorous scratching of the walls of the container. After standing for about two hours at room temperature (Note 8), practically all (Note 9) the arginine dinitro-naphtholsulfonate should have separated in crystalline, readily filterable form. The product is filtered by suction and washed first with three 100-cc. portions of a 0.5 per cent solution of dinitronaphtholsulfonic acid and then with two 25-cc. portions of 95 per cent ethyl alcohol. After being dried in air, the product weighs 113–118 g. (Note 10); it decomposes at 245–265° (Note 11).

One hundred grams (0.21 mole) of finely powdered arginine dinitro-naphtholsulfonate is added, all at once, to 230 cc. of cold 2 N sodium hydroxide; the salt dissolves readily on agitation. There is then added *without delay* (Note 12) 35 g. (0.33 mole) of benzaldehyde, in four portions and with vigorous shaking, each portion being accompanied by 75 cc. of ice-cold water. During this process, benzylidenearginine separates as a crystalline cake. The mixture is allowed to stand at 15–20° for one to two hours, whereupon the product is filtered (Note 13) and washed successively with two to four 50-cc. portions of ice-cold water, three 20-cc. portions of a mixture of 20 cc. of methyl alcohol and 40 cc. of ether, and finally two 50-cc. portions of ether (Note 5). It is then dried in air. The yield is 39–43 g., corresponding to about 44–48 g. from 500 g. of gelatine (Note 10).

Arginine Hydrochloride.—A suspension of 50 g. (0.18 mole) of benzylidenearginine in 39 cc. of 5 N hydrochloric acid is heated in a boiling water or steam bath for forty-five minutes, with occasional shaking. The mixture is allowed to cool and is freed of benzaldehyde by shaking with three 100–150 cc. portions of ether. The aqueous solution is filtered if necessary, decolorized with 3 g. of decolorizing carbon, filtered, and concentrated on the water bath at 70° under reduced pressure until crystallization sets in. The residue is transferred from the flask with the aid of 25 cc. of hot 70 per cent ethyl alcohol; the arginine hydrochloride is precipitated by adding 300 cc. of absolute alcohol. After filtering the product, a further small quantity of crystalline hydrochloride is obtained by adding 300 cc. of ether to the mother liquor. The combined (Note 14) yield amounts to 33–34 g. (88–90 per cent of the theoretical amount). It melts at 220° (corr.) and exhibits a rotation of $[\alpha]_D^{25°} = +12.2$ to 12.3° (5 per cent in water).

2. Notes

1. The quality of gelatine is technically defined on the basis of its physical properties, and different samples vary widely in chemical composition. In checking, the highest yields (9.7–10.3 per cent of the weight of gelatine taken) of benzylidenearginine were secured from the "Bactogelatine" of the Digestive Ferments Company.

2. The biuret reaction is generally found to be negative after five hours.

3. The third distillate generally contains only 1–2 g. of hydrogen chloride. In checking this preparation on a larger scale, it has been found convenient to add the water continuously below the surface of the boiling syrup; this modification, which constitutes a steam distillation under reduced pressure, brings about a more rapid removal of the excess hydrochloric acid.

4. Unless the temperature is held below 5°, difficulty is experienced in emulsifying the benzaldehyde.

5. Benzylidenearginine is quite insoluble in ether but appreciably soluble in methyl alcohol and in water. Attempts to recrystallize it from the latter solvents lead to a product of inferior quality, owing to decomposition in solution. Impure or contaminated samples may be purified by hydrolysis with hot hydrochloric acid and reprecipitation with benzaldehyde after neutralization.

6. The free dinitronaphtholsulfonic acid can be prepared readily from commercial Naphthol Yellow S by treating a filtered saturated solution of the dye with three volumes of concentrated hydrochloric acid. The crystals which separate are washed with cold 20 per cent hydrochloric acid and dried, first in air and finally in a vacuum desiccator over solid sodium hydroxide.

7. The boiling prevents the precipitation of arginine diflavianate and minimizes the separation of the flavianates of other amino acids.

8. Crystallization is occasionally delayed, particularly in first runs when traces of arginine flavianate are not available in the atmosphere for spontaneous inoculation. In such cases it may be necessary to chill the solution in the refrigerator with occasional vigorous scratching.

9. The mother liquor, on long standing in the icebox, may deposit a second crop of crystals which appear to consist largely of sodium dinitronaphtholsulfonate and yield no arginine on further treatment. The

filtrate thus obtained in Method 2 is suitable for the recovery of other amino acids, thereby differing from the corresponding mother liquor from Method 1.

10. This yield was obtained from a batch of gelatine from which 35–36 g. of benzylidenearginine was obtained by Method 1.

11. According to the literature,[1] pure arginine dinitronaphtholsulfonate melts at 258–260° with decomposition. The presence of moisture lowers the melting point considerably. If the mother liquor is allowed to stand for several days at 0–5°, some sodium dinitronaphtholsulfonate may crystallize.

12. Delay in adding the benzaldehyde must be avoided or sodium dinitronaphtholsulfonate may crystallize; the water is added to prevent this. The presence of the excess of benzaldehyde also appears to help prevent this crystallization.

13. A sintered-glass suction filter is advantageous for collecting and washing the benzylidenearginine.

14. The over-all loss involved in the various steps may be estimated from the following experiment: 5.0 g. of arginine nitrate was converted through the dinitronaphtholsulfonate into benzylidenearginine and then back into arginine nitrate, when 4.2 g. was recovered.

3. Methods of Preparation

Arginine has been precipitated (a) in the form of its silver derivative at pH 10; [2] (b) as its dinitronaphtholsulfonate which is then decomposed by means of 33 per cent sulfuric acid,[1] by the combined action of hot dilute sulfuric acid and butyl alcohol,[3] by cold concentrated hydrochloric acid followed by aniline,[4] or by barium hydroxide.[5] Arginine has likewise been precipitated (c) in the form of its benzylidene derivative from solutions rendered alkaline with barium hydroxide or sodium hydroxide.[6] It has been separated as such (d) from protein hydrolysates by electrolysis under controlled pH.[7]

In the present directions, Method 1 is essentially that developed by Bergmann and Zervas;[6] Method 2 forms a combination of methods

[1] Kossel and Gross, Z. physiol. Chem. **135**, 167 (1924).

[2] Kossel, ibid. **22**, 176 (1896–7); **25**, 165 (1898); Kossel and Kutscher, ibid. **31**, 165 (1900); Vickery and Leavenworth, J. Biol. Chem. **72**, 403 (1927); **75**, 115 (1927).

[3] Pratt, ibid. **67**, 351 (1926).

[4] Cox, ibid. **78**, 475 (1928).

[5] Felix and Dirr, Z. physiol. Chem. **176**, 38 (1928).

[6] Bergmann and Zervas, ibid. **152**, 282 (1926); **172**, 277 (1927).

[7] Foster and Schmidt, J. Biol. Chem. **56**, 545 (1923); J. Am. Chem. Soc. **48**, 1709 (1926); Cox, King, and Berg, J. Biol. Chem. **81**, 755 (1929).

(*b* and *c*) and thus at once affords a product of high purity while avoiding the mechanical difficulties involved in the complete removal of the dinitronaphtholsulfonic acid from its arginine salt.

AZELAIC ACID

$$\text{Castor Oil} \rightarrow CH_3(CH_2)_5CHOHCH_2CH{=}CH(CH_2)_7CO_2H$$

$$\xrightarrow[\text{(KMnO}_4)]{[O]} HO_2C(CH_2)_7CO_2H$$

Submitted by JULIAN W. HILL and W. L. McEWEN.
Checked by REYNOLD C. FUSON and CHARLES F. WOODWARD.

1. Procedure

FIVE HUNDRED grams of castor oil (Note 1) is added to a solution of 100 g. of potassium hydroxide in 1 l. of 95 per cent alcohol. The mixture is placed in a 3-l. flask equipped with a reflux condenser and is boiled for three hours. The solution is then poured into 3 l. of water and acidified by the addition of a solution of 100 cc. of concentrated sulfuric acid in 300 cc. of water. The acid which separates is washed twice with warm water, shaken intermittently for one hour with 100 g. of anhydrous magnesium sulfate, and then filtered with suction. The yield of crude ricinoleic acid thus obtained is 480 g. The acid should be oxidized at once (Note 2).

Two hundred and forty grams (0.8 mole) of the dried ricinoleic acid is dissolved in 1.6 l. of water containing 64 g. of potassium hydroxide. A 12-l. round-bottomed flask is equipped with a powerful mechanical stirrer, and in it are placed 625 g. (3.5 moles) of potassium permanganate and 7.5 l. of water at 35°. The mixture is stirred to facilitate solution of the permanganate, and, if necessary, heat is applied to maintain the temperature at 35°. When the permanganate has completely dissolved, the alkaline solution of ricinoleic acid is added in a single portion with vigorous stirring (Note 3). The temperature rises rapidly to about 75°. Stirring is continued for a half hour, or until a test portion added to water shows no permanganate color.

The oxidation mixture is now divided into two equal portions and each portion is treated as follows: To the mixture is added a solution of 200 g. of concentrated sulfuric acid in 600 cc. of water (Note 4). The mixture is heated on a steam bath for 15 minutes to coagulate the manganese dioxide which is filtered while still very hot (Note 5). After filtration the manganese dioxide is placed in a 4-l. beaker and boiled with 2 l. of water to dissolve any azelaic acid that may adhere to it. This mixture is filtered while hot, and the filtrate is added to the main portion.

The combined filtrates and washings for the two portions of the oxidation mixture are evaporated to a volume of about 4 l., and this solution is cooled in ice. The crystals which separate are filtered with suction, washed once with cold water, and dried. The yield is 70–80 g. of material having a melting point that may vary from 95 to 105°.

The crude substance is dissolved in 1.2 l. of boiling water, filtered with suction, and allowed to cool. The crystals are filtered, washed with water, and dried. There is obtained 48–55 g. of product (32–36 per cent of the theoretical amount, based upon the amount of crude ricinoleic acid taken for oxidation). The melting point of the purified azelaic acid is 104–106°.

2. Notes

1. The castor oil used was a commercial grade designated as "Crystal." It was obtained from the Baker Castor Oil Company, New York City.

2. If ricinoleic acid is allowed to stand, polymerization occurs; Baker and Ingold report [1] that the polymerized acid gives very poor yields in the oxidation by nitric acid.

3. At this point the mixture tends to froth quite badly, and, if stirring is not vigorous, material may be lost. Addition of a small quantity of ether or benzene may be resorted to but is unnecessary if stirring is efficient.

4. The acid must be added slowly and carefully to prevent too rapid evolution of carbon dioxide with consequent foaming. If possible, each of the two portions should be placed in a large container, such as a 12-l. flask.

5. For this purpose it is advisable to use three 20-cm. Büchner funnels supported in 2-l. filter flasks.

3. Methods of Preparation

Azelaic acid can be prepared by the oxidation of castor oil with nitric acid;[2] by the oxidation of ricinoleic acid with nitric acid[1] and with

[1] Baker and Ingold, J. Chem. Soc. **123**, 128 (1923); Verkade, Rec. trav. chim. **46**, 137 (1927).

[2] Arppe, Ann. **120**, 288 (1861); **124**, 86 (1862); Dale and Schorlemmer, ibid. **199**, 144 (1879); Kiliani, Ber. **54**, 469 (1921); Day, Kon, and Stevenson, J. Chem. Soc. **117**, 642 (1920); Böeseken and Lutgerhorst, Rec. trav. chim. **51**, 164 (1932).

alkaline permanganate;[3] by the oxidation of methyl oleate with alkaline permanganate;[4] by the ozonization of oleic acid and decomposition of the ozonide;[5] by the ozonization of methyl ricinoleate and decomposition of the ozonide;[6] by the action of carbon dioxide upon 1,7-heptamethylenemagnesium bromide;[7] by the hydrolysis of 1,7-dicyanoheptane;[8] and by the oxidation of dihydroxystearic acid with dichromate and sulfuric acid, the dihydroxy acid being prepared from oleic acid and hydrogen peroxide.[9]

AZLACTONE OF
α-BENZOYLAMINO-β-(3,4-DIMETHOXYPHENYL)-ACRYLIC ACID

[5(4)-Oxazolone, 2-phenyl-4-veratral-]

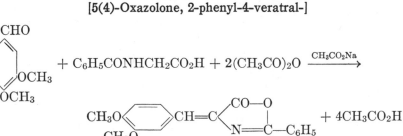

Submitted by JOHANNES S. BUCK and WALTER S. IDE.
Checked by JOHN R. JOHNSON and H. R. SNYDER.

1. Procedure

IN a 2-l. Erlenmeyer flask, a mixture of 160 g. (0.96 mole) of veratraldehyde (Note 1), 192 g. (1.07 moles) of powdered, dry hippuric acid (p. 328), 80 g. (0.98 mole) of powdered, freshly fused sodium acetate, and 300 g. (278 cc., 2.9 moles) of high-grade acetic anhydride is heated on an electric hot plate, with constant shaking. The mixture becomes almost solid, and then, as the temperature rises, it gradually liquefies and turns deep yellow in color (Note 2). As soon as the material has

[3] Maquenne, Bull. soc. chim. (3) 21, 1061 (1899).
[4] Armstrong and Hilditch, J. Soc. Chem. Ind. 44, 43T (1925).
[5] Harries and Tank, Ber. 40, 4556 (1907); Rieche, Ger. pat. 565,168 [C. A. 27, 1008 (1933)].
[6] Haller and Brochet, Compt. rend. 150, 500 (1910).
[7] v. Braun and Sobecki, Ber. 44, 1926 (1911).
[8] Dionneau, Ann. chim. (9) 3, 249 (1915).
[9] Bennett and Gudgeon, J. Chem. Soc. 1938, 1679.

liquefied completely the flask is transferred to a steam bath and heated for two hours. During this time a part of the product separates as deep yellow crystals. At the end of the heating 400 cc. of alcohol is added slowly to the contents of the flask. During this addition the flask is cooled slightly to moderate the vigor of the reaction. After allowing the reaction mixture to stand overnight, the yellow crystalline product is filtered with suction and washed on the filter with two 100-cc. portions of ice-cold alcohol and finally with two 100-cc. portions of boiling water. After drying, the product weighs 205–215 g. (69–73 per cent of the theoretical amount) and melts at 149–150°. This material is sufficiently pure for many purposes; it can be purified further by crystallization from hot benzene. In this way, using 1.2 l. of benzene, there is obtained 180–190 g. of the pure azlactone, melting at 151–152°.

2. Notes

1. The veratraldehyde obtained by methylating vanillin (p. 619) may be used without further purification.

2. The mixture should become completely liquid at a temperature of about 110°. Overheating should be avoided, since this causes the product to become red instead of bright yellow.

3. Methods of Preparation

The azlactones of α-benzoylaminocinnamic acids have always been prepared by the action of hippuric acid and acetic anhydride upon aromatic aldehydes,[1] usually in the presence of sodium acetate.[2] The procedure given here is essentially that of Kropp and Decker.[3]

[1] Plöchl, Ber. **16**, 2815 (1883).
[2] Erlenmeyer, Ann. **275**, 3 (1893).
[3] Kropp and Decker, Ber. **42**, 1184 (1909).

AZOXYBENZENE

$$4C_6H_5NO_2 + 3As_2O_3 + 18NaOH \rightarrow$$
$$2C_6H_5N{=}NC_6H_5 + 6Na_3AsO_4 + 9H_2O$$
$$\downarrow$$
$$O$$

Submitted by H. E. Bigelow and Albert Palmer
Checked by Henry Gilman and H. J. Harwood.

1. Procedure

Sodium arsenite is prepared by dissolving 226 g. (1.1 moles) of powdered arsenious oxide, made into a paste with a little water, in a solution of 275 g. (6.9 moles) of sodium hydroxide dissolved in 600 cc. of water. This solution, diluted with 600 cc. of water, is poured into a 2-l. three-necked flask provided with a reflux condenser and a mechanical stirrer, and 150 g. (125 cc., 1.2 moles) of freshly distilled nitrobenzene is added (Note 1).

The mixture is refluxed on an oil bath for eight hours with constant and vigorous stirring (Note 2). After removing the oil bath, the reaction mixture is allowed to cool to about 80°, while stirring is continued, and is then transferred to a separatory funnel previously heated to about the same temperature in an oven (Note 3).

The upper layer of oil is separated (Note 4), run at once into an open vessel, and washed with water to which a little hydrochloric acid has been added. Yellow crystals form at once (Note 5), and the yield of azoxybenzene melting at 35.5–36.5° is 102 g. (85 per cent of the theoretical amount) (Notes 6 and 7).

2. Notes

1. The excess of sodium arsenite and the eight-hour period of heating ensure the complete utilization of nitrobenzene. This makes it unnecessary to use steam distillation or other processes for the removal of unreacted nitrobenzene.

Crude nitrobenzene may be used, but a good grade is recommended. When crude nitrobenzene is used, a darker product having a slightly lower melting point is usually obtained.

2. The internal temperature should be about 104°, and the temperature of the bath should not greatly exceed 115°. A smaller flask might be used were it not for the danger of foaming which would result from

accidental stopping of the stirrer. If the reaction is interrupted, the oil bath must be removed even though the stirring has been stopped; otherwise, on resumption of stirring, the superheated material may be ejected through the condenser. With these precautions in mind it is unnecessary to have a continuous eight-hour period of heating.

Contrary to the general statement found in the literature, azoxybenzene is somewhat volatile with steam. Therefore, the presence of oil drops in the condenser at the end of the eight-hour period of refluxing is no criterion of unaltered nitrobenzene. Azoxybenzene is easily volatile with steam at 140–150°.

3. In this way the separation is effected at about 60°, and the danger of the solution's cooling to a point where sodium arsenate separates is avoided. Should the sodium arsenate separate because of undue cooling, the mixture is heated again, with stirring, until the arsenate redissolves. On dilution with sufficient water to keep the arsenate in solution, the oil settles to the bottom and does not separate readily from the liquid. Furthermore, the volume of solution is so large with such dilution that it is less easily handled.

4. The solution from which the original oil separated will yield, when diluted with an equal volume of water, a small additional quantity of azoxybenzene. This may be recovered by extraction with benzene, but the quantity of compound so obtained does not justify this extra procedure.

5. The presence of hydrochloric acid accelerates crystallization. Should crystallization be retarded, it is recommended that the oil be seeded with a crystal of azoxybenzene.

6. Recrystallization from 50 cc. of hot 95 per cent alcohol gives 72 g. of azoxybenzene. The recrystallized product melts at the same temperature, 35.5–36.5°, as the crude material but is distinctly lighter in color.

7. The following alternative procedure for preparing azoxybenzene is convenient.

In a 1-l. three-necked flask fitted with a reflux condenser and an efficient stirrer (p. 117) there are placed 60 g. of sodium hydroxide, 200 cc. of water, and 41 g. (34.2 cc., 0.33 mole) of nitrobenzene. The flask is immersed in a water bath kept at 55–60°, and 45 g. (0.23 mole) of dextrose is introduced in portions, with continuous stirring, in the course of one hour. The temperature of the bath is then raised to 100° and kept there for two hours. The hot mixture is poured into a 2-l. long-necked flask and steam-distilled to removed nitrobenzene and aniline. This requires some twenty minutes, during which time about 2 l. of distillate passes over. When the distillate is clear, the residue is poured into a beaker and cooled well in an ice bath. The azoxybenzene, which solidifies, is collected, the lumps are ground in a mortar, and the product is washed with water and dried. The yield of material melting at 34–35° is 26–27 g. (79–82 per cent of the theoretical amount). Crystallization from 15 cc. of methyl alcohol gives material melting at 35–35.5° with 90 per cent recovery. (NICHOLAS OPOLONICK, private communication. Checked by LOUIS F. FIESER and M. FIESER.)

3. Methods of Preparation

Azoxybenzene has been prepared by reduction of nitrobenzene with alcoholic potassium hydroxide,[1] with sodium amalgam,[2] with hydrogen in the presence of lead oxide,[3] with methyl alcohol and sodium hydroxide,[4] with sodium methoxide and methyl alcohol,[5] with lead suboxide in alkaline suspension,[6] with dextrose in alkaline suspension (see Note 7, above), and electrolytically;[7] by oxidation of azobenzene with chromic anhydride;[8] by treatment of β-phenylhydroxylamine with alkaline potassium permanganate,[9] with nitrobenzene,[10] with mineral acids,[11] and with mercury acetamide;[12] and by oxidation of aniline with hydrogen peroxide,[13] and with acid permanganate solution in the presence of formaldehyde.[14] The procedure described is a slight modification of one in the literature.[15]

[1] Zinin, J. prakt. Chem. (1) **36**, 98 (1845).

[2] Alexeyeff, Bull. soc. chim. (1) **1**, 325 (1864).

[3] Brown and Henke, U. S. pat. 1,451,489 [C. A. **17**, 1969 (1923)].

[4] Lachman, J. Am. Chem. Soc. **24**, 1180 (1902).

[5] Brühl, Ber. **37**, 2076 (1904). QDI D4

[6] Deutsche Gold- und Silber-Scheideanstalt vorm. Roessler, Ger. pat. 486,598 [C. A. **24**, 1389 (1930)].

[7] Löb, Ber. **33**, 2332 (1900); Ger. pat. 116,467 (Chem. Zentr. **1901**, I, 149).

[8] Wreden, Ber. **6**, 557 (1873).

[9] Reissert, ibid. **29**, 641 (1896).

[10] Bamberger and Renauld, ibid. **30**, 2278 (1897).

[11] Bamberger and Lagutt, ibid. **31**, 1501 (1898).

[12] Forster, J. Chem. Soc. **73**, 786 (1898).

[13] Prud'homme, Bull. soc. chim. (3) **7**, 622 (1892).

[14] Bamberger and Tschirner, Ber. **32**, 342 (1899).

[15] Loesner, J. prakt. Chem. (2) **50**, 564 (1894).

BARBITURIC ACID

$$CH_2(CO_2C_2H_5)_2 + CO(NH_2)_2 \xrightarrow{(C_2H_5ONa)} \begin{array}{c} HN\!-\!CO \\ | \qquad | \\ CO \qquad CH_2 \\ | \qquad | \\ HN\!-\!CO \end{array} + 2C_2H_5OH$$

Submitted by J. B. Dickey and A. R. Gray.
Checked by Reynold C. Fuson and W. E. Ross.

1. Procedure

In a 2-l. round-bottomed flask fitted with a reflux condenser protected by a calcium chloride tube, 11.5 g. (0.5 gram atom) of finely cut sodium is dissolved in 250 cc. of absolute alcohol. To this solution is added 80 g. (0.5 mole) of ethyl malonate followed by 30 g. (0.5 mole) of dry urea dissolved in 250 cc. of hot (70°) absolute alcohol. After being well shaken the mixture is refluxed for seven hours on an oil bath heated to 110°. A white solid separates rapidly. After the reaction is completed, 500 cc. of hot (50°) water is added and then enough hydrochloric acid (sp. gr. 1.18) to make the solution acidic (about 45 cc.). The resulting clear solution is filtered and cooled in an ice bath overnight. The white product is collected on a Büchner funnel, washed with 50 cc. of cold water, and then dried in an oven at 105–110° for three to four hours. The yield of barbituric acid is 46–50 g. (72–78 per cent of the theoretical amount).

2. Methods of Preparation

Barbituric acid has been prepared by the action of phosphorus oxychloride on malonic acid and urea;[1] by treating an acetic acid solution of urea and malonic acid with acetic anhydride;[2] from ethyl malonate and urea using sodium ethoxide as a condensing agent;[3] and from ethyl malonate and the sodium derivative of urea prepared from urea and sodium in liquid ammonia.[4]

The procedure described is an adaption of that of Michael.[3]

[1] Grimaux, Compt. rend. **87**, 752 (1878); Conrad and Guthzeit, Ber. **14**, 1643 (1881); Grimaux, Bull. soc. chim. (2) **31**, 146 (1879); Matignon, Ann. chim. phys. (6) **28**, 289 (1893).

[2] Biltz and Wittek, Ber. **54**, 1035 (1921).

[3] Michael, J. prakt. Chem. (2) **35**, 456 (1887); Tafel and Weinschenk, Ber. **33**, 3383 (1900); Gabriel and Colman, ibid. **37**, 3657 (1904).

[4] Jacobson, U. S. pat. 2,090,594 [C. A. **31**, 7068 (1937)].

BENZALPHTHALIDE

(Phthalide, 3-benzylidene-)

$$C_6H_4\underset{CO}{\overset{CO}{<}}O + C_6H_5CH_2CO_2H$$

$$\xrightarrow{CH_3CO_2Na} C_6H_4\underset{CO}{\overset{C=CHC_6H_5}{<}}O + CO_2 + H_2O$$

Submitted by RICHARD WEISS.
Checked by JOHN R. JOHNSON and H. R. SNYDER.

1. Procedure

IN a 500-cc. round-bottomed flask with a short neck (not longer than 3 cm.) are placed 100 g. (0.67 mole) of phthalic anhydride (Note 1), 110 g. (0.8 mole) of phenylacetic acid (Org. Syn. Coll. Vol. I, **1941**, 436), and 2.6 g. of freshly fused sodium acetate. A few chips of porous plate are added, and the flask is provided with a cork bearing a thermometer, which reaches almost to the bottom, and a wide, bent glass tube leading to a condenser. The tube ends just at the lower edge of the cork and does not protrude into the neck of the flask. The flask is imbedded up to the neck in a sand bath and is heated rapidly until the thermometer reaches 230°; then the temperature is raised slowly until the water produced in the reaction and some entrained organic matter pass out through the exit tube. The water is collected in a small vessel and its quantity noted from time to time in order to follow the progress of the reaction. The operation should be conducted so that the temperature rises from 230 to 240° in the course of about two hours. The reaction is maintained at 240° until the distillation of water ceases; this requires about one additional hour.

The flask now contains a brown mass covered with a film. The stopper is removed, and a test portion is taken out by means of a glass rod. The test portion is placed in a test tube or small beaker, treated with a little alcohol, and heated to boiling. When the reaction is complete, the material dissolves readily in the hot alcohol and crystallizes on cooling.

When this test has been found to be satisfactory, the flask is allowed to cool to 90–95°, and the product is dissolved in 400 cc. of boiling alcohol. The solution is filtered from insoluble matter and allowed to cool. The yellow crystals of benzalphthalide are filtered with suction and washed with 40–50 cc. of cold alcohol. The product weighs 115–116 g. and melts at 95–97°; for purification it is recrystallized from 370–380 cc. of alcohol. The yield of pure benzalphthalide, m.p. 100–101°, is 106–110 g. (71–74 per cent of the theoretical amount).

2. Note

1. A good grade of sublimed phthalic anhydride should be used (m.p. 129–131°); if this is not available the ordinary phthalic anhydride can be purified by sublimation.

3. Method of Preparation

Benzalphthalide has been prepared only from phthalic anhydride and phenylacetic acid in the presence of sodium acetate. The procedure given here is essentially that of Gabriel.[1]

BENZANTHRONE

(7-Benz[de]anthracene-7-one)

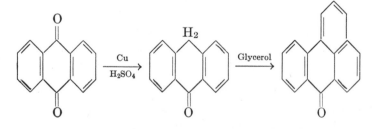

Submitted by L. C. MacLeod and C. F. H. Allen.
Checked by Louis F. Fieser and Max Tishler.

1. Procedure

In a 2-l. three-necked flask fitted with a mechanical stirrer and thermometer, 72 g. (0.35 mole) of anthraquinone (p. 554) is dissolved in 1060 g. of concentrated sulfuric acid by stirring at room temperature, and 42 cc. of water is then added to the red solution (Note 1). The flask is immersed up to the neck in an oil bath, and 48 g. (0.76 gram atom) of precipitated copper (Note 2) is added during one and one-half

[1] Gabriel, Ber. 18, 3470 (1885).

hours, the reaction mixture being kept at a temperature of 38–42°, if necessary by external heating, until all the copper has dissolved; this requires about three hours (Note 3).

A mixture of 96 g. (1.04 moles) of glycerol (Note 4) and 96 cc. of water is slowly introduced in the course of thirty minutes and the temperature is allowed to rise to 85–90°. The mixture is carefully heated to 120° during one and one-half hours, in such a way that the temperature rises uniformly at a rate of 1° every three minutes (Note 5). A temperature of 118–120° is maintained for an additional three-hour period; then the mixture is cooled to 70–80° and carefully poured with stirring into 4 l. of boiling water (Note 6). Spattering is avoided by pouring the acid mixture down the walls of the beaker while stirring. The suspension is boiled for a few minutes and preferably allowed to stand for several hours before being filtered.

The dark green benzanthrone is filtered on a large Büchner funnel, washed well with water, and boiled for thirty to forty minutes with 1.2 l. of 1 per cent sodium hydroxide solution. The product is filtered, washed free of the dark-colored liquor, and dried at 120°; weight, 67–71 g.; benzanthrone content, about 87 per cent. The crude material is boiled with 500 cc. of technical tetrachloroethane in which all but about 8 g. of a black char easily dissolves. The solution is boiled under reflux for fifteen minutes with 25 g. of decolorizing carbon, and then filtered while hot through a Büchner funnel directly into a 2-l. round-bottomed, long-necked flask, the residue being washed with hot tetrachloroethane (100–150 cc.) until the filtrate is colorless. After the addition of 400–500 cc. of hot water the solvent is removed by steam distillation, a process which requires but little time. The benzanthrone left as a residue is filtered and dried at 120°. The yield of yellow solid, which melts at 168–170° and is pure enough for many purposes, is 56–60 g. (70–75 per cent of the theoretical amount).

In order to secure a pure product the above material is dissolved in 175 cc. of tetrachloroethane by boiling and the solution is boiled under reflux for fifteen minutes with 12 g. of decolorizing carbon, and then filtered by suction into an Erlenmeyer flask, the charcoal being washed with about 50 cc. of hot solvent. The filtrate is kept hot, treated with 750 cc. of boiling alcohol, and set aside to crystallize. The benzanthrone separates as pure yellow needles melting at 170–171°; yield, 48–52 g. (60–65 per cent of the theoretical amount) (Note 7).

2. Notes

1. The solution of the anthraquinone is slower if the water is added at the outset.

2. The precipitated copper is prepared as on p. 446, using twice the quantities given.

3. The mixture becomes yellow-brown in color and some anthranol separates, but any unreacted copper can be seen on the bottom of the flask if the stirring is stopped for a few minutes.

4. The glycerol is a commercial, anhydrous product.

5. The heating must be done very carefully, and the temperature must never be allowed to rise above 120°. At higher temperatures much material is lost by charring.

6. A more granular and easily filterable product is obtained than when cold water is used.

7. On recovery of the tetrachloroethane by steam distillation of the mother liquor, a small additional quantity of material (5 g.) is obtained, but it is quite dark and of poor quality.

3. Methods of Preparation

Benzanthrone is commonly prepared by heating a reduction product of anthraquinone with sulfuric acid and glycerol,[1] or with a derivative of glycerol,[2] or with acrolein.[3] The anthraquinone is usually reduced in sulfuric acid solution, just prior to the reaction, by means of aniline sulfate,[1b] iron,[1c] or copper.[1d] However, the simultaneous reduction and condensation has been reported to give better yields.[4] Benzanthrone has also been prepared by dehydrogenating phenyl α-naphthyl ketone with aluminum or ferric chloride,[5] by dehydration of 1-phenylnaphthalene-2-carboxylic acid,[6] and by heating cinnamalanthrone with sodium-aluminum chloride.[7]

[1] (a) Bally, Ber. **38**, 194 (1905); Badische Anilin- und Soda-Fabrik, Ger. pat. 176,018 [Frdl. **8**, 372 (1905–7)]; (b) Bally and Scholl, Ber. **44**, 1665 (1911); (c) Iliinski, Russ. pat. 18,741 (Chem. Zentr. **1931**, II, 1759); (d) Caswell and Marshall, U. S. pat. 1,626,392 [C. A. **21**, 1992 (1927)]; (e) Bacharach and Cauliff, Jr., U. S. pat. 1,893,575 [C. A. **27**, 2163 (1933)].

[2] Badische Anilin- und Soda-Fabrik, Ger. pat. 204,354 [Frdl. **9**, 818 (1908–10)].

[3] Cross and Perkin, J. Chem. Soc. **1927**, 1297.

[4] Lukin, Org. Chem. Ind. (U. S. S. R.) **4**, 341 (1937) [C. A. **32**, 4977 (1938)].

[5] Scholl and Seer, Ann. **394**, 116 (1912); Monatsh. **33**, 1 (1912); Scholl, Ger. pat 239,671 [Frdl. **10**, 682 (1910–12)].

[6] Schaarschmidt, Ber. **50**, 295 (1917).

[7] I. G. Farbenind. A.-G., Ger. pat. 488,606 [Frdl. **16**, 1438 (1927–29)].

BENZIMIDAZOLE

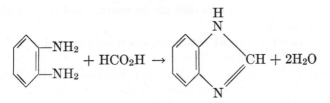

Submitted by E. C. WAGNER and W. H. MILLETT.
Checked by W. W. HARTMAN and G. W. SAWDEY.

1. Procedure

IN a 500-cc. round-bottomed flask 54 g. (0.5 mole) of o-phenylene-diamine (Note 1) is treated with 32 cc. (34.6 g.) of 90 per cent formic acid (0.75 mole) (Note 2). The mixture is heated in a water bath at 100° for two hours. After cooling, 10 per cent sodium hydroxide solution is added slowly, with thorough mixing by rotation of the flask, until the mixture is just alkaline to litmus. The crude benzimidazole is collected with suction in a 75-mm. Büchner funnel; ice-cold water is used to rinse all solid out of the reaction flask. The crude product is pressed thoroughly on the filter, washed with about 50 cc. of cold water, and then purified without previous drying (Note 3).

The benzimidazole is dissolved in 750 cc. of boiling water in a 1.5-l. beaker. The solution is digested for fifteen minutes with about 2 g. of Norite and filtered rapidly through a well-heated filter (Note 4). The filtrate is cooled to 10–15°, and the benzimidazole is filtered and washed with 50 cc. of cold water. The white (Note 5) product is dried at 100°. The melting point is 170–172°, and the yield is 49–50.5 g. (83–85 per cent of the theoretical amount) (Notes 6 and 7).

2. Notes

1. The o-phenylenediamine used was a good grade of commercial material, m.p. 99–101°. The hydrochloride can be used with or without addition of sodium formate. Directions for preparing o-phenylene-diamine are given on p. 501.

2. The yield is not greatly affected if the amount of formic acid is decreased almost to the theoretical, but a safe excess is recommended to ensure utilization of the o-phenylenediamine. Formic acid of considerably less than 90 per cent concentration will form benzimidazole; good yields were obtained with 40 per cent acid.

3. The crude benzimidazole, if dried at 100°, weighs 57.5–59 g. (97–99 per cent of the theoretical amount), melts at 167–168°, and is yellow tinged. This discoloration is difficult to remove and persists after two crystallizations (Note 5).

4. The solution is almost saturated when boiling, and crystallization begins at once on cooling. The filter must be thoroughly heated and filtration must be rapid, or crystallization will occur in the filter.

5. If the crystallized benzimidazole is discolored, the following treatment will yield a good product. The benzimidazole is dissolved in boiling water (13 cc. per gram), and a strong solution of potassium permanganate is added until the liquid becomes opaque owing to the precipitated brown oxide of manganese. To the hot mixture solid sodium bisulfite is added until clarification results. Decolorizing carbon is introduced, and the mixture is digested for fifteen minutes and filtered hot. The recovery is 90–92 per cent.

6. A small additional amount (2–2.5 g.) can be obtained by evaporation of the mother liquor to about 30 cc.

7. This is a general method of preparing benzimidazoles. Using an equivalent of acetic acid (45 g.) in place of formic acid, 2-methylbenzimidazole, m.p. 172–174°, can be prepared in 68 per cent yield.

3. Methods of Preparation

Benzimidazole has been prepared from o-phenylenediamine by the action of chloroform and alcoholic potassium hydroxide [1] and of formic acid,[2] and by the reduction of o-nitroformanilide.[3] Less serviceable methods include the interaction of o-phenylenediamine and dichloromethylformamidine,[4] or diphenylformamidine,[5] or formoacetic anhydride,[6] and the thermal decarboxylation of benzimidazole-2-carboxylic acid.[7] The procedure described was developed [8] from that of Wundt.[2]

The conversion of aliphatic acids to 2-alkylbenzimidazoles, by heating with o-phenylenediamine, has been proposed as a general method for preparing solid derivatives for identification.[9]

[1] Grassi-Cristaldi and Lambardi, Gazz. chim. ital. **25**, 225 (1895).

[2] Wundt, Ber. **11**, 826 (1878); Heller and Kühn, ibid. **37**, 3116 (1904); Pauly and Gundermann, ibid. **41**, 4012 (1908).

[3] Niementowski, ibid. **43**, 3018 (1910).

[4] Dains, ibid. **35**, 2503 (1902).

[5] Wagner, J. Org. Chem. **5**, 136 (1940).

[6] Béhal, Ger. pat. 115,334 [Frdl. **6**, 1280 (1900–02)].

[7] Bistrzycki and Przeworski, Ber. **45**, 3489 (1912).

[8] Wagner and Simons, J. Chem. Education **13**, 267 (1936).

[9] Seka and Müller, Monatsh. **57**, 97 (1931); Pool, Harwood, and Ralston, J. Am Chem. Soc. **59**, 178 (1937).

BENZOHYDROXAMIC ACID

$$C_6H_5CO_2C_2H_5 + H_2NOH + KOH$$
$$\rightarrow C_6H_5CONHOK + C_2H_5OH + H_2O$$

$$C_6H_5CONHOK + CH_3CO_2H \rightarrow C_6H_5CONHOH + CH_3CO_2K$$

Submitted by C. R. Hauser and W. B. Renfrow, Jr.
Checked by C. R. Noller and M. Synerholm.

1. Procedure

(A) *Potassium Benzohydroxamate.*—Separate solutions of 46.7 g. (0.67 mole) of hydroxylamine hydrochloride (Org. Syn. Coll. Vol. I, **1941**, 318) in 240 cc. of methyl alcohol, and of 56.1 g. (1 mole) of c.p. potassium hydroxide in 140 cc. of methyl alcohol, are prepared at the boiling point of the solvent. Both are cooled to 30–40° (Note 1), and the one containing alkali is added with shaking to the hydroxylamine solution; any excessive rise of temperature during the addition is prevented by occasional cooling in an ice bath. After all the alkali has been added, the mixture is allowed to stand in an ice bath for five minutes to ensure complete precipitation of potassium chloride. Fifty grams (0.33 mole) of ethyl benzoate is added with thorough shaking, and the mixture filtered immediately with suction. The residue in the funnel is washed with a little methyl alcohol. The filtrate is placed in an Erlenmeyer flask and allowed to stand at room temperature. Crystals begin to form within twenty minutes to three hours, depending upon the amount of supersaturation of the solution. After forty-eight hours the crystals are filtered, washed with a little absolute ethyl alcohol, and dried in air. The yield is 33–35 g. (57–60 per cent of the theoretical amount) (Notes 2, 3, and 4).

(B) *Benzohydroxamic Acid.*—A mixture of 35 g. (0.2 mole) of the potassium salt in 160 cc. of 1.25 N acetic acid is stirred and heated until a clear solution is obtained. The solution is allowed to cool to room temperature and finally chilled in an ice bath. Benzohydroxamic acid separates as white crystals. After filtering and drying, the product melts at 120–128° and weighs 25–26 g. (91–95 per cent of the theoretical amount). The crude material may be purified by dissolving it in 4.5 times its weight of hot ethyl acetate, filtering from a small amount of solid, and allowing the solution to cool to room temperature. The white crystals which separate are filtered, washed with a little benzene, and allowed to dry in air. The yield of recrystallized product, m.p. 125–128°, from 26 g. of crude material is 20 g. (77 per cent recovery) (Note 5).

2. Notes

1. The hydroxylamine hydrochloride solution should not be cooled too quickly or crystallization may occur before mixing. Exposure to atmospheric oxygen should be minimized after mixing the solutions, to avoid oxidation of the free hydroxylamine.

2. By concentrating the alcoholic mother liquors an additional 3–5 g. of the potassium salt may be obtained.

3. This salt can be used for the preparation of acyl derivatives; for example, when suspended in dioxane and treated with benzoyl chloride, dibenzohydroxamic acid is formed in excellent yield.

4. The potassium salts of p-methyl- and p-methoxybenzohydroxamic acid have been prepared by the submitters by this method in approximately the same yields.

5. This product has a neutralization equivalent of 137.5–138 (calculated 137.1) when determined as follows. Several drops of an alcoholic solution of 1,3,5-trinitrobenzene are added to 30–40 cc. of water in a flask, and 0.1 N alkali (about 0.5 cc.) is run in until a pink color is just produced. An accurately weighed sample (approximately 0.3 g.) of benzohydroxamic acid is then dissolved in the solution and titrated with standard alkali until the pink color is restored. The latter titer is used to calculate the neutralization equivalent.

3. Methods of Preparation

Benzohydroxamic acid has been prepared by the action of hydroxylamine on benzoyl chloride,[1] ethyl benzoate,[2, 3] or benzamide.[4]

[1] Lossen, Ann. **161,** 347 (1872); Jones and Hurd, J. Am. Chem. Soc. **43,** 2446 (1921)
[2] Renfrow and Hauser, ibid. **59,** 2312 (1937).
[3] Tiemann and Krüger, Ber. **18,** 740 (1885); Jeanrenaud, ibid. **22,** 1272 (1889).
[4] Hofmann, ibid. **22,** 2856 (1889).

BENZOIN ACETATE

$$C_6H_5CHOHCOC_6H_5 + (CH_3CO)_2O \rightarrow$$
$$C_6H_5CH(OCOCH_3)COC_6H_5 + CH_3CO_2H$$

Submitted by B. B. Corson and N. A. Saliani.
Checked by Frank C. Whitmore and Marion M. Whitmore.

1. Procedure

To a mixture of 212 g. (1 mole) of benzoin (Note 1), 200 cc. of glacial acetic acid, and 200 cc. (2.1 moles) of acetic anhydride, in a 1-l. beaker provided with a mechanical stirrer, is added slowly, with stirring, 20 cc. of concentrated c.p. sulfuric acid. This requires five minutes, during which the benzoin quickly dissolves and the temperature rises to about 50°. The beaker is placed on the steam bath for twenty minutes (Note 2). The mixture is allowed to cool somewhat, transferred to a large dropping funnel, and added *slowly* to 2.5 l. of water vigorously stirred in a 4-l. (1-gal.) crock during thirty minutes (Note 3). Stirring is continued for one hour. The mixture is filtered by suction on a 30-cm. Büchner funnel, and the crystals are sucked as dry as possible and spread on filter paper. After about two hours the crystals are transferred to a 1-l. beaker and warmed to about 60° with 400 cc. of 95 per cent ethyl alcohol. The clear solution is cooled with stirring to 5° and filtered by suction. The air-dried benzoin acetate, melting at 80–82°, weighs 220–230 g. (86 to 90 per cent of the theoretical amount). Another crystallization from 400 cc. of alcohol removes the slight yellow tinge and gives a product melting at 81.5–82.5°, with a loss of about 10 g.

2. Notes

1. The benzoin (Org. Syn. Coll. Vol. I, **1941**, 94) need not be recrystallized.
2. The mixture should not be heated longer or more vigorously.
3. If the product solidifies in lumps, the lumps must be removed, crushed to a paste in a large mortar, and returned to the mixture for stirring.

3. Method of Preparation

The only method of preparative interest is the acetylation of benzoin, either with acetyl chloride [1] or with acetic anhydride.[2] The melting point of benzoin acetate was reported by Zinin as "below 100°," by Jena and Limpricht as 75°, but by later investigators as 82–83°.

BENZOPHENONE OXIME

$$(C_6H_5)_2CO + H_2NOH \cdot HCl + NaOH \rightarrow$$
$$(C_6H_5)_2C{=}NOH + 2H_2O + NaCl$$

Submitted by ARTHUR LACHMAN.
Checked by C. R. NOLLER.

1. Procedure

A MIXTURE of 100 g. (0.55 mole) of benzophenone (Org. Syn. Coll. Vol. I, **1941**, 95), 60 g. (0.86 mole) of hydroxylamine hydrochloride (Org. Syn. Coll. Vol. I, **1941**, 318), 200 cc. of 95 per cent ethyl alcohol, and 40 cc. of water is placed in a 2-l. round-bottomed flask. To this is added in portions, with shaking, 110 g. (2.75 moles) of powdered sodium hydroxide. If the reaction becomes too vigorous, cooling with tap water may be necessary. After all the sodium hydroxide has been added, the flask is connected to a reflux condenser, heated to boiling, and refluxed for five minutes. After cooling, the contents are poured into a solution of 300 cc. of concentrated hydrochloric acid in 2 l. of water. The precipitate is filtered with suction, thoroughly washed with water, and dried (Note 1). The yield is 106–107 g. (98–99 per cent of the theoretical amount) of a product melting at 141–142°. On crystallizing 20 g. from 80 cc. of methyl alcohol, 13 g. of crystalline material of the same melting point is obtained (Note 2).

2. Notes

1. This crude material dried overnight at about 40° is practically pure and if used at once is satisfactory for the preparation of diphenylmethane imine hydrochloride (p. 234).

[1] Zinin, Ann. **104**, 120 (1857); Jena and Limpricht, ibid. **155**, 92 (1870); Päpcke. Ber. **21**, 1336 (1888).
[2] Francis and Keane, J. Chem. Soc. **99**, 346 (1911).

2. In the presence of oxygen and traces of moisture, benzophenone oxime is gradually converted into a mixture of benzophenone and nitric acid.[1] A good method of preserving this oxime is to dry it in a vacuum desiccator, fill the desiccator with pure carbon dioxide, re-evacuate, and fill again with carbon dioxide. The preparation may then be transferred to a bottle, also filled with carbon dioxide, and sealed against access of air.

3. Methods of Preparation

Benzophenone oxime has been prepared in quantity by treating an aqueous alcoholic mixture of benzophenone and hydroxylamine hydrochloride with hydrochloric acid,[2] with sodium carbonate,[3] with alcoholic potassium hydroxide,[4] or with aqueous sodium hydroxide.[5] It has also been obtained by treating bisnitrosylbenzohydryl with alcoholic potassium hydroxide,[6] and by the oxidation of α-aminodiphenylmethane with magnesium persulfate solution.[7]

BENZOPINACOL

$$2C_6H_5COC_6H_5 + CH_3CHOHCH_3$$

$$\xrightarrow{\text{(Sunlight)}} (C_6H_5)_2C\underset{OH}{\overset{|}{\rule{0pt}{0pt}}}\!\!\!-\!\!\!\underset{OH}{\overset{|}{\rule{0pt}{0pt}}}\!\!\!C(C_6H_5)_2 + CH_3COCH_3$$

Submitted by W. E. BACHMANN.
Checked by JOHN R. JOHNSON and H. R. SNYDER.

1. Procedure

A MIXTURE of 150 g. (0.82 mole) of benzophenone (Note 1), one drop of glacial acetic acid (Note 2), and 665 g. (850 cc., 11 moles) of isopropyl alcohol (Note 3) in a 1-l. round-bottomed flask is warmed to 45°. The flask is closed with a tight cork firmly wired or tied in place, and is

[1] Hollemann, Rec. trav. chim. **13**, 429 (1894); Lachman, J. Am. Chem. Soc. **46**, 1478 (1924).

[2] Beckmann, Ber. **19**, 989 (1886).

[3] Janny, ibid. **15**, 2782 (1882).

[4] Derick and Bornmann, J. Am. Chem. Soc. **35**, 1287 (1913).

[5] Lachman, ibid. **46**, 1481 (1924); **47**, 262 (1925).

[6] Behrend and Platner, Ann. **278**, 369 (1894).

[7] Bamberger and Seligman, Ber. **36**, 704 (1903).

supported in an inverted position in a tripod and exposed to direct sunlight. After three to five hours of bright sunshine crystals of benzopinacol begin to appear; after eight or ten days of exposure, depending upon the intensity of the light (Note 4), the flask is filled with crystals of benzopinacol. The solution is chilled in ice and the crystalline product is filtered with suction, washed with a small quantity of isopropyl alcohol, and allowed to dry in the air. The filtrate is reserved for subsequent reductions (see below). The yield of practically pure benzopinacol, m.p. 188–190° (Note 5), is 141–142 g. (93–94 per cent of the theoretical amount). The product is sufficiently pure for most purposes. It may be crystallized by dissolving it in 1 l. of hot benzene, filtering, and adding 400 cc. of hot ligroin (b.p. 90–100°) to the hot filtrate. After cooling in ice and filtering there is obtained 129–130 g. of purified product. The melting point is not changed by this purification.

To the isopropyl alcohol filtrate is added another 150-g. portion of benzophenone, and the solution is exposed to sunlight as in the first reduction. The benzopinacol which separates is filtered and dried. The yield in the second and subsequent runs is 142–143 g. (94–95 per cent of the calculated amount). This procedure can be repeated with the same filtrate until six or seven portions (900–1050 g.) of benzophenone have been reduced.

2. Notes

1. Although a practical grade of benzophenone can be used in this preparation, it is better to have material that has been recrystallized from alcohol. Directions for preparing benzophenone are given in Org. Syn. Coll. Vol. I, **1941**, 95.

2. No more than one drop of acetic acid should be used. The acid is added to ensure the removal of traces of alkali, which cause decomposition of the pinacol into benzophenone and benzohydrol.

3. If isopropyl alcohol is not available, absolute ethyl alcohol can be used. With ethyl alcohol the reaction is slower and a yellow solution is obtained; nevertheless, the crystals of benzopinacol are colorless.

4. About five clear bright days are required to complete the reduction. The reaction can be interrupted at any time, the crystals filtered, and the filtrate then exposed further.

5. Since the pinacol decomposes near its melting point the latter will vary with the rate of heating. The temperatures reported here were obtained by slow heating; if the tube is placed in a bath at 150° and heated rapidly, the observed melting or decomposition point is 193–195°.

3. Methods of Preparation

Benzopinacol has been prepared by the action of phenylmagnesium bromide on benzil [1] or methyl benzilate.[1] Usually it has been obtained by reduction of benzophenone, the reducing agents being zinc and sulfuric acid [2] or acetic acid,[3] aluminum amalgam,[4] and magnesium and magnesium iodide.[5] The present method is based on a study by Cohen [6] of the photochemical reaction discovered by Ciamician and Silber.[7]

β-BENZOPINACOLONE

$$(C_6H_5)_2C\text{---}C(C_6H_5)_2 \xrightarrow{(I_2)} C_6H_5COC(C_6H_5)_3 + H_2O$$
$$\underset{OH}{|} \quad \underset{OH}{|}$$

Submitted by W. E. BACHMANN.
Checked by JOHN R. JOHNSON and H. R. SNYDER.

1. Procedure

IN a 1-l. round-bottomed flask provided with a reflux condenser is placed a solution of 1 g. of iodine in 500 cc. of glacial acetic acid. One hundred grams (0.27 mole) of benzopinacol (Note 1) is added, and the flask is heated over a wire gauze, with shaking, until the solution boils gently. It is then refluxed for five minutes during which the solid benzopinacol disappears completely and a clear red solution is obtained (Note 2). The solution is transferred at once to a 1-l. beaker, and, upon cooling, the benzopinacolone separates in fine threads. The product is filtered with suction, washed with two or three 60-cc. portions of cold glacial acetic acid until colorless, and dried. The filtrate is reserved for subsequent preparations. The yield of practically pure benzopinacolone melting at 178–179° is 90–91 g. (95–96 per cent of the theoretical amount). If a purer product is desired the material may be dissolved in 450 cc. of hot benzene, filtered, and treated with 250 cc. of hot ligroin (b.p. 90–100°). After cooling in ice the benzopinacolone is filtered and dried. The purified product weighs 82–83 g. and melts at 179–180°.

[1] Acree, Ber. **37**, 2761 (1904).
[2] Linnemann, Ann. **133**, 26 (1865).
[3] Zagumenni, Ber. **14**, 1402 (1881).
[4] Cohen, Rec. trav. chim. **38**, 75 (1919).
[5] Gomberg and Bachmann, J. Am. Chem. Soc. **49**, 241 (1927).
[6] Cohen, Rec. trav. chim. **39**, 243 (1920).
[7] Ciamician and Silber, Ber. **33**, 2911 (1900).

To the acetic acid filtrate is added another 100-g. portion of benzo-pinacol and the reaction is carried out in the same way. The yield of benzopinacolone in the second and subsequent runs is 94–94.5 g. (98–99 per cent of the theoretical amount). This procedure can be repeated in the same filtrate until 500 g. of the pinacol has been rearranged.

2. Notes

1. The benzopinacol obtained by photochemical reduction of benzo-phenone (p. 71) may be used directly without purification.

2. Frequently the benzopinacolone begins to crystallize in the boiling solution during the last minute of heating.

3. Method of Preparation

β-Benzopinacolone has been prepared by rearrangement of benzo-pinacol. The rearrangement has been carried out by heating benzo-pinacol with benzoyl chloride,[1] with acetyl chloride,[2] with acetic acid at 180–200°,[2] with dilute sulfuric acid at 180–200°,[2] and with concentrated hydrochloric acid at 200°.[2] The present procedure is based on the method described by Gomberg and Bachmann.[3]

ε-BENZOYLAMINO-α-BROMOCAPROIC ACID

(Caproic acid, ε-benzamido-α-bromo-)

$$3C_6H_5CONH(CH_2)_5CO_2H + 3Br_2 + PBr_3$$
$$\rightarrow 3C_6H_5CONH(CH_2)_4\underset{\underset{Br}{|}}{C}HCOBr + P(OH)_3 + 3HBr$$

$$C_6H_5CONH(CH_2)_4\underset{\underset{Br}{|}}{C}HCOBr \xrightarrow{H_2O} C_6H_5CONH(CH_2)_4\underset{\underset{Br}{|}}{C}HCO_2H$$

Submitted by J. C. Eck and C. S. Marvel.
Checked by C. R. Noller and William Munich.

1. Procedure

An intimate mixture of 150 g. (0.64 mole) of dry ε-benzoylamino-caproic acid (p. 76) and 26.4 g. (0.85 gram atom) of dry red phosphorus

[1] Linnemann, Ann. **133**, 28 (1865).
[2] Thörner and Zincke, Ber. **10**, 1475 (1877).
[3] Gomberg and Bachmann, J. Am. Chem. Soc. **49**, 246 (1927).

is placed in a 1-l. three-necked flask provided with a separatory funnel, an air-cooled condenser connected through a calcium chloride tube to a water trap, and a mechanical stirrer (Note 1). The reaction flask is surrounded by an ice-salt mixture, the stirrer started, and 408 g. (131 cc., 2.55 moles) of dry bromine added dropwise from the separatory funnel. When all the bromine has been added the cooling bath is removed. The mixture is warmed slowly at first and finally heated on a steam bath, with stirring, until the bromine vapors have practically disappeared. The hot mixture is poured slowly into 400 cc. of water in a 1-l. beaker with hand stirring. The viscous acid bromide reacts with the water with the evolution of heat, and the solid acid is formed. The lumps are pulverized, the mixture is replaced in the original reaction flask, and the whole is treated with a slow stream of sulfur dioxide to remove excess bromine. The solid product is filtered on a Büchner funnel, washed with three 50-cc. portions of water, and air-dried. The crude material is dissolved in 250 cc. of hot 95 per cent alcohol, filtered, and poured with stirring into 1 l. of cold water (Note 2). After filtering and air-drying, 130–180 g. (64–89 per cent of the calculated amount) of acid melting at 162–165° is obtained.

2. Notes

1. The stirrer must be very powerful because the mixture becomes lumpy and finally very viscous. If the material agglomerates so badly that the stirrer will not operate, hand stirring may be necessary temporarily until the mass liquefies sufficiently to renew mechanical stirring. The checkers tried one run in which carbon tetrachloride was added to facilitate stirring, but, though it accomplished this purpose, the yield was only about half of that obtained without a liquid medium.

2. The crude product may be recrystallized from ethyl alcohol, but the melting point is the same as that of the product obtained by precipitation with water, and the yield is considerably less.

3. Method of Preparation

The above procedure [1] is essentially that of Braun.[2]

[1] Eck and Marvel, J. Biol. Chem. **106**, 387 (1934).
[2] Braun, Ber. **42**, 839 (1909).

ε-BENZOYLAMINOCAPROIC ACID

(Caproic acid, ε-benzamido-)

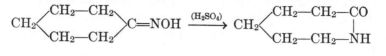

$$\xrightarrow{\text{H}_2\text{O}} \text{H}_2\text{N}(\text{CH}_2)_5\text{CO}_2\text{H} \xrightarrow{\text{C}_6\text{H}_5\text{COCl}} \text{C}_6\text{H}_5\text{CONH}(\text{CH}_2)_5\text{CO}_2\text{H}$$

Submitted by J. C. Eck and C. S. Marvel.
Checked by C. R. Noller and William Munich.

1. Procedure

(A) *Cyclohexanone Oxime.*—In a 5-l. flask, fitted with an efficient mechanical stirrer and an 8-mm. glass inlet tube reaching to within 5 cm. of the bottom of the flask, are placed 1.5 kg. of cracked ice and a solution of 182 g. (2.5 moles) of technical sodium nitrite (95 per cent) in 500 cc. of water (Note 1). The flask is placed in an ice-salt mixture, and a cold (−8°) solution of sodium bisulfite, prepared by saturating with sulfur dioxide a solution of 143 g. (1.35 moles) of anhydrous sodium carbonate in 600 cc. of water, is added. While the temperature is kept below 0°, a moderate stream of sulfur dioxide is passed into the mixture until it is acid to Congo red and then just enough longer to remove the dark color which appears shortly before the solution becomes acid.

To this solution are added 196 g. (2 moles) of technical cyclohexanone and 500 cc. of 85 per cent ethyl alcohol; the cooling bath is replaced by a steam bath, the stirrer started, and the mixture heated to 75°. The flask is then packed in mineral wool or other insulating material and allowed to cool slowly, with effective stirring, for forty-eight hours. The solution at room temperature is exactly neutralized to litmus with a 50 per cent solution of sodium hydroxide, with cooling and stirring. About 330 g. of sodium hydroxide solution is required.

The oily layer is separated and the aqueous solution extracted with two 200-cc. portions of ether. The oil and ether extracts are combined, the ether is removed, and the residue distilled from a 500-cc. modified Claisen flask having a 25-cm. fractionating side arm. The fraction boiling at 95–100° at 5 mm. weighs 170–190 g. and melts at 78–80°. This product is transferred to a large mortar, allowed to cool, and ground with 120 cc. of petroleum ether (b.p. 35–60°). After filtering with suction, and allowing the solvent to evaporate from the crystals, there is obtained 133–147 g. of cyclohexanone oxime (59–65 per cent of the calculated amount) melting at 86–88° (Note 2).

(B) ε-Benzoylaminocaproic Acid.—The rearrangement of 100 g. (0.88 mole) of pure cyclohexanone oxime (Note 3) is carried out in the following way. In a 1-l. beaker are placed a 10-g. portion of the oxime and 20 cc. of 85 per cent sulfuric acid (sp. gr. 1.783) (Note 4). The beaker is heated with a low flame and the contents are mixed with a rotary motion until bubbles first appear. The beaker is then removed from the flame immediately, and the violent reaction, which lasts a few seconds, is allowed to subside. The acid solution of ε-caprolactam is transferred to a 5-l. round-bottomed flask, and another 10-g. portion of the oxime is placed in the beaker and rearranged with sulfuric acid as before. The combined acid solution from the ten operations is diluted with 2.5 l. of water and boiled gently for one and one-half hours with 5 g. of decolorizing carbon. The solution is filtered and exactly neutralized to litmus with 50 per cent sodium hydroxide solution. About 510 g. of sodium hydroxide solution is usually required. The neutral solution is boiled for one-half hour with 5 g. of decolorizing carbon and filtered.

The filtrate is placed in a 5-l. flask fitted with a mechanical stirrer, cooled in an ice bath to 10°, and a solution of 55 g. (1.37 moles) of sodium hydroxide in 55 cc. of water added. The temperature being kept at 10°, 94 g. (0.67 mole) of benzoyl chloride is added dropwise from a separatory funnel over a period of thirty-five to forty minutes, with rapid stirring. The mixture is stirred for an hour longer, filtered, and placed in a 4-l. beaker. The cold filtrate is slowly acidified to Congo red by adding 10 per cent hydrochloric acid, with hand stirring (about 450 cc. of acid is required). The solid (Note 5) is filtered with suction, washed with water, and spread out to dry. When dry, the product is washed with two 100-cc. portions of petroleum ether (35–60°) to remove the admixed benzoic acid. The adhering petroleum ether is allowed to evaporate from the crystals, and the final drying is carried out in a vacuum desiccator over sulfuric acid. The yield of purified product, m.p. 77–80°, is 135–150 g. (65–72 per cent of the calculated amount, based on cyclohexanone oxime).

2. Notes

1. The sodium nitrite may be added directly to 2 kg. of ice, but, if this is done, the nitrite and ice should be mixed thoroughly outside the flask to prevent caking of the ice.

2. Cyclohexanone oxime can be prepared with better yields from the ketone, hydroxylamine hydrochloride (Org. Syn. Coll. Vol. I, 1941, 318),

and sodium carbonate according to the procedure given on p. 314. The preparation using hydroxylamine hydrochloride, however, is more expensive than that given above.

3. The rearrangement of the oxime is carried out in 10-g. portions and in a large open beaker because of the violence of the reaction. It is essential to use oxime of good quality or a product of inferior grade results.

4. Acid of this concentration may be prepared by mixing five volumes of concentrated sulfuric acid with one volume of water.

5. If the acid separates as an oil, it should be allowed to stand with occasional stirring until it solidifies.

3. Methods of Preparation

ε-Benzoylaminocaproic acid has been prepared by treatment of benzoylpiperidine with phosphorus pentachloride to form ε-benzoylaminoamyl chloride, conversion to the nitrile, and hydroylsis to the acid,[1] and by the procedure described above.[2]

Directions for the rearrangement of cyclohexanone oxime and isolation of ε-caprolactam and ε-aminocaproic acid are to be found on pp. 371 and 28.

[1] Braun, Ber. **42**, 839 (1909).
[2] Eck and Marvel, J. Biol. Chem. **106**, 387 (1934).

BENZOYLENE UREA

[2,4(1,3)-Quinazolinedione]

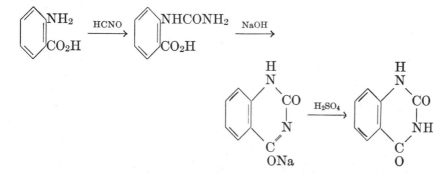

Submitted by N. A. LANGE and F. E. SHEIBLEY.
Checked by W. W. HARTMAN and J. B. DICKEY.

1. Procedure

IN a 3-l. beaker a mixture of 20 g. (0.146 mole) of anthranilic acid, 700 cc. of warm water (35°), and 11 cc. (11.6 g., 0.19 mole) of glacial acetic acid is stirred mechanically and allowed to cool to room temperature. A freshly prepared solution of 15 g. (0.185 mole) of potassium cyanate (Note 1) in 50 cc. of water is then added dropwise with stirring over a period of fifteen to twenty minutes (Note 2). The resulting pasty mixture is stirred for twenty minutes, and then 200 g. (5 moles) of flaked sodium hydroxide (Note 3) is added slowly in small portions. During this addition the reaction mixture is kept below 40° by cooling in a cold-water bath. A clear solution is obtained momentarily, but in a short time a fine granular precipitate of the hydrated monosodium salt of benzoylene urea precipitates. After the mixture has cooled overnight in an ice box, the precipitated sodium salt is collected on a Büchner funnel, using a hardened filter paper (Note 4). The colorless salt is dissolved in 1 l. of hot water (90–95°), and the solution is filtered and heated to boiling in a 3-l. beaker. The benzoylene urea is precipitated by adding dilute sulfuric acid (1 : 1) with vigorous stirring until the liquor is acid to litmus. The product separates as a hydrate which forms small, lustrous, colorless needles. The material is collected on a Büchner funnel, washed with 200 cc. of water, and dried in an oven at 100°. The yield is 19.5–20.5 g. (82–87 per cent of the theoretical amount) (Note 5).

2. Notes

1. The yield is highly dependent upon the quality of the potassium cyanate employed, and some samples were found worthless for the purpose. The yields given were realized using Eastman's regular grade of potassium cyanate.

2. If the addition is too rapid the odor of isocyanic acid (remindful of that of sulfur dioxide) becomes strong and the yield is diminished.

3. The commercial grade of flaked sodium hydroxide dissolves readily and is convenient to handle. Any silica present is not objectionable since it is removed by filtering the redissolved sodium salt before precipitating the product with acid.

4. Schleicher and Schüll's hardened filters (No. 575, 9 cm.) are satisfactory.

5. The melting point of benzoylene urea is above 350°.

3. Methods of Preparation

Benzoylene urea has been prepared by passing cyanogen into a solution of anthranilic acid in alcohol and hydrolyzing the resulting 2-ethoxy-4-ketodihydroquinazoline; [1] by fusing anthranilic acid with urea; [2,3] and by the action of aqueous cyanic acid on anthranilic acid. [2,3,4,5] The procedure described is adapted from that of Bogert and Scatchard [5] with several modifications.

[1] Griess, Ber. **2**, 415 (1869).

[2] Griess, J. prakt. Chem. (2) **5**, 371 (1872).

[3] Bogert and Scatchard, J. Am. Chem. Soc. **41**, 2056 (1919).

[4] Gabriel and Colman, Ber. **38**, 3561 (1905); Scott and Cohen, J. Chem. Soc. **119**, 664 (1921).

[5] Bogert and Scatchard, J. Am. Chem. Soc. **38**, 1611 (1916).

β-BENZOYLPROPIONIC ACID

(Propionic acid, β-benzoyl-)

$$\begin{array}{c} CH_2—CO \\ | \qquad\qquad >O + C_6H_6 \xrightarrow{\ AlCl_3\ } C_6H_5COCH_2CH_2CO_2H \\ CH_2—CO \end{array}$$

Submitted by L. F. SOMERVILLE and C. F. H. ALLEN.
Checked by REYNOLD C. FUSON and DON B. FORMAN.

1. Procedure

IN a 2-l. three-necked, round-bottomed flask fitted with a mechanical stirrer and two reflux condensers are placed 68 g. (0.68 mole) of succinic anhydride (p. 560) and 350 g. (4.5 moles) of dry, thiophene-free benzene (Note 1). The stirrer is started, and 200 g. (1.5 moles) of powdered, anhydrous aluminum chloride is added all at once. Hydrogen chloride is evolved and the mixture becomes hot (Note 2). It is heated in an oil bath and refluxed, with continued stirring, for half an hour (Note 3). The flask is then surrounded by cold water, and 300 cc. of water is slowly added from a dropping funnel inserted in the top of one of the condensers (Note 4). The excess benzene is removed by steam distillation, and the hot solution is at once poured into a 2-l. beaker. After the mixture is cold the liquid is decanted from the precipitated solid and acidified with concentrated hydrochloric acid (about 20 cc. is required); 5 to 15 g. of benzoylpropionic acid separates and is filtered (Note 5). The residual suspension in the beaker is boiled for five hours with 1.5 l. of water containing 360 g. of commercial soda ash; the resulting solution is filtered with suction, the filter cake washed with hot water, and the filtrate acidified with concentrated hydrochloric acid; about 300 cc. is required (Note 6). The precipitated benzoylpropionic acid is filtered and washed with hot water (Note 7). After drying for a day it weighs 95–100 g. (77–82 per cent of the theoretical amount) and melts at 111–113° (Note 8).

If the first, colored, precipitate (weight, 5–15 g.) is separated as suggested it is unnecessary to purify the remainder. The acid may be further purified, if desired, by dissolving in dilute sodium hydroxide and precipitating with concentrated hydrochloric acid, the first portion of the precipitate being collected separately. The pure product melts at 116° (Note 9).

2. Notes

1. Commercial benzene is shaken with concentrated sulfuric acid, then with water and dried, first with anhydrous calcium chloride and then over metallic sodium.

2. If the reaction does not start at once it is initiated by gentle heating.

3. The yield is not increased by longer heating.

4. The procedure from this point on can be simplified and improved as follows. After the addition of water to the aluminum chloride complex, 100 cc. of concentrated hydrochloric acid (sp. gr. 1.18) is added and the benzene is removed by steam distillation. The hot mixture is transferred to a 1-l. beaker, and the β-benzoylpropionic acid separates as a colorless oil which soon solidifies. After cooling to 0°, it is collected, washed with a cold mixture of 50 cc. of concentrated hydrochloric acid and 150 cc. of water, and then with 200 cc. of cold water. The crude acid is dissolved in a solution of 75 g. of anhydrous sodium carbonate in 500 cc. of water by boiling for fifteen minutes. The solution is filtered by suction and the small amount of aluminum hydroxide washed twice with 50-cc. portions of hot water. Four grams of charcoal is added to the hot filtrate; the solution is stirred for three to four minutes and then filtered with suction. The clear, colorless filtrate is transferred to a 2-l. beaker, cooled to 50–60°, and carefully acidified with 130 cc. of concentrated hydrochloric acid. After cooling to 0° in an ice-salt bath the acid is filtered, washed well with water, dried overnight at room temperature, and finally dried to constant weight at 40–50°. The yield is 110–115 g. (92–95 per cent of the theoretical amount). It melts at 114–115° and needs no further purification. (E. L. MARTIN and LOUIS F. FIESER, private communication.)

5. The amount of acid isolated at this point varies according to the length of time the solution is allowed to stand before acidification.

6. The first portion of the precipitated acid is usually colored; it is best to filter it separately. The remaining acid will then be colorless and very nearly pure (m.p. 114–115°).

7. By evaporating the filtrate to a small volume and extracting with ether, a further 3 g. of acid may be obtained.

8. The acid tenaciously retains traces of water to which apparently higher yields may be due. The weights given were obtained on material that had been dried overnight in a vacuum desiccator.

9. γ-Benzoylbutyric acid may also be prepared by this method. Glutaric anhydride (0.68 mole) dissolved in part of 350 g. of benzene is added to the rest of the benzene, in which the aluminum chloride is

suspended; the temperature is kept below 15° for one and one-half hours, including the time of addition; an 80–85 per cent yield of γ-benzoylbutyric acid melting at 125–126° is obtained.

3. Methods of Preparation

β-Benzoylpropionic acid can be prepared from succinic anhydride, benzene, and aluminum chloride,[1] or from succinic anhydride and phenylmagnesium bromide.[2] It has also been obtained by prolonged heating of cinnamic aldehyde, hydrocyanic acid, hydrochloric acid, and water;[3] by reduction of benzoylacrylic acid;[4] by the ketonic hydrolysis of benzoylsuccinic ester;[5] by heating phenylbromoparaconic, phenylbromoisoparaconic, or γ-phenyl-β,γ-dibromobutyric acids with water;[6] by the hydrolysis of phenacylbenzoylacetic ester;[7] and by heating phenacylmalonic acid.[8]

BENZYL PHTHALIMIDE

(Phthalimide, N-benzyl-)

$$2C_6H_4(CO)_2NH + 2C_6H_5CH_2Cl + K_2CO_3 \rightarrow$$
$$2C_6H_4(CO)_2NCH_2C_6H_5 + CO_2 + 2KCl + H_2O$$

Submitted by Richard H. F. Manske.
Checked by Henry Gilman and H. J. Harwood.

1. Procedure

An intimate mixture of 166 g. (1.2 moles) of anhydrous potassium carbonate (Note 1) and 294 g. (2 moles) of phthalimide is treated with 506 g. (4 moles) of benzyl chloride (Note 2), and the mixture is heated in an oil bath at 190° under a reflux condenser for three hours (Note 3). While the mixture is still hot, the excess benzyl chloride is removed by

[1] Burcker, Ann. chim. phys. (5) 26, 435 (1882); Kohler and Engelbrecht, J. Am. Chem. Soc. 41, 768 (1919).
[2] Komppa and Rohrmann, Ann. 509, 263 (1934).
[3] Matsmoto, Ber. 8, 1145 (1875); Peine, ibid. 17, 2114 (1884).
[4] von Pechmann, ibid. 15, 889 (1882).
[5] Perkin, J. Chem. Soc. 47, 276 (1885).
[6] Fittig and Leoni, Ann. 256, 81 (1890); Fittig, Obermüller, and Schiffer, ibid. 268, 74 (1892).
[7] Kapf and Paal, Ber. 21, 1487 (1888).
[8] Kues and Paal, ibid. 18, 3325 (1885).

steam distillation (Note 4). Near the end of this operation the benzyl phthalimide crystallizes. It is advisable to cool the mixture rapidly with very vigorous agitation so that the material is in as fine a state of division as possible. The solid is filtered on a large Büchner funnel, thoroughly washed with water, and drained as completely as possible by suction. It is then washed once with 400 cc. of 60 per cent alcohol and drained again. The yield of this product, melting at 100–110°, is 340–375 g. (72–79 per cent of the theoretical amount). It is conveniently purified by crystallizing from glacial acetic acid. The recovery in the crystallization is about 80 per cent, and the pure product melts at 116° (corr.) (Note 5).

2. Notes

1. The potassium carbonate is conveniently dehydrated by heating in a large basin over a moderate flame. It must be ground to a very fine powder and mixed with the phthalimide in a mortar.

2. A good grade of benzyl chloride having a boiling range of 3° was used.

3. There is no apparent advantage in using mechanical stirring.

4. The excess benzyl chloride is recovered from the distillate and dried with calcium chloride. About 200–300 g. is recovered.

5. Using the same general procedure trimethylene bromide furnishes γ-bromopropylphthalimide and, as a by-product, α,γ-diphthalimido-propane, while β-phenylethyl bromide gives β-phenylethylphthalimide.

3. Methods of Preparation

Benzyl phthalimide can be prepared from potassium phthalimide and benzyl chloride; [1] from phthalimide, potassium carbonate, and benzyl chloride; [2] and from phthalimide, sodium ethoxide, and benzyl chloride.[3] The two latter procedures avoid the preparation of potassium phthalimide. Phthalic anhydride, benzylamine, and glacial acetic acid on refluxing also furnish benzyl phthalimide.[4] The procedure described has been published.[2]

[1] Gabriel, Ber. **20**, 2227 (1887).
[2] Ing and Manske, J. Chem. Soc. **1926**, 2348.
[3] Weisz and Lányi, Magyar Chem. Folyóirat **39**, 153 (1933) [C. A. **28**, 5815 (1934)].
[4] Vanags, Acta Univ. Latviensis, Kim. Fakultat. Ser. 4, No. 8, 405 (1939) [C. A. **34**, 1983 (1940)].

BETAINE HYDRAZIDE HYDROCHLORIDE

Girard's Reagent

$$\left[\begin{array}{l} \text{Ammonium compounds, substituted.} \\ \text{(Carboxymethyl)trimethyl— \quad chloride, hydrazide} \end{array} \right]$$

$$(CH_3)_3N + ClCH_2CO_2C_2H_5 + H_2NNH_2 \rightarrow$$
$$(CH_3)_3\overset{|}{\underset{Cl}{N}}CH_2CONHNH_2 + C_2H_5OH$$

Submitted by ANDRÉ GIRARD.
Checked by LOUIS F. FIESER and ROBERT P. JACOBSEN.

1. Procedure

IN a 1-l. three-necked flask fitted with a stirrer, a thermometer, and an ice-cooled spiral condenser (Org. Syn. Coll. Vol. I, **1941,** 529) is placed a solution of 98 g. (84.5 cc., 0.8 mole) of ethyl chloroacetate (Note 2, p. 263) in 200 cc. of absolute alcohol. The solution is cooled to 0° by stirring in a salt-ice bath; and, after the stirrer is stopped, 74 cc. (49 g., 0.83 mole) of trimethylamine, measured after precooling to −5°, is added all at once. The exothermic reaction is controlled sufficiently by cooling so that the temperature of the mixture rises to 60° in the course of about one hour (Note 1). When there is no longer any heat effect, the mixture is allowed to stand at room temperature for twenty hours (without replenishment of the condenser ice).

The condenser is removed, the thermometer replaced by a dropping funnel, and 40 g. (0.8 mole) of 100 per cent hydrazine hydrate (Note 2) added, with stirring, in the course of ten to fifteen minutes. After being stirred for forty-five minutes longer, the solution is cooled slightly, and, unless crystallization of the reaction product starts spontaneously, the walls of the vessel are scratched with a glass rod to induce crystallization (Note 3). The product separates in fine, colorless needles. After being thoroughly cooled in an ice bath, the highly hygroscopic salt is collected quickly on a Büchner funnel, washed with 150 cc. of cold absolute alcohol, and pressed dry under a rubber dam. Dried in a vacuum desiccator over concentrated sulfuric acid, this material weighs 100–108 g. A further crop can be obtained after distilling 200–300 cc. of solvent from the mother liquor and washings at the pressure of the water pump. The total yield of salt, m.p. 175–180°, with decomposition, is 112–120 g (83.5–89.5 per cent of the theoretical amount) (Notes 4, 5, and 6).

2. Notes

1. Without external cooling the temperature rises to about 75° and it is difficult to avoid some loss of amine.

2. Hydrazine hydrate may be prepared by the ammonolysis of hydrazine sulfate (Org. Syn. Coll. Vol. I, 1941, 309) as described in Org. Syn. 21, 70.

The 42 per cent hydrazine hydrate solution supplied by the Eastman Kodak Company is too dilute for use as such, but may be concentrated by distillation with xylene.[1] A mixture of 144 cc. (150 g.) of the 42 per cent solution and 230 cc. of xylene is distilled from a 500-cc. flask through a 17-cm. Hempel column fitted into a cork covered with tin foil. After distillation of the xylene, with about 85 cc. of water, the residue yields on distillation 45–50 g. of 80–85 per cent hydrazine hydrate. This material, assayed best by titration with standard acid using methyl orange as indicator, may be used as such or concentrated further (see Note 6).

3. When more dilute hydrazine hydrate is used, the crystallization is slower, but it is not advisable to cool the solution thoroughly until crystals have begun to appear.

4. Although this material contains a small amount of the symmetrical dihydrazide, which is not easily eliminated on crystallization, it is entirely satisfactory as a reagent for the isolation of ketones. A purer product, m.p. 192°, with decomposition, can be obtained by adding the solution prepared from ethyl chloroacetate and trimethylamine to an alcoholic solution containing a considerable excess of hydrazine hydrate.

5. When stored in a dry, tightly stoppered container the reagent can be kept for long periods without deteriorating (odor), but samples withdrawn after some time are best recrystallized from absolute alcohol before use.

6. The yields obtained by the checkers when using 75 per cent and 50 per cent hydrazine hydrate were 78 per cent and 66 per cent, respectively, of the theoretical amount.

3. Method of Preparation

The above procedure is essentially that of Girard and Sandulesco.[2]

[1] Hurd and Bennett, J. Am. Chem. Soc. 51, 265 (1929).
[2] Girard and Sandulesco, Helv. Chim. Acta 19, 1095 (1936).

BROMAL

$$(CH_3CHO)_3 + 9Br_2 \rightarrow 3CBr_3CHO + 9HBr$$

Submitted by F. A. Long and J. W. Howard.
Checked by W. W. Hartman and G. L. Boomer.

1. Procedure

In a 2-l. three-necked, round-bottomed flask, fitted with a liquid-sealed mechanical stirrer, a dropping funnel, and an efficient reflux condenser, are placed 720 g. (230 cc., 4.5 moles) of bromine (Note 1) and 1.5 g. of sulfur (Note 2). A glass tube is connected to the top of the condenser to carry the evolved hydrogen bromide to a gas trap. Sixty-nine grams (69 cc., 0.52 mole) of dry paraldehyde (Note 1) is added slowly, with stirring, over a period of about four hours. The reaction proceeds under its own heat during the addition of the paraldehyde; subsequently the mixture is heated externally for two hours at 60–80°. The solution is distilled and a fraction collected over the range 155–175° (Note 3).

On redistillation under reduced pressure there is obtained 220–240 g. (52–57 per cent of the theoretical amount) of reddish yellow bromal, boiling at 59–62°/9 mm., or 71–74°/18 mm.

2. Notes

1. The bromine is dried by shaking with concentrated sulfuric acid; the paraldehyde is dried over calcium chloride.
2. The use of sulfur as a catalyst increases the yield 5–10 per cent and causes no trouble in the purification.
3. The fore-run amounts to 90–180 g. and consists mostly of bromine, bromoacetaldehyde, and dibromoacetaldehyde. An additional quantity of bromal may be obtained by treating this material with a small amount of bromine, heating for two hours at 60–80°, and distilling as before.

3. Methods of Preparation

Bromal has been prepared by brominating a solution of paraldehyde in ethyl acetate,[1] by passing bromine vapor through absolute alcohol,[2] and by treatment of chloral with a metallic bromide.[3]

[1] Pinner, Ann. **179**, 67 (1875).
[2] Schäffer, Ber. **4**, 366 (1871).
[3] Müller, U. S. pat. 2,057,964 [C. A. **31**, 112 (1937)].

BROMOACETONE

(2-Propanone, 1-bromo-)

$$CH_3COCH_3 + Br_2 \rightarrow CH_3COCH_2Br + HBr$$

Submitted by P. A. Levene.
Checked by Frank C. Whitmore and J. Pauline Hollingshead.

1. Procedure

A 5-l., three-necked, round-bottomed flask is provided with an efficient mechanical stirrer, a 48-cm. Allihn reflux condenser, a thermometer, and a 500-cc. separatory funnel, the stem of which reaches nearly to the bottom of the flask (Note 1).

Through the separatory funnel are introduced 1.6 l. of water, 500 cc. of c. p. acetone, and 372 cc. of glacial acetic acid. The stirrer is started and the temperature of the water bath is raised to 70–80°, so that the mixture in the flask is at about 65° (Note 2). Then 354 cc. (7.3 moles) of bromine is carefully added through the separatory funnel. The addition, which requires one to two hours, is so regulated as to prevent the accumulation of unreacted bromine (Note 3). As a rule the solution is decolorized in about twenty minutes after the bromine has been added. When the solution is decolorized, it is diluted with 800 cc. of cold water, cooled to 10°, made neutral to Congo red with about 1 kg. of solid anhydrous sodium carbonate, and the oil which separates is collected in a separatory funnel and dried with 80 g. of anhydrous calcium chloride. After drying, the oil is fractionated and the fraction boiling at 38–48°/13 mm. is collected. The yield is 470–480 g. (50–51 per cent of the theoretical amount). It may be used without further purification for the preparation of acetol (p. 5); but, if a purer product is desired, the above product is refractionated and the fraction boiling at 40–42°/13 mm. is collected. The yield is 400–410 g. (43–44 per cent of the theoretical amount).

The higher-boiling fraction contains a mixture of isomeric dibromoacetones.

2. Notes

1. The apparatus should be set up with the flask in a large container (such as a 14-qt. galvanized pail) to be used as a water bath, and under a well-ventilated hood, as both the bromine and bromoacetone are powerful irritants to the skin and mucous membranes.

2. It is necessary to warm the reaction mixture to this temperature to ensure a smooth reaction

3. It is not advisable to have too great an excess of bromine present at any time, as it sometimes reacts suddenly with great violence.

3. Methods of Preparation

Bromoacetone has been prepared by the electrolysis of a mixture of acetone and hydrobromic acid,[1] and by more orthodox methods of bromination: the addition of bromine to acetone dissolved in ten times its weight of water;[2] the addition of bromine to acetone in which marble is suspended;[3] the addition of bromine to acetone, water, and concentrated hydrochloric acid;[4] and the introduction of bromine by means of a current of air into cold acetone.[5]

A procedure similar to the one described above, except that the reaction mixture is illuminated with a powerful light, has been published.[6]

p-BROMOBENZALDEHYDE

(Benzaldehyde, *p*-bromo-)

$$p\text{-BrC}_6\text{H}_4\text{CH}_3 + 2\text{Br}_2 \rightarrow p\text{-BrC}_6\text{H}_4\text{CHBr}_2 + 2\text{HBr}$$

$$p\text{-BrC}_6\text{H}_4\text{CHBr}_2 + \text{H}_2\text{O} \rightarrow p\text{-BrC}_6\text{H}_4\text{CHO} + 2\text{HBr}$$

Submitted by GEORGE H. COLEMAN and G. E. HONEYWELL.
Checked by W. W. HARTMAN and A. J. SCHWADERER.

1. Procedure

IN a 1-l. three-necked flask fitted with a mechanical stirrer, a reflux condenser, a thermometer, and a dropping funnel is placed 100 g. (0.58 mole) of *p*-bromotoluene (Org. Syn. Coll. Vol. I, **1941**, 136). The stem of the dropping funnel and the thermometer should reach nearly to the bottom of the flask. The upper end of the condenser is connected to a gas absorption trap. The flask is heated with stirring in an oil bath

[1] Richard, Compt. rend. **133**, 879 (1901).
[2] Sokolowsky, Ber. **9**, 1687 (1876).
[3] Scholl and Matthaiopoulos, ibid. **29**, 1555 (1896).
[4] Hughes, Watson, and Yates, J. Chem. Soc. **1931**, 3322.
[5] Emmerling and Wagner, Ann. **204**, 29 (1880).
[6] Gohr and Thiekötter, Biochem. Z. **305**, 374 (1940).

until the temperature of the liquid reaches 105°. The liquid is illuminated with an unfrosted 150-watt tungsten lamp, and 197 g. (61.8 cc., 1.23 moles) of bromine is added slowly from the separatory funnel (Note 1). About one-half of the bromine is added during the first hour, during which time the temperature is kept at 105–110°. The rest is added during about two hours while the temperature is raised to 135°. When all the bromine has been added the temperature is raised slowly to 150°.

The crude product (Note 2) is transferred to a 2-l. flask and mixed thoroughly with 200 g. of powdered calcium carbonate. About 300 cc. of water is added and the mixture is heated cautiously (Note 3) and then refluxed for fifteen hours to effect hydrolysis. The product is then distilled in a rapid current of steam (Note 4); the distillate is collected in 500-cc. portions and cooled; and the p-bromobenzaldehyde is collected and dried in a desiccator. From the first liter of distillate 50–60 g. of p-bromobenzaldehyde melting at 55–57° is obtained. An additional 15–20 g. of product melting at 50–56° is obtained in about 2 l. more of distillate (Note 5). This may be purified through the bisulfite addition compound (Note 6) and yields 13–18 g. of product melting at 55–57°. The total yield of pure aldehyde is 65–75 g. (60–69 per cent of the theoretical amount).

2. Notes

1. The rate of addition of the bromine should be so regulated that a large excess of unreacted bromine does not accumulate in the reaction mixture. The amount of bromine present may be roughly estimated by the color of the solution and by the amount of bromine vapor carried into the condenser.

2. p-Bromobenzal bromide is a lachrymator and also produces a burning sensation on the skin. Washing the affected parts with alcohol gives relief.

3. In order to avoid breaking the flask the mixture is heated first on a water bath and then on a wire gauze over a flame with continuous shaking until the liquid begins to boil. The refluxing may then be continued without danger.

4. The inlet tube for steam should reach to the bottom of the distillation flask. A 16-mm. bulb on the end of this tube with four 0.8-mm. openings helps to ensure thorough mixing of the heavy residue. If this residue is not well stirred the aldehyde distils very slowly. It is well to connect the flask to the condenser through a large Hopkins still head in order to prevent the entrainment of foam during the distillation.

5. Five to ten grams of crude p-bromobenzoic acid can be obtained by acidifying the solution left in the distilling flask.

6. The material is triturated with saturated sodium bisulfite solution (2 cc. per gram), and after about three hours the pasty mixture is filtered with suction. The addition product is washed with absolute alcohol and then with ether and transferred to a flask fitted for steam distillation. Excess sodium carbonate solution is added and the aldehyde is distilled in a current of steam.

3. Methods of Preparation

p-Bromobenzaldehyde has been prepared by the oxidation of p-bromotoluene with chromyl chloride,[1] by saponification of the acetal from p-bromophenylmagnesium bromide and orthoformic ester,[2] by the oxidation of ethyl p-bromobenzyl ether with nitric acid,[3] by the oxidation of p-bromobenzyl bromide with lead nitrate,[4] by the hydrolysis of p-bromobenzal bromide in the presence of calcium carbonate,[5] and by the procedure given on p. 442 which involves oxidation of p-bromotoluene with chromium trioxide in the presence of acetic anhydride, followed by hydrolysis of the resulting p-bromobenzaldehyde diacetate.

β-BROMOETHYLAMINE HYDROBROMIDE

(Ethylamine, 2-bromo-, hydrobromide)

$$HOCH_2CH_2NH_2 + 2HBr \rightarrow BrCH_2CH_2NH_3Br + H_2O$$

Submitted by Frank Cortese.
Checked by C. S. Marvel and C. L. Fleming.

1. Procedure

One kilogram (16.4 moles) of ice-cold ethanolamine (Note 1) is added, through a dropping funnel, with mechanical stirring, to 7 l. (9.94 kg., 52 moles) of ice-cold hydrobromic acid (sp. gr. 1.42) (Note 2) contained in a 12-l. round-bottomed flask. The flask is attached to an efficient fractionating column and heated until 1850 cc. of distillate has been collected. The rate of heating is then diminished to a point at which

[1] Wörner, Ber. **29**, 153 (1896).
[2] Tschitschibabin, ibid. **37**, 188 (1904); Bodroux, Bull. soc. chim. (3) **31**, 587 (1904); Gattermann, Ann. **393**, 223 (1912).
[3] Errera, Gazz. chim. ital. **17**, 206 (1887).
[4] Jackson and White, Ber. **11**, 1043 (1878); Am. Chem. J. **3**, 32 (1881).
[5] Adams and Vollweiler, J. Am. Chem. Soc. **40**, 1738 (1918).

the liquid ceases to distil and merely refluxes. The heating under reflux is continued for one hour. At the end of this time, 700 cc. more is distilled, and the solution is again heated under reflux for one hour. This procedure is followed with 600-, 300-, 250-, 150-, 100-, and 50-cc. portions of distillate. The process may be interrupted at any time. The solution is finally heated under reflux for three hours, and 2.3 l. of crude hydrobromic acid distilled. The total volume of distillate, including that which is collected during refluxing, must not be less than 6270 cc. or more than 6330 cc.

The dark-colored residue is divided into two approximately equal portions, and each is poured, while still hot, into a 4-l. beaker. After the liquid has cooled to about 70°, 1650 cc. of acetone is added to each portion. The mixture is stirred well, so that as much as possible of the dark-colored solid is brought into contact with the acetone. After standing in the icebox overnight, the β-bromoethylamine hydrobromide is collected on a filter, washed with acetone until colorless (Note 3), and air-dried for about fifteen minutes. The filtrates are combined, concentrated to a volume of 1 l., and cooled. After seeding, a second crop of nearly pure material is obtained. By evaporation to a syrup, cooling, and seeding, a third crop of slightly colored material is obtained. (Note 4). The yield is about 2.8 kg. (83 per cent of the theoretical amount) (Note 5).

2. Notes

1. Commercial ethanolamine is fractionated in a glass apparatus, and the fraction boiling at 167–169° is used.

2. The hydrobromic acid must have a specific gravity of at least 1.42.

3. It may be advisable for effective washing to transfer the crude cake to a mortar and crush it.

4. All three crops are suitable for use in the synthesis of taurine described on p. 564.

5. The procedure has been applied to the preparation of the following β-dialkylaminoethyl bromide hydrobromides.

$$BrCH_2CH_2NHR_2Br$$

R	Yield, Per Cent	Heating Periods
CH_3	83	1–1.5 hr., 2 hr.
C_2H_5	80	1 hr., 2 hr.
n-C_3H_7	55	0.75 hr., 2 hr.
n-C_4H_9	20	1 hr., 3 hr.
n-C_4H_9	59	0.5 hr., 2 hr.

The final distillation was not always carried as far as called for by the procedure of Cortese. If a faint brown or violet color appeared in the distillate, or if white fumes were given off, the distillation was discontinued; further distillation seemed to cause decomposition. However, the color always came near the end point specified in the procedure above and the total volume of the distillates was never less than 95 per cent of that called for. Since the solubilities of the products in acetone range from very slight to extreme as the molecular weights increase, the extraction of the products requires variations in the amounts of acetone specified above. [LAWRENCE H. AMUNDSEN and KARL W. KRANTZ, private communication, and J. Am. Chem. Soc. **63**, 305 (1941).]

3. Methods of Preparation

β-Bromoethylamine hydrobromide has been prepared by the reaction of potassium phthalimide with ethylene bromide, followed by hydrolysis;[1] by the addition of hydrogen bromide to ethyleneimine;[2] from ethanolamine and hydrobromic acid;[3] and from ethanolamine and hydrogen bromide at 140–190°.[4] The preparation from ethyleneimine and hydrogen bromide is reported to succeed only if the imine is added to the acid.[5]

α-BROMOISOVALERIC ACID

(Isovaleric acid, α-bromo-)

$$(CH_3)_2CHCH(CO_2C_2H_5)_2 + 2KOH \rightarrow$$
$$(CH_3)_2CHCH(CO_2K)_2 + 2C_2H_5OH$$

$$(CH_3)_2CHCH(CO_2K)_2 + 2HCl \rightarrow (CH_3)_2CHCH(CO_2H)_2 + 2KCl$$

$$(CH_3)_2CHCH(CO_2H)_2 + Br_2 \rightarrow (CH_3)_2CHCBr(CO_2H)_2 + HBr$$

$$(CH_3)_2CHCBr(CO_2H)_2 \xrightarrow{\text{(Heat)}} (CH_3)_2CHCHBrCO_2H + CO_2$$

Submitted by C. S. MARVEL and V. DU VIGNEAUD.
Checked by FRANK C. WHITMORE and A. M. GRISWOLD.

1. Procedure

A SOLUTION of 200 g. (3.6 moles) of potassium hydroxide in 200 cc. of water is placed in a 2-l. flask fitted with a reflux condenser. The mixture is heated to about 80°, and 202 g. (1 mole) of isopropylmalonic ester (Note 1) is added through the condenser over a period of about

[1] Gabriel, Ber. **21**, 566 (1888).

[2] Gabriel, ibid. **21**, 1054 (1888); Gabriel and Stelzner, ibid. **28**, 2929 (1895).

[3] Gabriel, ibid. **50**, 826 (1917); Cortese, J. Am. Chem. Soc. **58**, 191 (1936).

[4] I. G. Farbenind. A.-G., Brit. pat. 468,387 [C. A. **31**, 8545 (1937)].

[5] Masters and Bogert, J. Am. Chem. Soc. **64**, 2710 (1942).

one hour. The mixture should be shaken well to prevent the formation of two layers. The saponification proceeds rapidly, forming a clear solution. The solution is transferred to a 20-cm. evaporating dish, the flask is rinsed with 50 cc. of water, and the solution and washings are evaporated practically to dryness on a steam bath (Note 2).

The residue is dissolved in 200 cc. of water, transferred to a 1-l. flask, and cooled to 0° in an ice-salt bath. A mixture of 400 cc. of concentrated hydrochloric acid (sp. gr. 1.19) and 200 g. of cracked ice is added slowly until the mixture is acid to Congo red. The temperature of the mixture must not rise above 10° (Note 3). Potassium chloride separates. The mixture is extracted with two 200-cc. portions and four 100-cc. portions of alcohol-free ether (Note 4) to remove the isopropylmalonic acid. The ether solution (Note 5) is placed in a flask fitted with a reflux condenser, and 160 g. (1 mole) of bromine is added gradually over a period of about two hours at such a rate that the ether boils gently (Note 6). When the bromination is complete, the ether solution is washed with 100 cc. of water to remove the hydrobromic acid, dried over 25 g. of calcium chloride, and freed from ether by distillation on the steam bath. The crude isopropylbromomalonic acid is heated in the distilling flask in an oil bath at 125–130° until no more carbon dioxide is evolved. It is then distilled under reduced pressure; the fraction distilling at 140–160°/40 mm. is collected separately and redistilled (Note 7). The yield of product boiling at 148–153°/40 mm. (125–130°/15 mm.) is 100–120 g. (55–66 per cent of the theoretical amount) (Note 8).

2. Notes

1. The isopropylmalonic ester was prepared by the method used for n-butylmalonic ester (Org. Syn. Coll. Vol. I, 1941, 250). The yield of product boiling at 132–135°/44 mm. was 70–75 per cent of the theoretical amount.

2. Unless the saponification is complete, the final product will contain ethyl α-bromoisovalerate, which will appear in the low-boiling fraction. If all the alcohol is not removed, some esterification will occur on acidification.

3. A rise in temperature favors the loss of carbon dioxide with the formation of isovaleric acid, which will escape bromination.

4. The ether used for extraction is first extracted with one-tenth its volume of saturated calcium chloride solution to remove the alcohol, which would otherwise cause partial esterification of the acid.

5. No special drying is necessary before the bromination, but the ether and aqueous layers should be separated carefully.

6. Usually the bromination starts easily. Sometimes, however, the mixture has to be heated after the first few drops of bromine are added. The mixture may have to be heated to complete the bromination.

7. The product on the first distillation does not have a constant boiling point, as some carbon dioxide is liberated from undecomposed isopropylbromomalonic acid. This cannot be avoided by preliminary heating, even at temperatures much higher than those used.

8. This is a general method for preparing α-bromo acids. By using exactly analogous directions α-bromo-*n*-caproic acid may be obtained in 65–70 per cent yields from *n*-butylmalonic ester; α-bromoisocaproic acid in 65–70 per cent yields from isobutylmalonic ester; and α-bromo-β-methylvaleric acid in 75–80 per cent yields from *sec.*-butylmalonic ester.

3. Methods of Preparation

α-Bromoisovaleric acid has been prepared from bromine and isovaleric acid alone,[1] or in the presence of phosphorus,[2] or phosphorus trichloride;[3] and by the action of heat on isopropylbromomalonic acid.[4]

BROMOMESITYLENE

(Mesitylene, 2-bromo-)

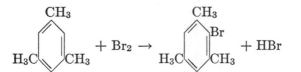

Submitted by LEE IRVIN SMITH.
Checked by ROGER ADAMS and H. A. STEARNS.

1. Procedure

In a 3-l. three-necked flask, provided with a short reflux condenser, a mechanical stirrer, and a separatory funnel, is placed a solution of 636 g. (5.3 moles) of mesitylene (Note 1) in 375–440 cc. of carbon tetrachloride. The flask is placed in an ice-salt bath, and when the temperature of the

[1] Ley and Popoff, Ann. **174,** 63 (1874).
[2] Schleicher, ibid. **267,** 115 (1892).
[3] Org. Syn. **20,** 106.
[4] Koenigs and Mylo, Ber. **41,** 4437 (1908).

reaction mixture is below 10° a solution of 900 g. (288 cc., 5.6 moles) of bromine in 565 cc. of carbon tetrachloride is added to the well-stirred solution. The bromination proceeds very readily, and the hydrogen bromide which is evolved is led off through the condenser and absorbed in water. The addition of the bromine solution requires about three hours, during which time the temperature is maintained at 10–15°.

After the addition of the bromine is complete, the reaction mixture is allowed to stand at room temperature for about one hour. It then has a light yellow color. The solution is washed with water and then with two 500-cc. portions of 20 per cent sodium hydroxide solution to remove any dissolved hydrobromic acid. The solution is dried over calcium chloride and filtered. The carbon tetrachloride is distilled through a good column until the temperature at the top of the column reaches about 120°. When the carbon tetrachloride has distilled, the oil is likely to turn dark and give off fumes.

The residue is added to a solution of 50 g. of sodium in about a liter of 95 per cent alcohol. The solution is boiled under a reflux condenser for about one hour (Note 2) and then allowed to stand overnight. The reaction mixture is diluted with about 6 l. of water and the two layers are separated. The aqueous layer is extracted with three or four 500-cc. portions of carbon tetrachloride (Note 3), and the extracts are added to the bromomesitylene. This solution is then washed thoroughly with water. The carbon tetrachloride solution is separated, dried over calcium chloride, and distilled. After the carbon tetrachloride is removed the bromomesitylene is fractionated carefully under reduced pressure from a modified Claisen flask. The fraction boiling at 105–107°/16–17 mm. (Note 4) is bromomesitylene. The yield is 840–870 g. (79–82 per cent of the theoretical amount). There is a small low-boiling portion (about 25 g.) and also a small high-boiling residue. The bromomesitylene obtained in this way gives no precipitate on standing twenty-four hours with alcoholic silver nitrate solution. It has a melting point of −1° to +1°.

2. Notes

1. The mesitylene was prepared as described in Org. Syn. Coll. Vol. I, **1941,** 341, and boiled at 58–59°/15 mm.

2. The treatment with sodium ethoxide removes any traces of side-chain halogen derivatives. Some care has to be exercised at this point as the solution may foam at the beginning of the heating period.

3. The carbon tetrachloride distilled from the crude bromomesitylene may be used for this purpose. The total volume of solution should be about 2 l.

4. Boiling points of bromomesitylene observed at different pressures were as follows: 132°/62 mm.; 139°/70 mm.; 146°/78 mm.; 157°/100 mm.

3. Method of Preparation

Bromomesitylene is always prepared by brominating mesitylene—a reaction that can be run in the dark,[1] or in daylight,[2] or using sulfur bromide and nitric acid,[3] or with metallic manganese as a catalyst in the absence of any solvent.[4]

o-BROMOPHENOL

(Phenol, *o*-bromo-)

$$C_6H_5OH + 2H_2SO_4 \rightarrow C_6H_3(OH)(SO_3H)_2(1,2,4) + 2H_2O$$
$$C_6H_3(OH)(SO_3H)_2 + 3NaOH \rightarrow C_6H_3(ONa)(SO_3Na)_2 + 3H_2O$$
$$C_6H_3(ONa)(SO_3Na)_2 + Br_2 \rightarrow C_6H_2(OH)(SO_3Na)_2(Br) + NaBr$$
$$C_6H_2(OH)(SO_3Na)_2(Br) + 2H_2SO_4$$
$$\rightarrow C_6H_2(OH)(SO_3H)_2(Br) + 2NaHSO_4$$

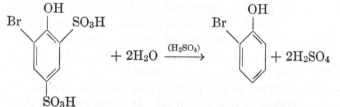

Submitted by RALPH C. HUSTON and MUREL M. BALLARD.
Checked by LOUIS F. FIESER and MAX TISHLER.

1. Procedure

IN a 3-l. three-necked flask is placed a mixture of 94 g. (1 mole) of phenol and 350 g. (190 cc., 3.5 moles) of concentrated sulfuric acid, and the mixture is heated on a boiling water bath for three hours with

[1] Schramm, Ber. **19**, 212 (1886).
[2] Fittig and Storer, Ann. **147**, 6 (1868); Smith and MacDougall, J. Am. Chem. Soc. **51**, 3002 (1929).
[3] Kalle and Company, Ger. pat. 123,746 [Frdl. **6**, 53 (1900–02)].
[4] Duke, Lewis, and Dunbar, Proc. S. Dakota Acad. Sci. **15**, 21 (1935) [C. A. **30**, 2556 (1936)].

constant mechanical stirring. At the end of this time the boiling water bath is replaced by an ice bath. When the reaction mixture has been cooled to room temperature it is made alkaline by the careful addition of a solution of 280 g. (7 moles) of sodium hydroxide in 700 cc. of water (Note 1). This must be done slowly and with good cooling to prevent boiling. A solid salt, which at first separates, largely dissolves at a later stage.

The alkaline solution is cooled to room temperature; with the stirrer still in constant operation, and after inserting a thermometer, 160 g. (1 mole) of bromine is added from a dropping funnel in the course of twenty to thirty minutes. During this operation the temperature is allowed to rise to 40–50°. Stirring is continued for one-half hour after all the bromine has been added. The solution should still be alkaline and should contain only a small amount of suspended material.

In order to evaporate the solution, the flask is then placed in an oil bath, which is brought to a temperature of 150–155°. As soon as solid material begins to separate, the mixture will bump badly unless a rather rapid current of air is passed through the reaction mixture. This has the further advantage of hastening the evaporation (Note 2). The heating is continued until a thick, pasty, gray mass is left as a residue, the process requiring thirty to forty minutes. The mixture is allowed to cool and then made strongly acid by the addition of 800 cc. of concentrated sulfuric acid. This must be done slowly and under a hood on account of the rapid evolution of hydrogen bromide.

The flask is then heated in an oil bath maintained at a temperature of 190–210° and the mixture is distilled with steam. The sulfonate groups are hydrolyzed in this process, and the bromophenol passes over as a heavy, colorless or pale yellow oil. In about one hour the distillate is clear. The product is extracted with ether, the ether is removed by distillation from the steam bath, and the residue is distilled at atmospheric pressure (Note 3). The fraction boiling at 194–200° represents practically pure o-bromophenol. The yield is 70–75 g. (40–43 per cent of the theoretical amount) (Note 4). o-Bromophenol is a colorless liquid with a very characteristic odor. It is rather unstable and decomposes on standing, becoming brown or red in color.

2. Notes

1. Too great an excess of water in the reaction mixture appears to result in the formation of higher bromination products. Insufficient water causes the reaction mixture to solidify during bromination, preventing efficient agitation.

2. A small amount of tribromophenol is eliminated in the evaporation, the substance being volatile with steam.

3. Distillation should be as rapid as possible, as the *o*-bromophenol is somewhat unstable and decomposes rapidly at the high temperature. Distillation at reduced pressure has not been found to offer much improvement.

4. The rather large residue of higher-boiling material probably contains higher bromination products of phenol.

3. Methods of Preparation

o-Bromophenol has been prepared by the bromination of phenol in various solvents and with various brominating agents,[1] and at high temperature in the absence of a solvent.[2] It has been obtained by the decarboxylation of 2-bromo-3-hydroxybenzoic acid,[3] by the diazotization of *o*-bromoaniline,[4] and from *o*-aminophenol by the Sandmeyer reaction.[5] The method given here, which is an adaptation of the procedure of Takagi and Kutani [6] for the preparation of 2-chlorophenol, has been improved by using nitrobenzene as a solvent in the bromination of the phenolsulfonic acid.[7]

[1] Hubner and Brenken, Ber. **6**, 171 (1873); Dinwiddie and Kastle, Am. Chem. J. **46**, 502 (1911); Skraup and Beifuss, Ber. **60**, 1077 (1927); Likhosherstov, J. Russ. Phys.-Chem. Soc. **61**, 1019 (1929) [C. A. **24**, 836 (1930)].

[2] E. Merck, Ger. pat. 76,597 [Frdl. **3**, 845 (1890–94)].

[3] Lellman and Grothmann, Ber. **17**, 2726 (1884).

[4] Fittig and Mayer, ibid. **8**, 362 (1875).

[5] Mendola and Streatfeild, J. Chem. Soc. **73**, 685 (1898).

[6] Takagi and Kutani, J. Pharm. Soc. Japan, No. 517, 260 (1925) [C. A. **20**, 2669 (1926)].

[7] Huston and Neeley, J. Am. Chem. Soc. **57**, 2176 (1935).

4-BROMORESORCINOL

(Resorcinol, 4-bromo-)

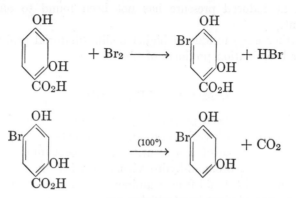

Submitted by R. B. SANDIN and R. A. McKEE.
Checked by W. W. HARTMAN and J. B. DICKEY.

1. Procedure

In a 1-l. flask fitted with a mechanical stirrer and a dropping funnel are placed 46.2 g. (0.3 mole) of 2,4-dihydroxybenzoic acid (β-resorcylic acid, p. 557) and 350 cc. of glacial acetic acid. After the stirrer is started the mixture is warmed until solution results (45°) and then is allowed to cool to 35°. Through the dropping funnel is added a solution of 48 g. (15 cc., 0.3 mole) of bromine in 240 cc. of glacial acetic acid with vigorous stirring over a period of about one hour. The temperature of the reaction mixture remains at 30–35°. When all the bromine is added the solution is poured into 5 l. of water, and the mixture is cooled to 0–5° and allowed to stand for several hours. The fine, white crystals of 2,4-dihydroxy-5-bromobenzoic acid are collected on a 10-cm. Büchner funnel and washed with about 500 cc. of cold water. The crude product, after air drying at room temperature, melts at 194–200° and weighs 55–60 g. For purification it is dissolved in 1.5 l. of boiling water, and the solution is refluxed for one hour (Note 1), filtered while hot, and cooled in an ice bath. The material which crystallizes is collected, washed with 100 cc. of cold water, and air dried. The yield of colorless 2,4-dihydroxy-5-bromobenzoic acid, melting at 206.5–208.5° (corr.), is 40–44 g. (57–63 per cent of the theoretical amount).

Thirty grams of purified 2,4-dihydroxy-5-bromobenzoic acid is refluxed for twenty-four hours with 375 cc. of water, and the resulting solution is filtered, cooled, and extracted with a 400-cc. and a 200-cc. portion of ether. The ether is removed by evaporation, and the 4-bromoresorcinol is dried on a steam bath. The yield of product melting at 100–102° (Note 2) is 22–22.5 g. (90–92 per cent of the theoretical amount).

2. Notes

1. The 2,4-dihydroxy-3,5-dibromobenzoic acid invariably present is in this way converted into the very soluble 2,4-dibromoresorcinol and removed. The monobromo acid is decarboxylated much more slowly.

2. Some samples were obtained with melting points ranging from 77° to 93°, but on dissolving the samples in chloroform and evaporating the solvent the values rose to 100–102°. This is not believed to be a process of purification.

3. Methods of Preparation

4-Bromoresorcinol has been prepared by the monobromination of resorcinol monobenzoate and subsequent hydrolysis;[1] from 2-bromo-5-aminophenol by the diazo reaction;[2] by treating resorcinol with dichlorourea and potassium bromide;[3] and by the bromination of 2,4-dihydroxybenzoic acid followed by decarboxylation.[4] The above procedure is based particularly upon the observations of Rice.[4]

[1] Fries and Lindemann, Ann. **404**, 61 (1914).

[2] Fries and Saftien, Ber. **59**, 1254 (1926).

[3] Likhosherstov, J. Gen. Chem. (U.S.S.R.) **3**, 172 (1933) [C. A. **28**, 1676 (1934)].

[4] Zehenter, Monatsh. **2**, 480 (1881); **8**, 293 (1887); von Hemmelmayr, ibid. **33**, 977 (1912); **34**, 374 (1913); Rice, J. Am. Chem. Soc. **48**, 3125 (1926); Davis and Harrington, ibid. **56**, 129 (1934).

1,3-BUTADIENE

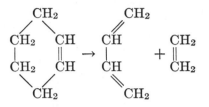

Submitted by E. B. HERSHBERG and JOHN R. RUHOFF.
Checked by W. H. CAROTHERS and J. HARMON.

1. Procedure

CYCLOHEXENE (Org. Syn. Coll. Vol. I, **1941**, 183, and p. 152 below) is boiled in the flask B, shown in Fig. 3, and the vapor is passed over a cracking element consisting of an expansible grid threaded with resistance ribbon L (Notes 1 and 2). The boiling flask B is supported over a 250-watt bowl heater provided with a rheostat, and the current in the cracking element is taken from a 115-volt a-c. or d-c. source and controlled by a second rheostat of 10-ampere capacity (Note 3). In order to trap any cyclohexene which passes the coil condenser, the gas-delivery tube D is connected to a tube leading close to the bottom of a 500-cc. distilling flask immersed in an ice bath. The exit tube of the flask is connected with a short section of rubber tubing to a receiver for condensing the butadiene, consisting of a large test tube with the entrance tube leading halfway to the bottom and an exit tube at the top for conducting the ethylene (saturated with butadiene) to a hood or outdoors. The receiver is cooled in a Dewar flask containing solid carbon dioxide and a eutectic mixture of equal parts by weight (or volume) of chloroform and carbon tetrachloride.

The flask B is two-thirds filled with cyclohexene, and, with the cooling water flowing, this is heated to vigorous boiling. When the vapor has displaced the air from the apparatus *completely* (Note 4), the current is turned on in the cracking unit. By adjusting both the rate of boiling and the current the ribbon is maintained at a bright red heat over its entire length. Very rapid refluxing is necessary in order to prevent undue carbonization on the filament and tar formation on the glass walls (Note 5). The generator can be run intermittently or until the charge is exhausted.

The butadiene collected is purified by a bulb-to-bulb distillation, the receiver in the cooling mixture being replaced by a similar container to which it is connected by means of rubber tubing and into which the

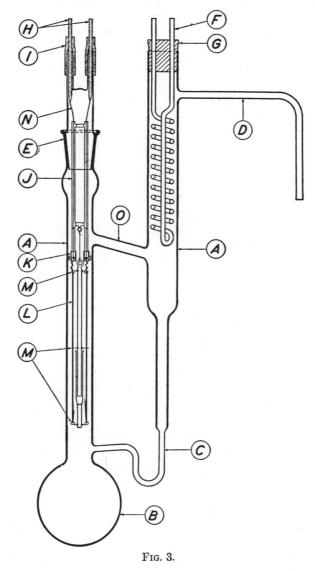

Fig. 3.

butadiene is allowed to distil. The product is quite satisfactory for most uses, as in the Diels-Alder reaction (Note 6). The cracking element uses about 500 watts at 8.7 amperes and produces 25–30 g. of butadiene per hour. The yield, based on the cyclohexene consumed

(Note 7) and on redistilled product, is 65–75 per cent of the theoretical amount. High-boiling residues accumulating in the boiling flask should be removed after preparing 100–150 g. of butadiene (Note 8).

2. Notes

1. *Apparatus.*—The unit shown in Fig. 3 is constructed of Pyrex glass, and the following is a summary of satisfactory dimensions for the various tubings (outer diameters) and of other specifications: A, 32-mm. tubing; B, 500-cc. flask; C, 7-mm. tubing; D, 8-mm. tubing; E, No. 35 standard taper joint; F, coil of $3/16$-in. copper tubing (see Note 2); G, cork stopper; H, $1/8$-in. brass rod; I, rubber tubing; J, 3-mm. glass tubing; K, brass electrical connectors; L, Chromel C resistance ribbon No. 37 B. and S. gauge, 1.9–2 ohms per foot, width $1/16$ in., length 56 to 60 in. (see below); M, tungsten wire-loop supports sealed into the suspended glass rod; the center supports are of 0.01-in. (dia.) wire, those at the top and bottom of 0.015-in. (dia.) wire; N, copper wire No. 22 B. and S. gauge; O, 12-mm. tubing.

The ribbon L of the cracking element is threaded between the tungsten loops M, of which there are five each at the top and bottom and eight at the center. The unit is suspended with a copper wire from the glass cross-support as shown. The lower end is free to drop down as the resistance wire expands; this prevents short-circuiting of the element.

The composition of the filament is of importance in determining the yield and the performance of the apparatus. With nickel-chromium alloys excessive carbonization occurs and the yield is poor. Much better results are obtained with the nickel-iron-chromium alloys called Chromel C and Nichrome Alloy Wire.

2. To provide for the proper functioning of the coil condenser even in warm weather it is advisable to increase the number of turns to 30–40, although the condenser as shown is satisfactory with tap water at 4–10°.

3. A slide-wire rheostat of 2-ampere capacity wound on a hollow enameled iron tube will carry the required current if a stream of cooling water is passed through the tube.

4. An explosion may occur if the filament is heated while an appreciable amount of air is still present. Furthermore, the heating element will burn out at once if an adequate supply of cyclohexene vapor is not supplied, since the current passing through the wire is far above the normal rating for air.

5. In case of excessive carbonization the rate of boiling should be increased or the filament temperature slightly lowered. In general the

vapor velocity should be as high as possible without exceeding the capacity of the copper condenser.

6. The crude product contains appreciable amounts of C_2, C_3, and C_6 fractions. The actual butadiene content lies between 82 and 88 per cent. If very pure material is desired the butadiene is converted into the tetrabromide, which is crystallized and reconverted to the hydrocarbon by means of zinc and alcohol.[1]

7. The cyclohexene collecting in the ice trap ordinarily is returned to the boiling flask; in determining the percentage conversion this was combined with any material left in the boiler and the pure starting material present recovered by fractionation.

8. The apparatus may be used also for the preparation of ketene from acetone (Org. Syn. Coll. Vol. I, 1941, 330).

3. Methods of Preparation

The methods for the preparation of gaseous products containing more or less butadiene are too numerous for profitable review here. Especially is this true since the development of commercial processes which are not particularly suitable for laboratory operation. The most satisfactory procedure for preparing butadiene in the laboratory is the pyrolysis of cyclohexene,[2] which has been shown to yield a product consisting essentially of butadiene.[1,3] For this pyrolysis, the apparatus described above and that described by Williams and Hurd[4] are available. Other laboratory preparations of butadiene start from butyl chloride,[5] 2,3-dibromobutane,[6] crotyl chloride,[6] and 1,3-butylene glycol.[7]

[1] Kistiakowsky, Ruhoff, Smith, and Vaughan, J. Am. Chem. Soc. 58, 146 (1936).

[2] Badische Anilin- und Soda-Fabrik, Ger. pat. 252,499 (Chem. Zentr. 1912, II, 1708).

[3] Zelinskiĭ, Mikhaĭlov, and Arbuzov, J. Gen. Chem. (U.S.S.R.) 4, 856 (1934) [C. A. 29, 2152 (1935)].

[4] Williams and Hurd, J. Org. Chem. 5, 122 (1940).

[5] Muskat and Northrup, J. Am. Chem. Soc. 52, 4043 (1930).

[6] Harries, Ann. 383, 176 (1911); Jacobson, J. Am. Chem. Soc. 54, 1545 (1932).

[7] Nagai, J. Soc. Chem. Ind. Japan 44, Suppl. binding 64 (1941) [C. A. 35, 3960 (1941)].

n-BUTYL BORATE

$$3C_4H_9OH + B(OH)_3 \rightarrow B(OC_4H_9)_3 + 3H_2O$$

Submitted by JOHN R. JOHNSON and S. W. TOMPKINS.
Checked by W. W. HARTMAN and J. B. DICKEY.

1. Procedure

IN a 2-l. round-bottomed flask, equipped with a 200-cc. dropping funnel and a 30-cm. column filled with glass beads (Note 1), and connected to a 40–50 cm. condenser, are placed 124 g. (2 moles) of boric acid, 666 g. (9 moles) of technical n-butyl alcohol, and a few chips of porous plate. The reaction mixture is heated to gentle boiling, and the rate of heating is adjusted so that 90–100 cc. of distillate is collected per hour. The temperature of the vapor at the top of the column remains constant at 91° over a period of three to three and one-half hours while the azeotropic mixture of n-butyl alcohol and water distils (Note 2). After two hours, the upper layer of n-butyl alcohol in the distillate is separated from the water, dried with a little anhydrous potassium carbonate or magnesium sulfate, and returned to the reaction mixture through the separatory funnel. Likewise, after the third hour of heating, the n-butyl alcohol in the distillate is separated, dried, and returned to the reaction flask.

During the third hour of heating, the temperature at the top of the column rises slowly as the removal of the water approaches completion. After the fourth hour, when the temperature of the distilling vapor has attained 110–112°, the heating is discontinued, and the reaction mixture is transferred to a 2-l. Claisen flask with the least possible exposure to atmospheric moisture (Note 3). The unreacted n-butyl alcohol is removed by distillation under reduced pressure (Note 4) until the thermometer registers a sudden rise in temperature. The receiver is then changed, and the main fraction of n-butyl borate distils almost entirely at 103–105°/8 mm. or 114–115°/15 mm. A negligible residue remains in the distilling flask. The weight of the distilled n-butyl borate, which contains a small amount of n-butyl alcohol, is 410–435 g. (89–94 per cent of the theoretical amount). From the aqueous distillate and the fore-run of the vacuum distillation, 190–210 g. of n-butyl alcohol is recovered.

The main fraction of n-butyl borate is redistilled from a Claisen flask provided with an indented fractionating column (Note 5), and the first

4–6 cc. of distillate is rejected. The purified product, b.p. 103–105°/8 mm. or 114–115°/15 mm., weighs 400–425 g. (87–92 per cent of the theoretical amount) (Note 6).

2. Notes

1. Any fractionating column of a reasonably effective type can be used. Columns longer than 30 cm. and of the most efficient types were found to give no better results than a simple Hempel column filled with glass beads or broken glass.

2. The distillate collected while the temperature remains at 91° separates into two layers; 100 cc. of this distillate contains 72 cc. of a supernatant layer of wet *n*-butyl alcohol.

3. Since *n*-butyl borate is hydrolyzed readily by atmospheric moisture, it is necessary to manipulate the reaction product so as to minimize exposure to the air.

4. *n*-Butyl borate can be purified by distillation at atmospheric pressure, but the separation from *n*-butyl alcohol is effected more readily under reduced pressure. *n*-Butyl borate is stated [1] to boil at 190°/200 mm. and at 230–235° under atmospheric pressure.

5. It is advantageous to use a Claisen flask with a 25–30 cm. side arm bearing a condenser jacket and connected to a device for collecting the fractions without interruption of the distillation or exposure to moist air.

6. *n*-Amyl borate can be prepared in a similar manner. From 792 g. (9 moles) of *n*-amyl alcohol and 124 g. (2 moles) of boric acid there is obtained 510–525 g. (93–96 per cent of the theoretical amount) of *n*-amyl borate, b.p. 146–148°/16 mm.; 210–215 g. of *n*-amyl alcohol is recovered. In this preparation the temperature of the distilling vapor remains at 95° during the first two hours and the distillate contains relatively more water (100 cc. of distillate contains 56 cc. of water and 44 cc. of *n*-amyl alcohol). After the second hour the temperature rises slowly to 136–137°. It is unnecessary, and not advantageous, to return the recovered *n*-amyl alcohol to the reaction mixture.

3. Methods of Preparation

The procedure described is essentially that disclosed in a patent.[1] *n*-Butyl borate can also be prepared by the action of *n*-butyl alcohol on boron triacetate or boric anhydride.

[1] W. J. Bannister, U. S. pat. 1,668,797 [C. A. **22**, 2172 (1928)].

n-BUTYL NITRITE

$$C_4H_9OH + HONO \xrightarrow{(H_2SO_4)} C_4H_9ONO + H_2O$$

Submitted by W. A. Noyes.
Checked by C. R. Noller and B. H. Wilcoxon.

1. Procedure

In a 3-l. three-necked, round-bottomed flask, fitted with a mechanical stirrer (Note 1), a separatory funnel extending to the bottom of the flask, and a thermometer, are placed 380 g. (5.5 moles) of c.p. sodium nitrite and 1.5 l. of water. The flask is surrounded by an ice-salt mixture, and the solution is stirred until the temperature falls to 0°. A mixture of 100 cc. of water, 136 cc. (250 g., 2.5 moles) of concentrated sulfuric acid (sp. gr. 1.84) (Note 2), and 457 cc. (370 g., 5 moles) of commercial n-butyl alcohol is cooled to 0° and by means of the separatory funnel is introduced slowly beneath the surface of the nitrite solution, with stirring. The alcohol solution is added slowly enough so that practically no gas is evolved, and the temperature is kept at ±1°. This usually requires from one and one-half to two hours.

The resulting mixture is allowed to stand in the ice-salt bath until it separates into layers, and the liquid layers are decanted from the sodium sulfate into a separatory funnel (Note 3). The lower aqueous layer is removed and the butyl nitrite layer washed twice with 50-cc. portions of a solution containing 2 g. of sodium bicarbonate and 25 g. of sodium chloride in 100 cc. of water. After drying over 20 g. of anhydrous sodium sulfate, the yield of practically pure butyl nitrite amounts to 420–440 g. (81–85 per cent of the theoretical amount) (Notes 4 and 5). If desired, the product may be distilled under reduced pressure when 98 per cent distils at 24–27°/43 mm. (Note 6). Butyl nitrite boils at 75° under atmospheric pressure, with some decomposition.

2. Notes

1. A stirrer capable of keeping solid material in motion and driven by a strong motor should be used because of the precipitation of large amounts of sodium sulfate towards the end of the reaction.

2. The concentration of sulfuric acid which is used keeps the butyl alcohol in solution but does not dissolve butyl nitrite.

3. If more butyl nitrite separates from the sodium sulfate after the first decantation, a second decantation is made. Care must be exercised

in handling butyl nitrite; inhalation of the vapor may cause severe headache and heart excitation.

4. The same procedure is quite satisfactory for runs of one-tenth this size. In small runs mechanical stirring is unnecessary since gentle rotation of the flask by hand gives good mixing.

5. Isoamyl nitrite may be made in the same manner and with approximately the same yields.

6. Butyl nitrite decomposes slowly on standing and should be kept in a cool place and used within a few days or, at most, a few weeks after it is prepared. A sample which stood for five months during a warm summer seemed to contain only 20–25 per cent of the original nitrite. The products of decomposition consist of oxides of nitrogen, water, butyl alcohol, and polymerization products of butyraldehyde.

3. Method of Preparation

Butyl nitrite has always been prepared by the action of nitrous acid on butyl alcohol.[1] The method described[2] is a modification of that of Wallach and Otto[3] for ethyl nitrite.

n-BUTYL PHOSPHATE *

$$3C_4H_9OH + POCl_3 + 3C_5H_5N \rightarrow PO(OC_4H_9)_3 + 3C_5H_5NHCl$$

Submitted by G. R. DUTTON and C. R. NOLLER.
Checked by JOHN R. JOHNSON and ANTHONY HUNT.

1. Procedure

IN a 2-l. round-bottomed flask, fitted with a reflux condenser, liquid-sealed mechanical stirrer, dropping funnel (Note 1), and thermometer, are placed 222 g. (274 cc., 3 moles) of dry *n*-butyl alcohol, 260 g. (265 cc., 3.3 moles) of pyridine, and 275 cc. of dry benzene (Note 2). The solution is stirred and the flask is cooled in an ice-salt mixture until the temperature has fallen to $-5°$. With efficient stirring (Note 3), 153 g. (91 cc., 1 mole) of phosphorus oxychloride (b.p. 106–107°) is added dropwise at such a rate that the temperature does not exceed 10°.

[1] Bertoni, Gazz. chim. ital. **18**, 434 (1888); Adams and Kamm, J. Am. Chem. Soc. **40**, 1285 (1918).

[2] Noyes, ibid. **55**, 3888 (1933).

[3] Wallach and Otto, Ann. **253**, 251 (1889).

* Commercially available; see p. v.

After the addition is completed the reaction mixture is heated slowly to the reflux temperature and held there for two hours. The mixture is cooled to room temperature, and 400–500 cc. of water is added to dissolve the pyridine hydrochloride (Note 4). The benzene layer is separated, washed with 100–150 cc. of water (Note 5), and dried over 20 g. of anhydrous sodium sulfate.

The benzene and other low-boiling materials are removed by distillation at 40–50 mm. pressure until the temperature of the distilling vapor reaches 90°. The n-butyl phosphate fraction is collected at 160–162°/15 mm., or 143–145°/8 mm., and weighs 190–200 g. (71–75 per cent of the theoretical amount) (Note 6).

2. Notes

1. The tip of the funnel should be placed sufficiently high above the surface of the reaction mixture to avoid encrustation with pyridine hydrochloride. It is advantageous to use a thermometer on which the scale above −5° is visible above the stopper; otherwise, the fog in the flask and the pyridine salt may obscure the scale.

2. The reactants and solvent were dried by distillation; fractions boiling over an interval of 1° were used.

3. The first 10–15 cc. of phosphorus oxychloride must be added very slowly to avoid vigorous reaction and overheating. It is essential to avoid an initial temperature so low that unreacted phosphorus oxychloride accumulates and then suddenly reacts with violence. The mechanical stirrer should be of such dimensions and operated at such speeds that the heat of reaction is dissipated rapidly without throwing solid material (and occluded reactants) against the upper walls of the flask.

4. About 50 per cent of the pyridine may be recovered by concentrating this aqueous solution over a steam bath, treating with strong caustic soda solution, and distilling the pyridine layer.

5. The solution should be neutral before distillation. The presence of hydrogen chloride promotes decomposition of phosphoric esters.[1] The benzene solution should not be washed with alkaline reagents, such as sodium carbonate solution, since alkaline reagents also cause decomposition during distillation.

6. This is a general method for preparing alkyl phosphates. Using a similar procedure, the n-propyl ester may be obtained in 60–65 per cent yields, the sec.-butyl ester in 40–45 per cent yields, and the n-amyl ester in 60–65 per cent yields.[2]

[1] Balarev, Z. anorg. allgem. Chem. 101, 227 (1917).
[2] Noller and Dutton, J. Am. Chem. Soc. 55, 424 (1933).

3. Methods of Preparation

n-Butyl phosphate has been prepared by the action of phosphorus pentachloride or oxychloride on butyl alcohol;[3] by the action of phosphorus oxychloride on aluminum butoxide[4] or sodium butoxide;[5] and by the oxidation of butyl phosphite.[6] The procedure described above[2] is similar to one which has been used for the preparation of alkyl phosphites.[7]

n-BUTYL SULFATE

$$(C_4H_9)_2SO_3 + SO_2Cl_2 \rightarrow C_4H_9OSO_2Cl + C_4H_9Cl + SO_2$$
$$C_4H_9OSO_2Cl + (C_4H_9)_2SO_3 \rightarrow (C_4H_9)_2SO_4 + C_4H_9Cl + SO_2$$

Submitted by C. M. SUTER and H. L. GERHART.
Checked by C. R. NOLLER and M. SYNERHOLM.

1. Procedure

IN a 2-l. three-necked flask, fitted with a dropping funnel, mercury-sealed stirrer, and condenser, is placed 625 g. (3.2 moles) of *n*-butyl sulfite (p. 112). The condenser is connected to a gas absorption trap, and 217 g. (131 cc., 1.6 moles) of sulfuryl chloride (Note 1) is added over a period of thirty minutes, with rapid stirring and cooling with tap water. The dropping funnel is then replaced by a thermometer dipping into the liquid, and the flask is heated slowly, with stirring, until the refluxing of the butyl chloride becomes vigorous (about 100–110°). Sulfur dioxide is evolved copiously during this time. The condenser is replaced with a 40-cm. Vigreux column and downward condenser; the temperature is gradually raised to 130–135° and kept there until no more sulfur dioxide or butyl chloride distils. The heating and distillation require about two hours. The residue of crude *n*-butyl sulfate is cooled to room temperature, and 100 cc. of saturated sodium carbonate solution is added. The mixture is stirred for about ten minutes, poured into a separatory funnel, and let stand for thirty minutes to allow the

[3] Nicolai, U. S. pat. 1,766,720 [C. A. **24**, 4053 (1930)]; Celluloid Corporation, Brit. pat. 455,014 [C. A. **31**, 1427 (1937)].

[4] Bannister, U. S. pat. 1,799,349 [C. A. **25**, 3014 (1931)].

[5] Evans, Davies, and Jones, J. Chem. Soc. **1930**, 1310.

[6] Chemische Fabrik von Heyden A.-G., Brit. pat. 398,659 [C. A. **28**, 1362 (1934)], and Ger. pat. 605,174 [C. A. **31**, 3066 (1937)].

[7] Milobendski and Sachnowski, Chemik Polski **15**, 34 (1917) [C. A. **13**, 2865 (1919)].

layers to separate. The upper layer is dried with calcium chloride, at least overnight, and then allowed to stand over 15 g. of anhydrous sodium or potassium carbonate for a day, with occasional shaking. The product is filtered into a 500-cc. modified Claisen flask with a 25-cm. fractionating side arm, and distilled from an oil bath. The first distillation gives 250–280 g. (74–83 per cent of the calculated amount) of n-butyl sulfate, b.p. 110–114°/4 mm., which has a sharp odor. Redistillation gives a pure material with a slight ester odor, boiling at 109–111° at 4 mm., with only a small mechanical loss (Note 2).

2. Notes

1. Commercial sulfuryl chloride was redistilled and the fraction boiling at 69–70° was used.
2. The submitters report that n-propyl sulfate may be prepared in similar fashion, the yield being about 66–70 per cent. The second distillation in this case should be done through a short column to remove higher-boiling material; n-propyl sulfate distils at 88–91° at 4 mm.

3. Methods of Preparation

n-Butyl sulfate has been prepared by the action of n-butyl chlorosulfonate upon n-butyl orthoformate or n-butyl sulfite.[1] It has been obtained also by oxidation of n-butyl sulfite with potassium permanganate in glacial acetic acid solution.[2] The method described above appears to be the most satisfactory for laboratory-scale preparations.

n-BUTYL SULFITE

$$2C_4H_9OH + SOCl_2 \rightarrow (C_4H_9)_2SO_3 + 2HCl$$

Submitted by C. M. Suter and H. L. Gerhart.
Checked by C. R. Noller and M. Synerholm.

1. Procedure

IN a 2-l. three-necked flask, fitted with a mercury-sealed stirrer, thermometer, condenser, and dropping funnel, is placed 684 g. (845 cc., 9.2

[1] Levaillant, Compt. rend. **197**, 648 (1933); Barkenbus and Owen, J. Am. Chem Soc. **56**, 1204 (1934).
[2] Evans, Ph.D. Dissertation, Northwestern University, 1935.

moles) of dry *n*-butyl alcohol (Note 1). The condenser is connected to a trap for absorbing hydrogen chloride, and 500 g. (305 cc., 4.2 moles) of thionyl chloride (Note 2) is added over a period of two hours, with stirring. The reaction mixture is kept at 35–45°, by immersing the flask in ice water, during the addition of the first half of the thionyl chloride (Note 3). After evolution of hydrogen chloride begins, the water bath is removed and a small flame applied to maintain this temperature. After all the thionyl chloride has been added the temperature is raised gradually to the boiling point, over a period of thirty minutes, to complete the reaction and remove the remainder of the hydrogen chloride. The reaction mixture is then transferred to a 1-l. modified Claisen flask with a 25-cm. fractionating side arm, and fractionated under diminished pressure. After a fore-run consisting largely of unchanged alcohol, there is obtained 625–689 g. (77–84 per cent of the calculated amount) of *n*-butyl sulfite, b.p. 109–115°/15 mm. Refractionation gives 585–674 g. (72–83 per cent) of a product distilling at 109–112° at 14 mm. (Note 4).

2. Notes

1. Commercial *n*-butyl alcohol was dried by distillation through a column; the fore-run and a small fraction collected after the vapors reached 117° were discarded.

2. Commercial thionyl chloride was redistilled, and the fraction boiling at 78–80° was employed.

3. Considerable heat is evolved until gas evolution begins, after which heat is absorbed.

4. The submitters report that *n*-propyl sulfite is obtained from *n*-propyl alcohol in somewhat smaller yields, by the same procedure.

3. Methods of Preparation

n-Butyl sulfite has been prepared from butyl alcohol and thionyl chloride by the method described,[1] and using anhydrous ether as a solvent together with dry pyridine to take up the hydrogen chloride evolved.[2] The pyridine method has been checked, but its advantages do not offset the extra expense of the ether and pyridine used. Butyl sulfite has also been made by the action of sulfur chloride on butyl alcohol.[3]

[1] Voss and Blanke, Ann. **485,** 258 (1931); Barkenbus and Owen, J. Am. Chem. Soc **56,** 1204 (1934).

[2] Gerrard, J. Chem. Soc. **1939,** 99.

[3] Bert, Compt. rend. **178,** 1827 (1924).

BUTYROIN

(4-Octanone, 5-hydroxy-)

$$2C_3H_7CO_2C_2H_5 + 4Na \longrightarrow \begin{array}{c} C_3H_7CONa \\ \| \\ C_3H_7CONa \end{array} + 2C_2H_5ONa$$

$$\begin{array}{ccc} C_3H_7CONa & C_3H_7COH & C_3H_7C{=}O \\ \| & \xrightarrow{H_2SO_4} \ \| & \rightarrow \quad | \\ C_3H_7CONa & C_3H_7COH & C_3H_7CHOH \end{array}$$

Submitted by JOHN M. SNELL and S. M. McELVAIN.
Checked by C. S. MARVEL and M. R. LEHMAN.

1. Procedure

IN a 3-l. three-necked, round-bottomed flask, fitted with a long reflux condenser and an efficient mechanical stirrer, are placed 92 g. (4 gram atoms) of clean metallic sodium and about 150 cc. of xylene. The sodium is finely powdered by heating the flask until the sodium melts and then cooling with very vigorous stirring. The cooled xylene is decanted, and the powdered sodium is thoroughly washed with four or five portions of dry, alcohol-free ether. About 1.2 l. of absolute ether is added (Note 1), and the flask is fitted with a reflux condenser, a 250-cc. separatory funnel, and a mechanical stirrer (Note 2).

The stirrer is started, and 232 g. (2 moles) of purified ethyl n-butyrate (Note 3) is slowly run in from the separatory funnel. It is advisable to add first a portion of about 25 cc.; the heat of reaction soon causes the ether to boil; the rest of the ester is then run in at such a rate that gentle ebullition is maintained. Stirring is continued until there is no further reaction and practically all the sodium has been converted into the voluminous yellow-white solid which begins to appear almost at once (Note 4).

The reaction flask is now surrounded by an ice bath, and the contents are vigorously stirred while a cooled solution of 210 g. of sulfuric acid (sp. gr. 1.84) in 350 cc. of water is carefully run in from the separatory funnel. The stirrer is now removed and the flask is allowed to stand in the ice bath until the lower layer of hydrated sodium sulfate ($Na_2SO_4 \cdot 10H_2O$) has solidified. The ether solution is decanted and the sodium sulfate crystals washed with 100–200 cc. of ether.

The combined solution and washings are shaken with about 100 cc. of 20 per cent sodium carbonate solution (Note 5) and are then dried over anhydrous potassium carbonate. The ether and alcohol are removed rapidly by distillation, and the residue is fractionated under reduced pressure in a 250-cc. modified Claisen flask (Note 6). The main fraction boils at 80–86° at 12 mm.; a fraction boiling up to about 15° above the main fraction should also be collected. The low-boiling and high-boiling fractions can be refractionated for recovery of a small additional amount of butyroin. The total yield is 94–101 g. (65–70 per cent of the theoretical amount) of product that is colored yellow by traces of the diketone (Notes 7 and 8).

2. Notes

1. The reaction can be run in benzene, but it is much slower than in ether.

2. For powdering the sodium a small, rapid stirrer is best; for stirring the reaction mixture, a fairly large, slower stirrer is best.

3. The ester is purified as follows: It is washed once with 10 per cent sodium carbonate solution and twice with an equal volume of saturated sodium chloride solution; it is then dried twenty-four hours over anhydrous potassium carbonate. The potassium carbonate is filtered and the ester allowed to stand overnight with about 2 per cent of its weight of phosphorus pentoxide. The ester is then distilled through a column directly from the phosphorus pentoxide; a fraction that distils over a range of 2 degrees or less should be taken. Lower yields of butyroin may be obtained from less carefully purified ester.

4. Addition of the ester requires one and one-half to two hours, and the mixture should then be refluxed an hour longer.

5. Small amounts of butyric acid and dipropylglycollic acid are present in the reaction mixture.

6. The ether should be removed rapidly and the distillation should not be too slow, for long heating favors the formation of a high-boiling by-product of unknown structure at the expense of the butyroin.

7. Usually the amount of diketone present is negligible. The diketone may be removed by shaking the butyroin vigorously from time to time during one hour with 100 cc. of a saturated sodium bisulfite solution, washing with strong sodium chloride solution, and then redistilling.

8. These directions have been used for the following acyloins:

Propionoin b.p. 60–65°/12 mm.; 50–55 per cent yield.
Isobutyroin b.p. 70–75°/14 mm.; 70–75 per cent yield.
Pivaloin m.p. 80–81°; b.p. 85–95°/12 mm.; 52–60 per cent yield

3. Methods of Preparation

Aliphatic acyloins can be obtained by the saponification of the reaction product of sodium on moist ethereal solutions of acid chlorides, the first product being the diester of the dienolic modification of the acyloin.[1] Of greater preparative interest, however, is the reaction between ethereal solutions of aliphatic esters and sodium [2] or potassium.[3]

2-CARBETHOXYCYCLOPENTANONE

(Cyclopentanecarboxylic acid, 2-oxo-, ethyl ester)

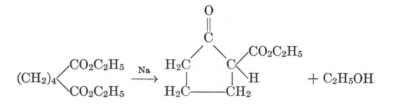

Submitted by P. S. PINKNEY.
Checked by LOUIS F. FIESER and T. L. JACOBS.

1. Procedure

A 3-l. three-necked, round-bottomed flask is fitted with a mercury-sealed mechanical stirrer (Note 1), a 250-cc. dropping funnel, and a reflux condenser protected from the air by means of a calcium chloride tube. In the flask are placed 23 g. (1 gram atom) of sodium and 250 cc. of dry toluene (Note 2). The stirrer is started, and 202 g. (1 mole) of ethyl adipate (p. 264) is added from the dropping funnel at such a rate that the addition is complete in about two hours. The reaction usually starts immediately on addition of the ethyl adipate. The temperature of the oil bath is maintained at 100–115° during the addition and for about five hours longer. Dry toluene is added through the condenser from time to time in order to keep the reaction mixture fluid enough for

[1] Klinger and Schmitz, Ber. **24**, 1273 (1891); Basse and Klinger, ibid. **31**, 1218 (1898); Anderlini, Gazz. chim. ital. **25** (II) 51, 128 (1895); Egorova, J. Russ. Phys.-Chem. Soc. **60**, 1199 (1928) [C. A. **23**, 2935 (1929)].

[2] Bouveault and Locquin, Bull. soc. chim. (3) **35**, 629 (1906); Feigl, Ber. **58**, 2299 (1925); Corson, Benson, and Goodwin, J. Am. Chem. Soc. **52**, 3988 (1930).

[3] Scheibler and Emden, Ann. **434**, 265 (1923).

efficient stirring (Note 3).   Between 750 cc. and 1 l. of toluene is added in this manner.

The reaction mixture is cooled in an ice bath and slowly poured into 1 l. of 10 per cent acetic acid cooled to 0° (ice-salt mixture).   The toluene layer is separated, washed once with water, twice with cooled 7 per cent sodium carbonate solution, and again with water.   The toluene is removed by distillation at ordinary pressure, and the residue is distilled under reduced pressure.   The yield is 115–127 g. (74–81 per cent of the theoretical amount) of a product boiling at 83–88°/5 mm. or 79–84°/3 mm. (Notes 4 and 5).

2. Notes

1. The Hershberg [1] stirrer, shown in part in Fig. 4, provides very efficient agitation of this or other pasty mixtures.   Two glass rings are

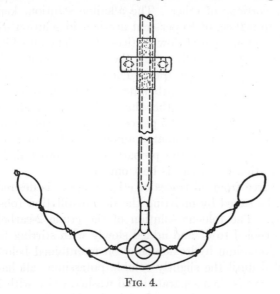

FIG. 4.

sealed to the end of a stirrer shaft at right angles to one another, and each is threaded with B. and S. No. 18 Chromel or Nichrome wire (in the drawing the wire is shown only for the lower ring; the upper wire is not provided with a cross brace).   The stirrer is easily introduced and removed through a narrow opening, and in operation it follows the contour of the flask.   It is convenient to use glass tubing for the stirrer shaft and to provide it with a pair of small ball bearings slipped on over

[1] Hershberg, Ind. Eng. Chem., Anal. Ed. **8**, 313 (1936).

short sections of rubber tubing (one of these bearings is shown in the drawing).

The yields reported were obtained using this stirrer; with various other stirrers it was seldom possible to duplicate the results.

2. The toluene is dried by distillation from sodium.

3. If the reaction mixture is allowed to become too thick for efficient stirring, or if the temperature of the oil bath is raised above 115–120°, the solid sodium derivative will cake on the sides of the flask. This makes the complete removal of the reaction mixture from the flask and the decomposition of the sodium derivative more difficult.

4. According to the literature,[2, 3] the product obtained in this manner may contain ethyl adipate. To remove this, the product is cooled to 0° and run slowly into 600 cc. of 10 per cent potassium hydroxide solution maintained at 0° with ice-salt. Water is added until the salt which separates has dissolved, and the cold alkaline solution is extracted twice with 200-cc. portions of ether. The alkaline solution, kept at 0°, is run slowly into 900 cc. of 10 per cent acetic acid solution with stirring, the temperature remaining below 1° (ice-salt). The oil which separates is taken up in 400 cc. of ether, and the aqueous solution is extracted with four 250-cc. portions of ether. The ether extract is washed twice with cold 7 per cent sodium carbonate solution and dried over sodium sulfate. After removal of the ether the residue is distilled, b.p. 79–81°/3 mm. The recovery is only 80–85 per cent, and in a well-conducted preparation the ethyl adipate eliminated amounts to less than 1 per cent of the total product. Unless the preparation has proceeded poorly the tedious purification ordinarily is best omitted.

If material free from all traces of ethyl adipate is desired, time and material can be saved by omitting the first distillation (observation of the checkers). The toluene solution of the crude 2-carbethoxycyclopentanone is cooled to 0° and added slowly with stirring to 300 cc. of 10 per cent potassium hydroxide solution maintained below 1°. Cold water is added until the slightly soluble potassium salt has dissolved. The toluene layer is then separated and washed twice with 150-cc. portions of cold, 10 per cent potassium hydroxide solution. After each washing, cold water is added to dissolve any solid which separates. The toluene solution, now very light yellow in color, is finally washed twice with 150-cc. portions of cold water. The aqueous solutions are combined, extracted with 250 cc. of ether, and treated as described for the alkaline solution above. The yield is 100–115 g. (64–74 per cent).

[2] Cornubert and Borrel, Bull. soc. chim. (4) **47,** 301 (1930).

[3] Bouveault, ibid. (3) **21,** 1019 (1899); Zelinsky and Ouchakoff, ibid. (4) **35,** 484 (1924).

5. The following procedure gives slightly better yields.

A 3-l. round-bottomed flask, which contains 50 g. of "molecular" sodium, is fitted with a reflux condenser protected from the air by means of a calcium chloride tube, and 1250 cc. of benzene, dried by distilling over sodium, is added. Three hundred and three grams (1.5 moles) of ethyl adipate is then added, in one lot, followed by 3 cc. of absolute alcohol. The flask is warmed on the steam bath until, after a few minutes, a vigorous reaction commences and a cake of the sodio-compound begins to separate. During this stage the flask is kept well shaken by hand. After the spontaneous reaction has abated, the mixture is refluxed on the steam bath overnight, then cooled in ice. The product is decomposed with ice and 6 N hydrochloric acid, the acid being added until Congo red paper is turned blue. The benzene layer is separated, and the aqueous layer is extracted once with 200 cc. of benzene. The united extract is washed with 200 cc. of 5 per cent sodium carbonate solution and 300 cc. of water. The solution is placed in a 3-l. distilling flask and the benzene and water are removed by distillation under ordinary pressure. The residue is fractionated under reduced pressure. The yield is 185–192 g. (79–82 per cent of the theoretical amount) of a product boiling at 108–111°/15 mm. On redistillation the product boils at 102°/11 mm.

The significant features of this procedure are the addition of alcohol, which eliminates or greatly reduces the induction period, and the excess of sodium, which contributes to the completeness of the reaction. The benzene, used as a solvent, may be replaced by petroleum ether (b.p. 60–80°).

The once-distilled carbethoxycyclopentanone is sufficiently pure for ordinary synthetic purposes; it gives a 93 per cent yield of the C-methyl derivative. (R. P. LINSTEAD and E. M. MEADE, private communication.[5] Checked by R. L. SHRINER and N. S. MOON.)

3. Methods of Preparation

2-Carbethoxycyclopentanone has been prepared from ethyl adipate by the action of sodium,[4, 2, 3, 5] sodamide,[6] and sodium ethoxide.[7] The method in the above procedure is based upon the work of Cornubert and Borrel.[2]

[4] van Rysselberge, Bull. acad. roy. Belg. (Sci.) [5] 12, 171 (1926); (Chem. Zentr. 1926, II, 1846).

[5] Linstead and Meade, J. Chem. Soc. 1934, 940.

[6] Bouveault and Locquin, Bull. soc. chim. (4) 3, 440 (1908).

[7] Wislicenus and Schwanhäusser, Ann. 297, 112 (1897).

CASEIN

Submitted by E. J. COHN and J. L. HENDRY.
Checked by H. T. CLARKE and W. M. KENNAN.

1. Procedure

To 1 l. of milk, from which the cream has been largely separated (Note 1), 0.05 M hydrochloric acid is slowly added, with stirring, through a capillary tube extending to the bottom of the beaker. The addition is continued until the solution attains a pH of 4.6 (Note 2). The end point is determined by withdrawing 5-cc. samples, diluting to 50 cc., adding methyl red, and matching against a buffered series (Note 3). Approximately 1 l. of acid is required; the separation of the casein is practically complete at this point. Three liters of water is then added, stirring is discontinued, and the flocculent precipitate of casein is allowed to settle in the refrigerator for twelve to twenty-four hours. The clear supernatant liquid which contains soluble proteins and salts is removed as completely as possible by siphoning; the precipitate is collected on a suction funnel and washed with cold distilled water until the washings are free of calcium (give no precipitate with ammonium oxalate).

The casein, which is contaminated with calcium phosphate and fats, is filtered to as small a volume as possible (about 500 cc.) and transferred to a 2-l. beaker. It is then treated with 0.1 M sodium hydroxide, the alkali being added slowly and with stirring through a capillary extending to the bottom of the beaker (Note 4). The addition of alkali is continued until the pH of the mixture reaches 6.3 (Note 5); 100–150 cc. of the alkali is required. The end point is determined by matching against a buffered series (Note 6), employing dibromo-o-cresolsulfonphthalcin ("bromocresol purple"). At this pH the casein is completely in solution in the form of its sodium salt; fats, calcium phosphate, and any calcium caseinate remain undissolved. Care must be taken not to add more alkali than is necessary to bring the pH to the above point (Note 4). The milky solution is filtered through a thick layer (10–15 mm.) of filter paper pulp tightly packed upon a suction funnel. The filtrate may be slightly opalescent; if it is less clear it is again filtered through a fresh layer of pulp.

The filtrate is brought to a pH of 4.6 with 0.05 M hydrochloric acid just as in the original precipitation, the necessary amount of acid being determined by titration of an aliquot portion, diluted fivefold, with 0.01 M hydrochloric acid; 220–250 cc. of 0.05 M acid is required. As the repreciptation progresses, the rate at which the acid is added is decreased in order to prevent precipitation at the tip of the capillary tube; vigorous mechanical stirring is, of course, essential. When the acidification is complete, 5 l. of cold distilled water is added and the flocculent precipitate allowed to settle in the refrigerator. After siphoning off the clear supernatant liquid, the casein is collected on a suction funnel, using hardened paper, washed with cold distilled water until free of chloride, sucked as dry as possible, and dried over calcium chloride in a vacuum desiccator. The yield is 23–29 g. of a colorless coherent product which may readily be pulverized in a mortar.

2. Notes

1. The cream is satisfactorily removed by allowing the milk to stand in a refrigerator overnight and siphoning off the lower layer.

2. Casein exists in milk in the form of a calcium derivative; pH 4.6 is the isoelectric point of free casein, which is soluble to the extent of only 0.11 g. per liter.[1]

3. Buffers for this range may be made up as follows:

0.1 M ACETIC ACID	0.1 M SODIUM ACETATE	pH
7.35 cc.	2.65 cc.	4.2
6.3	3.7	4.4
5.1	4.9	4.6
4.0	6.0	4.8
2.95	7.05	5.0

4. It is important to avoid a local excess of alkali, which would tend to denature the casein.[2]

5. At this pH sodium caseinate is largely dissolved, whereas calcium caseinate is largely undissolved.[3]

6. The buffer series may conveniently be prepared as follows:

DISODIUM PHOSPHATE (M/15)	MONOPOTASSIUM PHOSPHATE (M/15)	pH
0.78 cc.	9.22 cc.	5.8
1.2	8.8	6.0
1.85	8.15	6.2
2.65	7.35	6.4
3.75	6.25	6.6
5.0	5.0	6.8

[1] Cohn, J. Gen. Physiol. **4,** 697 (1922).
[2] Cohn and Hendry, ibid. **5,** 521 (1923).
[3] Loeb, ibid. **3,** 547 (1920–21).

3. Methods of Preparation

The precipitation of casein in its uncombined form by the addition to milk of one or another acid forms the basis of all methods of preparation. These differ widely, however, in the subsequent purification. In the method of Hammarsten,[4] just enough alkali is added to dissolve this casein completely. The alkalinity reached in this process somewhat modifies its physical properties but probably not its composition. In the method of Van Slyke and Bosworth [5] the last trace of calcium is removed by adding oxalate to an ammoniacal solution of the casein, but this procedure was shown to be unnecessary by Van Slyke and Baker.[6]

The present process is based in large part upon that of Van Slyke and Baker, the modifications depending upon the observation that casein forms far more soluble compounds with univalent than with bivalent bases at neutral reactions.

CELLOBIOSE

$$C_{12}H_{14}O_{11}(COCH_3)_8 + 8CH_3OH \xrightarrow{(CH_3ONa)} C_{12}H_{22}O_{11} + 8CH_3CO_2CH_3$$

Submitted by Géza Braun.
Checked by Reynold C. Fuson, William E. Ross, and William P. Campbell.

1. Procedure

In a 500-cc. three-necked flask, provided with a mercury-sealed stirrer and a calcium chloride tube, 68 g. (0.1 mole) of α-cellobiose octaacetate, m.p. 220–222° (p. 124), is suspended in 300 cc. of absolute methyl alcohol. A solution, prepared by dissolving 0.25 g. (0.01 gram atom) of sodium in 50 cc. of methyl alcohol, is added, and the mixture is stirred vigorously for one hour at room temperature (Note 1). The mixture becomes thin as the hydrolysis proceeds and the solvent acquires a slight color. After the time specified the crystalline solid is collected by suction filtration, washed with four 25-cc. portions of methyl alcohol, and dried at 40°. The weight of the nearly colorless crude cellobiose corresponds closely to the theoretical amount (34 g.). For purification it is

[4] Hammarsten, "Textbook of Physiological Chemistry," translation of 7th Ed., p. 619, John Wiley & Sons, 1911.

[5] Van Slyke and Bosworth, J. Biol. Chem. 14, 211 (1913).

[6] Van Slyke and Baker, ibid. 35, 127 (1918).

dissolved in 125 cc. of hot water containing a few drops of glacial acetic acid, and the solution is clarified with 1–2 g. of Norite and filtered by suction. The colorless filtrate is concentrated under reduced pressure to a small volume, continuing until a large portion of the cellobiose has crystallized, and the crystalline magma is washed into an Erlenmeyer flask with 100 cc. of methyl alcohol. The mixture is stirred well and allowed to stand for several hours for completion of the crystallization, and the sugar is collected on a Büchner funnel, washed with 25 cc. of methyl alcohol, and dried at 40°. The yield of pure cellobiose, $[\alpha]_D^{20°}$ + 34.8° (in 6 per cent aqueous solution), is 31 g. (91 per cent of the theoretical amount). On concentrating the mother liquor to a small volume and adding alcohol as before, 1 g. of equally pure product is obtained, making the total yield 94 per cent of the theoretical amount (Note 2).

2. Notes

1. The reaction may be carried out equally satisfactorily by shaking the mixture mechanically in a stoppered bottle.

2. Using an earlier procedure,[1] in which a solution of the octaacetate in chloroform is treated with sodium methoxide solution and then with water, the yields of pure cellobiose amounted to 67–79 per cent of the theoretical quantity.

3. Methods of Preparation

Cellobiose was prepared first by Skraup and König[2] by the saponification of the octaacetate with alcoholic potassium hydroxide, and the method was improved by Pringsheim and Merkatz.[3] Aqueous barium hydroxide also has been employed[4] for the purpose, and methyl alcoholic ammonia has been used extensively for the hydrolysis of carbohydrate acetates. The method of catalytic hydrolysis with a small quantity of sodium methoxide was introduced by Zemplén,[1] who considered the action to be due to the addition of the reagent to the ester-carbonyl groups of the sugar acetate and the decomposition of the addition compound by reaction with alcohol.[5] The present procedure, reported by Zemplén, Gerecs, and Hadácsy,[6] is a considerable improvement over the original method[1] (see Note 2).

[1] Zemplén, Ber. **59,** 1254 (1926).
[2] Skraup and König, ibid. **34,** 1115 (1901).
[3] Pringsheim and Merkatz, Z. physiol. Chem. **105,** 173 (1919).
[4] Abderhalden and Zemplén, ibid. **72,** 58 (1911).
[5] Zemplén and Kunz, Ber. **56,** 1705 (1923).
[6] Zemplén, Gerecs, and Hadácsy, ibid. **69,** 1827 (1936).

α-CELLOBIOSE OCTAACETATE

$$(C_{12}H_{20}O_{10})_n + nH_2O + 8n(CH_3CO)_2O \rightarrow$$
$$nC_{12}H_{14}O_{11}(COCH_3)_8 + 8nCH_3CO_2H$$

Submitted by Géza Braun.
Checked by Reynold C. Fuson, William E. Ross, and William P. Campbell.

1. Procedure

In a 1-l. wide-mouthed bottle with a glass stopper, 400 cc. of acetic anhydride is cooled to $-10°$ in a freezing mixture and 36 cc. of concentrated sulfuric acid is added at once with constant stirring. The temperature of the solution rises to about $20°$. The bottle is removed from the freezing mixture, and 20 g. of absorbent cotton (Note 1) is worked into the liquor immediately with a heavy glass rod. With constant stirring, the bottle is warmed in a water bath maintained at $60°$ until the temperature of the mixture reaches $45°$ (about ten minutes), and then it is removed and the temperature is kept from rising above $55°$ by suitable cooling with running water, stirring being continued throughout (Note 2). After about twenty minutes (total) the mixture becomes thin; at this point it is cooled to $50°$ and a second 20-g. portion of cotton is added with stirring, the temperature again being kept from rising above $55°$. The process of cooling to $50°$ and adding 20 g. of cotton is repeated at intervals of about ten minutes until 100 g. of cotton in all has been introduced. After the final addition stirring is continued until the mixture becomes thin (about ten minutes); then the bottle is stoppered and heated in a bath maintained at $50°$ for one hour, during which time the cotton dissolves completely to a thin, light brown syrup. The stoppered container is kept in an oven at $35°$ for seven days.

The solution darkens to a deep wine-red color, and α-cellobiose octaacetate begins to crystallize on the second day (Note 3). After seven days at $35°$ the semi-crystalline mass is stirred into 20 l. of cold water. With good stirring the flocculent precipitate of α-cellobiose octaacetate and cellulose-dextrin acetates soon becomes crystalline; after standing for one to two hours the solid is collected on a 12.5-cm. Büchner funnel, washed free from acid with cold water, and drained thoroughly. The moist product, weighing about 250 g., is triturated with 250 cc. of warm methyl alcohol, and, after cooling to room temperature, the undissolved solid is collected on a 7-cm. Büchner funnel, washed with three 50-cc. portions of methyl alcohol, and dried at $40°$. The yield of fairly pure α-cellobiose octaacetate is 69–74 g. For purification it is dissolved in

300 cc. of chloroform and the solution is filtered by suction into a dry receiver through a pad prepared by dusting Norite onto a Büchner funnel and washing the funnel with alcohol. The chloroform solution is filtered while the pad is still wet with alcohol, and the filter is washed at once with 100 cc. of chloroform without interruption in the filtration. The colorless filtrate is concentrated at reduced pressure until the acetate begins to crystallize (about 250 cc.), the crystals are redissolved by warming, and the solution is poured into 750 cc. of warm methyl alcohol. The acetate begins to crystallize at once as small needles which eventually form a thick paste. The mixture is cooled to 0° with stirring, and after about one hour the material is collected, washed with 100 cc. of methyl alcohol, and dried at 40°. The yield of colorless α-cellobiose octaacetate, m.p. 220–222°, $[\alpha]_D^{20°} + 41.6°$, is 65–69 g. (35–37 per cent of the theoretical amount, assuming that the cotton contains 10 per cent of moisture) (Notes 4, 5, and 6).

For further purification a solution of the material in 350 cc. of chloroform is clarified if necessary and poured into 750 cc. of methyl alcohol, and the mixture is cooled to 0°. The yield of the pure acetate, m.p. 225–226°, $[\alpha]_D^{20°} + 42.5°$, is 61–65 g. The constants are not altered by further crystallizations.

2. Notes

1. Pure commercial cotton or filter paper may be used. These materials ordinarily contain 7–10 per cent of moisture.

2. Effective stirring and careful control of the reaction are essential to obtain good yields. If left uncontrolled, the temperature may rise quickly above 100° with considerable decomposition.

3. The crystallization is accelerated by scratching the walls of the jar or by seeding. Seed is obtained easily by pouring a small portion of the reaction mixture into a large amount of water; the precipitate is separated, treated with alcohol, and dried at 40°.

4. In earlier experiments a temperature of 30–40° was maintained during the addition of the cotton (which then required more time), and the acetolysis was allowed to proceed at room temperature for six days. The yield of product melting at 220–222° was 50–56 g., but rose to 62–70 g. when the period of digestion was extended to eight days.

5. The quantities can be doubled without change in the procedure or the percentage yield.

6. Pringsheim's method [1] of isolating the acetate gives equally good results. Instead of pouring the acetolysis mixture into water, it is kept at 0° for one to two days and the crystalline material is collected on a

[1] Pringsheim and Merkatz, Z. physiol. Chem. **105, 173** (1919).

large Büchner funnel and washed with 50 cc. of ice-cold acetic anhydride. Cellulose-dextrin acetates and other by-products pass into the filtrate. The crude mass is digested with 200 cc. of warm methyl alcohol, collected, and crystallized from chloroform-alcohol as above. The yield is 65–68 g., m.p. 220–222°.

3. Methods of Preparation

The preparation of α-cellobiose octaacetate by the acetolysis of cellulose was discovered by Franchimont,[2] and the process has been studied carefully by a number of other investigators.[1,3,4,5] The observations of Freudenberg[4] and Klein[5] were particularly useful in developing the present procedure.

CHELIDONIC ACID

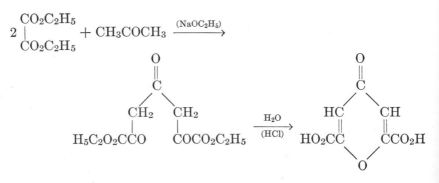

Submitted by E. RAYMOND RIEGEL and F. ZWILGMEYER.
Checked by REYNOLD C. FUSON and WILLIAM E. ROSS.

1. Procedure

IN a 1-l. round-bottomed flask, fitted with a reflux condenser protected by a calcium chloride tube, 46 g. (2 gram atoms) of sodium is dissolved in 600 cc. of absolute alcohol (Note 1). About one hour is required for the addition of the sodium, and another hour for its complete solution. Toward the end of the reaction the flask may be heated

[2] Franchimont, Ber. **12**, 1941 (1879).

[3] Skraup and König, Monatsh. **22**, 1011 (1901); Maquenne and Goodwin, Bull. soc. chim. (3) **31**, 854 (1904); Schliemann, Ann. **378**, 366 (1911); Ost, ibid. **398**, 332 (1913).

[4] Freudenberg, Ber. **54**, 767 (1921).

[5] Klein, Z. angew. Chem. **25**, 1409 (1912).

with a small smoky flame. While the sodium is being dissolved, the following materials are weighed in dry, stoppered containers: 58 g. (1 mole) of dry acetone (Note 2), 150 g. (1.03 moles) of freshly distilled ethyl oxalate (Org. Syn. Coll. Vol. I, **1941**, 261), and 160 g. (1.1 moles) of ethyl oxalate.

About half of the sodium ethoxide solution is poured into a 3-l. round-bottomed, three-necked flask provided with a liquid-sealed stirrer and a reflux condenser; the other half is kept warm by a small flame. The first half of the solution is allowed to cool until a solid begins to appear, then 58 g. of dry acetone mixed with 150 g. of ethyl oxalate is added at once and the stirrer is set in motion. Heat is evolved, and the liquid turns brown but remains clear. As soon as any turbidity appears, the other half of the hot sodium ethoxide solution is poured into the mixture together with 160 g. of ethyl oxalate, the two streams being allowed to mix as they flow into the flask. The liquid initially is clear and of a deep brown color, but after stirring for about thirty minutes the mixture becomes practically solid. The flask is then connected with a condenser for distillation and heated in an oil bath at 110° until 150 cc. of alcohol has distilled. The flask is protected by a calcium chloride tube and the reaction mixture is cooled to 20°. The sodium derivative is removed to a 3-l. beaker by means of a glass rod and treated with a mixture of 300 cc. of concentrated hydrochloric acid (sp. gr. 1.19) and 800 g. of cracked ice (Note 3). All lumps are carefully crushed, and the creamy yellow suspension of acetonedioxalic ester is collected on a 15-cm. Büchner funnel. The ester is removed from the filter, stirred with about 100 cc. of ice water, and again collected (Note 4). For hydrolysis the crude material is heated with 300 cc. of concentrated hydrochloric acid in a 5-l. flask (Note 5) on the steam bath for twenty hours. After cooling to 20° the solid hydrated acid is collected on a 10-cm. Büchner funnel, washed with two 50-cc. portions of ice water, and dried, first at 100° for two hours, and then at 160° to constant weight to remove the water of crystallization. The yield of product decomposing at 257° (corr.) is 140–145 g. (76–79 per cent of the theoretical amount).

2. Notes

1. A good grade of absolute alcohol should be used (Org. Syn. Coll. Vol. I, **1941**, 259).

2. The acetone is dried for several days over calcium chloride, filtered, and distilled. Some acetone is lost in this operation by combination with the calcium chloride.[1]

---

[1] Bagster, J. Chem. Soc. **111**, 494 (1917).

3. The temperature must be kept as low as possible during the neutralization, for any undue rise in temperature results in a darkening of the product.

4. The crude ester, after a further washing and after drying in a vacuum desiccator over sulfuric acid, melts at 98–100° and weighs 220 g. (85 per cent of the theoretical amount).

5. A large flask is used because the mixture froths seriously at first. If the frothing becomes troublesome it may be stopped by adding a little ether.

3. Methods of Preparation

Natural chelidonic acid is obtained from the herb celandine (*Chelidonium majus*). The synthesis from ethyl oxalate and acetone was first described by Claisen;[2] the process was simplified by Willstätter and Pummerer[3] and further improved by Ruzicka and Fornasir.[4] The present procedure is modeled after that of the last-mentioned investigators.

α-CHLOROANTHRAQUINONE

(Anthraquinone, 1-chloro-)

$$3C_6H_4\!\!<\!\!\begin{array}{c}CO\\CO\end{array}\!\!>\!\!C_6H_3SO_3K + NaClO_3 + 3HCl \rightarrow$$

$$3C_6H_4\!\!<\!\!\begin{array}{c}CO\\CO\end{array}\!\!>\!\!C_6H_3Cl(\alpha) + 3KHSO_4 + NaCl$$

Submitted by W. J. Scott and C. F. H. Allen.
Checked by Louis F. Fieser and E. B. Hershberg.

1. Procedure

A 2-l. three-necked flask fitted with a stirrer (Notes 1 and 2), condenser, and dropping funnel (Note 3) is mounted in the hood, and in it are placed 20 g. (0.061 mole) of potassium anthraquinone-α-sulfonate (p. 539), 500 cc. of water, and 85 cc. (1 mole) of concentrated hydrochloric acid. The solution is heated to boiling and stirred, while a solution of 20 g. (0.19 mole) of sodium chlorate (Note 4) in 100 cc. of water is added dropwise over a period of three hours (Note 5). The mixture

[2] Claisen, Ber. **24**, 111 (1891).
[3] Willstätter and Pummerer, ibid. **37**, 3744 (1904).
[4] Ruzicka and Fornasir, Helv. Chim. Acta **3**, 811 (1920).

is refluxed very slowly for an additional hour before the precipitated α-chloroanthraquinone is collected by suction filtration and washed free from acid with hot water (about 350 cc.). After drying *in vacuo* at 100°, the bright yellow product melts at 158–160° (corr.) and weighs 14.6–14.7 g. (97–98 per cent of the theoretical amount) (Notes 6 and 7).

2. Notes

1. Since the mixture tends to foam toward the end of the reaction, it is advisable to use an effective stirrer. A Hershberg stirrer (p. 117) of tantalum wire gave good service in the hands of the checkers; the metal was not appreciably attacked after repeated use.

2. Although a glass sleeve for the stirrer is fairly satisfactory, it is better to use a seal of the conventional type (Org. Syn. Coll. Vol. I, **1941,** 33, Fig. 2a) filled with water rather than mercury.

3. Since the usual dropping funnel has the disadvantage of requiring considerable attention from the operator, the checkers found it much more convenient to employ the device (of E. B. H.) shown in Fig. 5.

4. Potassium chlorate is less satisfactory because of its lower solubility.

5. If the addition is too rapid or the boiling too vigorous, chlorinating gases are lost through the condenser.

6. Crystallization of the product from 200 cc. of *n*-butyl alcohol gives 13.4 g. of material in the form of yellow needles, m.p. 161–162° (corr.). Larger amounts are conveniently crystallized from toluene, using 2 cc. per gram.

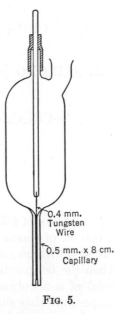

0.4 mm.
Tungsten
Wire

0.5 mm. x 8 cm.
Capillary

FIG. 5.

7. The melting point given for highly purified α-chloroanthraquinone is 162.5° (corr.). The possible contaminants include the β-isomer, 1,5-, and 1,8-dichloroanthraquinone, all of which melt at higher temperatures but depress the melting point of the α-monochloro compound.

3. Methods of Preparation

α-Chloroanthraquinone has been prepared from α-aminoanthraquinone by the diazo reaction,[1] by the action of thionyl chloride on potassium anthraquinone-α-sulfonate under pressure,[2] and by the above process.[3]

m-CHLOROBENZALDEHYDE

(Benzaldehyde, m-chloro-)

m-$NO_2C_6H_4CHO + 3SnCl_2 + 6HCl$
$$\rightarrow m\text{-}NH_2C_6H_4CHO + 3SnCl_4 + 2H_2O$$

m-$NH_2C_6H_4CHO + NaNO_2 + 2HCl$
$$\rightarrow m\text{-}ClN_2C_6H_4CHO + NaCl + 2H_2O$$

$$m\text{-}ClN_2C_6H_4CHO \xrightarrow{\text{(CuCl + HCl)}} m\text{-}ClC_6H_4CHO + N_2$$

Submitted by JOHANNES S. BUCK and WALTER S. IDE.
Checked by JOHN R. JOHNSON and PAUL W. VITTUM.

1. Procedure

A SOLUTION of 450 g. (2 moles) of stannous chloride crystals (Note 1) in 600 cc. of concentrated hydrochloric acid is placed in a 3-l. beaker provided with an efficient mechanical stirrer and cooled in an ice bath. When the temperature of the solution has fallen to +5°, 100 g. (0.66 mole) of m-nitrobenzaldehyde (Note 2) is added in one portion. The temperature rises slowly at first, reaching 25–30° in about five minutes, then rises very rapidly to about 100°. Stirring must be vigorous or the reaction mixture may be forced out of the beaker (Note 3). During the reaction the nitrobenzaldehyde dissolves, and an almost clear, red solution is obtained. The solution is cooled in an ice-salt mixture until the temperature has fallen to about +2°. During the cooling, orange-red crystals separate and a pasty suspension results.

[1] Bayer and Company, Ger. pat. 131,538 [Frdl. 6, 311 (1900–02)]; Groggins, "Unit Processes in Organic Syntheses," p. 175, McGraw-Hill Book Company, New York, 1935.

[2] Meister, Lucius, and Brüning, Ger. pat. 267,544 (Chem. Zentr. 1914, I, 89).

[3] Bayer and Company, Ger. pat. 205,195 (Chem. Zentr. 1909, I, 414); Badische Anilin- und Soda-Fabrik, Ger. pat. 228,876 (Chem. Zentr. 1911, I, 102); Ullmann and Ochsner, Ann. 381, 1 (1911).

A 250-cc. separatory funnel is fixed so that its stem extends below the surface of the pasty suspension. A solution of 46 g. (0.67 mole) of sodium nitrite in 150 cc. of water is placed in the funnel and is slowly added to the well-stirred mixture until it shows a positive starch-iodide test for nitrous acid. The temperature of the mixture is maintained between 0° and +5° (Note 4) throughout the addition of the nitrite solution, which requires about ninety minutes. Usually, all but 5–8 cc. of the nitrite solution must be added before a positive test for nitrous acid appears.

During the latter part of the diazotization of the aminobenzaldehyde, a hot solution of cuprous chloride is prepared. In a 5-l. round-bottomed flask, 189 g. (0.75 mole) of powdered copper sulfate crystals and 161 g. of sodium chloride are dissolved in 600 cc. of hot water, and to this solution is added a solution of 41 g. (0.22 mole) of sodium metabisulfite ($Na_2S_2O_5$) and 27 g. (0.67 mole) of sodium hydroxide in 300 cc. of water. The final temperature of the resulting cuprous chloride solution should be about 75°.

The diazonium solution is added to the hot cuprous chloride solution while the latter is shaken by hand but is not cooled. After the solutions are thoroughly mixed, 840 cc. of concentrated hydrochloric acid is added and the mixture is allowed to stand overnight. The reaction mixture is steam-distilled to separate the *m*-chlorobenzaldehyde, which is collected practically completely in the first 1.5 l. of distillate. The *m*-chlorobenzaldehyde is removed from the aqueous distillate by extraction with two 150-cc. portions of ether, and the ethereal solution is dried with 10–15 g. of anhydrous calcium chloride. After being decanted from the drying agent, the ether is distilled, and the residual liquid is distilled under diminished pressure. The *m*-chlorobenzaldehyde boils at 84–86°/8 mm., 107–109°/26 mm. (Note 5). The yield is 70–74 g. (75–79 per cent of the theoretical amount) (Note 6).

2. Notes

1. A chemically pure grade of stannous chloride crystals ($SnCl_2 \cdot 2H_2O$) was used. Lower yields were obtained when technical stannous chloride was used.

2. A practical grade of *m*-nitrobenzaldehyde was used; m.p. 52–55°.

3. During the vigorous reaction it is advisable to keep the cooling bath and the reaction mixture well stirred. Less satisfactory yields were obtained when the reaction was moderated by adding the nitrobenzaldehyde in several portions.

4. At temperatures below 0° the speed of diazotization is markedly decreased. Above +5° some decomposition of the diazonium salt takes place.

5. Since m-chlorobenzaldehyde is oxidized easily by atmospheric oxygen, it should be stored in a tightly corked or sealed container.

6. According to the submitters m-bromobenzaldehyde can be prepared by the same general procedure using, in place of cuprous chloride, a solution of cuprous bromide prepared from 189 g. of copper sulfate, 91 g. of sodium bromide, 41 g. of sodium metabisulfite, and 27 g. of sodium hydroxide. Instead of 840 cc. of concentrated hydrochloric acid, 200 cc. of 48 per cent hydrobromic acid is added after the diazonium solution has been mixed with the cuprous bromide. The m-bromobenzaldehyde boils at 93–98°/8 mm. The yield is 80 g. or 65 per cent of the theoretical amount.

It is reported, however, that m-bromobenzaldehyde prepared in this way may contain as much as 20 per cent of m-chlorobenzaldehyde. This contamination can be avoided by using stannous bromide as the reducing agent.

A solution of stannous bromide is prepared by heating 119 g. (1 gram atom) of mossy tin with 705 g. (4 moles) of 46 per cent hydrobromic acid for two hours on a steam bath, with mechanical stirring. The solution is cooled to 40°, and 50 g. (0.33 mole) of m-nitrobenzaldehyde is added in one portion, with continued stirring. The temperature rises from the heat of reaction and finally reaches about 105°. After heating for one-half hour longer on a steam bath, the reaction mixture is cooled to 0° and the aminobenzaldehyde diazotized by the gradual addition of 23 g. (0.33 mole) of sodium nitrite in 75 cc. of water. The diazonium solution is poured into a hot suspension of cuprous bromide, 100 cc. of 46 per cent hydrobromic acid is added, with stirring, and the mixture is allowed to stand overnight. The mixture is steam-distilled and the m-bromobenzaldehyde isolated by ether extraction and vacuum distillation; b.p. 90–92°/4 mm. The yield is 41 g. (67 per cent of the theoretical amount). (F. T. Tyson, private communication.)

3. Methods of Preparation

m-Chlorobenzaldehyde has been prepared by the chlorination of benzaldehyde [1] and by the oxidation of m-chlorobenzyl alcohol [2] and of m-chlorotoluene.[3] It is most conveniently prepared from m-nitrobenzal-

[1] Müller, Ger. pat. 30,329; 33,064 [Frdl. 1, 143, 146 (1877–87)].
[2] Mettler, Ber. 38, 2812 (1905).
[3] Law and Perkin, J. Chem. Soc. 93, 1636 (1908).

dehyde through *m*-aminobenzaldehyde and the diazonium reaction.[4] The procedure given above is essentially that described in the patent literature.[4]

p-CHLOROBENZALDEHYDE

(Benzaldehyde, *p*-chloro-)

$$p\text{-}ClC_6H_4CH_3 + 2Cl_2 \rightarrow p\text{-}ClC_6H_4CHCl_2 + 2HCl$$
$$p\text{-}ClC_6H_4CHCl_2 + H_2O \rightarrow p\text{-}ClC_6H_4CHO + 2HCl$$

Submitted by W. L. McEwen.
Checked by Henry Gilman and Chuan Liu.

1. Procedure

A 500-cc. two-necked, round-bottomed flask is provided with an air-cooled reflux condenser, 2 cm. in diameter and filled for a length of 60 cm. with 5–6 mm. glass pearls or rings (Note 1). Chlorine is to be introduced by means of a 4-mm. glass tube, inserted through a cork in a neck of the flask, extending close to the bottom of the flask and provided with a small bulb with fine perforations to break up the gas stream into small bubbles. The large quantities of hydrogen chloride formed can be disposed of by means of a gas absorption trap.

Into the tared flask are placed 126.5 g. (1 mole) of *p*-chlorotoluene (Org. Syn. Coll. Vol. I, **1941**, 170) and 3.8 g. of phosphorus pentachloride. The flask is heated in a bath kept at 160–170° (Note 2), and while illuminated with direct sunlight or with an unfrosted 100-watt tungsten lamp a rapid stream of chlorine is introduced directly from a cylinder until the gain in weight is 55–66 g. (Note 3).

The pale yellow or yellow-green product is then transferred to a 4-l. wide-mouthed bottle containing 400 cc. of concentrated sulfuric acid, and stirred vigorously (Hood) for five hours (Note 4). The viscous mixture is then transferred to a separatory funnel and allowed to stand overnight, after which the lower layer (Note 5) is run slowly, with stirring, into a 3-l. beaker three-quarters filled with cracked ice. The cream-colored solid obtained when the ice has melted is filtered by suction, washed with water, pressed dry on the funnel, and divided into three equal parts. Each portion is dissolved in a minimum of ether, and the ether solution is repeatedly shaken with 2 per cent sodium

[4] Meister, Lucius, and Brüning, Ger. pat. 31,842 [Frdl. **1**, 144 (1877–87)]; Erdmann and Schwechten, Ann. **260**, 59 (1890); Eichengrün and Einhorn, ibid. **262**, 135 (1891).

hydroxide solution until acidification of the washings gives no precipitate of *p*-chlorobenzoic acid (Note 6).

After removal of the ether by distillation on a steam bath, the residue is distilled under diminished pressure from a Claisen flask. The yield of *p*-chlorobenzaldehyde distilling at 108–111°/25 mm. and melting at 46–47° is 76–84 g. (54–60 per cent of the theoretical amount).

2. Notes

1. The glass packing reduces the tendency of the stream of hydrogen chloride to carry away *p*-chlorotoluene as a spray.

If only one run is to be made, good-quality corks are satisfactory. If several runs are made, it is recommended that the corks be impregnated with sodium silicate solution to prevent excessive corrosion by the hydrogen chloride and chlorine.

2. The bath contains either oil or graphite.

3. The time required was four and one-half hours. In a larger run by the submitter, in which 750 g. of *p*-chlorotoluene and 23 g. of phosphorus pentachloride were used, the time required for a gain in weight of 330–360 g. was six to ten hours.

4. Vigorous stirring is necessary to prevent undue foaming. Most of the hydrogen chloride is evolved early in stirring.

5. The waxy upper layer is discarded.

6. The yield of *p*-chlorobenzoic acid is about 20 g. From the larger runs, starting with 750 g. of *p*-chlorotoluene, the yield of acid averaged 260 g.

3. Methods of Preparation

p-Chlorobenzaldehyde can be prepared from *p*-chlorobenzyl chloride or *p*-chlorobenzyl bromide with aqueous lead nitrate;[1] from *p*-chlorotoluene and chromyl chloride;[2] by the hydrolysis of *p*-chlorobenzal chloride;[3] from *p*-aminobenzaldehyde by diazotization and subsequent treatment with cuprous chloride;[4] from *p*-chlorophenylmagnesium bromide and ethyl orthoformate;[5] from *p*-chlorobenzyl chloride with hexamethylenetetramine and subsequent hydrolysis;[6] by conversion of

[1] Beilstein and Kuhlberg, Ann. **147**, 352 (1867); Jackson and White, Am. Chem. J. **3**, 30 (1881).

[2] Law and Perkin, J. Chem. Soc. **93**, 1636 (1908).

[3] Erdmann and Kirchhoff, Ann. **247**, 368 (1888); Erdmann and Schwechten, ibid. **260**, 63 (1890); Kaeswurm, Ber. **19**, 742 (1886).

[4] Von Walther and Raetze, J. prakt. Chem. (2) **65**, 258 (1902).

[5] Bodroux, Bull. soc. chim. (3) **31**, 587 (1904).

[6] Mayer and English, Ann. **417**, 78 (1918).

p-chlorobenzonitrile to the iminochloride which is then hydrolyzed;[7] by the action of carbon monoxide and aluminum chloride on chlorobenzene;[8] and by the action of sodium carbonate on 2,5-dichlorobenzenesulfonyl-4-chlorobenzoylhydrazine.[9]

o-CHLOROBENZOIC ACID

(Benzoic acid, *o*-chloro-)

$$ClC_6H_4CH_3 + 2KMnO_4 \rightarrow ClC_6H_4CO_2K + 2MnO_2 + KOH + H_2O$$

Submitted by H. T. Clarke and E. R. Taylor.
Checked by Henry Gilman and J. H. McGlumphy.

1. Procedure

In a 12-l. flask fitted with stirrer and reflux condenser are placed 600 g. (3.8 moles) of potassium permanganate, 7 l. of water, and 200 g. (1.6 moles) of *o*-chlorotoluene (Org. Syn. Coll. Vol. I, **1941**, 170). The mixture is slowly heated to boiling (Note 1) with continual stirring until the permanganate color has disappeared. This requires three to four hours. The condenser is now set downward for distillation, and the mixture is distilled, with constant stirring, until no more oil passes over with the water. The unattacked *o*-chlorotoluene thus obtained amounts to 25–30 g. The hot mixture is filtered with suction and the cake of hydrated manganese dioxide washed with two 500-cc. portions of hot water. The combined filtrate is concentrated (Note 2) to about 3.5 l.; if it is not entirely clear it may be clarified by the use of 1–2 g. of decolorizing carbon. It is now, while still hot, acidified by cautiously adding 250 cc. of concentrated hydrochloric acid (sp. gr. 1.19) with continual agitation. When the mixture is cool the white precipitate of *o*-chlorobenzoic acid is filtered and washed with cold water. The dry weight is 163–167 g. (76–78 per cent of the theoretical amount, based on the amount of *o*-chlorotoluene actually oxidized) of a very nearly pure (Note 3) product melting at 137–138°. For purification this may be recrystallized from 600 cc. of toluene, when 135–140 g. of a product melting at 139–140° is obtained. Further crops can be obtained by concentrating the mother liquor.

[7] Stephen, J. Chem. Soc. **127**, 1874 (1925).

[8] Boehringer and Sons, Ger. pat. 281,212 (Chem. Zentr. **1915**, I, 178).

[9] McFadyen and Stevens, J. Chem. Soc. **1936**, 584.

2. Notes

1. If the mixture is heated too rapidly the reaction may be violent at the outset; it can be controlled by laying wet towels upon the upper part of the flask.

2. This concentration is satisfactorily carried out on the steam bath under reduced pressure (Org. Syn. Coll. Vol. I, **1941**, 427).

3. It is important to use pure o-chlorotoluene in this preparation: otherwise the o-chlorobenzoic acid may be contaminated with isomeric acids which are very difficult to remove. The o-chlorotoluene therefore should be prepared from pure o-toluidine or o-chlorotoluenesulfonic acid.

Directions for the purification of technical o-chlorobenzoic acid are to be found on p. 16.

3. Methods of Preparation

The only practical methods of preparing o-chlorobenzoic acid consist in the oxidation of o-chlorotoluene and the replacement of the amino group in anthranilic acid by a chlorine atom. Both methods have been fully discussed by Graebe,[1] who recommends the former for the preparation of relatively large quantities. The oxidation of o-chlorotoluene by permanganate was originally described by Emmerling.[2]

β-CHLOROETHYL METHYL SULFIDE

(Sulfide, β-chloroethyl methyl)

$$CH_3SCH_2CH_2OH + SOCl_2 \rightarrow CH_3SCH_2CH_2Cl + SO_2 + HCl$$

Submitted by W. R. KIRNER and WALLACE WINDUS.
Checked by H. T. CLARKE and S. GURIN.

1. Procedure

IN a 1-l. three-necked flask are mixed 150 g. (1.63 moles) of β-hydroxyethyl methyl sulfide (p. 345) (Note 1) and 200 g. of dry chloroform (Note 2). The flask is placed on a steam bath and is fitted with a dropping funnel, a mechanical stirrer, and a condenser. The condenser is fitted with a trap to remove the vapors of hydrogen chloride and sulfur dioxide. A solution of 204 g. (1.7 moles) (Note 3) of thionyl

[1] Graebe, Ann. **276**, 54 (1893).
[2] Emmerling, Ber. **8**, 880 (1875).

chloride in 135 cc. of dry chloroform is added dropwise to the β-hydroxy-ethyl methyl sulfide over a period of about two hours (Note 4). The reaction mixture is stirred vigorously during this addition and for about four hours after the addition is complete. The chloroform is distilled on the steam bath and the residue is distilled under reduced pressure. The yield is 135–153 g. (75–85 per cent of the theoretical amount) of a product boiling at 55–56°/30 mm. (Note 5).

2. Notes

1. Quantities of material seven times as large as the above may be used without decreasing the percentage yield of the product.
2. The chloroform is dried by distillation, and the fraction boiling at 60–61° is used.
3. The thionyl chloride is redistilled, and the fraction boiling over a two-degree range is employed.
4. The reaction mixture is heated once when about half the thionyl chloride has been added in order to keep the chloroform refluxing gently. Heating after the complete addition of the thionyl chloride is undesirable.
5. β-Chloroethyl methyl sulfide is a vesicant and must be handled with care. It boils at 140° under atmospheric pressure.

3. Method of Preparation

The above method is essentially that described in the literature.[1]

β-CHLOROPROPIONALDEHYDE ACETAL

(Propionaldehyde, β-chloro-, diethyl acetal)

$$CH_2{=}CHCHO + 2C_2H_5OH + HCl \rightarrow ClCH_2CH_2CH(OC_2H_5)_2 + H_2O$$

Submitted by E. J. WITZEMANN, WM. LLOYD EVANS, HENRY HASS, and
E. F. SCHROEDER.
Checked by FRANK C. WHITMORE and HARRY T. NEHER.

1. Procedure

IN a 3-l. round-bottomed flask, fitted with a mechanical stirrer and an inlet tube, is placed 200 g. (253 cc.) of absolute alcohol (Note 1). The flask is surrounded by an ice-salt bath, and the alcohol is saturated

[1] Kirner, J. Am. Chem. Soc. **50**, 2452 (1928).

with dry hydrogen chloride at 0°. The stem of a 200-cc. separatory funnel is fitted to a one-holed stopper inside a large short-stemmed funnel (Fig. 6). The space between the funnels is filled with finely cracked ice and water. In the separatory funnel is placed 112 g. (2 moles) of cold acrolein (Org. Syn. Coll. Vol. I, **1941,** 15) (Note 2).

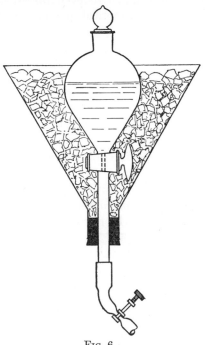

When the alcohol is saturated with hydrogen chloride, the generator is disconnected and the stem of the separatory funnel is connected to the inlet tube by a rubber tube provided with a screw clamp for adjusting the flow. The acrolein is added, with stirring, to the alcoholic hydrogen chloride solution at about 0°. The addition should require from one to two hours. After two layers have formed, the lower layer of acetal is separated and treated gradually with powdered sodium bicarbonate until all acid is neutralized (Note 3). The mixture is filtered. The filtrate is washed with two 50-cc. portions of ice water and dried over 10 g. of potassium carbonate for five to ten hours (Note 4). It is then filtered and distilled under reduced pressure. The yield of product boiling at 58–62°/8 mm. is 112 g. (34 per cent of the theoretical amount).

FIG. 6.

2. Notes

1. At least 99.5 per cent alcohol should be used (Org. Syn. Coll. Vol. I, **1941,** 259).

2. If the acrolein is not kept cold, the vapors become unbearable. The stopper of the separatory funnel should be provided with a fine glass capillary.

3. All acid must be removed before washing the product with water, because dilute acid hydrolyzes the acetal very readily.

4. It is recommended that the ice water be prepared from distilled water as the impurities in some samples of tap water lead to decomposition of the acetal. (Private communication from C. F. H. ALLEN.)

3. Methods of Preparation

β-Chloropropionaldehyde acetal has been prepared by the action of acrolein on alcoholic hydrogen chloride alone,[1] or in the presence of calcium chloride.[2]

CHOLESTANONE

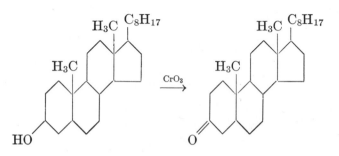

Submitted by WILLIAM F. BRUCE.
Checked by LOUIS F. FIESER and R. P. JACOBSEN.

1. Procedure

A SOLUTION of 50 g. (0.13 mole) of dihydrocholesterol (Note 1) in 500 cc. of benzene is added slowly with cooling (Note 2) to a solution of 68 g. (0.23 mole) of crystalline sodium dichromate, 50 cc. of glacial acetic acid, and 90 cc. of concentrated sulfuric acid in 300 cc. of water in a 3-l. flask. The mixture is agitated thoroughly in a shaking device or by efficient stirring (Note 3) for six hours at 25–30° (Note 4).

The benzene solution is separated and washed twice with 100 cc. of water, once with 200 cc. of 5 per cent potassium hydroxide, and twice with water. If the solution is not colorless it is clarified with 1 g. of Norite. The benzene is removed by distillation, and the resulting syrup is dissolved in 300 cc. of alcohol by heating. The solution on cooling deposits cholestanone as well-formed needles. The yield of collected, washed, and air-dried material, m.p. 129–130°, is 41.5–42 g. (83–84 per cent of the theoretical amount). The addition of 80 cc. of water to the filtrate gives about 2 g. of material melting at 125–126°.

[1] Alsberg, Jahresber. **1864**, 495; Wohl, Ber. **31**, 1797 (1898); Wohl and Emmerich, ibid. **33**, 2761 (1900); Brabant, Z. physiol. Chem. **86**, 208 (1913); Witzemann, J. Am. Chem. Soc. **36**, 1909 (1914); Spoehr and Young, Carnegie Inst. Washington Yearbook, **25**, 176 (1925–1926); Expt. Sta. Record, **57**, 817 (1927) [C. A. **22**, 2368 (1928)].

[2] Crawford and Kenyon, J. Chem. Soc. **1927**, 399.

2. Notes

1. Material melting at 140–141° (p. 191) is satisfactory. The presence of a trace of cholesterol is not objectionable since this is converted into acidic products which are removed in the course of the purification.

2. When the solutions are mixed without cooling the temperature rises to about 60° and the yield is somewhat less.

3. The checkers employed a Hershberg stirrer (p. 117).

4. The agitation may be continued twice as long without appreciable difference in yield. Six hours is regarded as the minimum time for the quantity specified.

3. Method of Preparation

Cholestanone has been prepared by the oxidation of dihydrocholesterol with chromic anhydride in acetic acid solution.[1] The yield is sometimes diminished as a result of the partial acetylation of the sterol.

CITRACONIC ANHYDRIDE AND CITRACONIC ACID

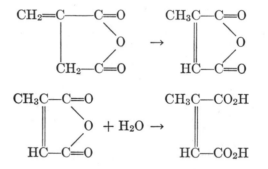

Submitted by R. L. Shriner, S. G. Ford, and L. J. Roll.
Checked by C. R. Noller.

1. Procedure

(A) *Citraconic Anhydride.*—Two hundred and fifty grams of itaconic anhydride (Note 1) is distilled rapidly at atmospheric pressure in a 500-cc. modified Claisen flask with a 15-cm. (6-in.) fractionating column (Note 2). The receivers for the distillate must be changed without

[1] Diels and Abderhalden, Ber. **39**, 884 (1906); Willstätter and Mayer, ibid. **41**, 2199 (1908); Vavon and Jakubowicz, Bull. soc. chim. (4) **53**, 584 (1933).

interrupting the distillation. The distillate passing over below 200° consists of water and other decomposition products. The fraction which distils at 200–215° consists of citraconic anhydride and is collected separately. The yield is 170–180 g. (68–72 per cent of the theoretical amount) of a product melting at 5.5–6°. On redistillation under reduced pressure there is obtained 155–165 g. (62–66 per cent of the theoretical amount) of a product which boils at 105–110°/22 mm. and melts at 7–8° (Note 3).

(B) *Citraconic Acid.*—To 22.4 g. (0.2 mole) of pure citraconic anhydride in a 100-cc. beaker is added from a pipet exactly 4 cc. (0.22 mole) of distilled water. The mixture is stirred on a hot plate until a homogeneous solution is formed, then covered with a watch glass and allowed to stand for forty-eight hours. At the end of this time the mixture has solidified completely. The yield is 26 g. of a product melting at 87–89°. For further purification it is finely ground in a mortar, washed with 50 cc. of cold benzene, dried in the air, and then dried for twenty-four hours in a vacuum desiccator over phosphorus pentoxide. This furnishes 24.4 g. (94 per cent of the theoretical amount) of citraconic acid which melts at 92–93°.

2. Notes

1. The crude itaconic anhydride obtained as described on p. 368 was used. Itaconic acid may be substituted for the anhydride.

2. The success of the preparation depends upon a rapid distillation and changing the receivers without interrupting the distillation. The best yields are obtained when the heating period is of short duration.

3. The crude citraconic anhydride contains a small amount of water, acetone, and citraconic acid. Vacuum distillation allows the removal of these impurities without materially decreasing the yield.

3. Methods of Preparation

Citraconic anhydride has been prepared by the distillation of citraconic acid and of citric acid.[1]

Citraconic acid has been obtained by distillation of citric acid,[2] of lactic acid,[3] and of hydroxypyrotartaric acid;[4] and by treating citric

[1] Anschütz, Ber. **14**, 2788 (1881).

[2] Crasso, Ann. **34**, 68 (1840); Kekulé, "Lehrbuch der organischen Chemie," **2**, 317 (1866); Wilm, Ann. **141**, 28 (1867).

[3] Engelhardt, ibid. **70**, 243, 246 (1849).

[4] Demarçay, Ber. **9**, 963 (1876).

acid with hydriodic acid.[5] A mixture of citraconic and itaconic acids is obtained by flowing a concentrated aqueous solution of citric acid into a heated evacuated vessel, distilling under reduced pressure the mixture of anhydrides formed, and allowing the mixture to react with water.[6]

COPPER CHROMITE CATALYST

$$\begin{matrix} Cu(NO_3)_2 \\ Ba(NO_3)_2 \\ (NH_4)_2CrO_4 \end{matrix} \xrightarrow{} \text{Complex chromates} \xrightarrow{\text{Ignition}} \begin{matrix} CuCr_2O_4 \\ BaCr_2O_4 \\ (\text{Note 1}) \end{matrix}$$

Submitted by W. A. Lazier and H. R. Arnold.
Checked by R. L. Shriner and J. F. Kaplan.

1. Procedure

A mixture of 26 g. (0.1 mole) of c.p. barium nitrate and 800 cc. of distilled water is warmed to 70°. After solution is complete 218 g. (0.9 mole) of c.p. copper nitrate trihydrate is added and the mixture stirred at 70° until a clear solution results (Note 2).

A solution of ammonium chromate is prepared by dissolving 126 g. (0.5 mole) of c.p. ammonium dichromate in 600 cc. of distilled water and adding 150 cc. of 28 per cent aqueous ammonia (sp. gr. 0.9) (Note 3). The warm solution of the nitrates is stirred (hand stirring is adequate) while the ammonium chromate solution is poured into it in a thin stream. Stirring is continued for a few minutes, after which the reddish brown precipitate of copper barium ammonium chromate is collected (Note 4) and pressed in a 16-cm. Büchner funnel, and dried at 110°. This dry precipitate is placed in a loosely covered nickel pan (Note 5), or one or two small porcelain casseroles covered with watch glasses, and heated in a muffle furnace for one hour at 350–450° (Note 6). At this point the yield of chromite should be about 160 g. The ignition residue is pulverized in a mortar to break up any hard lumps that may be present (Note 7) and then transferred to a 2-l. beaker containing 1.2 l. of 10 per cent acetic acid. After being stirred for ten minutes the mixture is allowed to settle. After about ten minutes, two-thirds or more of the spent acid solution is decanted and replaced by 1.2 l. of fresh 10 per cent acetic acid, and the extraction is repeated. The residue is washed by repeating the extraction procedure four times with 1.2 l. of distilled water

[5] Kämmerer, Ann. **139**, 269 (1866).
[6] Boehringer Sohn A.-G., Brit. pat. 452,460 [C. A. **31**, 1045 (1937)].

each time (Note 8). The insoluble portion is collected by filtering with suction on a Büchner funnel, dried at 110°, and ground in a mortar to a fine black powder (Note 9). The yield is 130–140 g. (Notes 10 and 11).

2. Notes

1. The reactions involved in this preparation cannot be expressed quantitatively in a simple equation. The process has been investigated by Gröger.[1]

2. Barium nitrate is sparingly soluble in cold water and even less soluble in copper nitrate solution. It is therefore necessary to heat the mixture in order to bring both salts into solution together.

3. The warm, freshly prepared ammonium chromate solution may be used for the precipitation at once or may be allowed to cool to room temperature. However, the solution prepared as indicated is supersaturated at room temperature and deposits crystals on standing. If a stock solution of ammonium chromate is to be held over from day to day, a portion of the ammonia should be withheld and added immediately before precipitation. Chromic acid may be used for the preparation of the ammonium chromate solution provided that the acid solution is kept cold while a sufficient quantity of ammonia is introduced below the surface.

4. The amount of ammonia used is calculated to give a mother liquid that is neutral to litmus, allowance being made for the basic character of the precipitate. While a deviation either way affects the yield adversely, it is better not to try to adjust the end point after precipitation, but to make a slight correction on the amount of ammonia to be used for the next batch. The mother liquor serves as its own indicator: if acid, the solution is yellow; if alkaline, a deep green.

5. A convenient-sized pan, 6 cm. deep, 19 cm. long, and 10 cm. wide, can be made from a sheet of nickel $\frac{1}{32}$ inch thick.

6. The ignition of copper ammonium chromate causes a spontaneous exothermic reaction which proceeds with a rapid evolution of gas. Although the size of the charge ignited at one time is unimportant, the retaining vessels should not be too full, else there will be considerable mechanical loss of product. It is also inadvisable to break up the lumps of dried precipitate, as the lumpy condition diminishes spraying of the chromite on ignition.

7. The unextracted catalyst should be a bluish black, friable powder. It is a satisfactory catalyst for the dehydrogenation of alcohols and for the less difficult hydrogenation reactions, such as the reduction of nitro

[1] Gröger, Z. anorg. Chem. **58**, 412 (1908); **76**, 30 (1912).

compounds. This mixture of copper chromite and copper oxide is somewhat less active and more susceptible of reduction to metallic copper than the catalyst from which the copper oxide has been removed by acid extraction.

8. If the washing is unduly prolonged, the catalyst tends to become colloidal and is difficult to separate either by decantation or filtration.

9. No special precautions are necessary in handling or storing the copper chromite catalyst, since it is unharmed by exposure to air or moisture.

10. Barium has been included as a catalyst component on account of its protective action against sulfate poisoning and its reported stabilization of the catalyst against reduction. Alternatively, the above procedure may be used for the preparation of a copper chromite catalyst containing no barium. In this case the barium nitrate is omitted and 242 g. (1 mole) of copper nitrate is used. All other details are the same as given above.

11. An alternative procedure described by Adkins [2] is as follows: Thirty-one grams of barium nitrate is dissolved in 820 cc. of distilled water which has been heated to 80°. To this hot solution is added 260 g. of copper nitrate trihydrate, and the mixture is stirred and heated until solution is complete. Meanwhile, 151 g. of ammonium dichromate is dissolved in 600 cc. of distilled water and 225 cc. of 28 per cent aqueous ammonia added. The hot nitrate solution is poured in a thin stream with stirring into the ammonium chromate solution. The orange precipitate is collected on a filter, pressed, and sucked as dry as possible. It is dried in an oven at 75–80° for twelve hours, pulverized, and divided into three portions. Each portion is decomposed separately in a large porcelain casserole (15-cm. diameter) by heating over a free flame. The mass should be heated just hot enough to cause decomposition to take place at the minimum temperature. During the decomposition the powder is stirred continuously with a steel spatula, and the heating is regulated so that the evolution of gases does not become violent. This is accomplished by heating only one side of the casserole and by increasing the rate of stirring when the decomposition starts to spread throughout the mass. During this process the color of the powder changes from orange to brown and finally to black. When the entire mass has become black the evolution of gases ceases, and the powder is removed from the casserole and allowed to cool. The combined product is then stirred for thirty minutes with 600 cc. of 10 per cent acetic acid solution, col-

[2] Adkins, "Reactions of Hydrogen with Organic Compounds over Copper-Chromium Oxide and Nickel Catalysts," p. 12, University of Wisconsin Press, Madison, 1937.

lected on a filter, washed by suspending it six times in 100-cc. portions of water, dried for twelve hours at 125°, and pulverized. The yield of catalyst is 170 g.

3. Methods of Preparation

Copper chromite has been made by the ignition of basic copper chromate at a red heat and by the thermal decomposition of copper ammonium chromate.[1] The procedure given here is a modification of the latter method[3] in which barium ammonium chromate is also incorporated.[4] Copper-chromium oxide hydrogenation catalysts have also been prepared by grinding or heating together copper oxide and chromium oxides, by the decomposition of copper ammonium chromium carbonates or copper-chromium nitrates,[2,4] and by the low-temperature ignition of copper ammonium chromates.[5]

COUPLING OF *o*-TOLIDINE AND CHICAGO ACID

Preparation of a Salt-Free Azo Dye

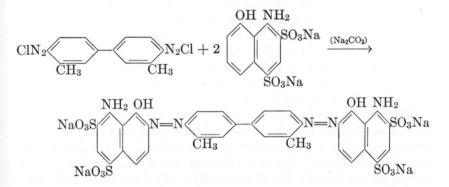

Submitted by J. L. Hartwell and Louis F. Fieser.
Checked by W. W. Hartman and G. L. Boomer.

1. Procedure

In a 1-l. beaker 21.2 g. (0.1 mole) of *o*-tolidine is made into a thin paste with 300 cc. of water and dissolved by the addition of 20 cc.

[3] Lazier, U. S. pats. 1,746,782 and 1,746,783 [C. A. **24**, 1649 (1930)]; U. S. pat. 1,964,000 [C. A. **28**, 5075 (1934)].

[4] Connor, Folkers, and Adkins, J. Am. Chem. Soc. **53**, 2012 (1931); **54**, 1138 (1932)

[5] Calingaert and Edgar, Ind. Eng. Chem. **26**, 878 (1934).

(23.6 g., 0.23 mole) of concentrated hydrochloric acid (sp. gr. 1.18), warming a little if necessary. The solution is cooled to 10° with ice, mechanical agitation is started, and 21 cc. (25 g., 0.24 mole) more of concentrated hydrochloric acid is added (Note 1). This causes partial separation of *o*-tolidine dihydrochloride. The temperature being kept at 10–15°, a solution of 14.5 g. (0.2 mole) of 95 per cent sodium nitrite in 40 cc. of water is run in rapidly, the last 10 per cent being held back and added slowly as needed to give a distinct positive test for nitrous acid on starch-iodide paper. This excess is maintained for one-half hour. Throughout the diazotization an excess of hydrochloric acid should be present as shown by a strong test on Congo red paper. When the diazotization is complete, the excess nitrous acid is eliminated by the careful addition of a small amount of *o*-tolidine hydrochloride solution, using starch-iodide paper to arrive at an exact balance (Note 2).

In a 3-l. beaker a paste is made from 82 g. (0.21 mole) of 88 per cent technical Chicago acid (1-amino-8-naphthol-2,4-disulfonic acid) (Note 3) and 500 cc. of water, and solution effected by the addition of 8 g. (0.2 mole) of sodium hydroxide in 30 cc. of water, testing with litmus toward the end and leaving the reaction still acid (Note 4). The solution is cooled to 18° by the addition of ice, and just before the coupling 35 g. (0.33 mole) of anhydrous sodium carbonate is added.

With vigorous mechanical agitation the diazonium salt solution is run into the Chicago acid solution rather rapidly. The blue dye which separates at first, gradually dissolves on further stirring. After one-half hour the solution is tested for alkalinity (Note 5) and for completeness of coupling (Notes 6 and 7).

After stirring for a total of two hours, the mixture is heated to 85°, 15 g. of decolorizing carbon is added, and the solution stirred for fifteen minutes and filtered. The filtrate is reheated to 85° with agitation and the volume estimated. For each 100 cc. of the solution 27 g. of hydrated sodium acetate (usually 450–500 g. in all) is added slowly, with vigorous stirring, in four or five portions (Note 8). The dye is thus caused to separate in an easily filterable form. On "spotting" a drop of the mixture on filter paper by means of a stirring rod, the dye should form a dark blue mass in the center of a red or violet rim (Note 9). The mixture, while still warm, is filtered with suction on a 20-cm. Büchner funnel, and as much as possible of the mother liquor is removed (Note 10). The filter cake is dissolved in 1.2 l. of water at 85° and resalted as above with about 300 g. of sodium acetate, in several portions, or until most of the dye is salted out. In a "spot" test on filter paper, little red should be left in the rim. The product is collected, redissolved in 1.2 l. of water at 85°, and salted again. This time enough sodium acetate should be added to give a small rim which shows no red impurity (Note 11). The product is collected, dried in an oven at 110° for at least twenty-four hours, ground, and passed through a 40-mesh sieve.

The sodium acetate is then extracted by digesting the sifted dye for ten to fifteen minutes with successive 500-cc. portions of boiling 95 per cent ethyl alcohol, filtering with suction after each extraction, and washing the product on the filter with a little alcohol (Note 12). After four alcohol extractions the filtrate should give a negative test for acetate. The product, which changes in appearance from bronze to green during the extraction, is dried in air at room temperature. The yield is 80–81 g. (83–84 per cent of the theoretical amount) of dye free from inorganic salts and organic impurities (Note 13).

2. Notes

1. If the hydrochloric acid is added all at once instead of in two portions as described, a solid will be obtained consisting of o-tolidine coated with its dihydrochloride and the diazotization will occur slowly.

2. If a clear solution is not obtained at this point, any solid may be removed by filtration.

3. Owing to the fact that this intermediate is a technical product and has been isolated by salting out, a knowledge of its purity is necessary. The compound is actually present as the acid sodium salt (molecular weight 341). A rapid method of analysis, sufficiently accurate for this preparation, is to titrate a known weight in dilute, strongly acid solution with 0.5 N sodium nitrite solution standardized against sulfanilic acid, using starch-iodide paper in obtaining the end points.

Some samples of technical Chicago acid may contain so much of an impurity giving rise to a red dye that the removal of this red dye by resalting, as recommended in Note 9, presents considerable difficulty. When this occurs it is advantageous to purify the Chicago acid in the following fashion. Seven hundred grams of a Chicago acid paste, found by titration to contain 26.6 per cent of 1-amino-8-naphthol-2,4-disulfonic acid, is dissolved by heating in 2 1. of water; the solution is filtered, heated to 70°, and salted out by the addition of sodium chloride (600–650 g.) until a thick paste is obtained. The precipitate is collected at 40°, pressed well by means of a rubber dam, dissolved in 2 1. of water, and resalted exactly as before. After drying for thirty-six hours in a vacuum oven at 75°, the nearly white product weighs 150–155 g. and contains 78 per cent of active reagent. The dye ("Tolidine-1824") obtained from this material showed no red impurity in the capillary test after the second salting. (Louis F. Fieser and M. E. Gross, private communication.)

4. Aminonaphthols rapidly turn dark in alkaline solution owing to oxidation. For this reason the solution is left faintly acid until just before coupling.

5. This test is made by removing a few drops of the coupling mixture by means of a stirring rod to one of the depressions in a white porcelain spot plate, salting out the dye by stirring with a little c.p. sodium chloride, and bringing a piece of litmus paper in contact with the mass.

6. The test for completeness of coupling involves testing for excess diazonium salt (a) and for excess Chicago acid (b):

(a) A drop or two of the coupling mixture is dissolved in about 20 cc. of water in a test tube and the solution divided equally in two test tubes. To one tube is added a few drops of a 10 per cent solution of Schaeffer's salt (2-naphthol-6-sulfonic acid) or R salt (2-naphthol-3,6-disulfonic acid) prepared from 0.5 g. of the acid, 5 cc. of water, and just enough sodium carbonate solution to dissolve the material. The other tube acts as a control. Both tubes are brought rapidly nearly to boiling and the colors compared. If too dark, both tubes are diluted equally with water until the solutions are light blue. The colors may be compared by pouring solution from each tube onto filter paper. If the color of the solution in the first tube is at all redder than that in the control tube, an excess of the diazonium compound is present in the coupling mixture.

(b) A few drops of the coupling mixture are salted out as in Note 5, the mass transferred to filter paper, and near the spot on the reverse side is placed a drop of solution of diazotized sulfanilic acid. A red or redviolet coloration at the meeting of the two spots shows the presence of excess Chicago acid.

7. Throughout the coupling an alkaline reaction should be maintained and sodium carbonate added if the alkalinity falls. If the mixture gives tests for both diazonium compound and Chicago acid, insufficient time has elapsed for complete coupling. A little more Chicago acid should be added (in alkaline solution) if a deficiency is shown.

8. The temperature of salting, speed of addition of salt, speed and duration of agitation, and alkalinity of the solution all affect the physical form of the dye and determine how well it will filter. The method described gives a product which filters rapidly.

9. The red impurity is another dye produced by coupling of the diazonium salt with an oxidation product which is present in the technical Chicago acid. It is usually easier to free the blue dye of this red impurity by resalting than to purify the Chicago acid. However, in the

case of Trypan blue (tolidine coupled with two moles of H acid) the red impurity cannot be removed by salting. It may be removed completely by extraction with alcohol, in which it is more soluble than the blue component, preferably by precipitating a hot saturated aqueous solution of the dye with at least six times its volume of alcohol.

10. In this and subsequent filtrations it is advantageous to use a rubber dam held in place by rubber bands. The use of the rubber dam is especially advantageous for a slow filtration continuing overnight.

11. The most sensitive test for purity of the dye is the capillary test,[1] depending on the fact that the red impurity is more soluble in water than the blue compound and less readily adsorbed by filter paper. An amount of press cake corresponding roughly to 0.05 g. of solid dye is dissolved in 200 cc. of water in a 400-cc. beaker, 0.5 g. of common salt is added, and the temperature raised to 95°. A 2 by 15 cm. strip of filter paper is immersed in the liquid and is supported on a stirring rod pushed through it and placed across the top of the beaker. The temperature is maintained at 95° for fifteen minutes and the filter paper removed. Any red impurity will show plainly above the blue. If purification is found to be incomplete, further resalting will produce a pure product.

12. The progress of the extraction may be followed by adding three drops of concentrated sulfuric acid to 10 cc. of the alcoholic filtrate in a test tube. If a white precipitate of sodium sulfate does not form at once, the solution is chilled in an ice bath. A precipitate will form with 0.002 g. of sodium acetate in 10 cc. of alcohol.

13. Using this procedure, other diamines such as benzidine and dianisidine may be diazotized and coupled with other aminonaphthols, such as S acid (1-amino-8-naphthol-4-sulfonic acid), J acid (2-amino-5-naphthol-7-sulfonic acid), Gamma acid (2-amino-8-naphthol-6-sulfonic acid), and H acid (1-amino-8-naphthol-3,6-disulfonic acid), or with simple naphthols such as NW acid (1-naphthol-4-sulfonic acid), Schaeffer's acid (2-naphthol-6-sulfonic acid), and R acid (2-naphthol-3,6-disulfonic acid). The procedure would be varied only in the manner of salting out the dyes, each of which would require a set of optimum conditions peculiar to itself (see Note 8). If the alkalinity of sodium acetate leads to a considerable increase in solubility of the dye, ammonium bromide may be used. The latter is extracted easily by hot ethyl alcohol or, better, by hot methyl alcohol.

Naphthylamines cannot be used as second components by this procedure, as coupling takes place in acid solution.

---

[1] Mulliken, "Identification of the Commercial Dyestuffs," pp. 10–11, John Wiley & Sons, New York, 1910.

3. Methods of Preparation

Dyes of the types described in the procedure and those mentioned in Note 13 are described in the patent literature.[2] The ordinary methods of preparation give products that contain salt and other impurities.

CYANOGEN BROMIDE

$$NaCN + Br_2 \rightarrow BrCN + NaBr$$

Submitted by W. W. HARTMAN and E. E. DREGER.
Checked by ROGER ADAMS and I. L. OZANNE.

1. Procedure

A 2-l. round-bottomed flask, surrounded by an ice-water bath and provided with a stirrer, a separatory funnel, and an outlet tube, is set up in a good hood. To the flask are added 500 g. (160 cc., 3.1 moles) of bromine and 50 cc. of water (Note 1). To the stirred mixture is added gradually a solution of 170 g. of sodium cyanide (3.5 moles) in 1.2 l. of warm water. The temperature of the reaction mixture is kept below 30°. When the reaction is complete (about two hours or less) the cyanogen bromide is distilled from a steam bath, using a 500-cc. flask (Note 2) as a receiver. The distillate is warmed with about 100 g. of anhydrous calcium chloride, filtered, and again distilled, preferably with the distilling flask used as a receiving vessel connected directly to the flask from which the cyanogen bromide is being distilled. The product boils at 60–62°. It is melted in the receiver (Note 3) (Hood) and poured into a warm tared bottle. The yield of white crystalline solid (Note 4) melting at 49–51° is 239–280 g. (73–85 per cent of the theoretical amount).

2. Notes

1. Water is used to decrease the volatilization of the bromine.
2. It is desirable to have the receiving flask close to the distilling flask because cyanogen bromide will clog a tube which is too small in diameter or too long.

[2] Bayer and Company, Ger. pats. 35,341 and 38,802 [Frdl. **1**, 469, 488 (1877–87)]; Cassella and Company, Ger. pat. 3949 [Frdl. **3**, 685 (1890–94)]; Badische Anilin- und Soda-Fabrik, Ger. pats. 57,327 and 75,469 [Frdl. **3**, 687, 690 (1890–94)].

3. Because of the toxic nature of the product it is best to wear a gas mask while transferring the molten product.

4. Cyanogen bromide does not keep well and may at times even become explosively unstable on standing. It is preferable to prepare it just before using.

3. Methods of Preparation

Cyanogen bromide has been prepared from an aqueous solution of potassium cyanide and bromine at $0°$;[1] by the action of bromine on an alkali cyanide in the presence of carbon tetrachloride and acetic acid;[2] and from bromine and moist mercuric cyanide.[3] Detailed directions have been published by Slotta for the preparation from bromine and aqueous potassium cyanide.[1]

CYCLOHEXYLBENZENE

(Cyclohexane, phenyl-)

$$H_2C \Big\langle \begin{matrix} CH_2-CH \\ CH_2-CH_2 \end{matrix} \Big\rangle CH + C_6H_6 \xrightarrow{(H_2SO_4)} H_2C \Big\langle \begin{matrix} CH_2-CH_2 \\ CH_2-CH_2 \end{matrix} \Big\rangle CHC_6H_5$$

Submitted by B. B. Corson and V. N. Ipatieff.
Checked by John R. Johnson and E. A. Cleveland.

1. Procedure

In a 1-l. three-necked flask equipped with a mechanical stirrer, dropping funnel, and thermometer are placed 468 g. (530 cc., 6 moles) of benzene and 92 g. (50 cc.) of concentrated sulfuric acid (sp. gr. 1.84). The mixture is cooled in an ice bath, and 164 g. (203 cc., 2 moles) of cyclohexene (Note 1) is added with stirring over a period of one and one-half hours, while the temperature is maintained between 5° and 10°. Stirring is continued for an additional hour after all the cyclohexene has been added.

The hydrocarbon layer is separated, cooled in ice, and washed with four 50-cc. portions of cold concentrated sulfuric acid (Note 2). The

[1] Langlois, Ann. chim. phys. (3) **61**, 482 (1861); Scholl, Ber. **29**, 1823 (1896); Baum, ibid. **41**, 523 (1908); Slotta, ibid. **67**, 1028 (1934).

[2] National Aniline and Chemical Company, U. S. pat. 1,938,324 [C. A. **28**, 1148 (1934)].

[3] Sérullas, Ann. chim. phys. (2) **34**, 100 (1827).

material is then washed twice with warm water (50°), twice with 3 per cent sodium hydroxide solution, and twice with pure water (Note 3). The hydrocarbon mixture is dried over anhydrous calcium chloride (Note 4) and subjected twice to fractional distillation, using a 30-cm. Vigreux or similar column; the cyclohexylbenzene is collected at 238–243° (Notes 5 and 6). The yield is 210–220 g. (65–68 per cent of the theoretical amount).

2. Notes

1. For the preparation of moderate amounts of cyclohexene the dehydration of cyclohexanol with 85 per cent phosphoric acid, according to the procedure of Dehn and Jackson, J. Am. Chem. Soc. **55**, 4285 (1933), is very convenient. Furthermore, very little carbonization occurs, in contrast with the sulfuric acid method described in Org. Syn. Coll. Vol. I, **1941**, 183, where there is much carbonization and the product is contaminated with sulfur dioxide.

In a 2-l. three-necked flask, carrying a separatory funnel and three-bulbed Wurtz column filled with broken glass tubing, is placed 200 g. of 85 per cent phosphoric acid. The column is attached to an *efficient* condenser leading to a receiver cooled in an ice bath, and the flask is heated in an oil bath at 165–170°. Through the funnel 1 kg. (10 moles) of practical cyclohexanol is dropped in over a period of four to five hours. After the addition has been completed the temperature of the bath is raised gradually to 200° and maintained at 200° for one-half hour. During the whole operation the temperature at the top of the column does not rise above 90°. The upper layer of the distillate is separated (salt may be added to break up emulsions) and dried with anhydrous magnesium sulfate; the lower aqueous layer is saved for reworking if desired; likewise the spent drying agent may be treated with water to recover admixed cyclohexene. The crude cyclohexene is distilled in an efficient column, and the fraction boiling at 81–83° is collected. The yield is 660–690 g. (79–84 per cent of the theoretical amount). The residue consists largely of cyclohexanol and may be recycled as described below.

An additional 25–30 g. of cyclohexene may be obtained by combining the residue from the distillation of the crude cyclohexene with the water layer from the original distillate and distilling with 25 g. of 85 per cent phosphoric acid. This distillate is added to the low-boiling fraction from the distillation of the crude cyclohexene and separated. The upper layer is dried with anhydrous magnesium sulfate and distilled as described above.

The phosphoric acid may be recovered by diluting with water and filtering, then evaporating with a little nitric acid to the proper concentration.

The same procedure when used with 86 g. (1 mole) of cyclopentanol and 15 cc. of 85 per cent phosphoric acid gave 55 g. (81 per cent of the theoretical amount) of cyclopentene, b.p. 44–45°. No attempt was made to recover the cyclopentanol. (OLIVER GRUMMITT and JOHN R. JOHNSON, private communication.)

2. The purpose of the sulfuric acid is to convert dicyclohexyl sulfate to cyclohexyl hydrogen sulfate, which is removed by the subsequent washing operations.

3. To avoid emulsification as much as possible it is advantageous to use warm water rather than cold, and dilute alkali rather than concentrated. The milkiness of the aqueous wash liquid represents only a very small loss of material.

4. It is well to allow suspended water to settle by standing overnight and to separate again before adding the drying agent.

5. In a typical preparation the fractions collected during the second distillation were as follows: 78–85°, 296 g.; 85–235°, 2 g.; 235–238°, 2 g.; 238–243°, 215 g.; 243–265°, 2 g.; residue above 265°, 46 g.

6. The distillation residue becomes semi-solid on cooling owing to the separation of 1,4-dicyclohexylbenzene. The latter may be recovered by filtering with suction, washing with methyl alcohol, and crystallizing from acetone (using 4 cc. of acetone per gram of the crude solid). The yield of purified dicyclohexylbenzene, m.p. 100–101°, is 15–24 g.

3. Methods of Preparation

Cyclohexylbenzene has been prepared by the hydrogenation of biphenyl [1] and of cyclohexenylbenzene; [2] by the reaction of cyclohexyl chloride [3] or bromide [4] with benzene in the presence of aluminum chloride; and from benzene and cyclohexene or cyclohexanol in the presence of aluminum chloride, [5] sulfuric acid, [6] or boron halides. [7]

[1] Eijkman, Chem. Weekblad 1, 7 (1903).

[2] Sabatier and Murat, Compt. rend. 154, 1390 (1912); Alder and Rickert, Ber. 71, 379 (1938).

[3] Kursanoff, Ann. 318, 309 (1901).

[4] Braun, Ber. 60, 1180 (1927).

[5] Bodroux, Ann. chim. (10) 11, 511 (1929); Berry and Reid, J. Am. Chem. Soc. 49, 3142 (1927); Corson and Ipatieff, ibid. 59, 645 (1937); Nametkin and Pokrovskaya, J. Gen. Chem. (U.S.S.R.) 7, 962 (1937) [C. A. 31, 5332 (1937)]; Tsukervanik and Sidorova, J. Gen. Chem. (U.S.S.R.) 7, 641 (1937) [C. A. 31, 5780 (1937)].

[6] Deschauer, Ger. pat. 515,177 (Chem. Zentr. 1931, I, 1829); Truffault, Compt. rend. 202, 1286 (1936); Corson and Ipatieff, J. Am. Chem. Soc. 59, 645 (1937).

[7] Hofmann and Wulff, Brit. pat. 307,802 (Chem. Zentr. 1929, II, 2101); McKenna and Sowa, J. Am. Chem. Soc. 59, 470 (1937).

DECAMETHYLENE GLYCOL

(1,10-Decanediol)

$$C_2H_5O_2C(CH_2)_8CO_2C_2H_5 + 8[H] \xrightarrow{(C_2H_5OH + Na)} HO(CH_2)_{10}OH + 2C_2H_5OH$$

Submitted by R. H. MANSKE.
Checked by W. H. CAROTHERS and W. L. McEWEN.

1. Procedure

To a solution of 65 g. (0.25 mole) of ethyl sebacate (p. 277) in 800 cc. of absolute ethyl alcohol (Note 1) contained in a 3-l. round-bottomed flask, to which is attached a 60-cm. bulbed reflux condenser protected by a calcium chloride drying tube, is added 70 g. (3 gram atoms) of sodium in large pieces in one lot. The somewhat vigorous reaction is easily kept under control by immersing the entire flask in a mixture of crushed ice and water. In a short time the reaction has subsided somewhat; the flask is then removed from the cooling mixture, and the reaction is allowed to proceed without external cooling. Reduction is completed by heating the mixture on a steam bath until all the sodium has dissolved. The partly cooled mixture is diluted with 300 cc. of water and heated on the steam bath until no more alcohol distils. The remaining small amount of alcohol is removed by gently applying suction from a water pump. The residue is diluted with about 600 cc. of hot water, and the mixture is allowed to cool without being disturbed. The separated oil solidifies to a solid cake from which the lower aqueous layer is easily decanted. The solid is washed once with a little cold water, drained as completely as possible, and dried by heating in the flask on a steam bath under reduced pressure. The residue is extracted with four successive 250-cc. portions of hot benzene. The united extract is clarified with a little charcoal, filtered, and concentrated to a volume of about 60 cc. About 200 cc. of alcohol is added; the solution is filtered, concentrated to about 60 cc., and mixed with an equal volume of hot benzene. On slow cooling the mixture sets to a solid mass of large crystals, which are filtered and washed with ether. The yield of this product, which melts at 72–74° (corr.), is 32–33 g. (73–75 per cent of the theoretical amount) (Notes 2, 3, and 4).

2. Notes

1. The alcohol must be perfectly dry, for any water present will cause immediate saponification of the ester with a consequent loss of yield. An excellent method for preparing completely anhydrous alcohol from alcohol of 99.5 per cent strength consists in dissolving 7 g. of sodium in 1 l. of the alcohol, adding 27.5 g. of a high-boiling ester such as ethyl phthalate, refluxing, and distilling.[1]

2. When several runs of the size indicated above or a single large run is made, the combined mother liquors may be evaporated free of solvent and the residue distilled under reduced pressure. The distillate on recrystallization from alcohol-benzene yields, however, only 5 to 7 grams of pure glycol per mole of sebacic ester.

3. The general method of Bouveault and Blanc has been used extensively for the preparation of glycols. Various modifications of detail have been suggested. Most of them are probably trivial or immaterial. In the experience of the checkers the rapid addition of the alcohol-ester mixture to the sodium gives results approximately equal to those described above, but mechanical stirring of the reaction mixture seriously reduces the yield.[2] The method used for isolating the glycol must be adapted to the properties of the glycol. At least for small runs the method of crystallization described here is the most suitable for decamethylene glycol. Lower members of the series are less readily crystallized. C. S. Marvel in a private communication has pointed out that for these glycols continuous ether extraction is the best method, and this has been used for hexamethylene glycol with success by the checkers.

Using the procedure described the checkers have prepared the following glycols.

		Yield, Per Cent
Heptamethylene glycol	b.p. 143–146°/8 mm.	88
Nonamethylene glycol	b.p. 147–150°/2 mm.	71
Undecamethylene glycol	m.p. 48–50°	57
Tridecamethylene glycol	m.p. 75–77°	88
Tetradecamethylene glycol	m.p. 83–85°	61
Octadecamethylene glycol	m.p. 96–98°	54

The heptamethylene glycol was separated by continuous ether extraction from the alkaline reduction solution after the latter had been diluted

[1] Smith, J. Chem. Soc. **1927**, 1288; Manske, J. Am. Chem. Soc. **53**, 1106 (1931).

[2] Bouveault and Blanc, Compt. rend. **137**, 329 (1903); Bull. soc. chim. (3) **31**, 1205 (1904); Ger. pat. 164,294 [Frdl. **8**, 1260 (1905–7)]; Franke and Kienberger, Monatsh. **33**, 1191 (1912); Chuit, Helv. Chim. Acta **9**, 264 (1926); Carothers, Hill, Kirby, and Jacobson, J. Am. Chem. Soc. **52**, 5287 (1930).

and distilled to remove the alcohol. The nonamethylene glycol was separated from the alkaline liquor by decantation (as above) and distilled. All the others were crystallized from benzene (without alcohol). Equally successful results have also been obtained with larger runs (e.g., 0.5 mole of ester).

4. Decamethylene glycol can also be prepared in excellent yields by the reduction of commercial butyl sebacate according to the procedure given for hexamethylene glycol on p. 325.

3. Methods of Preparation

Decamethylene glycol is best prepared by the reduction of sebacic esters either with sodium and an alcohol,[2] the Bouveault-Blanc procedure described above, or with hydrogen and a catalyst [3]—a procedure for which directions are given on p. 325. Both methods are applicable to the preparation of many other glycols.

Decamethylene glycol has also been prepared by the reduction of sebacamide with sodium and amyl alcohol [4] and by the reduction of methyl sebacate with sodium and liquid ammonia in absolute alcohol.[5]

DESOXYBENZOIN

$$3C_6H_5CH_2CO_2H + PCl_3 \longrightarrow 3C_6H_5CH_2COCl + H_3PO_3$$

$$C_6H_5CH_2COCl + C_6H_6 \xrightarrow{AlCl_3} C_6H_5CH_2COC_6H_5 + HCl$$

Submitted by C. F. H. ALLEN and W. E. BARKER.
Checked by C. S. MARVEL and TSE-TSING CHU.

1. Procedure

To 68 g. (0.5 mole) of phenylacetic acid (Note 1) in a 1-l. flask fitted with a reflux condenser and a system for absorbing hydrogen chloride, is added 35 g. (0.25 mole) of phosphorus trichloride. The mixture is heated on a steam bath for one hour. While the contents of the flask are still warm, 400 cc. of dry benzene is added. The benzene solution of phenylacetyl chloride is decanted from the residue of phosphorous

[3] Folkers and Adkins, ibid. **54**, 1146 (1932).

[4] Scheuble, Monatsh. **24**, 623 (1903); Scheuble and Loeble, ibid. **25**, 344 (1904); Alberti and Smieciuszewski, ibid. **27**, 411 (1906).

[5] Chablay, Compt. rend. **156**, 1021 (1913); Ann. chim. (9) **8**, 216 (1917).

acid onto 75 g. (0.56 mole) of anhydrous aluminum chloride in a dry, 1-l. flask which can be fitted to the same condenser. The reaction is vigorous at first and cooling is necessary. The mixture is refluxed for one hour on a steam bath, then cooled and poured into a mixture of 500 g. of cracked ice and 200 g. of concentrated hydrochloric acid. The benzene layer is separated, and the aqueous layer is extracted once with a mixture of 100 cc. of benzene and 100 cc. of ether (Note 2). The ether-benzene solution is washed once with 100 cc. of water (Note 3), and then dried over 40–50 g. of calcium chloride. The solution is filtered (Note 4) with suction into a 1-l. Claisen flask, and the solvent is removed by distillation under reduced pressure (Note 5); the residue consists of a brown oil which solidifies on cooling.

The crude material (91–92 g.) is purified by distillation under reduced pressure from a 250-cc. Claisen flask (Note 6). The product distils at 160°/5 mm. (172°/15 mm.; 200°/30 mm.) as a colorless oil which solidifies on cooling. The yield is 81–82 g. (82–83 per cent of the theoretical amount based on the phenylacetic acid used) of a product which melts at 53–54°. The product is recrystallized from methyl alcohol, using 4 cc. of solvent for each gram of product (Note 7); the yield is 55–56 g. of crystals melting at 55–56°. An additional 7 g. of crystals melting at 55–56° is obtained by cooling the filtrate in an ice-salt bath. On further cooling of the mother liquors, about 5 g. of crystals melting at 54–55° is obtained. The total yield of purified product is 67–70 g. (Note 8). Further recrystallization of the product from methyl alcohol does not raise the melting point above 55–56°.

2. Notes

1. Directions for preparing phenylacetic acid are given in Org. Syn. Coll. Vol. I, **1941**, 436. A very high grade of phenylacetic acid can also be obtained from companies supplying essential oils and perfumers' supplies, and some of the acid from these commercial sources is superior to that prepared by the Organic Syntheses procedure. Since the quality of the desoxybenzoin depends upon the quality of the phenylacetic acid used, it is important to employ a superior grade of acid.

2. A mixture of benzene and ether is used instead of ether alone because a more efficient separation of the two layers is obtained.

3. Washing with sodium hydroxide at this point does not improve the quality of the product but does, as a result of the formation of emulsions, cause an 8–10 per cent loss in yield.

4. It is better to remove the calcium chloride by filtration than by decantation even though the solution looks clear. Small particles of

calcium chloride and aluminum chloride not removed from the solution may cause bumping or even decomposition during the distillation.

5. Removal of the solvent by distillation under reduced pressure on the steam bath makes fractionation unnecessary during distillation of the product.

6. It is necessary to use a Claisen flask with a wide side-arm, since the desoxybenzoin may solidify and clog the apparatus. The distillation under reduced pressure is quiet if directions are followed carefully.

7. Methyl alcohol is the best solvent for purification. Desoxybenzoin tends to separate from ethyl alcohol as an oil.

8. Desoxybenzoin is somewhat unstable to light and consequently must be stored in dark bottles.

3. Methods of Preparation

Because of the availability of the starting materials, the most convenient methods of preparing desoxybenzoin are the Friedel-Crafts reaction,[1] described above, and the reduction of benzoin.[2] Desoxybenzoin can also be prepared, often with very good yields, by the treatment of bromostilbene with water in a sealed tube at 180–190°;[3] by the reduction of benzil;[4] by the action of zinc and hydrochloric acid on chlorobenzil;[5] from benzene, phenylacetic acid, and phosphorus pentoxide;[6] from benzoyl chloride and the magnesium halide derivative of sodium phenylacetate;[7] from benzylmagnesium chloride and benzamide;[8] and by alkaline hydrolysis of desylthioglycollic acid.[9]

[1] Graebe and Bungener, Ber. **12,** 1080 (1879).

[2] Kohler, Am. Chem. J. **36,** 182 (1906); Irvine and Weir, J. Chem. Soc. **91,** 1388 (1907); Kohler and Nygaard, J. Am. Chem. Soc. **52,** 4133 (1930); Ballard and Dehn, ibid. **54,** 3970 (1932).

[3] Limpricht and Schwanert, Ann. **155,** 60 (1870).

[4] Japp and Klingemann, J. Chem. Soc. **63,** 770 (1893).

[5] Thiele and Straus, Ann. **319,** 163 (1901).

[6] Zincke, Ber. **9,** 1771 (1876).

[7] Ivanoff and Nicoloff, Bull. soc. chim. (4) **51,** 1331 (1932).

[8] Jenkins, J. Am. Chem. Soc. **55,** 704 (1933).

[9] Behaghel and Schneider, Ber. **68,** 1588 (1935).

DESYL CHLORIDE

(Acetophenone, α-chloro-α-phenyl-)

$$C_6H_5CHOHCOC_6H_5 + SOCl_2 \xrightarrow{C_5H_5N} C_6H_5CHClCOC_6H_5 + SO_2 + HCl$$

Submitted by A. M. WARD.
Checked by C. S. MARVEL and TSE-TSING CHU.

1. Procedure

IN a 1-l. beaker are placed 100 g. (0.47 mole) of benzoin (Org. Syn. Coll. Vol. I, 1941, 94), and 50 g. (57 cc.) of pyridine. The mixture is heated until a solution is obtained, then cooled in an ice bath until solid. The mass is coarsely ground, and 75 g. (46 cc.; 0.63 mole) of thionyl chloride is added slowly with vigorous stirring and cooling in a water bath. After each addition of thionyl chloride, the reaction mixture becomes quite hot, and considerable amounts of sulfur dioxide and hydrogen chloride are evolved. At first the mass becomes pasty and then soon sets to a light yellow solid. After about an hour, water is added and the solid is coarsely ground and filtered. It is finely triturated twice with water, filtered by suction, and pressed as dry as possible. The white powder is dried to constant weight over sulfuric acid or calcium chloride. The yield of crude product is about 125 g. The compound is dissolved in 450 cc. of boiling 95 per cent alcohol (Note 1), filtered, and the filtrate cooled by running water. There is obtained 77 g. of colorless crystals which, after drying in the air, melt at 66–67°. On cooling the mother liquor in an ice-salt mixture, there is obtained an additional 9 g. of crystals melting at 65–66°. Further cooling of the filtrate yields no more product. The total yield is 80–86 g. (74–79 per cent of the theoretical amount) (Notes 2 and 3).

2. Notes

1. The product may be recrystallized from petroleum ether (b.p. 40–60°), but this solvent is less satisfactory for large amounts of material.

2. When the preparation is carried out with one-fifth these quantities the yield is only about 70 per cent of the theoretical amount.

3. Desyl chloride decomposes and becomes brown when exposed to sunlight, but is quite stable if kept in dark bottles.

3. Methods of Preparation

Desyl chloride has been prepared by the action of thionyl chloride on benzoin,[1] and on *l*-benzoin.[2] It has also been prepared by the action of hydrogen chloride on azibenzil.[3]

2,4-DIAMINOTOLUENE

(2,4-Toluenediamine)

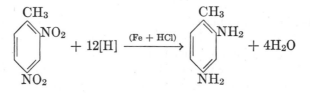

$$+ 12[H] \xrightarrow{(Fe + HCl)} + 4H_2O$$

Submitted by S. A. Mahood and P. V. L. Schaffner.
Checked by Roger Adams and P. R. Shildneck.

1. Procedure

In a 500-cc. three-necked flask, fitted with a reflux condenser and a mechanical stirrer (Note 1), are placed 45.5 g. (0.25 mole) of 2,4-dinitrotoluene (Note 2), 85 g. (1.5 moles) of iron (Note 3), and 100 cc. of 50 per cent (by weight) ethyl alcohol (Note 4). The mixture is heated to boiling on a water bath, the stirrer is started (Note 5), and a solution of 5.2 cc. (0.06 mole) of concentrated hydrochloric acid in 25 cc. of 50 per cent (by weight) ethyl alcohol is added slowly (Note 6). The mixture is refluxed for two hours after addition of the acid is complete. At the end of this time the apparatus is disconnected and the hot mixture is made just alkaline to litmus by the addition of the calculated amount of 15 per cent alcoholic potassium hydroxide solution (Note 7). Without allowing the mixture to cool, the iron is removed by filtration and the reaction flask is rinsed with two 50-cc. portions of 95 per cent ethyl alcohol; the same alcohol is used to wash the iron residue. To the filtrate is added 84 cc. of 6 N sulfuric acid; the normal sulfate of 2,4-diaminotoluene precipitates. The mixture is cooled to 25° and filtered by suction. The product is washed with two 25-cc. portions of 95 per

[1] Schroeter, Ber. **42**, 2348 (1909).
[2] McKenzie and Wren, J. Chem. Soc. **97**, 481 (1910).
[3] Curtius and Lang, J. prakt. Chem. (2) **44**, 547 (1891).

cent ethyl alcohol, dried in the air for two hours (Note 8) and then dried to constant weight at 110°. The yield is 49 g. (89 per cent of the theoretical amount) of a product which melts with decomposition at 249–251° (Note 9).

A solution of 20 g. of 2,4-diaminotoluene sulfate in 200 cc. of water at 60° is cooled to 40° (Note 10), and made alkaline to litmus with saturated sodium hydroxide solution (Note 11). To this solution is added 15 g. of the sulfate, which is dissolved by raising the temperature of the mixture to 55°. The solution is cooled to 40° and made slightly alkaline to litmus with saturated sodium hydroxide solution. The mixture is then cooled to 30° and filtered by suction. The remainder of the diaminotoluene sulfate (14 g.) is then dissolved in the filtrate by heating the mixture to 55°. The solution is cooled to 40° and again made alkaline to litmus with saturated sodium hydroxide solution. The mixture is cooled to 25°, and the diaminotoluene crystals are collected on a Büchner funnel. The entire product is dried to constant weight in a desiccator over calcium chloride. The yield is 26.5 g. (95 per cent of the theoretical amount based on the diaminotoluene sulfate used) of a product melting at 97–98.5°.

The crude diaminotoluene (Note 12) is dissolved in eight times its weight of benzene (212 g.) at 70°, and the solution is filtered quickly through a hot Büchner funnel (Note 13) with moderate suction (Note 14). The filtrate is cooled to 25°, and the mother liquor is decanted from the brown crystals. The mother liquor is concentrated to a volume of 25 cc. by distillation under atmospheric pressure and then cooled to 25°. The mother liquor is decanted from the diaminotoluene, and the entire product is dried in the air. The yield is 22.5 g. (81 per cent of the theoretical amount based on the diaminotoluene sulfate used; 74 per cent based on the dinitrotoluene used) of a product which melts at 98° (Note 15).

2. Notes

1. The mechanical stirrer should extend well to the bottom of the flask in order to prevent caking of the iron during the reduction.

2. The 2,4-dinitrotoluene used must be free from oil and have the correct melting point of 71°.

3. The iron used may be either iron powder or iron filings, but it should pass a 100-mesh sieve in order to give invariable results. If iron particles of larger size are used the reduction may be incomplete, although complete reduction was obtained with a sample of iron powder contaminated with filings.

4. The concentration of the alcoholic solution is important. Incomplete reduction was obtained with 95 per cent alcohol, as well as with methylated spirit, as suggested by West.[1] Complete reduction may be obtained in aqueous solution, but the product is difficult to isolate and is not always pure.

5. The stirrer should be run at such a speed that the iron particles do not settle or the iron will cake during the reduction. A speed of about 750 r.p.m. is satisfactory.

6. The reaction between the iron and hydrochloric acid is very vigorous, especially with iron powder, and the hydrochloric acid must be added very slowly at first. About one drop every ten seconds for the first ten minutes is satisfactory, and this may be increased to one drop every four seconds after this length of time has elapsed. The addition of the hydrochloric acid should require about thirty minutes.

7. The exact amount of alcoholic potassium hydroxide is determined previously by titration of a separate sample of 5.2 cc. of concentrated hydrochloric acid with the alcoholic potassium hydroxide.

8. Most of the alcohol and any aldehyde must be removed by evaporation at room temperature since heating the wet product at once to 110° always causes the formation of orange-colored impurities.

9. The same percentage yields of the sulfate were obtained using six times the amounts of materials.

10. The solution is cooled before addition of the alkali, as this is accompanied by a rise in temperature.

11. Saturated sodium hydroxide solution is used rather than more dilute solutions since dilution of the diaminotoluene solution is to be avoided as much as possible.

12. The diaminotoluene obtained by neutralizing the sulfate contains varying amounts of sodium sulfate and possibly some diaminotoluene sulfate, both of which are insoluble in benzene.

13. A hot Büchner funnel (70° or above) is used to prevent crystallization of the diaminotoluene in the funnel.

14. Moderate suction is used to prevent boiling of the benzene since this causes crystallization of the diaminotoluene in the funnel.

15. In an attempt to prepare a pure white product, the dry recrystallized material was distilled rapidly at atmospheric pressure using a short air condenser. A yellow product distilling at 292° was obtained. On distilling the recrystallized material at 148–150°/8 mm., a white product was obtained.

[1] West, J. Chem. Soc. **127**, 494 (1925).

3. Methods of Preparation

2,4-Diaminotoluene has been prepared by reduction of 2,4-dinitrotoluene with iron and acetic acid,[2] electrolytically,[3] or with hydrogen and Raney nickel;[4] and by reduction of 4-nitro-*o*-toluidine [5] or 2,4-dinitrobenzoyl chloride [6] with tin and hydrochloric acid.

DIAZOAMINOBENZENE

(Triazene, 1,3-diphenyl-)

$$C_6H_5NH_3Cl + NaNO_2 + HCl \rightarrow C_6H_5N_2Cl + 2H_2O + NaCl$$

$$C_6H_5NH_3Cl + CH_3CO_2Na \rightarrow NaCl + CH_3CO_2H + C_6H_5NH_2$$

$$C_6H_5NH_2 + C_6H_5N_2Cl + CH_3CO_2Na \rightarrow$$
$$C_6H_5N{=}NNHC_6H_5 + NaCl + CH_3CO_2H$$

Submitted by W. W. HARTMAN and J. B. DICKEY.
Checked by C. R. NOLLER and C. R. KEMP.

1. Procedure

IN a 5-l. flask fitted with a mechanical stirrer and a dropping funnel are placed 1 kg. of cracked ice, 1.5 l. of water, 279 g. (3 moles) of a technical grade of aniline, and 458 g. (388 cc., 4.5 moles) of concentrated hydrochloric acid (sp. gr. 1.18). The stirrer is started, and a solution of 109 g. (1.5 moles) of 95 per cent sodium nitrite in 250 cc. of water is added over a period of fifteen minutes. The reaction mixture is then stirred for fifteen minutes, and a solution of 422 g. (3.1 moles) of crystalline sodium acetate dissolved in 800 cc. of water is added over a period of five minutes. A yellow precipitate of diazoaminobenzene begins to form at once. Stirring is continued for forty-five minutes, keeping the temperature below 20° (Note 1). The yellow diazoaminobenzene is filtered on a 19-cm. Büchner funnel (Note 2), washed with 5 l. of cold water, and then sucked as dry as possible and spread out on a sheet of

---

[2] Hofmann, Jahresber. **1861**, 512.

[3] Hofer and Jakob, Ber. **41**, 3192 (1908).

[4] Albert and Ritchie, J. Proc. Roy. Soc. N. S. Wales **74**, 74 (1940) [C. A. **34**,7286 (1940)].

[5] Nölting and Collin, Ber. **17**, 268 (1884).

[6] Krasusky, J. Russ. Chem. Soc. **27**, 337 (1896) [Bull. soc. chim. (3) **16**, 370 (1896)].

paper to dry (Note 3). The product thus obtained is dissolved in 4 l. of boiling ligroin (b.p. 60–90°) (Note 4), filtered, and allowed to cool to room temperature and stand overnight. When crystallization is complete, the yellow crystals are filtered on a 19-cm. Büchner funnel, washed with 500 cc. of cold ligroin (b.p. 60–90°), and dried at room temperature. The yield of yellow crystals melting at 92–94° is 242–251 g. (82–85 per cent of the theoretical amount) (Note 5). If a product of greater purity is desired, the diazoaminobenzene is dissolved in 4 l. of boiling ligroin (b.p. 60–90°) and crystallized as before. The recrystallized diazoaminobenzene weighs 204–218 g. (69–73 per cent of the theoretical amount) and melts at 94–96° (Notes 6 and 7).

2. Notes

1. The temperature noted is not known to be the maximum temperature at which the reaction may be run.

2. A centrifuge of suitable size is preferable.

3. A rubber dam is fitted over the top of the Büchner funnel and held in place by rubber bands in order to remove as much of the water as possible.

4. Prolonged heating of the diazoaminobenzene with ligroin causes decomposition. For this reason it is well to heat the ligroin to boiling before it is added to the product to be crystallized. Solution is effected as rapidly as possible. If the crude diazoaminobenzene is not dry, a layer of water will separate at the bottom of the flask. This should be removed as completely as possible before filtering the hot ligroin solution.

5. An additional crop of crystals weighing 20–25 g. and melting at 79–83° can be obtained by evaporating the mother liquors to 1 l. and chilling in an ice bath.

6. The size of the run may be halved; a run with half quantities gave 125 g. of product.

7. Pure diazoaminobenzene is described as faintly yellow needles melting at 100°. A procedure for the purification of diazoaminobenzene and other aminoazo compounds has been described by Dwyer.[1]

3. Methods of Preparation

Diazoaminobenzene has been prepared by the action of sodium nitrite on aniline sulfate;[2] by the action of sodium nitrite on aniline hydrochloride;[3] by the action of sodium nitrite and sodium acetate on aniline

[1] Dwyer, J. Soc. Chem. Ind. **56,** 70T (1937).

[2] Staedel and Bauer, Ber. **19,** 1952 (1886).

[3] Martius, Zeit. für Chem. **1866,** 381; Curtius, Ber. **23,** 3033 (1890); Vaubel Chem. Ztg. **35,** 1238 (1911).

hydrochloride; [4] by the action of ammonium nitrate and hydrogen sulfide on aniline hydrochloride in the presence of iron; [5] by the action of sodium nitrite and potassium chromate or dichromate on aniline; [6] and from aniline and amyl nitrite.[7]

Diazoaminobenzene has also been prepared by the action of nitrous acid gas on aniline in alcohol; [8] by the action of silver nitrite on aniline hydrochloride; [9] and together with phenylurea by the action of nitrosophenylurea on aniline in methyl alcohol.[10] Niementowski and Roszkowski [11] have reported studies on the diazotization of aniline, aniline hydrochloride, and aniline sulfate with sodium nitrite and silver nitrite. The procedure described is adapted from that of Fischer.[4]

DIAZOMETHANE

(Methane, diazo-)

$$CH_3N(NO)CONH_2 + KOH \rightarrow CH_2N_2 + KCNO + 2H_2O$$

Submitted by F. ARNDT.
Checked by C. R. NOLLER and I. BERGSTEINSSON.

1. Procedure

Diazomethane is highly toxic. The utmost care is essential in the preparation and use of this material.

IN a 500-cc. round-bottomed flask are placed 60 cc. of 50 per cent aqueous potassium hydroxide solution and 200 cc. of ether. The mixture is cooled to 5°, and 20.6 g. (0.2 mole) of nitrosomethylurea (p. 461) is added with shaking. The flask is fitted with a condenser set for distillation. The lower end of the condenser carries an adapter passing through a two-holed rubber stopper and dipping below the surface of 40 cc. of ether contained in a 300-cc. Erlenmeyer flask and cooled in an

[4] Fischer, Ber. **17**, 641 (1884).

[5] Vaubel, Chem. Ztg. **37**, 637 (1913).

[6] Dwyer, Mellor, and Trikojus, J. Proc. Roy. Soc. N. S. Wales, **66**, 315 (1932) [C. A. **27**, 1331 (1933)].

[7] Meyer and Ambuhl, Ber. **8**, 1073 (1875).

[8] Griess, Ann. **121**, 257 (1862).

[9] Niementowski and Roszkowski, Z. physik. Chem. **22**, 158 (1897).

[10] Haager, Monatsh. **32**, 1089 (1911).

[11] Niementowski and Roszkowski, Z. physik. Chem. **22**, 145 (1897).

ice-salt mixture. The exit gases are passed through a second 40-cc. portion of ether likewise cooled below 0°. The reaction flask is placed in a water bath at 50° and brought to the boiling point of the ether with occasional shaking. The ether is distilled until it comes over colorless, which is usually the case after two-thirds of the ether has been distilled. *Under no circumstances should all the ether be distilled.* The combined ether solutions in the receiving flasks contain from 5.3 to 5.9 g. of diazomethane (63–70 per cent of the theoretical amount) (Notes 1 and 2), which is sufficiently dry for most purposes (Note 3).

If a dry solution of diazomethane is required, the ether solution is allowed to stand for three hours over pellets of pure potassium hydroxide (Note 4). For extremely dry solutions, further drying is effected with sodium wire.

2. Notes

1. For analysis an aliquot portion (about one-twentieth) of the solution is allowed to react at 0° with a solution of an accurately weighed sample of about 1.3 g. of pure benzoic acid in 50 cc. of absolute ether. The benzoic acid must be in excess as evidenced by the complete decolorization of the diazomethane solution. The unreacted benzoic acid is titrated with standard 0.2 N alkali.

2. The same procedure may be used for preparing two or three times the quantity obtained here.

3. The ether solution does not contain ammonia or methyl alcohol. It does contain traces of methylamine, but this is also present when diazomethane is prepared from nitrosomethylurethane.

If one does not require a pure, water-free solution, as is frequently the case when carrying out tests with small amounts of material, a simplified procedure may be used. To 100 cc. of ether is added 30 cc. of 40 per cent potassium hydroxide, and the mixture is cooled to 5°. To this, with continued cooling and shaking, is added 10 g. of finely powdered nitrosomethylurea in small portions over a period of one to two minutes. The deep yellow ether layer can be decanted readily; it contains about 2.8 g. of diazomethane, together with some dissolved impurities and water. The water may be removed by drying for three hours over pellets of pure potassium hydroxide. Solutions of diazomethane in benzene and other water-immiscible organic solvents may be prepared in the same way.

4. Broken sticks should not be used as the sharp corners facilitate the decomposition of the diazomethane.

3. Methods of Preparation

There are four methods of practical importance for the preparation of diazomethane: the action of alcoholic potassium hydroxide [1] or sodium dissolved in glycol [2] on nitrosomethylurethane; heating a mixture of potassium hydroxide, chloroform, hydrazine hydrate, and absolute alcohol; [3] the action of potassium hydroxide on nitrosomethylurea, [4] the method described above; and the action of alkoxides on the nitroso derivative of β-methylaminoisobutyl methyl ketone. [5] The choice of a method will usually depend upon the availability of the starting material. Directions for the preparation of the starting materials used in the first three methods are given in this volume; directions for preparing the nitroso derivative of β-methylaminoisobutyl methyl ketone and from it diazomethane will appear in a forthcoming volume of Organic Syntheses.

Arndt, Loewe, and Avan have discussed the merits of the various methods of preparing diazomethane, [6] as has Eistert. [7]

DIBENZALACETONE

(3-Pentanone, 1,5-diphenyl-)

$$2C_6H_5CHO + CH_3COCH_3 \xrightarrow{\text{(NaOH)}}$$
$$C_6H_5CH{=}CH{-}CO{-}CH{=}CHC_6H_5 + 2H_2O$$

Submitted by CHARLES R. CONARD and MORRIS A. DOLLIVER.
Checked by H. LOHSE and C. R. NOLLER.

1. Procedure

A COOLED solution of 100 g. of sodium hydroxide in 1 l. of water and 800 cc. of alcohol (Note 1) is placed in a 2-l. wide-mouthed glass jar which is surrounded with water and fitted with a mechanical stirrer. The solution is kept at about 20–25° and stirred vigorously (Note 2)

[1] v. Pechmann, Ber. **27**, 1888 (1894); **28**, 855 (1895).

[2] Meerwein and Burneleit, ibid. **61**, 1845 (1928).

[3] Staudinger and Kupfer, ibid. **45**, 505 (1912).

[4] Arndt and Amende, Angew. Chem. **43**, 444 (1930); Arndt and Scholz, ibid. **46**, 47 (1933).

[5] Adamson and Kenner, J. Chem. Soc. **1935**, 286; **1937**, 1551.

[6] Arndt, Loewe, and Avan, Ber. **73**, 606 (1940).

[7] Eistert, Angew. Chem. **54**, 99, 124 (1941).

while one-half of a mixture of 106 g. (1 mole) of benzaldehyde and 29 g. (0.5 mole) of acetone is added (Note 3). In about two or three minutes a yellow cloud forms which soon becomes a flocculent precipitate. After fifteen minutes the rest of the mixed reagents is added, and the container is rinsed with a little alcohol which is added to the mixture. Vigorous stirring is continued for one-half hour longer, and the mush is then filtered with suction on a large Büchner funnel. The product is thor· oughly washed with distilled water (Note 4) and then dried at room temperature to constant weight. The yield is 105–110 g. (90–94 per cent of the theoretical amount) (Note 5) of a product which melts at 104–107°.

The crude dibenzalacetone may be recrystallized from hot ethyl acetate, using 100 cc. of solvent for each 40 g. of material. The recovery in this purification is about 80 per cent; the purified product melts at 110–111°.

2. Notes

1. Sufficient alcohol is used to dissolve the benzaldehyde rapidly and to retain the benzalacetone in solution until it has had time to react with the second molecule of aldehyde. Lower concentrations of base slow up the formation of the dibenzalacetone and thus favor side reactions which yield a sticky product. Higher concentrations of base give added difficulty in washing. These concentrations were suggested by, and are approximately the same as, those used in the preparation of benzalacetophenone described in Org. Syn. Coll. Vol. I, **1941**, 78.

2. Only temperatures between 20 and 25° were tried; it was assumed that a change of temperature would have the same effect that it has in the preparation of benzalacetophenone mentioned above.

Stirring is essential, as it makes considerable difference in the uniformity of the product.

3. The benzaldehyde was u.s.p. quality which had been washed with sodium carbonate solution and distilled. Commercial c.p. acetone was used. The theoretical quantities are used, since an excess of benzaldehyde results in a sticky product while an excess of acetone favors the production of benzalacetone. The mixture is prepared before addition in order to ensure additions of equivalent quantities.

4. Since the product is practically insoluble in water, large amounts can be used in the washing. Sodium compounds are probably the chief impurities. The dried product contains some sodium carbonate which results from the failure to remove the sodium hydroxide completely. There remain also the impurities insoluble in water. However, the product is pure enough for use in most reactions.

5. If the mush is allowed to stand several hours, chilled, and filtered cold, a slightly larger yield is obtained, but this is not worth while. The filtrate may be used as a medium for a second run in which about 93 per cent of the theoretical yield is obtained. The melting point of the second product is slightly lower.

3. Methods of Preparation

Dibenzalacetone has been prepared by condensing benzaldehyde with acetone using as condensing agents dry hydrogen chloride,[1] 10 per cent sodium hydroxide solution,[2] and glacial acetic acid with sulfuric acid.[3] It has also been obtained by condensing benzalacetone with benzaldehyde in the presence of dilute sodium hydroxide.[4] Straus and Ecker [5] were the first to record the use of ethyl acetate for crystallization.

1,4-DIBENZOYLBUTANE

(1,6-Hexanedione, 1,6-diphenyl-)

$$
\begin{array}{l}
CH_2CH_2CO_2H \\
| \qquad\qquad\quad + 2SOCl_2 \longrightarrow \\
CH_2CH_2CO_2H
\end{array}
\qquad
\begin{array}{l}
CH_2CH_2COCl \\
| \qquad\qquad\quad + 2SO_2 + 2HCl \\
CH_2CH_2COCl
\end{array}
$$

$$
\begin{array}{l}
CH_2CH_2COCl \\
| \qquad\qquad\quad + 2C_6H_6 \xrightarrow{AlCl_3} \\
CH_2CH_2COCl
\end{array}
\qquad
\begin{array}{l}
CH_2CH_2COC_6H_5 \\
| \qquad\qquad\qquad + 2HCl \\
CH_2CH_2COC_6H_5
\end{array}
$$

Submitted by REYNOLD C. FUSON and JOSEPH T. WALKER.
Checked by W. H. CAROTHERS and W. L. McEWEN.

1. Procedure

ONE mole (146 g.) of adipic acid (Org. Syn. Coll. Vol. I, **1941**, 18), previously dried over sulfuric acid in a vacuum, is placed in a 1-l. round-bottomed flask equipped with a condenser, and 357 g. (217 cc., 3 moles) of thionyl chloride is added at once. The mixture is heated gently on a water bath held at a temperature of 50–60°. After about four hours, solution is complete and evolution of hydrogen chloride has ceased.

[1] Claisen and Claparède, Ber. **14**, 350 (1881).
[2] Schmidt, ibid. **14**, 1460 (1881); Claisen, ibid. **14**, 2470 (1881); Straus and Caspari, ibid. **40**, 2698 (1907).
[3] Claisen and Claparède, ibid. **14**, 2460 (1881).
[4] Claisen and Ponder, Ann. **223**, 141 (1884).
[5] Straus and Ecker, Ber. **39**, 2988 (1906).

The flask is now connected to a downward condenser and heated under diminished pressure by a water bath to remove any excess thionyl chloride. The light yellow residue of adipyl chloride is ready for use.

A mixture of 300 g. (2.25 moles) of anhydrous aluminum chloride and 1.5 l. (17 moles) of benzene (dried over sodium and distilled) is placed in a 3-l. three-necked flask equipped with a mercury-sealed stirrer, a reflux condenser, and a dropping funnel. The reaction mixture is cooled in an ice bath, and, with rapid stirring, the adipyl chloride is added through the dropping funnel at an *even rate* during the course of forty-five minutes (Note 1). The reaction mixture darkens slowly but does not become black. After the adipyl chloride has been added, the ice bath is removed and stirring is continued for two hours at room temperature (Note 2).

The solution is then poured slowly, with constant stirring, into a mixture of 1 kg. of cracked ice and 200 cc. of concentrated hydrochloric acid in a 5-l. flask. There should be a small quantity of ice remaining after decomposition is complete. When the ice has melted, the mixture of water, precipitated dibenzoylbutane, and benzene is divided into two equal portions, and 1–1.25 l. of benzene is added to each. The solid is dissolved by shaking and gentle warming on the steam bath; the benzene layers are separated and washed, first with an equal volume of dilute sodium carbonate solution, and then with water.

The benzene solution is placed in a 5-l. flask, and 3–3.5 l. of benzene is removed by distillation. The residual liquid is set aside and allowed to cool. The dibenzoylbutane which crystallizes after several hours is filtered; it melts at 104–107° (Note 3). An equal volume of ether is added to the light brown filtrate, and a second crop of crystals is obtained (Note 4).

The yield of crude product is 199–216 g. (75–81 per cent of the theoretical amount). The material may be recrystallized by dissolving it in 1 l. of hot 95 per cent ethyl alcohol. Upon cooling, crystals which melt at 106–107° separate. The yield of recrystallized product is 190–210 g.

2. Notes

1. It is very important to keep the reaction mixture cold during the addition of the acid chloride; otherwise there will be charring which will lead to a discolored product.

2. If the reaction mixture is allowed to stand at this point the yield is materially decreased.

3. The melting point and color of the crude product seem to be influenced by the quality of the aluminum chloride. The checkers used a

very pure aluminum chloride and found that the color of the reaction mixture was never darker than a bright orange, that the crude product was always colorless, and that the crude product always melted at 106–108°. If a less pure aluminum chloride is used the crude product is sometimes brown. This color may be completely removed by washing with a few cubic centimeters of cold ether, in which the diketone is only slightly soluble.

4. The checkers found that only 5–6 g. of material was precipitated by the addition of ether. For this reason it seems scarcely worth while to carry out this part of the procedure.

3. Methods of Preparation

Dibenzoylbutane has been prepared by the action of aluminum chloride on a mixture of benzene and adipyl chloride [1] or benzene and polymeric adipic anhydride.[2] It has also been obtained from adiponitrile and phenylmagnesium bromide,[3] and as a by-product in the action of zinc on α,β-dibromopropiophenone [4] and α,α'-dibromodibenzoylbutane.[5] Melting points between 102° [1] and 112° [4] have been recorded for the diketone.

1,2-DIBROMOCYCLOHEXANE

(Cyclohexane, 1,2-dibromo-)

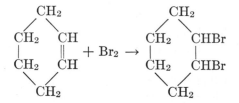

Submitted by H. R. SNYDER and L. A. BROOKS.
Checked by LEE IRVIN SMITH, R. T. ARNOLD, and JOHN RYAN.

1. Procedure

IN a 2-l. three-necked, round-bottomed flask, fitted with a 500-cc. separatory funnel, a mechanical stirrer, and a thermometer, is placed a solu-

[1] Etaix, Ann. chim. phys. (7) **9**, 372 (1896).
[2] Hill, J. Am. Chem. Soc. **54**, 4105 (1932).
[3] Compère, Bull. soc. chim. Belg. **44**, 523 (1935).
[4] Kohler, Am. Chem. J. **42**, 384 (1909).
[5] Fuson and Farlow, J. Am. Chem. Soc. **56**, 1593 (1934).

tion of 123 g. (1.5 moles) of cyclohexene (Note 1) in a mixture of 300 cc. of carbon tetrachloride and 15 cc. of absolute alcohol. The flask is surrounded by an ice-salt bath. The stirrer is started, and, when the temperature has reached −5°, a solution of 210 g. (67 cc., 1.3 moles) of bromine in 145 cc. of carbon tetrachloride is added from the separatory funnel at such a rate that the temperature of the reaction mixture does not exceed −1° (Note 2). The addition requires about three hours.

When the bromine has been added the contents of the flask are transferred directly to a 1-l. modified Claisen flask and the carbon tetrachloride and excess cyclohexene are distilled from a water bath (Notes 3, 4, and 5). The water bath is replaced by an oil bath and the product distilled under reduced pressure. There is a small low-boiling fraction, and then pure dibromocyclohexane distils at 99–103°/16 mm. (108–112°/25 mm.). The yield is 303 g. (95 per cent of the theoretical amount) (Notes 6 and 7).

2. Notes

1. Cyclohexene boiling over a two-degree range is satisfactory for this preparation. Directions for preparing cyclohexene are given in Org. Syn. Coll. Vol. I, **1941,** 183 and on p. 152 above.

2. Unless the temperature is controlled carefully, the yield is poor because of substitution reactions. Even at this low temperature some substitution occurs unless the excess of cyclohexene is used.

3. The three-necked flask may be rinsed with a little carbon tetrachloride.

4. The dibromide decomposes on continued exposure to the air and becomes very dark. Hence the product should be distilled at once.

5. The low-boiling distillate contains a trace of the dibromide as shown by the fact that it darkens on exposure to the air.

6. The product is stored best in sealed bottles with as little exposure to the air as possible.

7. A product which will not darken may be obtained by the following purification: The crude dibromide is shaken for five minutes with about one-third its volume of 20 per cent ethyl alcoholic potassium hydroxide. The mixture is diluted with its own volume of water and the organic layer is washed free of alkali, dried, and distilled. Material so treated will stay clear indefinitely. The loss in the purification is about 10 per cent. (WM. VON EGGERS DOERING and ALDRICH DURANT, JR., private communication.)

3. Method of Preparation

1,2-Dibromocyclohexane has been prepared from cyclohexene by the addition of bromine in chloroform,[1] carbon tetrachloride,[2] ether,[3] glacial acetic acid,[4] aqueous sodium bromide,[5] and in the absence of a solvent.[6] The bromination in carbon tetrachloride containing a little alcohol is more satisfactory than the bromination in carbon tetrachloride which is described in an earlier volume.[2]

2,6-DIBROMO-4-NITROPHENOL

(Phenol, 2,6-dibromo-4-nitro-)

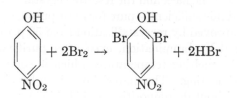

Submitted by W. W. HARTMAN and J. B. DICKEY.
Checked by LOUIS F. FIESER and C. H. FISHER.

1. Procedure

IN a 5-l. round-bottomed flask, fitted with a liquid-sealed mechanical stirrer, a dropping funnel, and a tube leading to a gas trap to carry off the hydrogen bromide, 278 g. (2 moles) of p-nitrophenol (m.p. 112–113°) is dissolved in 830 cc. of glacial acetic acid. To this solution at room temperature is added, dropwise with stirring during the course of three hours, a solution of 750 g. (240 cc., 4.7 moles) of bromine in 700 cc. of glacial acetic acid. After the addition of the bromine the reaction mixture is stirred for one-half hour and then warmed on the steam bath (internal temperature about 85°) for one hour in order to remove as much of the excess bromine as possible. The last traces of bromine are

[1] Baeyer, Ann. **278**, 108 (1894); Hofmann and Damm, Mitt. schlesischen Kohlenforsch. Kaiser Wilhelm Ges. **2**, 97 (1925) [C. A. **22**, 1249 (1928)]; Rothstein, Ann. chim. (10) **14**, 542 (1930).

[2] Coffey, Rec. trav. chim. **42**, 398 (1923); Greengard, Org. Syn. **12**, 27.

[3] Fortey, J. Chem. Soc. **73**, 948 (1898).

[4] Harries and Splawa-Neyman, Ber. **42**, 695 (1909).

[5] Markownikoff, Ann. **302**, 29 (1898); Swarts, Bull. soc. chim. Belg. **46**, 13 (1937);

[6] Truffault, Bull. soc. chim. (5) **1**, 398 (1934).

removed by passing a stream of air into the reaction mixture, which then has a yellow or brown color. The mixture is treated with 1.1 l. of cold water, stirred until cool (Note 1), and allowed to stand in ice, or in an ice chest, overnight. The pale yellow crystalline product is collected on a 19-cm. Büchner funnel and washed first with 500 cc. of 50 per cent aqueous acetic acid and then thoroughly with water. It is dried in an oven at 40–60° or in a vacuum desiccator over sodium hydroxide. The yield is 570–583 g. (96–98 per cent of the theoretical amount) of a nearly colorless product melting with decomposition at 138–140° (Note 2).

2. Notes

1. If the solution is stirred during cooling the product is less likely to cake on the walls of the flask and the resulting crystals are easier to wash.

2. This material is sufficiently pure for most purposes, and the quality is not greatly improved by recrystallization from 50 per cent acetic acid. Samples prepared by bromination, even after purification, invariably decompose at the melting temperature, which is somewhat dependent upon the rate of heating. The material obtained by nitration has been found to melt without decomposition at 144–145°.[1]

3. Methods of Preparation

2,6-Dibromo-4-nitrophenol has been prepared by the nitration of 2,6-dibromophenol [1] or of dibromophenolsulfonic acid,[2] and by the action of nitric acid on 2,6-dibromo-4-nitrosophenol [3] or on 2,4,6-tribromophenol.[4] It has been obtained from the corresponding ethyl ether [5] and by the action of bromine on p-nitrosophenol,[6] 4,6-dibromo-2-nitrophenol,[7] 5-nitro-2-hydroxybenzoic acid,[8] and 5-nitro-2-hydroxybenzenesulfonic acid.[9] The dibromination of p-nitrophenol [10, 11] has been car-

[1] Pope and Wood, J. Chem. Soc. 101, 1828 (1912).

[2] Armstrong and Brown, ibid. 25, 859 (1872); Contardi and Ciocca, Gazz. chim. ital. 63, 878 (1933).

[3] Kehrmann, Ber. 21, 3318 (1888); Forster and Robertson, J. Chem. Soc. 79, 688 (1901).

[4] Raiford and Heyl, Am. Chem. J. 43, 395 (1910).

[5] Jackson and Fiske, Ber. 35, 1132 (1902); Am. Chem. J. 30, 60 (1903).

[6] van Erp, Rec. trav. chim. 30, 290 (1911).

[7] Armstrong, J. Chem. Soc. 28, 522 (1875); Ling, ibid. 51, 147 (1887).

[8] Lellmann and Grothmann, Ber. 17, 2731 (1884).

[9] Post, Ann. 205, 94 (1880).

[10] Brunck, Z. Chem. 1867, 204; ibid. 1868, 323; Vaubel, J. prakt. Chem. (2) 49, 544 (1894).

[11] Möhlau and Uhlmann, Ann. 289, 94 (1896).

ried out in sulfuric acid solution [12] and in the presence of aluminum chloride.[13]   The method described here is essentially that of Möhlau and Uhlmann.[11]

2,6-DIBROMOQUINONE-4-CHLOROIMIDE

(Quinonimine, 2,6-dibromo-N-chloro-)

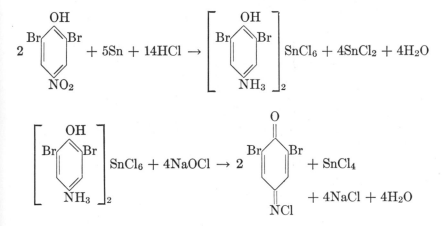

Submitted by W. W. HARTMAN, J. B. DICKEY, and J. G. STAMPFLI.
Checked by LOUIS F. FIESER and C. H. FISHER.

1. Procedure

(A) 2,6-*Dibromo-4-aminophenol Chlorostannate.*—In a 5-l. round-bottomed flask is placed 148.5 g. (0.5 mole) of 2,6-dibromo-4-nitrophenol (p. 173) with 300 cc. (3.7 moles) of concentrated hydrochloric acid (sp. gr. 1.19), 300 cc. of water, and 185 g. (1.56 gram atoms) of mossy tin. Three cubic centimeters of capryl alcohol is added to control the foaming, and the mixture is heated in the open flask with stirring on the steam bath until the reaction starts.   The reaction may proceed vigorously at the outset, and it is well to heat cautiously during the initial stages. Hydrochloric acid and water are added from time to time, and foaming can be controlled by the addition of a part of the water.   A total of 520 cc. (6.4 moles) of concentrated hydrochloric acid and 900 cc. of water are added during the course of the reaction.   When the first,

[12] Datta and Bhoumik, J. Am. Chem. Soc. **43**, 310 (1921).
[13] Bodroux, Compt. rend. **126**, 1285 (1898); Bull. soc. chim. (3) **19**, 759 (1898).

vigorous reaction is over the mixture is heated strongly until all the dibromonitrophenol has dissolved, and the hot solution (at about 85°) is filtered through a layer of Norite on a hot Büchner funnel. The filtrate, which usually is colorless, is cooled to 0° with stirring for two hours, or allowed to stand in a cool place overnight. The product, which separates in the form of colorless or slightly yellow needles, is collected on a Büchner funnel and washed with cold dilute hydrochloric acid (one volume of concentrated acid to one volume of water). The material is usually colorless and may be used directly for the reaction which follows (Note 1). After drying in an oven at 50–60°, or in a vacuum desiccator over sodium hydroxide, the tin salt weighs 214–220 g. The theoretical weight, based on the formula assumed above, is 217 g.

(B) 2,6-*Dibromoquinone*-4-*chloroimide.*—The chlorostannate is conveniently oxidized in two batches (Note 2). In a 3-l. flask is placed a solution of 115 g. (2.9 moles) of sodium hydroxide in 175 cc. of water, 1 kg. of cracked ice is added, and 108 g. (1.52 moles) of chlorine gas is passed into the mixture. About 80 per cent of the ice is melted by the operation. In a 5-l. flask 110 g. (0.127 mole) of the tin salt of 2,6-dibromo-4-aminophenol is dissolved in 1.2 l. of water and 12 cc. of concentrated hydrochloric acid. Solution is effected by warming the mixture to 40–50°, after which it is cooled to 15–17° and 600 g. of ice is added. The sodium hypochlorite solution is then added all at once with vigorous stirring (Hood). A yellow precipitate of 2,6-dibromoquinone-4-chloroimide separates immediately, and chlorine is evolved. As soon as the sodium hypochlorite solution has been stirred in, 120 cc. of concentrated hydrochloric acid is added in order to keep the tin salts in solution (Note 3). The fine, yellow precipitate is filtered under the hood on a Büchner funnel (Note 4) and washed with 1.5 l. of 5 per cent hydrochloric acid to remove tin salts and chlorine. The product is dried on a glass tray at 30–40° (Note 5), or in a vacuum desiccator over sodium hydroxide. From two such batches the yield of chloroimide melting at 80–82° is 126–130 g. (84–87 per cent of the theoretical amount based on the weight of 2,6-dibromo-4-nitrophenol used in Part A).

2. Notes

1. If the product has an appreciable yellow color, it should be recrystallized from a mixture of 150 cc. of concentrated hydrochloric acid and 375 cc. of water, heating to 85–90°. The yield of recrystallized product is 180–190 g. By evaporating the mother liquor to one-third the original volume under reduced pressure, an additional 15–25 g. of tin salt can be recovered.

2. As much as 700 g. of tin salt can be oxidized in one run by using a 70-l. crock for the reaction vessel.

3. The particle size of the chloroimide can be increased by stirring for one hour at this point.

4. The first liter of filtrate is refiltered until it is clear.

5. Care must be taken in drying as one larger run decomposed violently in an open tray when the temperature was about 60°.

3. Method of Preparation

2,6-Dibromoquinone-4-chloroimide has been prepared by the action of sodium hypochlorite on 2,6-dibromo-4-aminophenol in the form of the chlorostannate [1] or as the hydrochloride.[2]

α,β-DIBROMOSUCCINIC ACID

(Succinic acid, α,β-dibromo-)

$$HO_2CCH{=}CHCO_2H + Br_2 \rightarrow HO_2CCHBrCHBrCO_2H$$

Submitted by HERBERT S. RHINESMITH.
Checked by REYNOLD C. FUSON and W. E. ROSS.

1. Procedure

IN a 2-l. three-necked, round-bottomed flask, equipped with a mechanical stirrer (Note 1), dropping funnel, and Friedrichs condenser [1] (Note 2), are placed 200 g. (1.7 moles) of fumaric acid (Note 3) and 400 g. of water (Note 4). The materials are thoroughly mixed until the fumaric acid has been completely wet by the water. The resulting thick, viscous mass is then stirred vigorously (Note 5) and brought to boiling by heating on a wire gauze with a Bunsen flame (Note 6).

Two hundred and seventy-six grams (94.3 cc., 1.7 moles) of bromine (Note 7) is now added as rapidly as possible through the dropping funnel, the rate of addition being so controlled that the Friedrichs condenser is

[1] Möhlau, Ber. 16, 2845 (1883); Friedländer and Stange, ibid. 26, 2262 (1893); Möhlau and Uhlmann, Ann. 289, 94 (1896); Mikhaĭlov, Trans. Inst. Pure Chem. Reagents No. 16, 83 (1939) [C. A. 34, 3707 (1940)].

[2] Gibbs, J. Biol. Chem. 72, 653 (1927).

α,β-DIBROMOSUCCINIC ACID

[1] Friedrichs, Z. angew. Chem. 33 (I) 30 (1920).

continuously about half full of the refluxing liquid (Note 8). This operation takes about one hour (Note 9). After about 100 g. of bromine has been added, the dibromosuccinic acid forms rapidly and separates in tiny white needles. At the completion of the reaction there should be a slight excess of bromine, as indicated by the red color of the solution. Occasionally 5–10 g. of bromine has to be added at this point to ensure an excess.

The reaction flask is now surrounded with ice water and cooled to 10°, with stirring. The product is then collected on a large Büchner funnel, and washed with cold water to remove the bromine liquor. The filtrate may be discarded, as it contains only impurities. The material is dried overnight at room temperature and need not be recrystallized; the yield is 343–400 g. (72–84 per cent of the theoretical amount).

2. Notes

1. A heavy stirrer with as large a paddle as possible is used, in order to rotate the mass of crystals formed during the course of the reaction. A mercury seal is unnecessary, but it is advisable to have the stirrer bearing extend beneath the surface of the liquid.

2. Glass connections or rubber stoppers should be used throughout, as corks are rapidly disintegrated by the hot bromine.

3. Commercial fumaric acid ("practical") is sufficiently pure for this preparation. Directions for preparing fumaric acid are given on p. 302.

4. Any larger amount of water leads to the formation of monobromomalic acid, tartaric acid, and compounds of unknown composition.[2]

5. Vigorous stirring is essential to obtain good yields.

6. It is necessary to keep the reaction mixture boiling throughout the entire course of the reaction. During the addition of the bromine, however, the size of the flame should be reduced considerably, because the reaction is exothermic.

7. The apparatus should be set up under a hood, or the top of the condenser connected to a gas absorption trap for the removal of bromine vapor, small amounts of which escape continually under the conditions of the experiment.

8. By this procedure most of the unchanged bromine is washed back into the flask, so that the amount escaping from the top of the condenser is kept at a minimum.

9. If the bromine is added over a much longer period of time, the yield is materially decreased.

[2] Kekulé, Ann. Spl. Bd. 1, 338 (1861).

3. Methods of Preparation

α,β-Dibromosuccinic acid may be prepared by heating succinic acid with bromine and water in a closed tube at 180°;[3] by heating succinic acid, red phosphorus, and bromine in a closed tube at 140°;[4] by heating fumaric acid with 2 moles of bromine in acetic acid for seven hours in a sealed tube at 100°;[5] from fumaric acid, bromine, and water at 100° under pressure;[6] and by the method described above.

DI-*n*-BUTYLCARBINOL

(5-Nonanol)

$$C_4H_9Br + Mg \rightarrow C_4H_9MgBr$$

$$2C_4H_9MgBr + HCO_2C_2H_5 \rightarrow (C_4H_9)_2CHOMgBr + C_2H_5OMgBr$$

$$2(C_4H_9)_2CHOMgBr + H_2SO_4 \rightarrow 2(C_4H_9)_2CHOH + MgBr_2 + MgSO_4$$

Submitted by G. H. COLEMAN and DAVID CRAIG.
Checked by JOHN R. JOHNSON and H. B. STEVENSON.

1. Procedure

IN a 3-l. three-necked, round-bottomed flask, fitted with a 500-cc. separatory funnel, a liquid-sealed mechanical stirrer, and a reflux condenser, are placed 36.5 g. (1.5 gram atoms) of magnesium turnings and 500 cc. of absolute ether. A solution of 206 g. (1.5 moles) of *n*-butyl bromide (Org. Syn. Coll. Vol. I, **1941**, 28, 37) in 250 cc. of absolute ether is placed in the separatory funnel. The stirrer is started, and 10–15 cc. of the bromide solution is allowed to flow into the flask from the funnel; the reaction generally begins within a few minutes (Note 1). As soon as refluxing is vigorous, the flask is surrounded by ice and water and the rate of addition of the bromide is adjusted so that moderate refluxing occurs. After all the solution has been added (thirty to forty minutes), the cooling bath is removed. Stirring is continued for fifteen minutes longer, after which only a small residue of unreacted magnesium remains.

[3] Kekulé, Ann. **117**, 120 (1861); Ann. Spl. Bd. **1**, 338 (1861); Bourgoin, Bull. soc chim. (2) **19**, 148 (1873).

[4] Gorodetzky and Hell, Ber. **21**, 1729 (1888).

[5] Michael, J. prakt. Chem. (2) **52**, 289 (1895).

[6] Kekulé, Ann. Spl. Bd. **1**, 129 (1861); Baeyer, Ber. **18**, 674 (1885).

The flask is cooled in an ice bath and a solution of 55.5 g. (0.75 mole) of pure ethyl formate (Note 2) in 100 cc. of absolute ether is placed in the separatory funnel. The stirrer is started and the ethyl formate solution is added at such a rate that the ether refluxes gently. This addition requires about one-half hour. The cooling bath is then removed and stirring is continued for ten minutes.

With vigorous stirring (Note 3), 100 cc. of water is added through the separatory funnel at such a rate that rapid refluxing occurs. Following this, a cold solution of 85 g. (46 cc., 0.85 mole) of concentrated sulfuric acid in 400 cc. of water is added. After the addition of the acid, the two layers become practically clear. A large part of the ethereal layer is decanted into a 1-l. round-bottomed flask, and the remainder, together with the aqueous layer, is transferred to a separatory funnel. The solid remaining in the flask is washed with two 25-cc. portions of ether, which are added to the material in the separatory funnel. The ethereal layer is separated and combined with the decanted portion. The flask is fitted with an efficient fractionating column, and the ether is distilled from a steam bath until the temperature of the vapor reaches about 50°. To the residual impure carbinol (Note 4) is added 75 cc. of 15 per cent aqueous potassium hydroxide solution and the flask is fitted with a reflux condenser. The mixture is boiled vigorously under reflux for three hours, after which the purified carbinol is removed by steam distillation, the volume in the flask being kept at 250–300 cc. The distillate is collected in a separatory funnel so that the lower aqueous layer can be drawn off periodically. The distillation is complete when about 1.5 l. of water has been collected.

The upper layer of di-n-butylcarbinol is separated and allowed to stand over 10 g. of anhydrous potassium carbonate for one hour. The liquid is decanted into a 500-cc. Claisen flask, and the residual potassium carbonate is washed with three 10-cc. portions of dry ether, which are added to the material in the distilling flask. After removing a small fraction of low-boiling material, there is obtained 90–92 g. (83–85 per cent of the theoretical amount) of pure di-n-butylcarbinol, b.p. 97–98°/ 20 mm. (Note 5).

2. Notes

1. The reaction between the ethereal solution of n-butyl bromide and the magnesium generally starts without any assistance; if necessary, a small amount of a previously prepared Grignard reagent or a crystal of iodine may be used to start the reaction.

2. It is best to use freshly distilled ethyl formate, which may be purified in the following way: To 100 g. of commercial ethyl formate is added 15 g. of anhydrous potassium carbonate, and the mixture is allowed to

stand for one hour with occasional shaking. The ester is decanted into a dry 200-cc. flask, and 5 g. of phosphorus pentoxide is added. The flask is provided with an efficient fractionating column, and the ethyl formate is distilled into a receiver protected from atmospheric moisture. A fraction boiling at 53–54° was used in this preparation.

3. During the addition of the water it is necessary to stir efficiently so that the solid which is produced will be precipitated in a finely divided form and not in large aggregates.

4. The formic ester of di-n-butylcarbinol is present as an impurity in the crude product and is hydrolyzed by refluxing with potassium hydroxide solution.

5. Di-n-butylcarbinol can be distilled at atmospheric pressure without appreciable decomposition (b.p. 193–194°/743 mm.) but it is preferable to distil under diminished pressure. The following boiling points were observed under various pressures: 97°/20 mm., 104°/30 mm., 109°/40 mm., 117°/60 mm., 130°/100 mm.

3. Methods of Preparation

Di-n-butylcarbinol has been prepared by the action of n-butylmagnesium bromide upon n-valeraldehyde [1] and upon ethyl formate.[1,2] It has also been obtained by the catalytic hydrogenation of di-n-butyl ketone in the presence of platinum.[3]

DICHLOROACETIC ACID

(Acetic acid, dichloro-)

$$2CCl_3CH(OH)_2 + 2CaCO_3 \xrightarrow{\text{(NaCN)}}$$
$$(CHCl_2CO_2)_2Ca + 2CO_2 + 2H_2O + CaCl_2$$
$$(CHCl_2CO_2)_2Ca + 2HCl \rightarrow CaCl_2 + 2CHCl_2CO_2H$$

Submitted by ARTHUR C. COPE, JOHN R. CLARK, and RALPH CONNOR.
Checked by R. L. SHRINER and NEIL S. MOON.

1. Procedure

A SOLUTION of 250 g. (1.5 moles) of U.S.P. chloral hydrate in 450 cc. of warm water (50–60°) is placed in a 3-l. round-bottomed flask bearing

[1] Malengreau, Bull. acad. roy. Belg. cl. sci. 1906, 802 [C. A. 1, 1970 (1907)].
[2] Dillon and Lucas, J. Am. Chem. Soc. 50, 1713 (1928).
[3] Vavon and Ivanoff, Compt. rend. 177, 453 (1923).

a reflux condenser and thermometer (Note 1). The condenser is temporarily removed and 152.5 g. (1.52 moles) of precipitated calcium carbonate added; this is followed by 2 cc. of amyl alcohol (Note 2) and a solution of 10 g. of technical sodium cyanide in 25 cc. of water. Although the reaction is exothermic, the reaction mixture is heated with a low flame so that it reaches 75° in about ten minutes; at this point heating is discontinued. The temperature continues to rise to 80–85° during five to ten minutes and then drops. As soon as the temperature begins to fall the solution is heated to boiling and refluxed for twenty minutes. The mixture is then cooled to 0–5° in an ice bath, acidified with 215 cc. of concentrated hydrochloric acid (sp. gr. 1.18), and extracted with five 100-cc. portions of ether (Note 3). The combined ether extracts are dried with 20 g. of anhydrous sodium sulfate, the ether is removed by distillation from a steam bath, and the residue is distilled in vacuum from a Claisen flask with a fractionating side arm (Note 4). The yield of dichloroacetic acid, b.p. 99–104°/23 mm., is 172–180 g. (88–92 per cent of the theoretical amount) (Note 5).

2. Notes

1. The amount of hydrogen cyanide evolved is small, and the reaction may be carried out in a hood without any special device for removing this gas. The use of mechanical stirring does not improve the results.

2. Amyl alcohol is added to decrease the amount of foaming.

3. The emulsion which often forms during the ether extraction may be broken by filtering through a fluted filter or with suction.

4. The product decomposes when distilled at atmospheric pressure.

5. The preparation has been carried out with equally good results using double the quantities given above.

3. Methods of Preparation

Dichloroacetic acid has been prepared by the chlorination of acetic [1] or chloroacetic [2] acid, by hydrolysis of pentachloroethane,[3] from trichloroacetic acid by electrolytic reduction [4] or the action of copper,[5]

[1] Müller, Ann. 133, 159 (1865); Dow Chemical Company, U. S. pat. 1,921,717 [C. A. 27, 5084 (1933)].

[2] Maumené, Compt. rend. 59, 84 (1864).

[3] Alais, Froges, and Camargue, Fr. pat. 773,623 [C. A. 29, 1437 (1935)].

[4] Brand, Ger. pat. 246,661 [C. A. 6, 2496 (1912)].

[5] Doughty and Black, J. Am. Chem. Soc. 47, 1091 (1925); Doughty and Derge, ibid. 53, 1594 (1931).

and by the action of alkali cyanides on chloral hydrate.[6] The method described here is essentially that of Delépine.[7]

β-DIETHYLAMINOETHYL ALCOHOL

(Ethanol, 2-diethylamino-)

$$(C_2H_5)_2NH + ClCH_2CH_2OH + NaOH$$
$$\rightarrow (C_2H_5)_2NCH_2CH_2OH + NaCl + H_2O$$

Submitted by W. W. HARTMAN.
Checked by W. H. CAROTHERS and W. L. McEWEN.

1. Procedure

IN a 2-l. flask provided with a reflux condenser and a dropping funnel is placed 380 g. (5.2 moles) of diethylamine (b.p. 52–60°). The diethylamine is heated to boiling over a steam bath, and 320 g. (4 moles) of ethylene chlorohydrin is added from the dropping funnel during the course of about one hour. Heating is then continued for eight hours more. The reaction mixture is allowed to cool, and a solution of 230 g. of sodium hydroxide in 350 cc. of water is added fairly rapidly with constant shaking. Two layers form immediately, and sodium chloride is precipitated. The latter is dissolved by the addition of 400 cc. of water, and then 500 cc. of benzene is added and the mixture is stirred mechanically for five minutes. The benzene layer is separated and the aqueous layer is extracted three times more, 500 cc. of benzene being used for each extraction. The combined benzene extracts are dried over about 100 g. of solid potassium carbonate, with mechanical stirring until the turbidity of the solution has disappeared. The solution is distilled from a 3-l. flask provided with a 50-cm. column (packed with glass or Carborundum) and a thermometer dipping in the liquid. Distillation is continued until the temperature of the liquid reaches 100° and that at the top of the column is 85°. The residue is transferred to a 1-l. Claisen flask having a 30-cm. column, and is distilled under reduced pressure. Cuts are taken at 45°/20 mm., 45–64°/18 mm., and 64–65°/18 mm. The last fraction amounts to about 290 g. The first two fractions are redistilled and more β-diethylaminoethyl alcohol is obtained (Note 1). The total yield is 320–330 g. (68–70 per cent of the theoretical amount).

[6] Wallach, Ann. **173**, 288 (1874); Pucher, J. Am. Chem. Soc. **42**, 2251 (1920); Chattaway and Irving, J. Chem. Soc. **1929**, 1038.

[7] Delépine, Bull. soc. chim. (4) **45**, 827 (1929).

2. Note

1. The physical properties of β-diethylaminoethyl alcohol are described in detail by Headlee, Collett, and Lazzell.[1]

3. Methods of Preparation

β-Diethylaminoethyl alcohol has been prepared by reduction of diethylaminoacetic ester with sodium and alcohol;[2] by the action of ethylene chlorohydrin on diethylamine;[3] by the action of ethylene oxide on diethylamine;[1] and from ethanolamine and ethyl sulfate.[4]

DIETHYL ZINC

(Zinc, diethyl-)

$$2C_2H_5I + 2C_2H_5Br + 4Zn \rightarrow 2Zn(C_2H_5)_2 + ZnI_2 + ZnBr_2$$

Submitted by C. R. NOLLER.
Checked by HENRY GILMAN, HARRIET A. SOUTHGATE, and C. MEHLTRETTER.

1. Procedure

IN a 500-cc. round-bottomed flask, provided with a reflux condenser (Note 1) and a heavy stirrer (Note 2), is placed 130 g. (approximately 2 gram atoms) of zinc-copper couple (Note 3). To this is added a mixture of 78 g. (0.5 mole) of ethyl iodide and 54.4 g. (0.5 mole) of ethyl bromide (Note 4); the stirrer is started and the mixture heated to refluxing. After one-half hour of refluxing, the reaction starts (Note 5), as is evidenced by the greatly increased rate of refluxing, and the flame is removed. If the reaction becomes too vigorous, the flask is cooled with ice water, but only to the point where the reaction is again under control (Note 6). At the end of one-half hour from the time the flame is removed, the reaction is usually over. The flask is allowed to cool

[1] Horne and Shriner, J. Am. Chem. Soc. **54**, 2928 (1932); Headlee, Collett, and Lazzell, ibid. **55**, 1066 (1933).

[2] Gault, Compt. rend. **145**, 126 (1907); Bull. soc. chim. (4) **3**, 369 (1908).

[3] Ladenburg, Ber. **14**, 1878 (1881); Soderman and Johnson, J. Am. Chem. Soc. **47**, 1394 (1925).

[4] Carbide and Carbon Chemicals Corporation, Fr. pat. 792,046 [C. A. **30**, 4176 (1936)].

to room temperature (Note 7), connected with a distilling head, condenser, and receiver, and the contents are distilled under reduced pressure (Note 8) directly from the reaction flask into a 200-cc. round-bottomed flask immersed in an ice-salt mixture (Note 9). At the end of the distillation, dry carbon dioxide or purified nitrogen is admitted to the apparatus (Note 10). The yield of crude material, which is sufficiently pure for most purposes, is 53–55 g. (86–89 per cent of the theoretical amount).

For purification, a 30-cm. fractionating column of the Vigreux type is fitted with a condenser and receiving flask provided with a vent. The extension of this vent is covered with a test tube carrying another small tube for admitting carbon dioxide; the end of the test tube is then loosely plugged by a piece of cotton wool (Note 11). The apparatus is swept out with carbon dioxide, and the round-bottomed flask containing the once-distilled diethyl zinc is connected to the column. The diethyl zinc is then redistilled at atmospheric pressure. The yield of material boiling at 115–120° is 50–52 g. (81–84 per cent of the theoretical amount) (Notes 5 and 12).

2. Notes

1. The use of dry apparatus and the exclusion of atmospheric moisture are essential to smooth starting of the reaction. Consequently, the top of the condenser is either provided with a drying tube containing calcium chloride or attached to a trap [1] to exclude the atmosphere.

2. A slow-moving stirrer that fits the bottom and sides of the flask is necessary in order to keep the zinc thoroughly agitated.

3. The zinc-copper couple may be prepared by either of the following methods:

(a) A mixture of 120 g. of zinc dust and 10 g. of powdered copper oxide in a 200-cc. round-bottomed flask is heated gently over a free flame in a current of hydrogen with stirring or rotation of the flask until the copper oxide is reduced and a uniform gray mixture is obtained. The temperature should be kept just below the point of fusion during the heating.

(b) A zinc-copper alloy containing 5 to 8 per cent copper is prepared by melting zinc with clean brass turnings and casting into bars. This is turned into fine shavings and is ready for use.

When only small quantities of zinc alkyls are to be prepared, method (a) may be used to advantage. For large quantities it is believed that

---

[1] Gilman and Hewlett, Rec. trav. chim. **48**, 1124 (1929).

the zinc-copper alloy turnings (b) are more convenient. Zinc-copper couple prepared by method (a) must be used at once or stored in an atmosphere of dry inert gas, as it rapidly deteriorates on exposure to moist air.

The method recommended by Noyes [2c] does not seem to yield so active a couple as either of the above methods, and, although it gives entirely satisfactory results if pure iodides are used, the yields with a mixture of iodides and bromides are lower.

4. The use of a mixture of alkyl iodide and alkyl bromide is not only less expensive but the reaction is not so vigorous as when the alkyl iodide alone is used. Directions for preparing ethyl bromide are given in Org. Syn. Coll. Vol. I, **1941**, 29, 36.

5. The submitter observed that reaction generally started in twenty to forty minutes after refluxing began. The checkers found that the time for reaction to set in was nearer one and one-quarter hours. Unless all precautions to exclude moisture are exercised, the reaction may not start for several hours. In such cases, after the preliminary period of stirring and heating externally, the mixture can be allowed to stand at room temperature without stirring. Under these conditions the reaction always started spontaneously (sometimes after five hours) and required no attention, for when reaction did set in the refluxing was not sufficiently vigorous to require cooling. When the reaction is slow in starting, the yield of diethyl zinc is somewhat lower.

6. If the mixture is cooled too much, the reaction may stop entirely and is difficult to start again. If this happens the yield will be considerably lower.

7. On cooling, the reaction mass solidifies (probably on account of the re-formation of RZnX) and may be briefly exposed to air without danger. The solid material is much less reactive than diethyl zinc.

8. The pressure should be below 30 mm. of mercury or decomposition will occur.

9. Either a special adapter with a side arm for applying suction may be used or a piece of right-angled tubing may be inserted in the same stopper with the condenser tube.

10. When the distillation is carried out under reduced pressure there is no need to sweep out the apparatus with inert gas. At other times, exposure to air must be prevented by using an atmosphere of carbon dioxide or nitrogen.

[2] (a) Gladstone and Tribe, J. Chem. Soc. **26**, 445, 678, 961 (1873).

(b) Lachman, Am. Chem. J. **19**, 410 (1897); **24**, 31 (1900).

(c) Noyes, "Organic Chemistry for the Laboratory," 3rd Ed., p. 61, 1916.

(d) Renshaw and Greenlaw, J. Am. Chem. Soc. **42**, 1472 (1920).

The checkers carried out this distillation by heating the flask in an oil bath (about 200°) at a pressure of about 8 mm. The distillation was complete in about one hour. The flask can also be heated directly with a full Bunsen flame.

11. This vent can be replaced by a mercury seal formed by connecting a double right-angled glass tube from the upper part of the receiving flask to a container with mercury. The tube is only slightly immersed in the mercury. When a mercury seal is used, the receiver must be removed when distillation is completed in order that mercury may not be pulled into the receiver when the system commences to cool.

12. Using the same procedure, the yields and boiling points of higher zinc alkyls are as follows: di-n-propyl zinc, 85–86 per cent, b.p. 39–40°/9 mm.; di-n-butyl zinc, 78–79 per cent, b.p. 81–82°/9 mm.; di-isoamyl zinc, 50–55 per cent, b.p. 100–103°/12 mm. The higher zinc alkyls should always be distilled under reduced pressure.

3. Methods of Preparation

Diethyl zinc has been prepared from zinc and mercury diethyl,[3] and from ethyl bromide and zinc-copper couple using a special catalyst.[4] Diethyl zinc is usually prepared by the action of ethyl iodide on specially treated zinc,[5] the zinc-copper couple being most useful for this purpose.[2] The procedure described above has been published.[6] Arylzinc halides, zinc dialkyls, and zinc diaryls may be prepared by the action of the Grignard reagent on anhydrous zinc chloride in ether solution.[7]

Attention is called to an improved apparatus for the preparation, purification, and use of zinc ethyl.[8]

[3] Frankland and Duppa, J. Chem. Soc. 17, 31 (1864).
[4] Job and Reich, Bull. soc. chim. (4) 33, 1424 (1923).
[5] Frankland, Ann. 85, 360 (1853).
[6] Noller, J. Am. Chem. Soc. 51, 594 (1929).
[7] Houben-Weyl, "Methoden der organischen Chemie," 2nd Ed., 1924, Vol. 4, pp. 754, 901; Blaise, Bull. soc. chim. (4) 9, I–XXVI (1911); Gilman and Brown, J. Am. Chem. Soc. 52, 4482 (Note 7) (1930).
[8] McCleary and Degering, Proc. Indiana Acad. Sci. 43, 127 (1934) [C. A. 28, 7245 (1934)].

4,4'-DIFLUOROBIPHENYL

(Biphenyl, 4,4'-difluoro-)

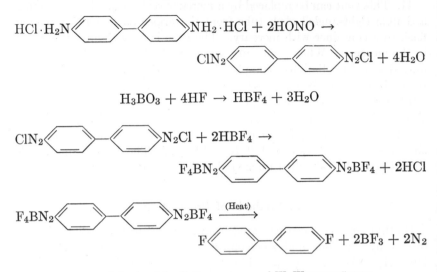

$$H_3BO_3 + 4HF \rightarrow HBF_4 + 3H_2O$$

Submitted by G. Schiemann and W. Winkelmüller.
Checked by W. W. Hartman, J. R. Byers, and J. B. Dickey.

1. Procedure

A MIXTURE of 280 g. (1.5 moles) of commercial benzidine and 880 cc. (10.2 moles) of concentrated hydrochloric acid (sp. gr. 1.18) is placed in a 5-l. round-bottomed flask and warmed on a steam bath for one to two hours, with occasional shaking, to form the dihydrochloride. The flask is then equipped with a mechanical stirrer and a dropping funnel, and cooled, with stirring, to −10° in an ice-salt bath. When this temperature has been reached, the benzidine dihydrochloride is tetrazotized over a period of two hours with a solution of 232 g. (3.2 moles) of 95 per cent sodium nitrite in 800 cc. of water, until a faint test for nitrous acid with starch-iodide paper is obtained after twenty minutes. During this reaction, the temperature is kept below −5°.

Concurrently with the tetrazotization, 104 g. (1.68 moles) of boric acid is dissolved in 222 g. (6.66 moles) of 60 per cent hydrofluoric acid. The solution is made in a 1-l. beaker, which has been coated inside with paraffin, and is cooled in an ice bath. The boric acid is added slowly in small portions, and the mixture is stirred with a lead rod. It is necessary to keep the temperature below 20–25° in order that the heat of solution may not melt the paraffin from the walls of the beaker (Note 1). The ice-cold fluoboric acid solution is added rather rapidly, with stirring, to the finished tetrazo solution, the temperature being kept below 10°. A thick paste of 4,4'-biphenylene-bis-diazonium borofluoride forms. The mixture is stirred at 10° for twenty to thirty minutes. It is then collected on a 19-cm. Büchner funnel and washed consecutively with about 200 cc. of cold water, 200 cc. of cold commercial methyl alcohol, and 200 cc. of commercial ether; the cake is sucked as dry as possible between washings. It is then dried in a vacuum desiccator over concentrated sulfuric acid (sp. gr. 1.84). The yield of the dry solid is 393–400 g. (68–69 per cent of the theoretical amount). The product decomposes at 135–137°.

A 1-l. distilling flask with a wide side arm may be used for the decomposition of the tetrazonium borofluoride. A 500-cc. distilling flask is fastened directly to the side arm of the decomposition flask and cooled with running water. To the side arm of the receiver is connected a rubber tube which is placed over 2 l. of water in a 5-l. flask (Note 2). The solid to be decomposed (Note 3) is placed in the decomposition flask and heated at the upper edge with a Bunsen burner. When white fumes begin to be evolved, the burner is removed and the decomposition permitted to continue spontaneously. More heat is applied as needed. Finally vigorous heating is employed to ensure complete decomposition. Some 4,4'-difluorobiphenyl is collected in the receiver, but the larger portion remains in the decomposition flask, from which it is recovered by steam-distilling the black residue. A second steam distillation gives a pure white compound melting at 88–89°, after drying in an oven at 60–70°. When 153 g. of the tetrazonium borofluoride is decomposed in this manner, 61–62 g. of 4,4'-difluorobiphenyl is obtained (80–81.5 per cent of the theoretical amount, based on the tetrazonium borofluoride; 54–56 per cent based on the benzidine used).

2. Notes

1. A small lead jar is excellent for preparing the solution. By means of a lead stirrer of the usual shape, mechanical stirring may be used. The stirrer should be thrust through a hole in a lead cover of sufficient size to prevent splashing of the hydrofluoric acid.

Hydrofluoric acid produces extremely painful burns. Exposed parts of the body must be protected when working with this material. Compare, Note 3, p. 297. Instead of preparing fluoboric acid, 355.5 g. of commercial 40 per cent fluoboric acid may be used.

2. A convenient apparatus is made by connecting a 1-l. round-bottomed flask by means of a bent tube of large diameter (2 cm.) to a second 1-l. round-bottomed flask containing 500 cc. of water. The second flask is equipped with an exit tube, and the gases which do not dissolve in the water are led into a hood.

3. It is very necessary that the tetrazonium borofluoride be dried completely. If the solid is wet, the decomposition proceeds very vigorously. There is formed at the same time a product of higher melting point (160°) as well as some tar. These products, although not volatile with steam, lower the yield of the 4,4'-difluorobiphenyl very materially.

3. Methods of Preparation

4,4'-Difluorobiphenyl has been prepared from 4,4'-biphenyl-bis-diazonium piperidide (by diazotizing benzidine and coupling with piperidine) and concentrated hydrofluoric acid;[1] by the action of sodium on p-fluorobromobenzene in ether;[2] from benzidine by tetrazotization and decomposing the biphenyl-bis-diazonium salt with concentrated hydrofluoric acid;[3] by the above method in the presence of ferric chloride;[4] and by the prolonged contact of the vapors of fluorobenzene with a red-hot wire.[5] The method described here is the most satisfactory for use in the laboratory and is an improvement on the method of Balz and Schiemann.[6]

[1] Wallach, Ann. **235,** 271 (1886).
[2] Wallach and Heusler, ibid. **243,** 244 (1888).
[3] Valentiner and Schwarz, Ger. pat. 96,153 [Frdl. **5,** 910 (1897–1900)].
[4] Valentiner and Schwarz, Ger. pat. 186,005 [Frdl. **8,** 1237 (1905–07)].
[5] Meyer and Hofmann, Monatsh. **38,** 149 (1917).
[6] Balz and Schiemann, Ber. **60,** 1189 (1927).

DIHYDROCHOLESTEROL

(β-Cholestanol)

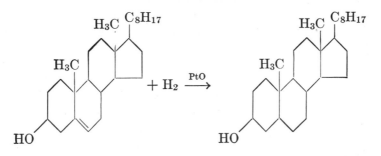

Submitted by (A) W. F. BRUCE.
(B) J. O. RALLS.
Checked by LOUIS F. FIESER, R. P. JACOBSEN, and M. S. NEWMAN.

1. Procedure

(A) *From Cholesterol.*—One hundred grams (0.26 mole) of commercial cholesterol is crystallized from 250 cc. of glacial acetic acid, using 1 g. of Norite if required, and the purified material (Note 1) is transferred, conveniently without being dried, to a hydrogenation vessel equipped with a thermometer and a heating device (Note 2). Three hundred cubic centimeters of purified glacial acetic acid (Note 3) and 0.5 g. of platinum oxide are added, and the hydrogenation is conducted at 65–75° at a slight positive pressure. The total amount of hydrogen usually is absorbed in two to four hours (Note 4). After the hydrogen has been replaced by air the solution is filtered hot and the product is obtained by crystallization and concentration. The total yield of air-dried, partially acetylated dihydrocholesterol, m.p. 130–135°, is 85–90 g.

Unless a specially purified product (see below) is desired, the crude material is heated for three hours on the steam bath with 400 cc. of alcohol and a solution of 25 g. of sodium hydroxide in 100 cc. of water. After cooling, the product is collected, washed, and crystallized from 500 cc. of alcohol. The yield is 75–80 g. (75–80 per cent of the theoretical amount), and a well-dried sample (Note 5) melts at 140–141°.

(B) *From Cholesteryl Acetate* (Note 6).—Five grams of cholesteryl acetate (Note 7) and 0.1 g. of platinum oxide are suspended in 25 cc. of absolute ether and 50 cc. of purified glacial acetic acid (Note 3), and the hydrogenation is conducted at room temperature at a slight positive pressure. The reaction is complete in ten to fifty minutes. The solution is filtered, using ether to dissolve any crystallized material, and, after removing the solvent by distillation at reduced pressure, the residue is either saponified as above or purified in the following manner.

Purification by the Method of Anderson and Nabenhauer.[1]—A solution of 20 g. of crude, partially or completely acetylated dihydrocholesterol in 200 cc. of' carbon tetrachloride is placed in a separatory funnel and treated with 100 cc. of acetic anhydride. About 5 cc. of concentrated sulfuric acid is added dropwise through the stem of the inverted funnel with cooling and shaking until there is no further increase in color. A blue or green color develops, the intensity depending on the amount of cholesterol present in the sample. After fifteen to twenty minutes about 10 cc. of water is added, by drops and with cooling and gentle shaking until two distinct layers form. The carbon tetrachloride solution (upper layer) is separated and washed free of acid with sodium chloride or sodium carbonate solution (pure water gives emulsions). After drying with sodium sulfate, the solvent is removed by distillation at diminished pressure and the residue is saponified as above with alcoholic alkali and crystallized from alcohol. The purified dihydrocholesterol weighs 12–14 g. and melts, after thorough drying (Note 5), at 142–143°. It gives a faint Liebermann-Burchard reaction (Note 8) only after ten to fifteen minutes.

2. Notes

1. The dry weight of the crystallized material is 90–95 g. Some samples may require recrystallization.

2. A suitable arrangement for heating the hydrogenation vessel is described in Org. Syn. Coll. Vol. I, **1941**, 61. An alternative arrangement is the following: a round-bottomed long-necked flask is supported at the top by a two-piece clamp with a loosened checknut, connected below to an eccentric, and heated in motion by means of a stationary microburner.

3. Glacial acetic acid is purified by boiling for one hour with 5 g. of potassium permanganate per liter and distilling.

4. The catalyst sometimes loses its activity when about half of the theoretical amount of hydrogen has been absorbed, probably because of the poisoning action of impurities not removed from the commercial cholesterol. If this happens the addition of one or two 0.2-g. portions of catalyst usually suffices to bring the reaction practically to completion.

[1] Anderson and Nabenhauer, J. Am. Chem. Soc. **46**, 1957 (1924).

5. The sterol forms a hydrate from which the water is eliminated only after thorough drying, as in vacuum at 100°.

6. The acetyl derivative is more easily reduced than the free sterol.

7. Cholesteryl acetate is prepared by boiling for one hour a solution of 5 g. of cholesterol in 7.5 cc. of acetic anhydride, cooling, filtering, and washing the crystalline product with cold methanol. The yield of material melting at 114–115° is 5 g.

8. This test for cholesterol is made by dissolving about 5 mg. of the material in 2 cc. of carbon tetrachloride, and adding 1 cc. of acetic anhydride and 3–4 drops of concentrated sulfuric acid. Cholesterol gives rise to a succession of color changes.

3. Methods of Preparation

Dihydrocholesterol has been prepared by the reduction of cholestenone with sodium and amyl alcohol [2] and by the hydrogenation of cholesterol. In the presence of platinum black or platinum oxide, yields varying from 6.5 per cent to 40 per cent have been obtained in ether,[3] acetone,[4] ethyl acetate,[5] and acetic acid.[6]

[2] Diels and Abderhalden, Ber. **39**, 889 (1906); Diels and Stamm, ibid. **45,** 2230 (1912); Neuberg, ibid. **39**, 1155 (1906).

[3] Willstätter and Mayer, ibid. **41**, 2200 (1908); Dorée, J. Chem. Soc. **95**, 644 (1909); Boehm, Biochem. Z. **33**, 474 (1911); Windaus and Uibrig, Ber. **47**, 2386 (1914); Ellis and Gardner, Biochem. J. **12**, 72 (1918); Anderson, J. Biol. Chem. **71**, 411 (1926); Vavon and Jakubowicz, Bull. soc. chim. (4) **53**, 584 (1933); Ruzicka, Brüngger, Eichenberger, and Meyer, Helv. Chim. Acta **17**, 1407 (1934).

[4] Nord, Biochem. Z. **99**, 265 (1919).

[5] Shriner and Ko, J. Biol. Chem. **80**, 6 (1928).

[6] v. Fürth and Felsenreich, Biochem. Z. **69**, 420 (1915).

3,4-DIHYDRO-1,2-NAPHTHALIC ANHYDRIDE

(1,2-Naphthalenedicarboxylic anhydride, 3,4-dihydro-)

(A) $C_6H_5CH_2CH_2CH_2CO_2C_2H_5 + (CO_2C_2H_5)_2 \xrightarrow{\text{(NaOC}_2\text{H}_5)}$

$C_6H_5CH_2CH_2CH(CO_2C_2H_5)COCO_2C_2H_5 + C_2H_5OH$

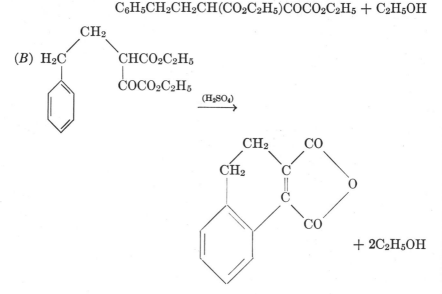

(B)

$+ 2C_2H_5OH$

Submitted by E. B. HERSHBERG and LOUIS F. FIESER.
Checked by C. R. NOLLER and S. KINSMAN.

1. Procedure

(A) *Ester Condensation.*—In a 1-l. round-bottomed flask, fitted with a reflux condenser equipped with a dropping funnel and a calcium chloride tube, is placed a suspension of 6.1 g. (0.27 mole) of powdered sodium (Note 1) in 150 cc. of absolute ether. A solution of 12.6 g. (0.27 mole) of absolute ethyl alcohol (Note 2) and 50 cc. of absolute ether is added, and the mixture is allowed to stand overnight to complete the reaction. To the suspension of sodium ethoxide, 57 g. (0.39 mole) of ethyl oxalate (Org. Syn. Coll. Vol. I, **1941,** 261), diluted with 50 cc. of ether, is added in portions. After the spontaneous reaction subsides, the pale yellow solution is allowed to stand for one-half hour, and 50 g. (0.26 mole) of ethyl γ-phenylbutyrate (Note 3), diluted with 50 cc. of absolute ether, is added. The mixture is refluxed gently for twenty-four hours (Note 4).

The deep red solution is cooled in an ice bath and neutralized by the addition, with shaking, of an ice-cold solution of 15 cc. of concentrated sulfuric acid in 200 cc. of water. The ether layer is separated, washed with water, and dried over sodium sulfate. The ether is removed by dropping the solution from a separatory funnel the stem of which extends to the bottom of an evacuated Claisen flask heated on the steam bath. The residue is a pale yellow oil consisting of a mixture of ethyl α-ethoxalyl-γ-phenylbutyrate and unchanged ethyl oxalate (Note 5).

(B) *Cyclization.*—The above oil is poured slowly into 500 cc. of concentrated sulfuric acid, the temperature being kept at 20–25° by cooling in an ice bath. After standing for one and one-half hours at 20–25°, the deep red solution is poured on 3 l. of ice and water. The anhydride, precipitated as a pale yellow solid, is collected and washed thoroughly with water. Dried *in vacuo* at 25°, the material weighs 40–45 g. and melts at 117–122°. Distillation under diminished pressure gives a light yellow product, m.p. 122–124°. The yield is 38–42 g. (73–81 per cent of the theoretical amount). This material is suitable for most purposes. Crystallization from 100 cc. of benzene with the addition of 75 cc. of ligroin (b.p. 60–80°) gives 34–41 g. of pale yellow prisms, m.p. 125–126° (Note 6).

2. Notes

1. The powdered sodium for this preparation may be prepared as in Org. Syn. Coll. Vol. I, **1941**, 252, or according to the following procedure for potassium, using xylene instead of toluene. (With certain other γ-arylbutyric esters it is better to use potassium.) Commercial potassium is cleaned by melting it under toluene, and 10.4 g. (0.27 mole) of the metal and 150 cc. of dry toluene are placed in a 1-l. flask. After the liquid is heated to boiling on a hot plate, the flask is removed and closed with a ground-glass stopper carrying a sealed-on stopcock. Apparatus with interchangeable ground joints is essential. After one shake with the stopcock open to relieve superheating, the stopcock is closed, and the flask is shaken quickly and vigorously to powder the metal. The mixture is allowed to cool undisturbed, and nitrogen is admitted. The stopper is replaced by a distilling head carrying a 500-cc. flask, into which the toluene can be decanted. The powdered metal is washed several times with absolute ether and finally covered with ether (150 cc.) and converted into the ethoxide with 12.6 g. of alcohol diluted with 150 cc. of ether. Traces of potassium in the wash liquors are destroyed safely by treatment under reflux with alcohol diluted with ether.

2. The alcohol was dried according to Org. Syn. Coll. Vol. I, **1941**, 251, Note 1.

3. Ethyl γ-phenylbutyrate is prepared in 85–88 per cent yields by refluxing for three hours a mixture of 50 g. of γ-phenylbutyric acid (p. 499), 150 cc. of alcohol dried over lime, and 5 g. of concentrated sulfuric acid. The ester is isolated by distilling 100 cc. of the alcohol under reduced pressure from a steam bath, diluting the residue with 200 cc. of water, separating, and extracting the aqueous layer twice with 50-cc. portions of ether. The combined ester and ether layers are dried with sodium sulfate, the ether removed, and the residue distilled under diminished pressure; the portion boiling at 144–147°/19 mm. is collected.

4. With potassium ethoxide the reaction is complete in twelve hours.

5. The keto ester decomposes on distillation, even under diminished pressure.

6. For the cyclization of the keto esters from γ-naphthylbutyric esters it is advisable to use 80 per cent sulfuric acid, and to heat the mixture, with stirring, at 70–80° for one-half hour.

3. Method of Preparation

The above procedure is a modification [1] of the method of von Auwers and Möller.[2]

2,6-DIIODO-*p*-NITROANILINE

(Aniline, 2,6-diiodo-4-nitro-)

$$p\text{-}NO_2C_6H_4NH_2 + 2ICl \rightarrow \underset{NO_2}{\overset{NH_2}{\underset{I}{\bigcirc}}I} + 2HCl$$

Submitted by R. B. Sandin, W. V. Drake, and Frank Leger.
Checked by Frank C. Whitmore and Marion M. Whitmore.

1. Procedure

One hundred thirty-eight grams (1 mole) of *p*-nitroaniline (Eastman Technical grade) is dissolved in 370 cc. of boiling glacial acetic acid in a 2-l. three-necked flask provided with a mechanical stirrer, a reflux condenser, and a dropping funnel. The burner is removed, and a mixture of 325 g. (2 moles) of iodine monochloride (Note 1) in 100 cc. of glacial acetic acid is added slowly from the dropping funnel with rapid

[1] Fieser and Hershberg, J. Am. Chem. Soc. **57**, 1851 (1935).
[2] von Auwers and Möller, J. prakt. Chem. (2) **109**, 137 (1925).

stirring during thirty minutes. Considerable heat is evolved. The mixture is heated on a rapidly boiling water bath for two hours and then transferred to a 1-l. beaker and allowed to cool. The solidified mixture is treated with 100 cc. of glacial acetic acid and any hard lumps are crushed thoroughly with a flat glass stopper (Note 2). The mixture is transferred to a large Büchner funnel and filtered by suction, two 25-cc. portions of glacial acetic acid being used to wash the last of the crystals into the funnel. The dark mother liquor is removed as completely as possible by suction. The crystals are returned to the beaker, thoroughly stirred with 200 cc. of cold glacial acetic acid, and transferred to the suction filter again. The beaker is rinsed with two 25-cc. portions of glacial acetic acid, and the crystals are sucked as dry as possible. The suction is shut off, and the crystals are wetted with 50 cc. of ether. The suction is again used to remove the ether. The product is air dried to constant weight (about twenty-four hours). The yield of air-dried diiodo-*p*-nitroaniline melting at 243–245° is 220–250 g. (56–64 per cent of the theoretical amount) (Note 3).

2. Notes

1. The iodine monochloride is made by passing dry chlorine gas into 254 g. (1 mole) of iodine in a tared 500-cc. Erlenmeyer flask until the gain in weight is 71 g. Frequent shaking is necessary. The iodine monochloride is used directly without distillation. Compare p. 344.

2. It is advisable to wear rubber gloves during these operations with glacial acetic acid.

3. It is reported that the following modifications increase the yield to 86 per cent: The reaction mixture is refluxed for two hours in an oil bath, cooled, filtered, and sucked as dry of solvent as possible. The cake is then made into a paste with 600 cc. of hot water, a little sodium bisulfite is added in order to remove excess iodine, and the product is filtered and dried. 2,6-Diiodo-4-nitroaniline can be crystallized from nitrobenzene followed by washing with alcohol; it then melts at 249–250°. [CARL NIEMANN and C. E. REDEMANN, private communication, and J. Am. Chem. Soc. 63, 1550 (1941)].

3. Method of Preparation

2,6-Diiodo-*p*-nitroaniline has been made by the action of iodine chloride on a chloroform solution [1] and also on a glacial acetic acid solution [2] of *p*-nitroaniline.

[1] Michael and Norton, Ber. 11, 113 (1878).
[2] Willgerodt and Arnold, ibid. 34, 3344 (1901).

2,4-DIMETHYL-5-CARBETHOXYPYRROLE

(2-Pyrrolecarboxylic acid, 3, 5-dimethyl-, ethyl ester)

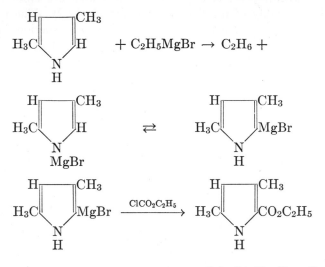

Submitted by HANS FISCHER.
Checked by C. R. NOLLER.

1. Procedure

IN a 1-l. round-bottomed, three-necked flask fitted with an efficient reflux condenser, liquid-sealed stirrer, and dropping funnel is placed 13 g. (0.53 gram atom) of magnesium turnings. A few cubic centimeters of a solution of 60 g. (42 cc., 0.55 mole) of pure ethyl bromide in 50 cc. of absolute ether is added and the stirrer started (Note 1). When the bromide begins to react 200 cc. of absolute ether is added, and then the balance of the bromide solution is run in as fast as the refluxing permits (about one-half hour). After allowing fifteen minutes for the completion of the reaction, a solution of 40 g. (0.42 mole) of 2,4-dimethylpyrrole (p. 217) in 100 cc. of absolute ether is added in the course of twenty minutes (Note 2) and the mixture is refluxed for one-half hour on the steam bath.

The reaction mixture is cooled to room temperature, and a solution of 58 g. (51 cc., 0.53 mole) of freshly distilled ethyl chloroformate (b.p. 92.5–93.5°) in 100 cc. of absolute ether is added dropwise in the course of one-half hour (Note 3). The mixture is heated on the steam bath for one and one-half hours and then allowed to stand overnight at room temperature.

The flask is placed in an ice-salt mixture, and the contents are decomposed by the gradual addition of 300 cc. of saturated ammonium chloride solution and 100 cc. of water (Note 4). The aqueous layer is removed by means of a 1.5-l. separatory funnel, and sufficient ether is added to dissolve the yellow precipitate. The total volume of ether solution is about 1 l. This is washed with two 200-cc. portions of water, and the three aqueous layers are extracted consecutively with a 100-cc. portion of ether. The combined ether solution is dried over 30 g. of anhydrous sodium sulfate, concentrated on the steam bath to a volume of about 200 cc., and cooled to room temperature. The product which crystallizes is collected with suction and washed with two 25-cc. portions of ether. The yield is 35–38 g. of light yellow product, m.p. 122–123°. The ether is completely removed from the combined filtrates by heating on the steam bath, and the black oil is allowed to stand overnight. The semi-solid mass is filtered with suction and washed with a minimum amount of cold ether. In this way an additional 6–7 g. of yellow material is obtained which melts at 119–121°.

The combined crude material is crystallized from 75 cc. of 95 per cent alcohol and yields 37–39 g. of slightly colored material, m.p. 123–124°. A second crystallization from alcohol gives 34–36 g. of colorless product melting at the same temperature. By the systematic working of the alcoholic mother liquors, an additional 5–6 g. of pure material is obtained, making the total yield 40–41 g. (57–58 per cent of the theoretical amount).

2. Notes

1. Stirring is continued without interruption throughout the preparation to the point where the mixture is allowed to stand overnight.

2. The reaction is not exothermic, but the large volume of ethane evolved necessitates the regulated addition of the solution of dimethylpyrrole.

3. Considerable heat is produced during the addition of about two-thirds of the solution, after which the addition may be more rapid.

4. The first third of the ammonium chloride solution must be added quite slowly with frequent and thorough shaking.

3. Methods of Preparation

The most convenient laboratory method for the preparation of 2,4-dimethyl-5-carbethoxypyrrole is that given above.[1] A cheaper method

[1] Fischer, Weiss, and Schubert, Ber. **56**, 1199 (1923); Ingraffia, Gazz. chim. ital. **63**, 584 (1933).

of obtaining large quantities of the material consists in the partial hydrolysis of 2,4-dimethyl-3,5-dicarbethoxypyrrole with sulfuric acid, followed by decarboxylation.[2]   The ester has been obtained also by the alcoholysis of 5-trichloroaceto-2,4-dimethylpyrrole in the presence of sodium ethoxide.[3]   The free acid has been obtained from 1-[2,4-dimethylpyrrole-5]-2,4-dimethylpyrrole-5-carboxylic acid [4] and from 2,4-dimethylpyrrole-5-aldehyde.[5]

5,5-DIMETHYL-1,3-CYCLOHEXANEDIONE

(1,3-Cyclohexanedione, 5,5-dimethyl-)

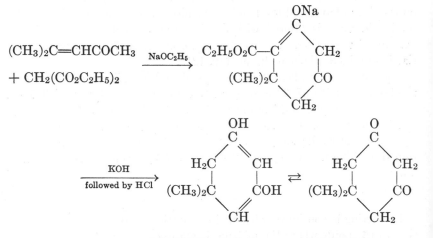

Submitted by R. L. Shriner and H. R. Todd.
Checked by W. H. Carothers and W. L. McEwen.

1. Procedure

In a dry 2-l. three-necked, round-bottomed flask, fitted with a liquid-sealed stirrer, a 500-cc. dropping funnel, and an efficient reflux condenser protected at the top with a calcium chloride tube, is placed 400 cc. of absolute alcohol.   Through the condenser tube is added 23 g. (1 gram atom) of clean sodium at such a rate that the solution is kept at the

[2] Fischer and Walach, Ber. **58**, 2820 (1925).

[3] Houben and Fischer, ibid. **64**, 2639 (1931).

[4] Magnanini, ibid. **22**, 38 (1889).

[5] Alessandri, Atti accad. Lincei **24** (II) 199 (1915) [C. A. **10**, 1350 (1916)]; Alessandri and Passerini, Gazz. chim. ital. **51** (I), 277 (1921).

boiling temperature. After the sodium has dissolved completely, 170 g. (1.06 moles) of ethyl malonate is added, and then 100 g. (1.02 moles) of mesityl oxide (Note 1) is added slowly through the dropping funnel. The solution is refluxed with constant stirring for two hours, after which a solution of 125 g. (2.2 moles) of potassium hydroxide in 575 cc. of water is added and the mixture is stirred and refluxed again on the water bath for six hours.

The mixture while still hot is made just acid to litmus with dilute hydrochloric acid (1 volume concentrated acid to 2 volumes water; sp. gr. 1.055); about 550 cc. is required. The flask is fitted with a condenser set for distillation, and as much alcohol as possible (about 550 cc.) is distilled by heating on a water bath.

The residue in the flask is boiled with about 15 g. of Norite (Note 2), filtered, and the treatment with the decolorizing charcoal repeated. The residue is again neutralized to litmus with dilute hydrochloric acid (about 150 cc.) and again boiled with charcoal. The hot, neutral or alkaline, yellow filtrate is finally made distinctly acid to methyl orange with additional dilute hydrochloric acid (50 to 100 cc.), boiled for a few minutes, and allowed to cool, whereupon the methone crystallizes. The product is filtered by suction from the acid liquid, washed with ice-cold water, and dried in the air. The yield is 96–122 g. (67–85 per cent of the theoretical amount) (Note 3).

2. Notes

1. The yield of methone depends on the purity of the mesityl oxide (Org. Syn. Coll. Vol. I, **1941**, 345), which should be freshly distilled, the fraction boiling at 126–131° being collected.

2. Care is necessary when adding the decolorizing charcoal or the hot acid solution may foam vigorously owing to the liberation of carbon dioxide. A large container should be used, and the charcoal should be added very slowly.

3. The submitters in carrying out this preparation invariably obtained yields ranging from 120 to 128 g. of a product melting between 145° and 147°. Crystallization from about 1 l. of acetone gave 100 g. (70 per cent) of pure white material melting at 147°. The checkers (using mesityl oxide purchased from the Eastman Kodak Co. and freshly distilled) obtained yields ranging from 96 to 112 g., but their product melted at 147–148° and recrystallization failed to raise the melting point. The melting point of methone is given in the literature as 148–150°.

3. Method of Preparation

5,5-Dimethyldihydroresorcinol, methone, dimedone, has always been made from mesityl oxide and malonic ester. The procedure given above is that of Vorländer.[1]

2,4-DIMETHYL-3,5-DICARBETHOXYPYRROLE

(2,4-Pyrroledicarboxylic acid, 3,5-dimethyl-, diethyl ester)

$$CH_3COCH_2CO_2C_2H_5 + HNO_2 \rightarrow CH_3COCH(NO)CO_2C_2H_5 + H_2O$$

$$CH_3COCH(NO)CO_2C_2H_5 + 4[H] \xrightarrow{(Zn + CH_3CO_2H)} CH_3COCH(NH_2)CO_2C_2H_5 + H_2O$$

$$\begin{matrix} CH_3COH \\ \| \\ C_2H_5O_2CCNH_2 \end{matrix} + \begin{matrix} HCCO_2C_2H_5 \\ \| \\ HOCCH_3 \end{matrix} \rightarrow \begin{matrix} CH_3 \quad CO_2C_2H_5 \\ \\ C_2H_5O_2C \quad CH_3 \\ N \\ H \end{matrix} + 2H_2O$$

Submitted by Hans Fischer.
Checked by C. R. Noller.

1. Procedure

In a 3-l. three-necked, round-bottomed flask, fitted with a liquid-sealed mechanical stirrer and dropping funnel, are placed 390 g. (3 moles) of ethyl acetoacetate (Org. Syn. Coll. Vol. I, **1941**, 235) and 900 cc. of glacial acetic acid. The solution is cooled in an efficient freezing mixture to 5°, and a cold solution of 107 g. (1.47 moles) of 95 per cent sodium nitrite in 150 cc. of water is added dropwise with vigorous stirring at such a rate that the temperature remains between 5° and 7°. With efficient cooling about one-half hour is required to add the nitrite. The mixture is stirred for one-half hour longer and then allowed to stand for four hours, during which time it warms up to room temperature.

The separatory funnel is replaced by a wide-bore condenser, and the third neck of the flask is fitted with a stopper. The solution is stirred

[1] Vorländer and Erig, Ann. **294**, 314 (1897); Vorländer, Z. anal. Chem. **77**, 245 (1929); Z. angew. Chem. **42**, 46 (1929); Chavanne, Miller, and Cornet, Bull. soc. chim. Belg. **40**, 673 (1931).

and portions of 196 g. (3 gram atoms) of zinc dust (Note 1) are added quickly through the third neck of the flask until the liquid boils and then frequently enough to keep it boiling (Note 2). After the addition has been completed, the mixture is heated by a burner and refluxed for one hour (Notes 3 and 4). While still hot the contents of the flask are decanted from the remaining zinc into a crock containing 10 l. of water which is being vigorously stirred. The zinc residue is washed with two 50-cc. portions of hot glacial acetic acid which are also decanted into the water. After standing overnight, the crude product is filtered by suction, washed on the filter with two 500-cc. portions of water, and dried in air to constant weight. The yield is 205–230 g. (57–64 per cent of the theoretical amount) of material melting at 126–130°. On recrystallizing a 50-g. portion from 100 cc. of 95 per cent alcohol and washing twice with 20-cc. portions of cold alcohol, there is obtained 38.5 g. of pale yellow crystals melting at 136–137° (Note 5).

2. Notes

1. The zinc dust should be at least 80 per cent pure, and an amount equivalent to 196 g. of 100 per cent material should be used. An excess of zinc dust up to 3.5 gram atoms has been used without changing the yield.

2. Four portions of approximately 15 g. each are required to bring the solution to the boiling point, after which the remainder is added in about 5-g. portions over a period of three-quarters of an hour. Great care must be taken not to add too much zinc dust at first as the mixture foams badly. It is well to have a bath of ice and water and wet towels handy in order to control the reaction if it should become too violent.

3. For reasons unknown some runs behave entirely differently from others. Occasionally the foaming suddenly stops after about half of the zinc has been added and the remainder can be added much more rapidly, causing the solution to boil vigorously. During the period of external heating of the solution the mixture becomes thick and gummy and as much as 300 cc. of acetic acid must be added before stirring can be continued. The lower yields reported are obtained when this occurs.

4. It is reported that the yield is increased to 75 per cent by adding sodium acetate to the zinc dust reduction mixture to form a complex with the zinc acetate and so increase its solubility, and by adding the solution of the nitroso compound to the mixture of zinc dust, sodium acetate, ethyl acetoacetate, and glacial acetic acid, instead of adding the zinc dust last.[1]

[1] Corwin and Quattlebaum, Jr., J. Am. Chem. Soc. **58**, 1083 (1936).

5. The crude product turns pink when exposed to light whereas the recrystallized product is quite stable. A somewhat lighter product is obtained if 2 g. of decolorizing carbon is used during the crystallization, but this has no effect on the melting point.

3. Method of Preparation

The above method, which has always been used for the preparation of 2,4-dimethyl-3,5-dicarbethoxypyrrole, is essentially that of Knorr.[2]

DIMETHYLGLYOXIME

(Glyoxime, dimethyl-)

$$C_2H_5OH + HNO_2 \xrightarrow{(NaNO_2 + H_2SO_4)} C_2H_5ONO + H_2O$$

$$C_2H_5ONO + CH_3COCH_2CH_3 \rightarrow CH_3COC(NOH)CH_3 + C_2H_5OH$$

$$\begin{array}{c} CH_3CO \\ | \\ CH_3C{=}NOH \end{array} + NaO_3SNHOH \rightarrow \begin{array}{c} CH_3C{=}NOH \\ | \\ CH_3C{=}NOH \end{array} + NaHSO_4$$

Submitted by W. L. SEMON and V. R. DAMERELL.
Checked by HENRY GILMAN and R. E. FOTHERGILL.

1. Procedure

(A) Ethyl Nitrite.—Two solutions are prepared. Solution I contains 620 g. (9 moles) of sodium nitrite (650 g. of technical 95 per cent quality), 210 g. (4.6 moles) of alcohol (285 cc. of 90 per cent denatured alcohol, or its equivalent), and water to make a total volume of 2.5 l. Solution II contains 440 g. of sulfuric acid (255 cc. of sp. gr. 1.84) and 210 g. of alcohol, diluted with water to 2.5 l. Ethyl nitrite may be generated continuously in gaseous form by allowing solution II to flow into solution I.

The gas can be conveniently made by putting solution II in a 2.5-l. bottle fitted with a two-holed rubber stopper (Fig. 7) and provided with glass tubing as indicated in the diagram. The rate of flow of solution II into a 6-l. bottle containing solution I is regulated by a screw clamp. The ethyl nitrite generated flows out of the second opening in the stopper of the lower bottle. The stoppers to both bottles should be wired on.

[2] Knorr, Ann. **236**, 318 (1886).

A mechanical stirrer in the lower bottle is helpful in securing a steady evolution of the gas (Note 1).

(B) *Biacetyl Monoxime.*—In a 2-l. three-necked flask provided with a condenser, a thermometer, and an inlet tube for ethyl nitrite, and arranged for external cooling, is placed 620 g. of commercial methyl ethyl ketone which has been dried with, and filtered from, 75 g. of anhydrous copper sulfate. Forty cubic centimeters of hydrochloric acid (sp. gr. 1.19) is added (Note 2) and the temperature raised to 40°. The ethyl nitrite from the preceding preparation is now bubbled in, the temperature being maintained between 40° and 55°. After all the ethyl nitrite has been passed in (Note 3) the crude product may be used in the preparation of dimethylglyoxime by distilling off the alcohol formed in the reaction until the temperature of the liquid reaches 90° (Note 4).

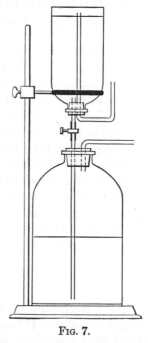

If pure biacetyl monoxime is desired, the crude product is neutralized with about 35 cc. of concentrated aqueous ammonia and diluted with half its volume of water. It is then distilled to remove alcohol, etc., until the distillate is no longer inflammable. The receiver is changed, and the contents are now rapidly distilled, using superheated steam. Almost all the biacetyl monoxime comes over in the first 5 l. of distillate. One to 1.5 kg. of salt is dissolved in the distillate, which is then cooled

FIG. 7.

to 0°. The solid biacetyl monoxime crystallizes and may be filtered. The yield is 480–520 g. The product can be further purified, if desired, by recrystallizing from water (Note 5).

(C) *Sodium Hydroxylamine Monosulfonate.*—In a 12-l. crock are mixed 5 kg. of shaved ice (Note 6) and 569 g. of sodium nitrite (594 g. of the technical 95 per cent quality). Into this is stirred a suspension of sodium bisulfite containing 1.1 kg. of available sulfur dioxide (about 1775 g. of technical bisulfite) in 750 cc. of water. Then, beneath the surface of the solution is added, with constant stirring, 150 cc. of glacial acetic acid (Note 7) followed by a mixture of 550 cc. of concentrated hydrochloric acid (sp. gr. 1.19) with 400 g. of shaved ice (Note 8). The entire solution should always remain below 0°, further ice being added if there is a tendency for the temperature to rise. The solution should now become acid to Congo red paper and contain 6 or more moles of the sodium hydroxylamine disulfonate, which rapidly hydrolyzes to an acid solution of the monosulfonate.

(D) *Dimethylglyoxime.*—The crude biacetyl monoxime remaining after distilling to 90°, and containing about 5 moles of biacetyl monoxime, is added to the sodium hydroxylamine monosulfonate solution (which has been filtered to remove any sediment) and which is contained in a 15-l. flask. It is heated to 70° and allowed to remain warm (with occasional stirring) for several hours (Note 9). The dimethylglyoxime separates in crystals which can be filtered from the solution as soon as it has become cold (Note 10). The crystals are washed with cold water until free of sulfate. The yield of compound melting at 238–240° is 540–575 g.

Recrystallization is unnecessary since the product is pure white and free from tarry material (Note 5).

(E) *Sodium Dimethylglyoximate.*—To a solution of 75 g. of sodium hydroxide in 300 cc. of water is added, with stirring, 100 g. of dimethylglyoxime (Note 11). Heat is applied to effect solution (Note 12), and the mixture is filtered from any slight residue. The solution is poured while hot into 500 cc. of 95 per cent alcohol. After cooling to 5°, with stirring, the crystals which form are filtered, then suspended in 150 cc. of alcohol and again filtered, and finally dried at 25° until the solid has no odor of alcohol (Note 13). The yield is 213–230 g. (81–88 per cent of the theoretical amount) of the octahydrate.

2. Notes

1. If a mechanical stirrer is not used attention is necessary to prevent a layer of solution II from collecting at the bottom of solution I, since, under these conditions, a little shaking produces too violent an evolution of gas.

Because ethyl nitrite is harmful if inhaled continuously, the reaction should be run in a hood or out of doors.

2. The methyl ethyl ketone should not be allowed to stand any length of time between the addition of the hydrochloric acid and the treatment with ethyl nitrite, inasmuch as the acid causes a condensation of the ketone with itself, thereby lowering the yield of biacetyl monoxime.

3. The ethyl nitrite may be passed into the solution as rapidly as possible, provided that the temperature does not exceed 55°. Very complete absorption takes place, and the time of addition is about one and one-half hours.

4. This distillation for the removal of alcohol and excess methyl ethyl ketone should not stop before the temperature of the liquid reaches 90° or continue after this temperature is attained. If the distillation is

stopped too soon, the alcohol that was not removed later increases the solubility of the dimethylglyoxime; too high a temperature causes the formation of tarry products. Both factors reduce the yield.

5. About 70–80 per cent of the biacetyl monoxime is recovered on crystallization. Practically all the monoxime left in the mother liquor can be recovered by steam distillation. The recrystallized and dried compound melts at 76.5°. Purification by distillation is not recommended.

Biacetyl monoxime turns brown rapidly after preparation; it must not be allowed to stand before adding it to the hydroxylamine monosulfonate in the preparation of dimethylglyoxime if this final product is to be obtained in a colorless condition.

6. Shaved ice is used in place of crushed ice because of more efficient cooling. The use of snow is advised, if it is available.

7. The acetic acid helps to buffer the solution and maintain a low hydrogen-ion concentration, which is favorable for a good yield.

8. The acid may be added as rapidly as desired so long as the temperature remains below 0° and no evolution of gas takes place. The entire addition should require not more than fifteen minutes.

9. The mixture is warmed to hasten the hydrolysis of disulfonate to monosulfonate, and also to increase the solubility of the biacetyl monoxime in the monosulfonate solution.

10. The dimethylglyoxime should not be filtered until the mixture has become perfectly cool, because dimethylglyoxime is slightly soluble in a warm acid solution.

11. Crude or discolored dimethylglyoxime may be used since the impurities remain in solution in the aqueous alcohol.

12. The solution of dimethylglyoxime in sodium hydroxide should not be boiled, since prolonged heating causes decomposition.

13. The sodium salt may be further purified if desired by dissolving in water and reprecipitating with alcohol. Large crystals may be secured by crystallizing slowly from water.

Too long drying, or drying at too high a temperature, partially dehydrates the salt, with the result that it dissolves much more slowly in water and is therefore less desirable for making solutions for use in qualitative or quantitative analysis. Sodium dimethylglyoximate is extremely soluble in water. A 3 per cent ($0.1\ M$) aqueous solution is suggested to replace the 1 per cent alcoholic solution now used in analytical work.

3. Methods of Preparation

Biacetyl monoxime has been prepared by the action of amyl nitrite on methyl ethyl ketone using sodium hydroxide [1] or hydrochloric acid [2] as a condensing agent, and by melting nitrosolevulinic acid.[3]

Dimethylglyoxime has been prepared by the action of hydrochloric acid upon biacetyl monoxime; [4] by the action of hydroxylamine on biacetyl; [5] by the action of hydroxylamine on biacetyl monoxime; [6] and by the action of sodium hydroxylamine monosulfonate on biacetyl monoxime.[7] It is formed in small amounts, together with ethylnitrolic acid, by the action of oxides of nitrogen upon methyl ethyl ketone.[8]

sym.-DIMETHYLHYDRAZINE DIHYDROCHLORIDE

(Hydrazine, 1,2-dimethyl-, dihydrochloride)

$$2C_6H_5COCl + N_2H_4 \cdot H_2SO_4 \xrightarrow{\text{(NaOH)}} C_6H_5CONHNHCOC_6H_5$$

$$C_6H_5CONHNHCOC_6H_5 + 2(CH_3)_2SO_4 + 2NaOH \rightarrow$$
$$C_6H_5CON(CH_3)N(CH_3)COC_6H_5 + 2CH_3SO_4Na + 2H_2O$$

$$C_6H_5CON(CH_3)N(CH_3)COC_6H_5 + 2HCl + 2H_2O \rightarrow$$
$$CH_3NHNHCH_3 \cdot 2HCl + 2C_6H_5CO_2H$$

Submitted by H. H. HATT.
Checked by REYNOLD C. FUSON and ELLSWORTH ELLINGBOE.

1. Procedure

(A) *Dibenzoylhydrazine.*—In a 2-l. flask, provided with a mechanical stirrer (Note 1) and cooled in a bath of cold water, are placed a solution of 48 g. (1.2 moles) of sodium hydroxide in 500 cc. of water and 65 g. (0.5 mole) of hydrazine sulfate (Org. Syn. Coll. Vol. I, **1941**, 309).

[1] Claisen, Ber. **38**, 696 (1905).

[2] Diels and Farkas, ibid. **43**, 1957 (1910); Biltz, Z. anal. Chem. **48**, 164 (1909); Semon and Damerell, J. Am. Chem. Soc. **47**, 2038 (1925).

[3] Thal, Ber. **25**, 1720 (1892).

[4] Schramm, ibid. **16**, 180 (1883); Johlin, J. Am. Chem. Soc. **36**, 1218 (1914).

[5] Fittig, Daimler, and Keller, Ann. **249**, 182 (1888).

[6] Tschugaeff, Ber. **38**, 2520 (1905); Gandarin, J. prakt. Chem. (2) **77**, 414 (1908); Biltz, Z. anal. Chem. **48**, 164 (1909); Adams and Kamm, J. Am. Chem. Soc. **40**, 1281 (1918).

[7] Semon and Damerell, ibid. **46**, 1290 (1924); **47**, 2033 (1925).

[8] Behrend and Tryller, Ann. **283**, 244 (1894).

With stirring, 145 g. (120 cc., 1.03 moles) of freshly distilled benzoyl chloride and 120 cc. of an aqueous solution containing 45 g. (1.1 moles) of sodium hydroxide are added slowly from separate dropping funnels. The benzoyl chloride is added over a period of one and one-half hours; the alkali is added slightly faster. After both additions are completed the mixture is stirred for two hours longer and then saturated with carbon dioxide (Note 2). The dibenzoylhydrazine is filtered with suction, pressed thoroughly, and ground to a paste with 50 per cent aqueous acetone. The material is filtered with suction, washed with water, and pressed as dry as possible. The crude product is dissolved in about 650 cc. of boiling glacial acetic acid, from which, on cooling, the dibenzoylhydrazine separates as a mass of fine white needles. These are filtered with suction, washed with cold water, and dried by heating under reduced pressure on a water bath, in a slow current of air. This first fraction, m.p. 234–238°, amounts to 80–90 g. (66–75 per cent of the theoretical amount) and is practically pure. Small quantities of less pure material can be obtained by concentration of the mother liquor.

(B) *Dibenzoyldimethylhydrazine.*—The following operations should be performed under a hood. In a 2-l. three-necked flask, provided with a mechanical stirrer, a thermometer, and two dropping funnels, are placed 80 g. (0.33 mole) of dibenzoylhydrazine, 10 g. (0.25 mole) of sodium hydroxide, and 600 cc. of water. The mixture is maintained at about 90° (Note 3) by heating on a water bath. With stirring, 320 g. (240 cc., 2.54 moles) of methyl sulfate (Note 4) and 250 cc. of an aqueous solution containing 125 g. (3.1 moles) of sodium hydroxide are added from separate dropping funnels. The methyl sulfate is added in 10-cc. portions at five-minute intervals, the sodium hydroxide solution at a rate that maintains the reaction mixture slightly alkaline. When markedly alkaline the mixture has a distinctly yellow color; it is best to maintain a degree of alkalinity slightly less than that required to produce this color.

During the additions, which require about two hours, the dibenzoyldimethylhydrazine separates as a supernatant liquid (Note 5); the semisolid material thrown up on the sides of the flask must be washed down from time to time. After being heated for one-half hour longer the mixture is allowed to cool. The solid dibenzoyldimethylhydrazine is collected, crushed with water, filtered, and dissolved in 100 cc. of chloroform. The solution is filtered from insoluble impurities and dried over sodium sulfate; the chloroform is removed by heating on a water bath, finally under reduced pressure. The solid residue melts at 77–84° and is sufficiently pure for further use. The yield is 77–83 g. (86–93 per cent of the theoretical amount) (Note 6).

(C) *Dimethylhydrazine Dihydrochloride.*—In a 2-l. flask, a mixture of 67 g. (0.25 mole) of dibenzoyldimethylhydrazine and 250 cc. of 32 per cent hydrochloric acid (sp. gr. 1.16) is refluxed gently for two hours, under a hood. To remove the benzoic acid, the mixture is steam-distilled until 10 l. of distillate has collected (Note 7); the residual liquor is evaporated to dryness under reduced pressure on a water bath. The crystalline dihydrochloride is treated with 25 cc. of absolute ethyl alcohol, and the mixture is evaporated to dryness under reduced pressure. This treatment is repeated, and the dihydrochloride is crushed with a mixture of 25 cc. of absolute alcohol and 2–3 cc. of concentrated hydrochloric acid, filtered, and washed with 10–15 cc. of cold absolute alcohol. The first fraction of dimethylhydrazine dihydrochloride, after being dried in a vacuum desiccator, weighs 22–23 g. By evaporation of the mother liquors and further treatment with 5–6 cc. of absolute alcohol and a little hydrochloric acid, a second fraction of 3 g. of the hydrochloride is obtained. Repetition of this procedure yields a third fraction amounting to about 0.5 g. (Note 8). The total yield is 25–26 g. (75–78 per cent of the theoretical amount).

2. Notes

1. The reaction flask need neither be corked nor placed under a hood if a glass tube, connected to a filter pump, is passed a short way into the neck of the flask.

2. Most of the dibenzoylhydrazine separates during the reaction as a white precipitate; if the mixture is not saturated with carbon dixoide, about 3 g. of the product remains in solution.

3. The temperature may be determined by a thermometer sheathed in a glass tube, which passes through the stopper and dips into the liquid. A section of rubber tubing slipped over the upper end of the thermometer serves to hold it in place. This device also provides a convenient method of testing the alkalinity of the solution.

4. Since methyl sulfate is very toxic, care should be exercised to avoid spilling the liquid or inhaling the vapor of the reaction mixture. Ammonia is a specific antidote.

5. The stated quantities of methyl sulfate and sodium hydroxide are merely approximations of the amounts actually required. The methyl sulfate is added at the stated rate until all the solid dibenzoylhydrazine has disappeared, and then another 25 cc. is added. Usually less than the stated amount of alkali is required.

6. The dibenzoyldimethylhydrazine can be obtained in a purer state by treating the cold, filtered chloroform solution with a 1 : 1 mixture of ether and petroleum ether (b.p. 40–60°) until crystallization begins and then adding slowly a moderate excess of petroleum ether. The hydrazine separates in small white needles, m.p. 84–87°.

7. It is found that this treatment removes the benzoic acid almost completely. Alternatively, the benzoic acid can be extracted with a 1 : 1 benzene-ether mixture.

8. The mono- and dihydrochlorides are very hygroscopic and should not be exposed to air for any length of time. Solutions of the dihydrochloride lose hydrochloric acid on evaporation, and the dihydrochloride obtained from solutions so treated tends to be oily. In recrystallization a little concentrated hydrochloric acid is added to the alcoholic solution. The dihydrochloride may be crystallized by dissolving in boiling absolute ethyl alcohol (1 g. requires 20 cc. of alcohol), then adding a little concentrated hydrochloric acid and, after cooling slightly, about one-fifth of the volume of ether. The dimethylhydrazine dihydrochloride separates as a white crystalline powder, melting at 165–167°.

3. Methods of Preparation

Symmetrical dimethylhydrazine has been obtained by heating the methiodide of 1-methylpyrazole with potassium hydroxide.[1] It has usually been prepared by methylation of diformylhydrazine and subsequent hydrolysis with hydrochloric acid.[2] The present method is based on the observation of Folpmers that dibenzoylhydrazine may be similarly employed.[3]

unsym.-DIMETHYLHYDRAZINE HYDROCHLORIDE

(Hydrazine, 1,1-dimethyl-, hydrochloride)

$$(CH_3)_2NH \cdot HCl + HONO \longrightarrow (CH_3)_2NNO + HCl + H_2O$$

$$(CH_3)_2NNO + 4[H] \xrightarrow{(Zn + CH_3CO_2H)} (CH_3)_2NNH_2 + H_2O$$

$$(CH_3)_2NNH_2 + HCl \longrightarrow (CH_3)_2NNH_2 \cdot HCl$$

Submitted by H. H. HATT.
Checked by W. W. HARTMAN and W. D. PETERSON.

1. Procedure

(A) *Nitrosodimethylamine.*—In a 2-l. round-bottomed flask provided with a mechanical stirrer are placed 245 g. (3 moles) of dimethylamine

[1] Knorr and Köhler, Ber. **39**, 3257 (1906).

[2] Harries and Klamt, ibid. **28**, 503 (1895); Harries and Haga, ibid. **31**, 56 (1898); Thiele, ibid. **42**, 2575 (1909).

[3] Folpmers, Rec. trav. chim. **34**, 34 (1915).

hydrochloride, 120 cc. of water, and 10 cc. of approximately 2 N hydrochloric acid. The resulting solution is stirred vigorously and maintained at 70–75° by heating on a water bath, while 235 g. (3.23 moles) of 95 per cent sodium nitrite suspended in 150 cc. of water is added from a dropping funnel over a period of an hour. The reaction mixture is tested frequently and maintained barely acid to litmus by further 1-cc. additions of 2 N hydrochloric acid when necessary (about 30–35 cc. of acid is required). Stirring and heating are continued for two hours after all the sodium nitrite has been added.

The flask is arranged for distillation, and the reaction mixture is distilled under slightly diminished pressure on a water bath until the residue is practically dry. To the residue 100 cc. of water is added, and the process of distillation to dryness is repeated. The distillates are combined and saturated with potassium carbonate (about 300 g. is required); the upper layer of dimethylnitrosoamine is removed, and the water layer is extracted with three 140-cc. portions of ether. The combined nitrosoamine and ethereal extracts are dried over anhydrous potassium carbonate and distilled through a 30-cm. fractionating column. The yield of product boiling at 149–150°/755 mm. is 195–200 g. (88–90 per cent of the theoretical amount). Nitrosodimethylamine is a yellow oil which darkens in bright light.

(B) *unsym.-Dimethylhydrazine Hydrochloride.*—In a 5-l. round-bottomed flask, provided with a mechanical stirrer, dropping funnel, and thermometer, are put 200 g. (2.7 moles) of nitrosodimethylamine, 3 l. of water, and 650 g. (10 gram atoms) of 100 per cent zinc dust, or an equivalent amount of lower-quality material (Note 1). While the mixture is stirred and maintained at 25–30° by immersion in a water bath, 1 l. (14 moles) of 85 per cent acetic acid is added from the dropping funnel over a period of two hours. Subsequently the reaction mixture is heated for one hour at 60°, allowed to cool, and the excess zinc dust filtered and washed with a little water. The aqueous liquors are combined and transferred to a 12-l. flask arranged for steam distillation. The flask is fitted with a dropping funnel, and the steam inlet is provided with a trap. A large filtering or distilling flask serves as the receiver, and the side tube is connected with two absorption flasks, each containing 1 : 1 hydrochloric acid (Note 2). The aqueous liquors are made distinctly alkaline by adding a concentrated solution of 1 kg. of sodium hydroxide through the dropping funnel, and the mixture is then steam-distilled until a test portion of the distillate shows only a faint reduction with Fehling's solution. Usually about 5–6 l. of distillate suffices to remove the dimethylhydrazine. The distillation is more rapid if a free flame is placed under the distillation flask.

The aqueous distillate is treated with 650 cc. of concentrated hydrochloric acid and concentrated on a steam bath, under reduced pressure, until the residual liquor becomes a syrupy mass (Note 3). This is desiccated further by adding 150 cc. of absolute ethyl alcohol and evaporating under reduced pressure. Two or three such treatments with alcohol dry the crystalline material sufficiently so that it no longer sticks to the sides of the flask. The crude product is dried in a vacuum desiccator over calcium chloride. The pale yellow, dry, crystalline solid weighs 200–215 g. (77–83 per cent of the theoretical amount). The dried product may be purified by dissolving it in an equal weight of boiling absolute ethyl alcohol and then chilling in an ice bath. The yield of pure white crystals melting at 81–82° is 180–190 g. (69–73 per cent of the theoretical amount) (Note 4).

2. Notes

1. The approximate strength of the zinc dust should be known in order to ensure complete reduction.

2. If this step is carried out under a hood, the absorption flasks are unnecessary. During the first minutes of the subsequent steam distillation, volatile nitrogen bases, consisting chiefly of ammonia and methylamine, are driven over.

3. A concentrated aqueous solution of the free hydrazine may be obtained from the syrupy residue by allowing it to drop onto a large excess of solid sodium hydroxide and distilling until the temperature reaches 100°. To obtain the anhydrous base, the concentrated aqueous solution is redistilled after standing over potassium hydroxide, the base is collected over barium oxide and after several days' standing is distilled again. The free hydrazine boils at 62–65°/765 mm.; it is extremely hygroscopic and attacks cork and rubber.

4. The product is the monohydrochloride of the base.

3. Methods of Preparation

The preparation of unsymmetrical dimethylhydrazine by reduction of nitrosodimethylamine was described by Fischer [1] and by Renouf.[2] Methylation of hydrazine,[3] reduction of nitrodimethylamine,[4] and the action of aminopersulfuric acid on dimethylamine [5] also furnish unsymmetrical dimethylhydrazine.

[1] Fischer, Ber. **8,** 1587 (1875).
[2] Renouf, ibid. **13,** 2170 (1880).
[3] Harries and Haga, ibid. **31,** 56 (1898).
[4] Franchimont, Rec. trav. chim. **3,** 427 (1884); Backer, ibid. **31,** 150 (1912).
[5] Sommer and Schulz, Ger. pat. 338,609 [Frdl. **13,** 203 (1916–21)].

2,6-DIMETHYLPYRIDINE

(2,6-Lutidine)

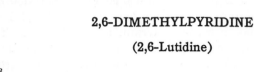

$$2\ \underset{\substack{|\\CH_2\\|\\CO_2C_2H_5}}{\overset{\substack{CH_3\\|}}{C}}=O \quad + NH_3 + CH_2O$$

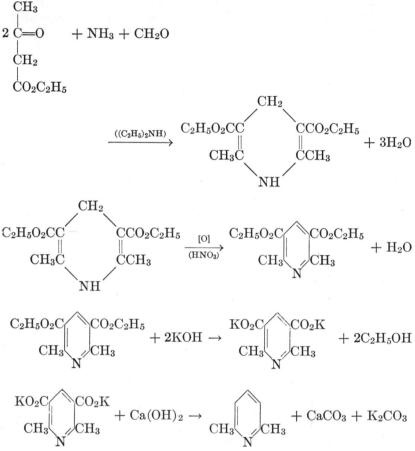

Submitted by ALVIN SINGER and S. M. McELVAIN.
Checked by REYNOLD C. FUSON and CHARLES F. WOODWARD.

1. Procedure

To 500 g. (3.85 moles) of freshly distilled ethyl acetoacetate in a 1-l. flask, set in ice and well cooled, are added 152 g. (2 moles) of 40 per cent aqueous formaldehyde solution and 20–25 drops of diethylamine. The flask and contents are kept cold for six hours and are then allowed to

stand at room temperature for forty to forty-five hours. At the end of
this time two layers are present, a lower oily layer and an upper aqueous
layer. The layers are separated, and the aqueous layer is extracted with
50 cc. of ether. The ether solution is added to the oily layer, and the
resulting solution is dried over 30 g. of calcium chloride. The ether is
then removed by distillation on a steam bath. The residue, amounting
to approximately 500 g., is diluted with an equal volume of alcohol and
is thoroughly cooled in an ice bath. Ammonia is then passed into the
mixture until the solution is saturated. This requires from four to eight
hours, and during this time the flask is kept packed in ice. The ammoniacal alcoholic solution is allowed to stand at room temperature for
forty to forty-five hours. Most of the alcohol is now evaporated; the
residue is cooled, and the solid 1,4-dihydro-3,5-dicarbethoxy-2,6-di-
methylpyridine is removed from the remaining alcohol on a suction filter.
The dried ester melts at 175–180° and weighs 410–435 g. (84–89 per cent
of the theoretical amount).

To 200 g. (0.79 mole) of the ester in a 5-l. flask is added a mixture of
270 g. of water, 72 g. of concentrated nitric acid (sp. gr. 1.42), and 78 g.
of concentrated sulfuric acid. The flask is then very cautiously heated,
and the contents are kept in a swirling motion by a slow shaking of the
flask by hand. The oxidation is accompanied by considerable foaming;
if the heating is too rapid, part of the reaction mixture may be lost by
excessive frothing. After the foaming has subsided, the reaction mixture is again warmed cautiously until the liquid assumes a deep red
color. The entire oxidation is carried out in ten to fifteen minutes.
After the liquid has ceased boiling, it is treated with 500 cc. of water
and 500 g. of finely chopped ice. The resulting solution is made strongly
alkaline by the gradual addition of ammonium hydroxide (sp. gr. 0.90).
The precipitated 3,5-dicarbethoxy-2,6-dimethylpyridine is filtered with
suction, dried on a porous plate, and then distilled (Note 1). The yield
of product boiling at 170–172°/8 mm. is 115–130 g. (58–65 per cent of
the theoretical amount based on the dihydro ester).

A solution of 130 g. (0.52 mole) of this ester in 400 cc. of ethyl alcohol
is placed in a two-necked 2-l. flask, carrying a dropping funnel and a
reflux condenser, and is heated to boiling. Then one-third of a solution
(Note 2) of 78.5 g. (1.4 moles) of potassium hydroxide in 400 cc. of
alcohol is added from the dropping funnel, and the alcoholic solution is
boiled until it becomes clear. Then a second third of the alkali solution
is added, and the reaction mixture is again boiled until any precipitate
disappears. Finally, the last third of the alcoholic potassium hydroxide
solution is added. The addition of the alkali requires about twenty
minutes. The reaction mixture is then boiled for forty minutes longer.

The contents of the flask while still hot are poured into a 30-cm. evaporating dish and the alcohol is evaporated on a steam bath. The dry salt is pulverized and thoroughly mixed with 390 g. of calcium oxide, placed in a 2-l. copper retort (Note 3), and heated with the full flame of a Meker burner. The distillate is placed in a distilling flask and heated on a steam bath; all material distilling under 90° is removed and discarded. The residue is then allowed to stand over solid potassium hydroxide for twelve hours and is finally fractionated. The dimethylpyridine distils at 142–144°/743 mm. The yield is 35–36 g. (63–65 per cent of the theoretical amount based on the 3,5-dicarbethoxy-2,6-dimethylpyridine, or 30–36 per cent based on the original ethyl acetoacetate).

2. Notes

1. Before the pressure is reduced, the 3,5-dicarbethoxy-2,6-dimethylpyridine should be melted by immersing the distilling flask in boiling water. If this is not done, considerable foaming takes place during the distillation.

2. If the entire amount of alcoholic potassium hydroxide solution is added at this point, there is precipitated a colorless solid, which does not dissolve even on prolonged heating.

3. A metal retort is very desirable for this decomposition since glass flasks soften at the temperature necessary for the reaction.

3. Methods of Preparation

2,6-Dimethylpyridine has been isolated from the basic fraction of coal tar [1] and also from the bone oil fraction distilling at 139–142°.[2] It has been prepared from ethyl acetopyruvate and ethyl β-aminocrotonate.[3] A procedure for the oxidation of the dihydroester and the saponification and decarboxylation of the product, similar to the procedure given above, has been published.[4]

[1] Lunge and Rosenberg, Ber. **20**, 129 (1887); Heap, Jones, and Speakman, J. Am. Chem. Soc. **43**, 1936 (1921); Komatsu and Mohri, J. Chem. Soc. Japan **52**, 722 (1931) [C. A. **26**, 4936 (1932)].

[2] Ladenberg and Roth, Ber. **18**, 51 (1885).

[3] Mumm and Hüneke, ibid. **50**, 1568 (1917).

[4] Oparina, Karasina, and Smirnov, J. Applied Chem. (U.S.S.R.) **11**, 965 (1938) [C. A. **33**, 1732 (1939)].

2,4-DIMETHYLPYRROLE

(Pyrrole, 2,4-dimethyl-)

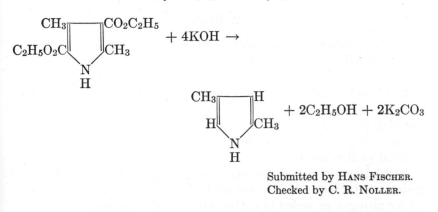

$+ 4KOH \rightarrow$

$+ 2C_2H_5OH + 2K_2CO_3$

Submitted by HANS FISCHER.
Checked by C. R. NOLLER.

1. Procedure

A SOLUTION of 270 g. (4.8 moles) of potassium hydroxide in 150 cc. of water is prepared in a 3-l. round-bottomed flask, 120 g. (0.5 mole) of crude 2,4-dimethyl-3,5-dicarbethoxypyrrole (p. 202) and a pinch of sand are added, and the whole is mixed thoroughly by shaking. The flask is fitted with a reflux condenser, and the mixture is heated in an oil bath at 130° for two to three hours with occasional shaking until the thick paste has become partially liquefied owing to the formation of dimethylpyrrole.

The flask is next fitted for distillation with superheated steam and with a separatory funnel for the introduction of water into the center of the flask. A 3-l. round-bottomed flask fitted with a vertical condenser is used as a receiver (Org. Syn. Coll. Vol. I, 1941, 479). The temperature of the oil bath is raised to 160°, and superheated steam at 220–250° is introduced. The temperature of the oil bath is then gradually raised to 200°. If foaming becomes too great, a few drops of water are added from the separatory funnel, care being taken that the water does not strike the hot glass walls (Note 1). Steam distillation is continued until no more dimethylpyrrole comes over. This takes from one to two hours, and the distillate amounts to 2.5–3 l. The distillate is extracted once with 200 cc. of ether and three times with 100-cc. portions, and the extract is dried for two hours over 20 g. of anhydrous potassium carbonate. The ether is removed by distillation from a 100-cc. modified Claisen flask having a 15-cm. fractionating side arm, the solution being added gradually through a separatory funnel. After the ether is removed the residue is distilled and the fraction boiling at 160–165° is collected. The yield is 27 to 30 g. (57–63 per cent of the theoretical amount) (Notes 2 and 3).

2. Notes

1. If the contents of the flask cake or become semi-solid, the temperature of the oil bath should be lowered and the rate of flow of superheated steam gradually decreased.

2. There is practically no fore-run, but if repeated or larger batches are made it is possible to obtain about 2 g. more of product per run from the higher-boiling fractions. If recrystallized ester is used a higher yield of dimethylpyrrole with less high-boiling products is obtained but the over-all yield is lower.

3. 2,4-Dimethylpyrrole is very readily oxidized in the air to a red resinous substance. If it is not used immediately, it should be stored under nitrogen or sealed in a glass vial under a vacuum.

3. Methods of Preparation

2,4-Dimethylpyrrole has been obtained by a large number of reactions, but the method of Knorr [1] and the condensation of acetone with aminoacetone [2] are the only ones of preparative interest. The procedure given above is one modification of Knorr's method. In another modification, the pyrrole ester and alkali are heated in a steel bomb and the product is distilled in an inert atmosphere; a yield of 95 per cent is reported.[3]

[1] Knorr, Ann. **236**, 326 (1886).
[2] Piloty and Hirsch, ibid. **395**, 65 (1913).
[3] Corwin and Krieble, J. Am. Chem. Soc. **63**, 1830 (1941).

2,5-DIMETHYLPYRROLE

(Pyrrole, 2,5-dimethyl-)

$$CH_3COCH_2CH_2COCH_3 + NH_3 \rightarrow$$ 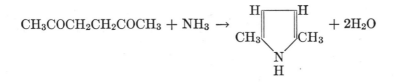 $$+ 2H_2O$$

Submitted by D. M. Young and C. F. H. Allen.
Checked by John R. Johnson and H. B. Stevenson.

1. Procedure

In a 500-cc. Erlenmeyer flask, fitted with an air-cooled reflux condenser of large bore (Note 1), are placed 100 g. (0.88 mole) of acetonylacetone (Note 2) and 200 g. (1.75 moles) of ammonium carbonate, in lumps. The mixture is heated in an oil bath at 100° until effervescence stops; this requires sixty to ninety minutes. The air-cooled condenser is replaced by a water-cooled condenser of large bore, and the mixture is refluxed gently at 115° (bath temperature) for thirty minutes longer (Notes 1 and 3). The mixture is cooled, and the upper, yellow layer of the pyrrole is separated (Note 4). The lower layer is extracted with 15 cc. of chloroform, which is added to the crude pyrrole. The whole is dried over anhydrous calcium chloride, in a tightly stoppered container which has been swept out with nitrogen beforehand.

The material is transferred (Note 4) to a modified Claisen flask having a fractionating side arm, and the chloroform is distilled completely under reduced pressure (without being condensed). The dimethylpyrrole is collected at 51–53°/8 mm. or 78–80°/25 mm.; only a small residue (4–5 g., b.p. 80–85°/25 mm.) remains. The product weighs 68–72 g. (81–86 per cent of the theoretical amount) and is quite pure (Note 5); it should be stored in an inert atmosphere in a tightly sealed, dark glass container.

2. Notes

1. Sublimed ammonium carbonate must be prevented from blocking the condenser. From time to time the sublimate is pushed back into the reaction mixture by means of a stout glass rod.

2. Acetonylacetone may be prepared conveniently by hydrolysis of 2,5-dimethylfuran. The following procedure is based upon directions

supplied to the checkers by Mr. Gwyn Benson, of the Shawinigan Chemicals, Ltd. In a 500-cc. flask are placed 125 g. (1.2 moles) of 2,5-dimethylfuran (b.p. 93–96°), 60 g. of water, 50 g. of glacial acetic acid, and 3 cc. of 10 per cent sulfuric acid. The mixture is boiled gently for thirty-six hours, 1 g. of sodium acetate crystals is added to convert the sulfuric acid to sodium sulfate, and the material is distilled at atmospheric pressure through a fractionating column. The fraction boiling up to 155° is removed, and the residual liquid is distilled under reduced pressure. The acetonylacetone is collected at 78–79°/15 mm., or 88–89°/25 mm., and weighs 128–133 g. (86–90 per cent of the theoretical amount).

3. At the end of the first and second periods of heating it is well to dissolve the sublimed ammonium carbonate, by pouring 5–10 cc. of hot water back and forth through the condenser, and return the solution to the reaction mixture.

4. In this and all the subsequent operations, the dimethylpyrrole should be manipulated with a minimum exposure to atmospheric oxygen. The distilling apparatus should be swept out with nitrogen at the start and nitrogen admitted, instead of air, whenever the vacuum is released.

5. Since 2,5-dimethylpyrrole does not form solid derivatives, the refractive index may be used as a criterion of purity. The value for a freshly distilled specimen is $n_D^{22°}$ 1.500. On standing the material gradually turns red and the refractive index rises. This change is accelerated by air and light.

3. Methods of Preparation

2,5-Dimethylpyrrole has been prepared by heating acetonylacetone with ammonium acetate in glacial acetic acid [1] or with alcoholic ammonia in a sealed tube at 150°.[2] It has also been prepared by hydrolysis and decarboxylation of 2,5-dimethyl-3,4-dicarbethoxypyrrole [3] and the monocarbethoxy compound.[4]

[1] Ajello and Cusmano, Gazz. chim. ital. **69**, 210 (1939).

[2] Paal, Ber. **18**, 2254 (1885).

[3] Knorr, ibid. **18**, 299, 1565 (1885).

[4] Timoshevskaya, J. Gen. Chem. (U.S.S.R.) **9**, 766 (1939) [C. A. **34**, 423 (1940)].

2,4-DINITROANILINE

(Aniline, 2,4-dinitro-)

$$(NO_2)_2C_6H_3Cl + 2NH_3 \rightarrow (NO_2)_2C_6H_3NH_2 + NH_4Cl$$

Submitted by F. B. WELLS and C. F. H. ALLEN.
Checked by REYNOLD C. FUSON and CHAN MANN LU.

1. Procedure

A WIDE-MOUTHED 250-cc. flask (Note 1) containing a mixture of 50 g. (0.25 mole) of technical 2,4-dinitrochlorobenzene (Note 2) and 18 g. (0.23 mole) of ammonium acetate is half immersed in an oil bath. The flask is fitted with a reflux condenser and an inlet tube, the lower end of which is at least 2 cm. wide (to prevent clogging) and which nearly touches the surface of the reaction mixture. During the operation ammonia gas is introduced through a bubble counter which contains a small amount of strong potassium hydroxide solution (12 g. of potassium hydroxide in 10 cc. of water).

The oil bath is heated to 170° and maintained at that temperature for six hours during which time ammonia gas is passed through at the rate of three to four bubbles per second. After the mixture has cooled, the solid is broken up by means of a glass rod and mixed with 100 cc. of water; the mixture is then heated to boiling and filtered while hot. The residue is dissolved in 500 cc. of boiling alcohol, and water is added (about 150 cc.) until the solution becomes turbid. Heat is applied until the turbidity disappears and then the solution is allowed to cool. After standing overnight, the crystals are filtered and dried. The yield is 31–35 g. (68–76 per cent of the theoretical amount) of 2,4-dinitroaniline, melting at 175–177° (Notes 3 and 4). For further purification the product is recrystallized in the same manner from alcohol and water, using 20 cc. of alcohol per gram of solid. Ninety per cent of the crude material is recovered as recrystallized product melting sharply at 180°.

2. Notes

1. The neck of the flask must be large enough to admit the wide inlet tube. A 250-cc. extraction flask is most convenient.
2. The technical dinitrochlorobenzene used had a freezing point of 45°. The higher yield given is obtained using a product once recrystallized from alcohol (m.p. 48°).

3. Some specimens of technical dinitrochlorobenzene used contained undetermined impurities that formed double compounds with the dinitroaniline. These remained in the filtrate. It was not worth the trouble to recover the small amount of amine thus lost.

4. The value of this method is that no autoclave is required. It was not satisfactory when applied to the nitrochlorobenzenes.

3. Methods of Preparation

2,4-Dinitroaniline has been prepared by heating dinitrochlorobenzene and ammonia under pressure; [1] by heating 1,2,4-trinitrobenzene with concentrated ammonia; [2] by heating dinitrochlorobenzene with urea [3] or ammonium acetate; [4] by hydrolysis of the dinitroacetanilide obtained when dinitrochlorobenzene and acetamide are heated at 200–210°; [5] by heating 2,4-dinitroanisole or 2,4-dinitroanisic acid with aqueous or alcoholic ammonia under pressure; [6] by heating 2,4-dinitrophenol and aqueous ammonia under pressure; [7] and by the rearrangement of p-nitrophenylnitroamine spontaneously [8] or with concentrated sulfuric acid at 0°. [9]

[1] Clemm, J. prakt. Chem. (2) **1**, 170 (1870); Willgerodt, Ber. **9**, 979 (1876).

[2] Hepp, Ann. **215**, 362 (1882).

[3] Rohm and Haas Company, U. S. pat. 1,752,998 [C. A. **24**, 2468 (1930)].

[4] Soc. chim. de la Grande-Paroisse, Brit. pat. 169,688 [C. A. **16**, 721 (1922)].

[5] Kym, Ber. **32**, 3539 (1899).

[6] Salkowski, Ann. **174**, 263 (1874).

[7] Barr, Ber. **21**, 1542 (1888).

[8] Bamberger and Dietrich, ibid. **30**, 1253 (1897).

[9] Hoff, Ann. **311**, 98 (1900).

2,4-DINITROBENZALDEHYDE

(Benzaldehyde, 2,4-dinitro-)

$$C_6H_5N(CH_3)_2 + 2HCl + NaNO_2 \rightarrow$$
$$ONC_6H_4N(CH_3)_2 \cdot HCl + NaCl + H_2O$$

$$ONC_6H_4N(CH_3)_2 \cdot HCl + CH_3C_6H_3(NO_2)_2 + Na_2CO_3 \rightarrow$$
$$(CH_3)_2NC_6H_4N{=}CHC_6H_3(NO_2)_2 + NaCl + NaHCO_3 + H_2O$$

$$(CH_3)_2NC_6H_4N{=}CHC_6H_3(NO_2)_2 + H_2O + HCl \rightarrow$$
$$C_6H_3(NO_2)_2CHO + (CH_3)_2NC_6H_4NH_2 \cdot HCl$$

Submitted by G. M. Bennett and E. V. Bell.
Checked by J. B. Conant and R. E. Schultz.

1. Procedure

A solution of 300 g. (2.5 moles) of technical dimethylaniline in 1050 cc. of concentrated hydrochloric acid is placed in a large jar or crock and finely divided ice is added until the temperature has fallen to 5°. The contents of the jar are then stirred mechanically, and a solution of 180 g. (2.6 moles) of sodium nitrite in 300 cc. of water is slowly added from a separating funnel, the stem of which dips beneath the surface of the liquid. The addition takes one hour, and the temperature is kept below 8° by the addition of ice if necessary. When all the nitrite has been added the mixture is allowed to stand one hour and then filtered. The solid p-nitrosodimethylaniline hydrochloride is washed with 400 cc. of 1 : 1 hydrochloric acid and then with 100 cc. of alcohol. After drying in air, it weighs 370–410 g. (80–89 per cent of the theoretical amount) (Note 1).

In a 3-l. flask a mixture of 330 g. (1.8 moles) of air-dried p-nitrosodimethylaniline hydrochloride (Note 2) with 100 g. of anhydrous sodium carbonate and 1.5 l. of 95 per cent alcohol is heated on the steam bath for thirty minutes and filtered to remove sodium chloride and unchanged carbonate. The filtrate is placed with 300 g. (1.6 moles) of 2,4-dinitrotoluene (m.p. 70°) in a 5-l. flask provided with a mechanical stirrer and a reflux condenser and heated on the steam bath with stirring for five hours. When the contents of the flask are cold, the condensation product is collected on a Büchner funnel and well drained by means of a suction pump and a piece of "rubber dam" fastened over the top of the funnel. The rubber is drawn down onto the solid by the suction, thus forcing the liquid out. The dark green solid is washed by transferring it to a large beaker, stirring it with 1 l. of 95 per cent alcohol on the steam bath for half an hour, cooling, and filtering again, using the rubber dam. The dinitrobenzylidene-p-aminodimethylaniline is placed while still damp in a large round-bottomed flask containing 500 cc. of a mixture of equal volumes of concentrated hydrochloric acid and water, and steam is blown in by a tube reaching to the lowest part of the vessel. Vigorous injection of steam is continued for fifteen minutes after the liquid has reached a temperature of 105° (Note 3).

When cold the aqueous layer is decanted (or filtered) from the solidified product, and the process of steam-agitation with acid is repeated, another 500 cc. of dilute acid being added (Note 4). Finally the solid is filtered, washed with water, and dried *in vacuo*. The crude dinitrobenzaldehyde so obtained weighs 170–210 g. and melts at 40–50°. It is purified by heating in two lots each with 8 l. of naphtha (b.p. 90–110°) in a large flask on the steam bath with mechanical stirring for three hours, decanting the solution (temperature of liquid 80°) and allowing the aldehyde to crystallize. The purified dinitrobenzaldehyde, air-dried, weighs 79–104 g. (24–32 per cent of the theoretical amount) and melts at 69–71° (Note 5). From the mother liquors impure material may be recovered by evaporation or by distillation in steam, from which by a similar process of crystallization a further 20 g. of aldehyde of m.p. 66° can be separated.

2. Notes

1. It is reported that p-nitrosodiethylaniline hydrochloride can be prepared by a similar procedure and that the free base can be obtained from the hydrochloride in the following way: The reaction mixture containing the p-nitrosodiethylaniline hydrochloride prepared from 50 g. of once-distilled diethylaniline is stirred and kept at 5° while 580 cc. of 2 N sodium carbonate solution is added dropwise. The addition requires about two hours, after which the color of the reaction mixture changes to dark green and a dark green solid separates. Stirring is continued for ten minutes, then the precipitate is allowed to settle for five minutes and is filtered with suction. The precipitate is washed with three 25-cc. portions of distilled water, added dropwise by means of a pipet, then with three 20-cc. portions of a 1 : 1 alcohol-ether mixture, also added dropwise. The p-nitrosodiethylaniline is dried in a desiccator over calcium chloride. The material weighs 57 g. (95 per cent of the theoretical amount). If a very pure sample is desired, the nitrosoamine should be crystallized from petroleum ether. (M. Q. DOJA and A. MOKEET, private communication.)

2. Attempts to use technical nitrosodimethylaniline in place of the freshly prepared material were not successful.

3. It is essential that the temperature reach the maximum (about 105°) and be kept at this point for sufficient time during the hydrolysis.

4. The acid aqueous liquors contain *p*-aminodimethylaniline and might be utilized in the preparation of quinoneimine dyes.

5. 2,4-Dinitrobenzaldehyde is a useful reagent since it forms crystalline condensation products with amines and with substances having a reactive methylene or methyl group.[1]

3. Methods of Preparation

Dinitrobenzaldehyde was first prepared from dinitrotoluene.[1,2,3] It has also been prepared from 2,4-dinitrobenzylaniline by oxidation to the benzylidene compound and subsequent hydrolysis,[4,5] and by oxidation of dinitrobenzyl alcohol.[5]

p-DINITROBENZENE

(Benzene, *p*-dinitro-)

$$p\text{-}NO_2C_6H_4NH_2 + HNO_2 + HBF_4 \rightarrow p\text{-}NO_2C_6H_4N_2BF_4 + 2H_2O$$

$$p\text{-}NO_2C_6H_4N_2BF_4 + NaNO_2 \xrightarrow{\text{(Cu)}} p\text{-}C_6H_4(NO_2)_2 + N_2 + NaBF_4$$

Submitted by E. B. STARKEY.
Checked by LEE IRVIN SMITH and H. E. UNGNADE.

1. Procedure

THIRTY-FOUR grams (0.25 mole) of *p*-nitroaniline is dissolved in 110 cc. of fluoboric acid solution (Note 1) in a 400-cc. beaker. The beaker is placed in an ice bath and the solution stirred with an efficient stirrer. A cold solution of 17 g. (0.25 mole) of sodium nitrite in 34 cc. of water is added dropwise. When the addition is complete, the mixture is stirred for a few minutes and filtered by suction on a sintered glass filter. The solid diazonium fluoborate is washed once with 25–30 cc. of cold

[1] Bennett and Pratt, J. Chem. Soc. **1929**, 1465.

[2] Sachs and Kempf, Ber. **35**, 1224 (1902); Sachs, Ger. pat. 121,745 [Frdl. **6**, 1047 (1900–02)].

[3] Müller, Ber. **42**, 3695 (1909).

[4] Sachs and Everding, ibid. **35**, 1237 (1902).

[5] Cohn and Friedländer, ibid. **35**, 1266 (1902).

fluoboric acid, twice with 95 per cent alcohol, and several times with ether (Note 2). The product weighs 56–59 g. (95–99 per cent of the theoretical amount).

Two hundred grams of sodium nitrite is dissolved in 400 cc. of water in a 2-l. beaker, and 40 g. of copper powder is added (Note 3). The mixture is stirred with an efficient stirrer (Note 4), and a suspension of the *p*-nitrophenyldiazonium fluoborate in 200 cc. of water is added slowly. Much frothing occurs, and 4–5 cc. of ether is added from time to time to break the foam. The reaction is complete when all the diazonium compound has been added. The product is filtered with suction, washed several times with water, twice with dilute sodium hydroxide solution, and again with water. The solid is dried in an oven at 110°, powdered, and extracted with 300-cc., 200-cc., and 150-cc. portions of boiling benzene. The benzene is evaporated on a water bath, and the residue is crystallized from 120–150 cc. of boiling glacial acetic acid. The resulting reddish yellow crystals, melting at 172–173°, weigh 28–34.5 g. (67–82 per cent yield) (Note 5). Recrystallization from alcohol yields pale yellow crystals melting at 173°.

2. Notes

1. Fluoboric acid is made by adding 184 g. of boric acid slowly, with constant stirring, to 450 g. of hydrofluoric acid (48–52 per cent) in a copper, lead, or silver-plated container placed in an ice bath. *Hydrofluoric acid causes very painful burns. Exposed parts of the body must be protected when working with this material.* Compare Note 3, p. 297. Fluoboric acid (40 per cent) is now a commercial product.

2. A sintered glass filter should be used for filtering, and the fluoborate stirred well on the filter with each washing, before suction is applied. The diazonium fluoborate is stable and may be dried in a vacuum desiccator over phosphorus pentoxide.

3. The copper used was "Copper metal, precipitated powder."

4. In the decomposition reaction, efficient stirring is quite essential. An off-center stirrer is best suited for the purpose. The reaction should require about two hours. Less efficient stirring and a shorter reaction time cause the formation of an impure product which is not readily purified by crystallization.

5. *o*-Dinitrobenzene may be prepared from *o*-nitroaniline by the same general method. From 34 g. (0.25 mole) of the amine there is obtained 38 g. (63 per cent yield) of dry *o*-nitrobenzenediazonium fluoborate. After carrying out the reaction of the diazonium fluoborate with sodium nitrite and copper, the *o*-dinitrobenzene is separated by steam distilla-

tion instead of filtration and extraction. A little paraffin is added to diminish the troublesome foaming and creeping during distillation, and a special flask (Note 6) designed for such operations may be used advantageously. The crystalline product is filtered from the steam distillate and recrystallized from alcohol. The yield of *o*-dinitrobenzene, m.p. 116–116.5°, is 14–16 g. (33–38 per cent of the theoretical amount).

6. A Claisen flask modified as shown in Fig. 8 is used in the synthetic laboratories of the Eastman Kodak Company for the distillation of

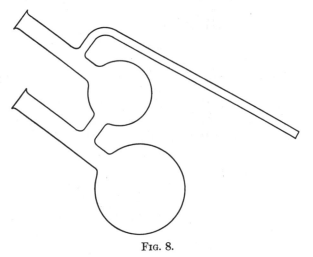

FIG. 8.

liquids which foam and creep badly. The large bulb inserted in the side arm should have a capacity about one-third that of the distillation flask. During distillation the flask is supported at the angle shown in the figure.

3. Methods of Preparation

p-Dinitrobenzene has been prepared from *p*-nitrosonitrobenzene by treatment with nitric acid;[1] from *p*-nitroaniline by the Sandmeyer reaction;[2] and by the oxidation of *p*-nitroaniline in concentrated sulfuric acid with ammonium persulfate.[3] *o*-Dinitrobenzene has been prepared in similar fashion from *o*-nitroaniline by the Sandmeyer reaction[4] and by oxidation with persulfate.[3] The preparation described above has been published.[5]

[1] Bamberger and Hübner, Ber. **36**, 3808 (1903).
[2] Meisenheimer and Patzig, ibid. **39**, 2528 (1906).
[3] Witt and Kopetschni, ibid. **45**, 1134 (1912).
[4] Körner and Contardi, Atti accad. Lincei (5) **23**, I, 283 (1914) [C. A. **8**, 3020 (1914)].
[5] Starkey, J. Am. Chem. Soc. **59**, 1479 (1937).

2,4-DINITROPHENYLHYDRAZINE

[Hydrazine, (2,4-dinitrophenyl)-]

$$C_6H_3(NO_2)_2Cl + 3CH_3CO_2K + NH_2NH_2 \cdot H_2SO_4$$
$$\rightarrow C_6H_3(NO_2)_2NHNH_2 + K_2SO_4 + KCl + 3CH_3CO_2H$$

Submitted by C. F. H. ALLEN.
Checked by REYNOLD C. FUSON and MARK W. FARLOW.

1. Procedure

THIRTY-FIVE grams (0.27 mole) of hydrazine sulfate (Org. Syn. Coll. Vol. I, **1941**, 309) is suspended in 125 cc. of hot water in a 400-cc. beaker and stirred by hand during the addition of 85 g. (0.87 mole) of potassium acetate (Note 1). The mixture is boiled five minutes and then cooled to about 70°; 75 cc. of alcohol is added, and the solid is filtered with suction and washed with 75 cc. of hot alcohol. The filtered hydrazine solution is saved for the next step.

In a 1-l. flask fitted with a stirrer and reflux condenser, 50.5 g. (0.25 mole) of technical 2,4-dinitrochlorobenzene is dissolved in 250 cc. of alcohol; the hydrazine solution is added, and the mixture is refluxed with stirring for an hour. Most of the product separates during the first ten minutes (Note 2); it is cooled well, filtered, and washed, once with 50 cc. of warm alcohol (60°) to remove unchanged halide and then with 50 cc. of hot water. The solid weighs 30 g. and melts at 190–192° with evolution of gas (Note 3); it is pure enough for most purposes. By distilling half the alcohol from the filtrate a less pure second crop is obtained; this is recrystallized from n-butyl alcohol (30 cc. per g.) (Note 4). The total yield is 40–42 g. (81–85 per cent of the theoretical amount) (Notes 5–10).

2. Notes

1. An equivalent amount of sodium acetate may be substituted for the potassium acetate.

2. Considerable heat is evolved during the separation.

3. The melting point is not sharp; in the capillary tube the sample shrinks about 10° below the melting point.

4. For recrystallization, n-butyl alcohol is the best solvent in spite of the large amount required, but tetralin, pyridine, or dioxane (10 cc. per g.) may be used where large quantities are involved. Fortunately, most of the material as prepared does not need further purification.

5. Complete evaporation of the filtrate yields a gummy residue; the amount of dinitrochlorobenzene present is too small to justify recovery.

6. By substituting 10 g. of sodium hydroxide for every 35 g. of potassium acetate and boiling for five minutes without filtering the salt, a 70 per cent yield of dinitrophenylhydrazine results; the quality is not quite so good as that obtained by the use of potassium acetate.

7. A slightly higher yield is obtained by starting with hydrazine hydrate.

8. 2,4-Dinitrophenylhydrazine is used in qualitative organic analysis for preparing solid derivatives of carbonyl compounds.[1,2]

9. By this procedure 2,6-dinitrophenylhydrazine can be prepared from 2,6-dinitrochlorobenzene; picryl chloride gives 2,4,6-trinitrophenylhydrazine.

10. These directions have been used equally successfully with twice, and with five times, the amounts specified.

3. Methods of Preparation

2,4-Dinitrophenylhydrazine has been prepared from hydrazine hydrate and 2,4-dinitrochlorobenzene [3,2] or 2,4-dinitrobromobenzene,[4] and from the same halogen compounds and hydrazine acetate.[1]

1,4-DIPHENYLBUTADIENE

(Bistyryl)

$$C_6H_5CH_2CO_2H + C_6H_5CH{=}CHCHO \xrightarrow[(CH_3CO)_2O]{(PbO)}$$

$$C_6H_5CH{=}CHCH{=}CHC_6H_5 + CO_2 + H_2O$$

Submitted by B. B. CORSON.
Checked by C. R. NOLLER and J. F. CARSON.

1. Procedure

IN a 1-l. round-bottomed flask equipped with a reflux condenser and protected by a calcium chloride tube are placed 150 g. (1.1 moles) of

[1] Allen, J. Am. Chem. Soc. 52, 2955 (1930).

[2] Brady and Elsmie, Analyst 51, 77 (1926); Brady, J. Chem. Soc. 1931, 757.

[3] Purgotti, Gazz. chim. ital. 24 (I) 555 (1894).

[4] Curtius and Dedichen, J. prakt. chem. (2) 50, 258 (1894).

phenylacetic acid (Org. Syn. Coll. Vol. I, **1941,** 436), 147 g. (1.1 moles) of freshly distilled cinnamic aldehyde, 122 g. of litharge, and 155 cc. of acetic anhydride. The mixture is boiled for five hours (Note 1), poured while still hot into a beaker, and allowed to stand overnight. The semi-solid contents are stirred to a mush, filtered with suction in a large Büchner funnel, and pressed dry. The solid is washed on the funnel with two 35-cc. portions of ethyl alcohol, the material being stirred thoroughly before suction is applied. The cake is transferred to a beaker, stirred to a mush with 50 cc. of alcohol, and again filtered with suction. The material is washed with another 50-cc. portion of alcohol in the same manner (Note 2). The product, which is light yellow to tan in color, weighs 62–67 g. (27–29 per cent of the theoretical amount) (Note 3) and melts at 149.5–153.5°.

For purification the material is dissolved in 300 cc. of hot benzene; the solution is boiled three minutes with 5 g. of decolorizing carbon, and filtered hot, with gentle suction, through a warm Büchner funnel. The benzene filtrate is treated with 500 cc. of hot ethyl alcohol, heated to boiling, and then cooled to 10° in an ice bath, with shaking. After the crystals are filtered with suction and pressed thoroughly, 50 cc. of ethyl alcohol is allowed to soak into the cake and suction is applied again. The recrystallized product weighs 52–57 g. (23–25 per cent of the theoretical amount) (Note 4) and melts at 152.5–153.5°. This product is the trans-trans form of the diene.

2. Notes

1. During the first half hour, the flask should be heated gently and shaken several times to facilitate solution of the lead oxide.

2. Thorough washing is essential.

3. About 8 g. of crude hydrocarbon can be recovered from the original mother liquor, but the recovery is tedious and is not recommended.

4. An additional quantity of about 5.5 g. can be obtained by evaporating the mother liquor to 25 cc. and allowing to crystallize.

3. Methods of Preparation

The only method of preparative interest is the condensation of phenylacetic acid and cinnamic aldehyde.[1] The above procedure follows the method of Kuhn and Winterstein.[2] Bistyryl has been obtained also

[1] Thiele and Schleussner, Ann. **306,** 198 (1899).
[2] Kuhn and Winterstein, Helv. Chim. Acta **11,** 103 (1928).

from styrylmagnesium bromide and cupric chloride,[3] or azobenzene,[4] and by the reduction of β-bromostyrene with hydrazine in the presence of palladium.[5]

4,5-DIPHENYLGLYOXALONE

[2(5)-Imidazolone, 4,5-diphenyl-]

$$C_6H_5CHOHCOC_6H_5 + NH_2CONH_2 \rightarrow \begin{array}{c} C_6H_5CH—NH \\ | \\ C_6H_5C=N \end{array} CO + 2H_2O$$

Submitted by B. B. Corson and Emeline Freeborn.
Checked by Frank C. Whitmore and Marion M. Whitmore.

1. Procedure

A MIXTURE of 212 g. (1 mole) of benzoin (Note 1), 110 g. (1.8 moles) of urea and 800 cc. of glacial acetic acid in a 2-l. round-bottomed flask is heated under a reflux condenser for seven hours (Note 2). The hot solution is quickly poured into a 2-l. beaker (Note 3) and allowed to stand at least three hours. The cold mixture is then transferred by means of a large wooden spoon or spatula to a 30-cm. Büchner funnel and sucked as dry as possible with the help of a rubber dam. The crystals are pressed down during the suction filtration. The filtrate is discarded. The crystals are returned to the beaker, stirred mechanically with 500 cc. of ether for thirty minutes, filtered again by suction (Note 4), and spread out to dry at least overnight (Note 5). The product is dissolved by heating with 1 l. of glacial acetic acid in a 2-l. round-bottomed flask attached to a reflux condenser (Note 6), and the clear solution is poured, with mechanical stirring, into 2.5 l. of water in a 4-l. (1-gal.) crock. Stirring is continued for thirty minutes (Note 7). The mixture is filtered on a 30-cm. Büchner funnel and pressed and sucked as dry as possible. The crystals are transferred to the 2-l. beaker and stirred mechanically for ten minutes with 1 l. of water. After filtration, the washing with water is repeated. The product is then returned to the beaker, stirred with 750 cc. of ether, and filtered again. The material is then air-dried to constant weight. The yield of white fluffy crystals of diphenylglyoxalone melting at 330–335° (corr.) (Note 8) is 220–230 g. (93–97 per cent of the theoretical amount).

[3] Sakellarios and Kyrimis, Ber. **57**, 325 (1924); see also Gilman and Parker, J. Am. Chem. Soc. **46**, 2827 (1924).

[4] Gilman and Pickens, ibid. **47**, 2410 (1925).

[5] Busch and Weber, J. prakt. Chem. (2) **146**, 54 (1936).

2. Notes

1. The benzoin (Org. Syn. Coll. Vol. I, **1941,** 94) need not be re-crystallized.

2. At first the color is reddish orange but later it changes to dark yellow.

3. If left in the flask the product solidifies too much to be removed readily.

4. Unreacted benzoin is removed by the ether, in which diphenylgly-oxalone is only sparingly soluble.

5. The product is difficult to dry completely.

6. This step is not successful if less than 1 l. of glacial acetic acid is used.

7. The water removes unreacted urea.

8. The melting point is conveniently taken on the surface of mercury heated in a test tube with the thermometer dipping in the mercury.

3. Method of Preparation

The only method of preparative interest is the interaction between benzoin and urea in acetic acid solution as described by Biltz[1] and Chattaway.[2]

DIPHENYLMETHANE

(Methane, diphenyl-)

$$C_6H_5CH_2Cl + C_6H_6 \xrightarrow{\text{(Aluminum amalgam)}} (C_6H_5)_2CH_2 + HCl$$

Submitted by W. W. HARTMAN and ROSS PHILLIPS.
Checked by REYNOLD C. FUSON and S. H. BABCOCK.

1. Procedure

IN a 5-l. flask, provided with a reflux condenser and an S-tube attached to a dropping funnel, are placed 2 kg. (2.3 l., 25.6 moles) of benzene, which has been dried by distilling until the distillate comes over clear, and 10 g. of amalgamated aluminum turnings (Note 1). The benzene is heated to boiling on a steam bath, the steam is turned off, and 500 g. (3.96 moles) of benzyl chloride is added at such a rate as to cause the

[1] Biltz, Ann. **368,** 173 (1909).
[2] Chattaway and Coulson, J. Chem. Soc. **1928,** 1363.

solution to boil (Note 2). The hydrogen chloride is absorbed in water or allowed to pass out-of-doors. When all the benzyl chloride has been added (one hour), the mixture is warmed for ten to fifteen minutes, or until the evolution of hydrogen chloride ceases. When cool, the benzene solution of diphenylmethane is decanted from the small amount of tarry material (Note 3) and washed with 5 per cent sodium hydroxide solution and then with water. After a partial drying with calcium chloride, the benzene is distilled from a steam bath and the residue fractionated under diminished pressure. The fore-run is collected up to 125°/10 mm., the main product at 125–130°/10 mm., and an after-run up to 150°/10 mm. (Note 4). Redistillation of the fore-run and after-run yields a small amount of material which is added to the main fraction. The latter is chilled and a small amount of oil is decanted from the crystals. The yield of material melting at 24–25° is 330–350 g. (49.5–52.5 per cent of the theoretical amount).

2. Notes

1. Amalgamated aluminum is prepared as follows: Aluminum turnings, freed from any oil by washing with ether, are stirred with a 5 per cent mercuric chloride solution for a few minutes and then washed quickly with water followed by methyl alcohol. The amalgamated aluminum is used at once.

2. At times the reaction is slow in starting. Not more than 50–60 g. of benzyl chloride is added at first, and the mixture is heated until the evolution of hydrochloric acid indicates that the reaction is under way. If too much benzyl chloride is present when the reaction starts, the contents of the flask may boil over.

3. Succeeding batches may be started in the same flask containing the aluminum turnings and the trace of tarry material without the addition of further catalyst; these batches do not show an induction period.

4. The material in the residue and high-boiling fraction may be partially converted into diphenylmethane by heating with one-third its weight of aluminum chloride and five times its weight of benzene.

3. Methods of Preparation

Diphenylmethane can be prepared from benzene and benzyl chloride with aluminum chloride,[1] hydrogen fluoride,[2] beryllium chloride,[3] the

[1] Friedel and Balsohn, Bull. soc. chim. (2) **33**, 337 (1880).
[2] Simons and Archer, J. Am. Chem. Soc. **61**, 1521 (1939).
[3] Bredereck, Lehmann, Schönfeld, and Fritzsche, Ber. **72**, 1414 (1939).

double salt of aluminum and sodium chlorides,[4] zinc dust,[5] zinc chloride,[6] or aluminum amalgam [7] as a condensing agent. The procedure described above is a slight modification of the last method.

Benzene and benzyl alcohol furnish diphenylmethane on treatment with boron fluoride,[8] hydrogen fluoride,[9] or beryllium chloride.[3] Diphenylmethane has also been prepared from benzene, methylene chloride, and aluminum chloride,[10] and from benzene, formaldehyde, ethanol, and concentrated sulfuric acid.[11] The reduction of benzophenone to diphenylmethane has been effected by hydriodic acid and phosphorus,[12] sodium and alcohol,[13] and fusion with zinc chloride and sodium chloride.[14] The condensation of benzylmagnesium chloride and benzene to diphenylmethane can be brought about by small amounts of magnesium and water.[15]

DIPHENYLMETHANE IMINE HYDROCHLORIDE

(Benzohydrylideneimine, hydrochloride)

$$8(C_6H_5)_2C\!\!=\!\!NOH \rightarrow$$
$$4(C_6H_5)_2CO + 4(C_6H_5)_2C\!\!=\!\!NH + N_2 + 2H_2O + 2NO$$
$$(C_6H_5)_2C\!\!=\!\!NH + HCl \rightarrow (C_6H_5)_2C\!\!=\!\!NH \cdot HCl$$

Submitted by ARTHUR LACHMAN.
Checked by C. R. NOLLER.

1. Procedure

A PIECE of glass tubing 80 cm. long and 2 cm. in internal diameter is sealed at one end and loosely packed with 49 g. (0.25 mole) of benzophenone oxime (p. 70). The tube is supported in a nearly horizontal position, dipping slightly toward the closed end, and connected to a small

[4] Norris and Klemka, J. Am. Chem. Soc. 62, 1432 (1940).
[5] Zincke, Ann. 159, 374 (1871).
[6] Friedel and Crafts, Ann. chim. phys. (6) 1, 478 (1884).
[7] Hirst and Cohen, J. Chem. Soc. 67, 827 (1895).
[8] McKenna and Sowa, J. Am. Chem. Soc. 59, 470 (1937).
[9] Simons and Archer, ibid. 62, 1623 (1940).
[10] Friedel and Crafts, Bull. soc. chim. (2) 41, 324 (1884); Schwarz, Ber. 14, 1526 (1881).
[11] Blythe and Company, Ltd., Brit. pat. 446,450 [C. A. 30, 6760 (1936)].
[12] Graebe, Ber. 7, 1624 (1874).
[13] Klages and Allendorff, ibid. 31, 999 (1898).
[14] Clar, ibid. 72, 1645 (1939).
[15] Kharasch, Goldberg, and Mayo, J. Am. Chem. Soc. 60, 2004 (1938).

filter flask by means of rubber stoppers and a piece of glass tubing bent at a right angle. The system is evacuated by means of a water pump, and dry carbon dioxide is admitted; the system is re-evacuated and carbon dioxide is admitted again. The oxime is then heated with a free flame, beginning at the upper end and heating at one spot until decomposition takes place before heating a further portion (Note 1). When all the oxime has been decomposed, the mixture which has collected at the closed end of the tube (Note 2) is heated strongly for a short time to complete the decomposition, and allowed to cool. Suction is again applied to the tube, and the condensed water is driven out by gently warming the tube. The liquid is then transferred to a small distilling flask and distilled at a pressure of about 20 mm. The distillate, consisting of a mixture of benzophenone and diphenylmethane imine, is dissolved in 400 cc. of ligroin (60–90°), and the imine hydrochloride is precipitated by bubbling in dry hydrogen chloride. The salt is filtered with suction (Note 3), washed with a little ligroin, dried, and preserved in a dry atmosphere (Note 4). It sublimes without decomposition at 230–250° (Note 5). The yield is 16–18 g. (59–66 per cent of the theoretical amount) (Note 6).

2. Notes

1. When the heating is carefully done, no material is carried out of the tube with the gases.

2. Care should be taken to prevent any drops of water that have condensed in the cool portion of the tube from running back and mixing with the liquid that is being heated.

3. Benzophenone may be recovered from the filtrate.

4. In moist air the hydrochloride is changed to a mixture of benzophenone and ammonium chloride. The free base on standing in air gives off ammonia and gradually deposits crystals of benzophenone.

5. The imine salt can be converted into the free base by the method of Hantzsch and Kraft, which involves treating a solution of the salt in chloroform with dry ammonia.

6. The yield depends greatly on the quality of the benzophenone oxime. If this contains moisture or has been exposed to moist air, especially in a closed space, for any length of time, the yield is markedly decreased.

3. Methods of Preparation

Diphenylmethane imine (or its hydrochloride) has been obtained by heating diphenyldichloromethane with urethane at 130°;[1] by the action

[1] Hantzsch and Kraft, Ber. **24**, 3516 (1891).

of ammonia on diphenyldibromomethane;[2] by treating benzophenone-chloroimide with phosphorus pentachloride in ether solution,[3] or dry hydrogen chloride in ligroin solution;[4] by the action of phenylmagnesium bromide on N-bromobenzamide,[2] benzonitrile,[5] cyanogen bromide,[6] cyanogen chloride,[7] and alkyl thiocyanates.[8] It has also been obtained by passing a mixture of ammonia and benzophenone vapor over thorium oxide at 380–390°;[9] by the catalytic reduction of benzophenone oxime using hydrogen and a nickel catalyst in absolute alcohol solution at ordinary temperature and pressure;[10] by passing a mixture of hydrogen and benzophenone oxime vapor over reduced copper at 200°;[11] and by the action of sodium ethoxide on N-monochlorodiphenylmethylamine in alcoholic solution.[12] The method described here has been previously published by Lachman.[13]

β,β-DIPHENYLPROPIOPHENONE

(Propiophenone, β,β-diphenyl-)

$$C_6H_5CH{=}CHCOC_6H_5 + C_6H_6 \xrightarrow{(AlCl_3)} (C_6H_5)_2CHCH_2COC_6H_5$$

Submitted by P. R. SHILDNECK.
Checked by C. R. NOLLER and C. R. KEMP.

1. Procedure

IN a 3-l. round-bottomed, three-necked flask fitted with a liquid-sealed mechanical stirrer, a thermometer, and a 500-cc. separatory funnel are placed 1.7 l. of dry benzene and 160 g. (1.2 moles) of powdered, anhydrous aluminum chloride (Note 1). The mixture is cooled to 10° by

[2] Moore, ibid. **43,** 564 (1910).
[3] Vosburgh, J. Am. Chem. Soc. **38,** 2095 (1916).
[4] Peterson, Am. Chem. J. **46,** 331 (1911).
[5] Moureu and Mignonac, Compt. rend. **156,** 1806 (1913); Ann. chim. (9) **14,** 336 (1920).
[6] Grignard, Bellet, and Courtot, ibid. (9) **4,** 34 (1915).
[7] Grignard, Bellet, and Courtot, ibid. (9) **12,** 379 (1919).
[8] Adams, Bramlet, and Tendick, J. Am. Chem. Soc. **42,** 2372 (1920).
[9] Mignonac, Compt. rend. **169,** 239 (1919).
[10] Mignonac, ibid. **170,** 938 (1920).
[11] Yamaguchi, Bull. Chem. Soc. Japan **1,** 35 (1926) [C. A. **21,** 75 (1927)].
[12] Hellerman and Sanders, J. Am. Chem. Soc. **49,** 1742 (1927).
[13] Lachman, ibid. **46,** 1477 (1924).

means of an ice-water bath and maintained at 10–20° during the addition of a solution of 120 g. (0.58 mole) of benzalacetophenone (Note 2) in 300 cc. of dry benzene. This addition requires about thirty minutes. The cooling bath is then removed and stirring continued at room temperature until all the dense, yellow precipitate formed at first has gone into solution (Note 3). The reaction is complete after stirring for an additional hour.

The bulk of the brown-colored benzene solution is decanted into a cold mixture of 100 cc. of concentrated hydrochloric acid and 1.5 l. of water in a 5-l. round-bottomed flask. The remainder is filtered on a Büchner funnel, and the lumps of aluminum chloride are washed with two 100-cc. portions of benzene. The filtrates are added to the main solution and the whole is washed thoroughly with the dilute acid. If the layers do not separate readily the mixture is filtered with gentle suction and the water separated by siphoning. The solution is washed twice with 1.5-l. portions of water and filtered again if necessary.

The clear, light yellow benzene solution is subjected to rapid steam distillation (Note 4) in the same 5-l. flask, and when no more benzene passes over the flask is cooled under the tap with shaking. The residual oil solidifies to light brown pellets. These are collected, separated from water as much as possible, and dissolved in 2250 cc. of boiling alcohol. Five grams of decolorizing carbon is added; the hot solution is filtered with suction and allowed to cool. The best results are obtained if the alcoholic solution is stirred slowly with a mechanical stirrer while cooling to room temperature. Stirring is stopped when the mixture becomes semi-solid, and the mass is then allowed to stand for twenty-four hours. The fine, colorless needles are filtered on a 15-cm. Büchner funnel and pressed as dry as possible. The yield of thoroughly air-dried, colorless material melting at 91–92° (Note 5) amounts to 125–140 g. (76–85 per cent of the theoretical amount) (Note 6).

2. Notes

1. Trial runs demonstrated that one mole of benzalacetophenone required at least two moles of anhydrous aluminum chloride to complete the reaction at room temperature. When less aluminum chloride was used the yellow addition product failed to dissolve entirely even after stirring for twenty-four hours, and the yield was decreased.

2. The benzalacetophenone (Org. Syn. Coll. Vol. I, **1941**, 78) must be quite pure (m.p. 55–56°) and, in particular, free from benzaldehyde.

3. A change in appearance of the mixture is very noticeable at the end of the reaction. The yellow color rapidly gives way to a dark brown.

A perfectly clear solution is not produced as the aluminum chloride remains in suspension.

4. If the benzene solution is steam-distilled directly from the acid mixture the crude product is darker in color, more difficult to crystallize, and less pure after crystallization.

5. The melting point of β,β-diphenylpropiophenone is given in the literature [1,2,3] as 96°. The product melting at 91–92° was recrystallized to constant melting point from alcohol and from ligroin, but the melting point remained at 92° (corr.). The oxime and the monobromoketone prepared according to Kohler [3] were found to melt at 133° (corr.) and 166° (corr.), respectively.

6. An additional 12–16 g. of less pure product, melting at 88–91°, may be obtained by concentrating the mother liquor to 400 cc.

3. Methods of Preparation

β,β-Diphenylpropiophenone has been prepared from benzalacetophenone with phenylmagnesium bromide,[1] and a number of other phenylmetallic compounds; [2] by the condensation of benzalacetophenone and benzene with concentrated sulfuric acid [3] or with aluminum chloride,[4] and by the action of aluminum chloride on a mixture of benzene and the hydrochloride of benzalacetophenone.[4] The procedure described here is essentially that of Vorländer and Friedberg.[4]

DIPHENYL SELENIDE

(Phenyl selenide)

$$C_6H_5NH_2 + NaNO_2 + 2HCl \rightarrow C_6H_5N_2Cl + NaCl + 2H_2O$$
$$2C_6H_5N_2Cl + K_2Se_x \rightarrow (C_6H_5)_2Se + 2KCl + (x - 1)Se + 2N_2$$

Submitted by HENRY M. LEICESTER.
Checked by W. W. HARTMAN and R. H. BULLARD.

1. Procedure

IN a 500-cc. beaker are placed 360 g. (6.4 moles) of powdered potassium hydroxide and 240 g. (3 gram atoms) of black powdered selenium,

[1] Kohler, Am. Chem. J. **29**, 352 (1903).
[2] Gilman and Kirby, J. Am. Chem. Soc. **63**, 2047 (1941).
[3] Kohler, Am. Chem. J. **31**, 642 (1904).
[4] Vorländer and Friedberg, Ber. **56**, 1144 (1923).

which have been previously ground together in a mortar. This mixture is heated (Note 1) in an oil bath at 140–150° until a thick, dark red liquid is formed (Note 2), and then it is added (Note 3) in small portions to 400 cc. of ice water in a 5-l. flask. The solution is kept in an ice bath until used.

To 375 cc. (4.3 moles) of hydrochloric acid (sp. gr. 1.18) and 200 g. of ice is added 139.6 g. (1.5 moles) of aniline. The resulting solution is diazotized with a solution of 103.5 g. (1.5 moles) of c.p. sodium nitrite, ice being added to the reaction mixture, as necessary, in order to keep the temperature below 5°. The final volume of the diazotized solution is about 1 l. This solution is added in a slow stream from a dropping funnel to the potassium selenide solution, which is vigorously stirred with a mechanical stirrer. When all the diazotized solution has been added, the red aqueous solution is decanted from the dark oil which forms and is heated to boiling (Note 4). It is then poured back on the oil, the mixture is well stirred (Note 5), 200 cc. of chloroform is added, and the selenium is collected on a filter and washed with a little more chloroform (Note 6). After the chloroform layer is separated, the aqueous layer is again extracted with 200 cc. of chloroform. The combined extracts are then distilled, the diphenyl selenide being collected from 300 to 315°. The yield of yellow oil of rather unpleasant odor is 138–150 g. (79–86 per cent of the theoretical amount) (Note 7). If a purer product is desired, this material can be distilled under diminished pressure. It boils at 165–167°/12 mm.

2. Notes

1. Although no toxic gases are evolved in this reaction, the mixture has a rather unpleasant odor, so that it is best to use a hood throughout.

2. If the potassium hydroxide and selenium are absolutely dry, a thick paste is formed. The addition of a few cubic centimeters of water will form the dark red liquid mentioned.

3. If the mass is allowed to cool and solidify, it becomes very hard and is difficult to break up and dissolve.

4. Unless the aqueous layer is heated separately, the beaker containing the viscous oil always cracks, no matter how well the liquid is stirred.

5. This treatment converts the selenium from the red, colloidal form into the more easily filterable black modification.

6. The amount of selenium recovered is 101–115 g.

7. When smaller amounts of the selenide are prepared, the yield drops to 70–75 per cent. The method may be applied equally well to the

preparation of the selenides from the three toluidines, giving yields of 50–70 per cent.

3. Methods of Preparation

The method described here is a modification of that of Schoeller.[1] Diphenyl selenide has also been prepared from diazotized aniline and alkali monoselenides;[2] by the Friedel-Crafts reaction with benzene and selenium tetrachloride[3] or selenium dioxide;[4] from diphenyl sulfone and selenium;[5] from phenylmagnesium bromide and selenium,[6] selenium dichloride,[7] selenium oxychloride,[7] or selenium dibromide;[8] and from sodium selenophenoxide and bromobenzene.[9]

DIPHENYLSELENIUM DICHLORIDE AND TRIPHENYLSELENONIUM CHLORIDE

(Selenium compounds, diphenyl- dichloride)
(Selenonium compounds, triphenyl- chloride)

(A) $\qquad$ $(C_6H_5)_2Se \xrightarrow{\text{(HNO}_3 + \text{HCl)}} (C_6H_5)_2SeCl_2$

(B) $\quad (C_6H_5)_2SeCl_2 + C_6H_6 \xrightarrow{\text{(AlCl}_3)} (C_6H_5)_3SeCl + HCl$

Submitted by HENRY M. LEICESTER.
Checked by W. W. HARTMAN and R. H. BULLARD.

1. Procedure

(A) *Diphenylselenium Dichloride.*—One hundred twenty-five grams (0.53 mole) of diphenyl selenide (p. 238) is added in portions to 250 cc. (4 moles) of nitric acid (sp. gr. 1.42) in a 1.5-l. beaker. Hydrochloric acid (sp. gr. 1.18) is then added until precipitation is complete. About 170 cc. (2 moles) of acid is required. The mixture is then diluted with

[1] Schoeller, Ber. **52**, 1517 (1919).

[2] Lesser and Weiss, ibid. **47**, 2521 (1914).

[3] Krafft and Kaschau, ibid. **29**, 428 (1896); Bradt and Green, J. Org. Chem. **1**, 540 (1937).

[4] Lyons and Bradt, Ber. **60**, 60 (1927).

[5] Krafft and Vorster, ibid. **26**, 2817 (1893); Krafft and Lyons, ibid. **27**, 1761 (1894)

[6] Taboury, Ann. chim. (8) **15**, 35 (1908).

[7] Strecker and Willing, Ber. **48**, 196 (1915).

[8] Pieroni and Balduzzi, Gazz. chim. ital. **45** (II) 106 (1915).

[9] Foster and Brown, J. Am. Chem. Soc. **50**, 1182 (1928).

500 cc. of water, and the yellow precipitate is separated by filtration and air-dried. The crude product is purified by extracting with 500 cc. of boiling benzene. The crystals which separate on cooling are collected on a filter and the filtrate is used for a further extraction. Three such treatments are necessary for a complete crystallization. The yield of yellow needles, decomposing at 187–188°, is 137–141 g. (85–87 per cent of the theoretical amount).

(B) *Triphenylselenonium Chloride.*—To 87 g. (100 cc., 1.1 moles) of benzene in a 1-l. three-necked flask provided with a mechanical stirrer, is added 30 g. (0.22 mole) of anhydrous aluminum chloride. The suspension is cooled in an ice bath, and to it is added, with stirring, 40 g. (0.13 mole) of diphenylselenium dichloride in portions of approximately 1 g. at a time over a period of twenty-five minutes. Before each addition the temperature should be below 10° to prevent the final product from becoming dark. When the addition is complete, the reaction mixture is allowed to stand for three hours at room temperature, and then 200 cc. of water is added cautiously (Note 1). The benzene layer is separated and discarded. If the water layer is colored, further extractions with benzene will remove most of the color without reducing the yield. The water layer is then extracted three times with 50-cc. portions of chloroform. The combined extracts are concentrated to a volume of 40 cc. and treated with 120 cc. of ether (Note 2). A yellow oil precipitates and solidifies almost at once to a white powder. The triphenylselenonium chloride is collected on a filter and recrystallized from 300 cc. of methyl ethyl ketone to which 20 cc. of water has been added (Note 3). The yield of anhydrous product, after drying at 100°, is 30 g. (67 per cent of the theoretical amount) (Note 4).

2. Notes

1. Much heat is evolved during the first part of the hydrolysis.
2. Triphenylselenonium chloride may be precipitated from aqueous solution as the zinc chloride double salt.[1]
3. Triphenylselenonium chloride is soluble in anhydrous methyl ethyl ketone only to the extent of 2 g. in 300 cc. Using the water-methyl ethyl ketone mixture, the substance crystallizes with two molecules of water. This can be removed by heating for half an hour at 100°, but is again taken up from moist solvents or moist air.
4. This method can also be used for the preparation of p-tolyl- or of mixed phenyl-p-tolylselenonium salts.

[1] Crowell and Bradt, J. Am. Chem. Soc. **55**, 1500 (1933).

3. Methods of Preparation

Diphenylselenium dichloride has been prepared from diphenyl selenide [2,3] by the action of chlorine and by treatment first with nitric acid and then hydrochloric acid. Triphenylselenonium chloride has been prepared by fusing together diphenylmercury and diphenylselenium dichloride,[4] and by the action of diphenylselenium dichloride on benzene in the presence of aluminum chloride.[3]

DIPHENYL SULFIDE

(Phenyl sulfide)

$$2C_6H_6 + S_2Cl_2 \xrightarrow{(AlCl_3)} C_6H_5SC_6H_5 + S + 2HCl$$

Submitted by W. W. HARTMAN, L. A. SMITH, and J. B. DICKEY.
Checked by REYNOLD C. FUSON and S. H. BABCOCK.

1. Procedure

IN a 5-l. three-necked, round-bottomed flask, fitted with a mechanical stirrer, dropping funnel, and a condenser connected to an apparatus for removing hydrogen chloride, are placed 858 g. (980 cc., 11 moles) of dry benzene (Note 1) and 464 g. (3.48 moles) of aluminum chloride. The reaction mixture is cooled in an ice bath to 10°, and then 405.1 g. (3 moles) of commercial sulfur chloride in 390 g. (450 cc., 5 moles) of benzene is added, with stirring, over a period of one hour, the temperature being kept at about 10°. The reaction begins at once as evidenced by the evolution of hydrogen chloride and the separation of a yellow viscous aluminum chloride complex. When all the sulfur chloride has been added, the reaction mixture is removed from the ice bath, stirred at room temperature for two hours, and then heated at 30° until practically no hydrogen chloride is evolved (one hour). The mixture is then poured on 1 kg. of cracked ice, and, when hydrolysis is complete, the benzene layer is separated from the water layer by means of a separatory funnel. The benzene is distilled on a steam bath, and the resulting dark-colored oil is cooled to 0° and filtered through a Büchner funnel to remove the sulfur which separates. The residue is dissolved in 500

[2] Foster and Brown, ibid. **50**, 1182 (1928).

[3] Leicester and Bergstrom, ibid. **51**, 3587 (1929).

[4] Leicester and Bergstrom, ibid. **53**, 4428 (1931); Leicester, ibid. **57**, 1901 (1935).

cc. of commercial methyl alcohol, and the solution is cooled to 0°. Stirring is continued for three hours, and the precipitated sulfur is removed as before. The alcohol is removed on a steam bath, and the residue is distilled from a 1-l. modified Claisen flask with a water-cooled side-arm receiver. After a small amount of low-boiling product passes over, there is obtained 470–490 g. (Note 2) of a yellow liquid boiling at 155–170°/18 mm. The material thus obtained is heated for one hour on a steam bath, with stirring, with 70 g. of zinc dust and 200 g. of 40 per cent sodium hydroxide solution (Note 3). The diphenyl sulfide is then separated from the sodium hydroxide, washed with two 500-cc. portions of water, dried over anhydrous sodium sulfate, and distilled. The yield of colorless diphenyl sulfide boiling at 162–163°/18 mm. is 450–464 g. (81–83 per cent of the theoretical amount).

2. Notes

1. The benzene can be dried by distilling on a steam bath until the distillate is no longer milky. About 15 per cent of the benzene is distilled.
2. On distillation of the residue in the distillation flask a fraction boiling at 170–200°/18 mm. is obtained. Crystallization of this material from methyl alcohol yields 8–10 g. of thianthrene, melting at 155–156°.
3. It is necessary to treat the diphenyl sulfide as described in order to obtain a colorless product.

3. Methods of Preparation

Diphenyl sulfide can best be prepared by treating benzene and aluminum chloride with sulfur chloride,[1] sulfur dichloride,[2] or sulfur.[3] In addition to diphenyl sulfide, traces of thiophenol and varying amounts of thianthrene are found.

[1] Böeseken, Rec. trav. chim. 24, 209 (1905); Böeseken and Waterman, ibid. 29, 319 (1910); Böeseken and Koning, ibid. 30, 116 (1911); Genvresse, Bull. soc. chim. (3) 15, 409 (1896); Hartman, Smith, and Dickey, Ind. Eng. Chem. 24, 1317 (1932).
[2] Böeseken, Rec. trav. chim. 24, 217 (1905); Böeseken and Koning, ibid. 30, 116 (1911).
[3] Friedel and Crafts, Ann. chim. phys. (6) 14, 437 (1888); Böeseken, Rec. trav. chim. 24, 17, 219 (1905); Dougherty and Hammond, J. Am. Chem. Soc. 57, 117 (1935).

DIPHENYL TRIKETONE

(Propanetrione, diphenyl-)

$$C_6H_5COCH_2COC_6H_5 + 2Br_2 \rightarrow C_6H_5COCBr_2COC_6H_5 + 2HBr$$

$$C_6H_5COCBr_2COC_6H_5 + 2CH_3CO_2Na + 2H_2O$$
$$\rightarrow C_6H_5COC(OH)_2COC_6H_5 + 2NaBr + 2CH_3CO_2H$$

$$C_6H_5COC(OH)_2COC_6H_5 \xrightarrow{\text{Heat}} C_6H_5COCOCOC_6H_5 + H_2O$$

Submitted by Lucius A. Bigelow and Roy S. Hanslick.
Checked by W. W. Hartman and Lloyd A. Smith.

1. Procedure

(A) *Dibenzoyldibromomethane.*—In a 1-l. three-necked flask, equipped with a mechanical stirrer, a dropping funnel, and a thermometer, are placed 56 g. (0.25 mole) of dibenzoylmethane (Org. Syn. Coll. Vol. I, 1941, 205; Org. Syn. 20, 32) and 14 cc. of chloroform. The flask is surrounded by an ice bath, the stirrer is started, and a solution of 28.5 cc. (88 g., 0.55 mole) of dry bromine (Note 1) in 230 cc. of chloroform is added slowly from the dropping funnel during a period of about thirty minutes. The temperature of the mixture should not exceed 15° during the bromination; the hydrogen bromide evolved is continuously removed by aspirating a gentle stream of air over the surface of the solution. After all the bromine has been added, stirring is continued for about fifteen minutes. The solution is then transferred to a distilling flask, and the solvent is completely removed under diminished pressure at room temperature (Note 2). The slightly colored residue is crystallized from 125 cc. of hot 95 per cent ethyl alcohol. The dibenzoyldibromomethane is obtained in the form of white crystals melting at 94–95°. The yield is 72.4 g. (76 per cent of the theoretical amount) (Note 3).

(B) *Diphenyl Triketone Hydrate.*—A solution of 34.3 g. (0.42 mole) of fused sodium acetate in 142 cc. of hot glacial acetic acid is prepared in a 1-l. round-bottomed flask; 72.4 g. (0.19 mole) of dibenzoyldibromomethane is added, and the mixture is refluxed until the precipitation of sodium bromide ceases (one and one-half to two hours). The mixture is then cooled to room temperature and diluted with 150–200 cc. of water with constant shaking to dissolve the inorganic salt and to precipitate the triketone hydrate, which separates as a white, curdy mass (Note 4). This is separated by filtration, washed well with water, and dried in an oven at 60°. The melting point varies from 65 to 90°, depending upon the extent of dehydration that occurs during the drying operation. The yield is 41.5 g. (86 per cent of the theoretical amount based on the dibenzoyldibromomethane).

(C) *Diphenyl Triketone.*—The 41.5 g. (0.16 mole) of triketone hydrate is distilled in vacuum from a Claisen flask heated by means of a sand bath. A distilling flask is used as a receiver, and no condenser is necessary. The neck of the receiving flask must be warmed, however, to prevent clogging of the apparatus by crystallization of the distillate. The anhydrous triketone distils at 174–176°/2 mm. as a reddish oil that solidifies to a light yellow, crystalline mass. This is dissolved in 70 cc. of hot ligroin (b.p. 90–120°); it separates on cooling in light yellow needles which melt at 68–70°. The yield is 35 g. (91 per cent of the theoretical amount based on the triketone hydrate; 59 per cent based on the dibenzoylmethane) (Note 5).

2. Notes

1. The bromine should be dried by washing it with concentrated sulfuric acid.

2. If the solvent is removed by heat at ordinary pressure, there is a decrease both in the yield and the purity of the product.

3. If the recrystallization is omitted, a lower yield of the triketone hydrate is obtained.

4. If the hydrate separates in part in an oily condition, it may be dissolved in a small amount of glacial acetic acid and reprecipitated with an equal volume of water.

5. The triketone is hygroscopic and must be kept in a vacuum desiccator or sealed tube.

3. Methods of Preparation

The two general methods for the preparation of diphenyltriketone involve the treatment of dibenzoylmethane with bromine [1] and with oxides of nitrogen.[2] De Neufville and v. Pechmann, who originated the first method, recommended [1] conversion of the diketone to dibenzoylbromomethane and transformation of this to the acetate and then to dibenzoylbromocarbinol acetate which was split to the triketone. This sequence has been used with success by Kohler and Erickson [3] to prepare the triketone in good yields, but the present method, mentioned but not described by De Neufville and v. Pechmann,[1] is simpler and more direct.

[1] De Neufville and v. Pechmann, Ber. **23**, 3375, 3379 (1890).

[2] Wieland and Bloch, ibid. **37**, 1524, 1531 (1904).

[3] Kohler and Erickson, J. Am. Chem. Soc. **53**, 2308 (1931).

n-DODECYL BROMIDE

(Dodecane, 1-bromo-)

$$C_{12}H_{25}OH + HBr \rightarrow C_{12}H_{25}Br + H_2O$$

Submitted by E. EMMET REID, JOHN R. RUHOFF, and ROBERT E. BURNETT.
Checked by W. H. CAROTHERS and W. L. McEWEN.

1. Procedure

IN a 500-cc. distilling flask, fitted with a thermometer and an inlet tube leading to the bottom (Note 1), is placed 186 g. (1 mole) of n-dodecyl alcohol (Note 2). An adapter, one end of which is immersed in about 75 cc. of water contained in a 125-cc. Erlenmeyer flask, is attached to the side arm of the flask. All connections are of rubber. The alcohol is heated to 100°, and dry hydrogen bromide (Note 3) is passed in at 100–120° (Note 4) until no more absorption occurs (Note 5). The crude bromide, together with any of the product that has been carried over into the receiving flask, is transferred to a separatory funnel, separated from the aqueous hydrobromic acid formed during the reaction, and shaken with one-third its volume of concentrated sulfuric acid (Note 6). The lower acid layer is drawn off and discarded (Note 7). The residual bromide is mixed with an equal volume of 50 per cent methyl alcohol (Note 8), and aqueous ammonia is added with intermittent shaking until the solution is alkaline to phenolphthalein. The lower bromide layer is drawn off and washed once with an equal volume of 50 per cent methyl alcohol. It is then dried with calcium chloride, filtered, and distilled. The yield of product boiling at 199.5–201.5°/100 mm. or 134–135°/6 mm. is 220 g. (88 per cent of the theoretical amount) (Note 9).

2. Notes

1. In order to obtain more efficient absorption of the hydrogen bromide, a small bulb is blown on the end of the inlet tube, and a number of pin holes are made in it by means of a small, white-hot tungsten wire.

2. The dodecyl alcohol used was obtained by the fractionation of "Lorol"; its boiling point was 192.5–193.5°/100 mm. or 151–152°/21 mm. If the alcohol is not of good quality, the yield is somewhat lower. Dodecyl alcohol may also be prepared according to the procedure given on p. 372.

3. The hydrogen bromide is conveniently prepared by the direct combination of hydrogen and bromine (p. 338). An excess of hydrogen is to be avoided since it causes loss of product by volatilization.

4. The heat of the reaction maintains the alcohol at this temperature until the preparation is nearly completed.

5. For each mole of alcohol about 1.5 moles of hydrogen bromide is required, of which 1 mole is used to convert the alcohol to the bromide and approximately 0.5 mole to saturate the water formed in the reaction. The rate of addition should be regulated so as to require not less than an hour and a half. The Erlenmeyer flask that serves as a receiver should be weighed with the water in it before it is put in place. When the reaction is complete, the receiver gains weight rapidly and becomes warm owing to the heat of solution of the hydrogen bromide in the water.

6. The crude bromide must be shaken well with the sulfuric acid. The function of the sulfuric acid appears to be to convert any free alcohol to the acid sulfate, which is then soluble in 50 per cent methyl alcohol and ammonia.

7. Care must be taken that the separation of the two layers in this and subsequent washings is complete. Failure to observe this precaution is usually the cause of a low yield.

8. The use of methyl alcohol prevents, to a large extent, the formation of troublesome emulsions. Less than 0.1 g. of dodecyl bromide dissolves in 100 cc. of 50 per cent methyl alcohol at room temperature.

9. The authors have prepared other bromides by this method with the yields indicated below:

BROMIDE	YIELD, %	SOLUBILITY OF BROMIDES IN METHYL ALCOHOL
Cyclohexyl	72–75	Less than 1 g. in 100 cc. of 65% methyl alcohol
n-Heptyl	87–90	Less than 0.5 g. in 100 cc. of 50% methyl alcohol
Tetradecyl	88–89	Less than 0.1 g. in 100 cc. of 50% methyl alcohol
Octadecyl	90–91	Practically insoluble in 90% methyl alcohol

Obviously for the lower bromides it is desirable to use no more methyl alcohol than is necessary to prevent the formation of an emulsion. A convenient method is to place the water, phenolphthalein, and crude bromide in a separatory funnel, and add ammonia until the mixture becomes pink. Methyl alcohol is then added in small portions until the emulsion is broken and two layers separate with a distinct boundary after the mixture has been agitated.

3. Methods of Preparation

The above method for preparing *n*-dodecyl (lauryl) bromide is an adaptation of that of Ruzicka [1] and has been published.[2] It is thought

[1] Ruzicka, Stoll, and Schinz, Helv. Chim. Acta **11**, 685 (1928).
[2] Ruhoff, Burnett, and Reid, J. Am. Chem. Soc. **56**, 2784 (1934).

to present some advantages in ease of manipulation and quality of product over the older method involving the action of aqueous hydrobromic acid on the alcohol in the presence of sulfuric acid.[3]

DURENE

(By-products, penta- and hexamethylbenzene)

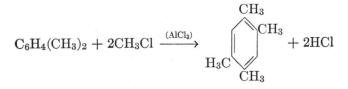

$$C_6H_4(CH_3)_2 + 2CH_3Cl \xrightarrow{(AlCl_3)} \qquad + 2HCl$$

Submitted by Lee Irvin Smith.
Checked by Frank C. Whitmore and Thos. E. Hollingshead.

1. Procedure

A 5-l. flask, mounted on a steam bath, is fitted with a wide (Note 1) inlet tube reaching to the bottom of the flask, a reflux condenser, and a delivery tube running from the top of the reflux condenser and reaching to the bottom of a heavy glass cylinder containing a column of mercury about 10 cm. high. All stoppers and joints of the apparatus must fit tightly and be wired, as the reaction is carried out under a slight pressure. In the flask are placed 3180 g. (3.7 l., 30 moles) of xylene (Note 2) and 1 kg. of anhydrous aluminum chloride (Note 2). The inlet tube of the flask is then connected to a tank of methyl chloride or to a methyl chloride generator (Note 3), the steam is turned on in the bath, and a fairly rapid stream of methyl chloride is passed into the mixture (Note 4). At first there will be a rapid absorption, and the stream of gas must be regulated so that the mercury in the pressure tube does not suck back. The hydrogen chloride formed may be conveniently disposed of by a trap. As the reaction slows down, the pressure increases until both hydrogen chloride and methyl chloride begin to escape through the mercury. At this point the current of incoming gas should be cut down so that undue loss of methyl chloride is avoided. About one hundred hours is required for the completion of the reaction (Note 5).

When the reaction is completed, the steam is turned off and the flask is disconnected and loosely stoppered. After standing overnight the re-

[3] Kamm and Marvel, ibid. **42**, 299 (1920); Org. Syn. Coll. Vol I, **1941**, 29.

action mixture is decomposed by pouring it slowly onto 5 kg. of chopped ice. The greenish oil which separates from the aqueous layer is removed, dried over calcium chloride, filtered and fractionated twice, using a round-bottomed flask fitted with an efficient column and an air condenser (Note 6).

The fractions are cut as follows:

FRACTION	TEMP. RANGE	DISTILLATE	YIELD, g.
I and II	Up to 150°	Benzene, xylene	Little
III	150 to 180	Trimethyl benzenes	570
IV	180 to 205	Tetramethyl benzenes	2075
V	Above 205	Mostly pentamethylbenzene	815

Typical yields of the various fractions are given in the table; they may vary 10 to 20 per cent from these figures, owing to slight differences in procedure and materials, and especially to differences in the quality of the aluminum chloride employed. The more efficient the aluminum chloride, the greater is the percentage of the higher-boiling fractions.

The tetramethylbenzene fraction (IV) is rich in durene, which may be frozen out and filtered because of its relatively high melting point (80°). To isolate the durene, fraction IV is thoroughly chilled in an ice-calcium chloride pack, and filtered through a cold filter (Note 7), using good suction and pressing down the solid compactly. When no more liquid drips through the cold filter, the filtration apparatus is allowed to come gradually to room temperature (Note 7) and the suction is continued as long as any liquid drips through, after which the solid is removed and bottled. The yield is 540 to 610 g.

Fraction III may be methylated to the tetramethylbenzene stage by heating on the steam bath with 100 g. of anhydrous aluminum chloride and passing in 225 g. of methyl chloride. The filtrates from the durene likewise yield more durene when heated on the steam bath with 50 g. of aluminum chloride. The products are worked up in the usual way; that is, they are decomposed by pouring onto twice their weight of chopped ice, separated from the aqueous layer, fractionated twice, and the durene frozen out as before. By conducting one methylation of the trimethylbenzene fraction, and one treatment of the durene filtrates, the combined yield of crude durene will average 1 to 1.4 kg. (25–35 per cent of the theoretical amount based on the original 30 moles of xylene).

To purify the durene, 200 g. of the crude product is placed in a 1-l. round-bottomed flask fitted with a reflux condenser, and melted in a water bath at 95°; 200 cc. of warm (50°) 95 per cent ethyl alcohol is then added through the top of the condenser, and the mixture is carefully heated until homogeneous. The solution is filtered on a hot-water funnel, allowed to stand tightly covered (Note 8) in a fairly warm place (35°) overnight, cooled to about 0°, and filtered on a suction filter. The product (about 169 g.) melts at 74–78°. A second recrystallization yields 149 g. melting at 77–79°. A third recrystallization yields 140 g. having a melting point of 79–80°. The alcoholic filtrates are fractionated, and the crude durene obtained is worked over with the isomers of durene.

Pentamethylbenzene.—Fraction V, and xylene which is methylated beyond the tetramethylbenzene stage, may be worked up for pentamethylbenzene. If xylene is to be methylated to obtain pentamethylbenzene, one more mole of methyl chloride should be used, and the mixture of xylene and anhydrous aluminum chloride should be methylated for one hundred and ten hours instead of the one hundred used for durene. Otherwise the procedure is exactly the same as for durene. The reaction mixture is decomposed and fractionated in the usual way, and the material boiling above 205° (fraction V) is separated into three fractions:

FRACTION	TEMPERATURE RANGE	
VI	205–215°	Tetra- and pentamethylbenzene
VII	215–235	Mostly pentamethylbenzene
VIII	Residue above 235	Hexamethylbenzene and tars

The pentamethylbenzene obtained in this way is nearly pure, and one recrystallization from 95 per cent alcohol or from a mixture of equal volumes of alcohol and benzene gives a snow-white product, but the product generally melts over too wide a range for practical purposes. However, if fraction VII is refractionated under diminished pressure and the fraction boiling at 123–133°/22 mm. (practically all at 127–129°) is collected and recrystallized as in the following paragraph, a product melting quite sharply at 52° (true m.p. 53°) is obtained.

Six hundred grams of crude pentamethylbenzene is heated to 100° and poured slowly with stirring into 1 l. of 95 per cent ethyl alcohol heated to 70° in a 2-l. beaker, and allowed to stand overnight at a temperature of approximately 30°. The crystals formed are collected on a suction filter and dried at room temperature overnight on a porous plate. The yield is about 250 g. (Note 9).

Hexamethylbenzene.—Fraction VIII is fractionated in 250-cc. batches in a Claisen flask at 20 mm. pressure, the following fractions being collected:

FRACTION	TEMPERATURE RANGE	
IX	80–110°	Mostly tetra- and pentamethyl benzenes
X	110–135	Mostly pentamethylbenzene
XI	135–170	Hexamethylbenzene
XII	Residue above 170	Hexamethylbenzene and tars

This fractionation may be carried out in an ordinary Claisen flask, but there is some difficulty in maintaining the desired pressure owing to the solvent action of the hydrocarbons on the rubber stopper. This difficulty may be avoided by the use of a Claisen flask with very long necks and a wide side tube. The material should be distilled fast enough to prevent it from solidifying in the column and side tube. Prolonged heating of hexamethylbenzene also causes a considerable amount of decomposition to tars.

The method for the production of large amounts of hexamethylbenzene is the rapid methylation of pentamethylbenzene or the durene filtrates. A mixture of 378 g. of pentamethylbenzene and 200 g. of anhydrous aluminum chloride is heated on an oil bath at 190–200° and a rapid stream of dry methyl chloride is bubbled through for three to four hours, using the same apparatus as for the preparation of durene. The mixture is allowed to stand overnight at room temperature. One liter of hot xylene is added to dissolve the solidified material, and the reaction mixture is decomposed by pouring it onto 3 kg. of chopped ice. The resulting oil is separated, and the xylene and other low-boiling material are removed by distillation under reduced pressure (Note 10). The fractions are divided as above. Two refractionations and two recrystallizations (Note 11) give 98–121 g. of white crystals, melting at 157–161° (Note 12).

2. Notes

1. The inner tube of a condenser makes a good inlet tube. It should be placed so that the wide end is inside the large flask.

2. The xylene should be a good, colorless laboratory grade, b.p. 135–140°. Any moisture present may be removed by distilling and discarding the first 10 per cent of the distillate. The best aluminum chloride available should be used, for the methylation is very unsatisfactory if the catalyst is of an inferior grade. It should be in small pieces but need not be powdered.

3. The methyl chloride generator consists of a 5-l. flask resting on a sand bath and fitted with a reflux condenser, with a delivery tube running from the top of the condenser to a train of wash bottles, two containing water and two containing concentrated sulfuric acid, with three

safety bottles, one at each end of the train, and one between the water and sulfuric acid bottles. To charge for about 45 moles (theoretical) of methyl chloride: 200 g. of water and 2.2 kg. (1.2 l.) of concentrated sulfuric acid are placed in the flask, and 1.4 kg. (1760 cc.) of methyl alcohol is added, with cooling, at such a rate that the temperature does not rise above 70°. Then 2.4 kg. of sodium chloride is added, the apparatus is tightly connected, and the flask is heated on the sand bath so that the gas is evolved at a fairly rapid rate. It has been found in practice that, using materials of the commercial grade, the yield of methyl chloride is about 55 to 65 per cent of the theoretical amount, so that about double the calculated quantities must be used. This means that the generator has to be charged three times in order to convert 30 moles of xylene to tetramethylbenzene. If a tank of methyl chloride is available, 65–70 moles of methyl chloride should be used for this same amount of xylene. The tank should be weighed before starting and the reaction stopped when the tank has lost the proper amount in weight.

4. Experiments have shown that the rapid current of methyl chloride furnishes sufficient stirring.

5. The normal time of one hundred hours can be shortened by increasing the amount of aluminum chloride. The product is then very viscous, however, and rather difficult to handle in large amounts.

6. Two systematic fractionations (not redistillations) with a good column are absolutely necessary in order to obtain good separations. The more efficient the column the better.

7. The material should be filtered through a large Büchner funnel, which is immersed in a freezing mixture as long as any liquid drips through. It is stated in the literature that the first filtrate obtained in this way is mostly isodurene (1,2,3,5-), whereas the second filtrate, obtained as the material warms slowly to room temperature, is pseudodurene or prehnitene (1,2,3,4-), m.p. −4°.

8. Durene, being quite volatile, should not be allowed to remain exposed to the air any longer than necessary. It is also quite volatile with alcohol, and the mother liquors resulting from the recrystallizations should be distilled: the alcoholic distillate is used for further recrystallizations, and the residues may be worked up for durene by heating with aluminum chloride.

9. The melting point of pentamethylbenzene is only slightly affected by recrystallization, because most of the impurity is hexamethylbenzene, which can be removed only by fractionation.

10. Hexamethylbenzene decomposes when heated very strongly for any length of time. Therefore better results are obtained if the distillations are carried out under reduced pressure.

11. Small amounts of impurities greatly influence the melting point of hexamethylbenzene, and several recrystallizations of a fraction of close boiling range are necessary in order to prepare a sharply melting product.

12. Small amounts (25 g. or less) of hexamethylbenzene which is nearly pure are best recrystallized from ethyl alcohol. It requires about 600 cc. of boiling alcohol to dissolve 25 g., but on cooling 20 g. of pure product will result. Ether and benzene dissolve the substance much more readily, and larger amounts of materials are best recrystallized from either of these solvents, or from a mixture of one of them with alcohol. One hundred and twenty-five grams of the hexamethylbenzene distillate which has been refractionated is melted and poured slowly with stirring into 1.5 l. of 95 per cent alcohol. A small amount which remains undissolved may be brought into solution by adding about 300 cc. of hot benzene, the beaker being heated on a steam bath and the mixture stirred constantly until all is dissolved. The solution is allowed to stand overnight at approximately 25°. The crystals are filtered by suction and washed with enough 95 per cent alcohol to moisten thoroughly (about 25 cc.). After drying, the crystals weigh approximately 112 g. and melt at 155–159°.

3. Methods of Preparation

Durene, pentamethylbenzene and hexamethylbenzene have usually been prepared from benzene or one of its methylated derivatives by the Friedel-Crafts synthesis.[1] Durene has been made from bromine derivatives of methylated benzenes by the Fittig reaction.[2] It has been obtained in 20 per cent yield by passing methyl alcohol and acetone vapors over heated aluminum oxide [3] and in 45 per cent yield by the chloromethylation of xylene and reduction of the chloromethylated products.[4] Hexamethylbenzene has been obtained by the action of zinc chloride on methyl alcohol [5] or on acetone.[6] The method described in the procedure above has been published.[7]

[1] Ador and Rilliet, Ber. **12**, 331 (1879); Friedel and Crafts, Compt. rend. **91**, 257 (1880); Ann. chim. phys. (6) **1**, 449 (1884); Jacobsen, Ber. **14**, 2629 (1881); **18**, 338 (1885); **20**, 896 (1887); Anschütz, Ann. **235**, 185 (1886); Claus and Foecking, Ber. **20**, 3097 (1887); Beaurepaire, Bull. soc. chim. (2) **50**, 676 (1888).

[2] Jannasch and Fittig, Z. Chem. **13**, 161 (1870); Jannasch, Ber. **7**, 692 (1874); **10**, 1354 (1877).

[3] Reckleben and Scheiber, ibid. **46**, 2363 (1913).

[4] v. Braun and Nelles, ibid. **67**, 1098 (1934).

[5] LeBel and Greene, Compt. rend. **87**, 260 (1878).

[6] Greene, ibid. **87**, 931 (1878).

[7] Smith and Dobrovolny, J. Am. Chem. Soc. **48**, 1413 (1926).

DUROQUINONE

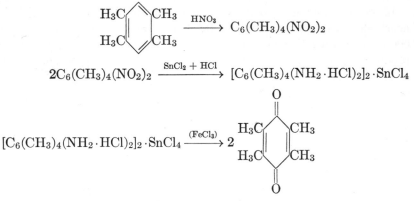

Submitted by LEE IRVIN SMITH.
Checked by FRANK C. WHITMORE and THOS. E. HOLLINGSHEAD.

1. Procedure

(A) *Dinitrodurene.*—A solution of 13.4 g. (0.1 mole) (Notes 1 and 2) of durene (p. 248) in 100 cc. of chloroform is added to 75 cc. of concentrated sulfuric acid in an 800-cc. beaker provided with a thermometer and an efficient mechanical stirrer. The mixture is cooled to 10°, and 16 g. (10.7 cc.) of fuming nitric acid (sp. gr. 1.5) (Note 3) is added drop by drop, with stirring, from a 125-cc. separatory funnel, the mixture being cooled in an ice-salt bath and the nitric acid added at such a rate that the temperature does not rise above 50° (about fifteen minutes is required for the addition). As soon as all the acid has been added the mixture is poured into a separatory funnel, the sulfuric acid layer is removed, and the upper chloroform layer is immediately (Note 4) run into 500 cc. of 10 per cent sodium carbonate solution. The sulfuric acid layer is discarded because it contains very little dinitrodurene. Four portions are nitrated, and the combined chloroform solutions are washed twice with 2.5 per cent sodium carbonate solution, dried overnight with 30 g. of anhydrous calcium chloride, filtered, and the chloroform distilled until crystals of dinitrodurene begin to appear. At this point four times the volume of hot 95 per cent ethyl alcohol is added (about 500 cc.), and the resulting mixture is cooled to 10°. The solid is filtered and washed twice with 50 cc. of cold (10°) 95 per cent ethyl alcohol. The yield from four nitrations is 82.5–84 g. (92–94 per cent of the theoretical amount) of a product melting at 207–208° (Note 5).

(B) *Reduction of Dinitrodurene.*—A solution of 90 g. of dinitrodurene in 1 l. of glacial acetic acid is boiled in a 12-l. flask (Note 6); 700 g. of stannous chloride is dissolved in 800 cc. of concentrated hydrochloric acid and heated to boiling. The heat is removed from the acetic acid solution of the nitro compound, and the stannous chloride solution is poured very carefully (during about ten minutes) into the dinitrodurene solution. The reaction is complete in fifteen minutes, and, as the solution cools, the stannic chloride compound of the diamine begins to crystallize. The reaction mixture is cooled to 10° in an ice-water bath, and the solid is filtered by suction, washed twice with 50 cc. of 95 per cent ethyl alcohol and twice with 50 cc. of ether, and dried. The filtrates from the tin compound contain very little of the reduction product and may be discarded. The composition of this compound is $[C_6(CH_3)_4(NH_2 \cdot HCl)_2]_2 \cdot SnCl_4$, and it crystallizes from the reaction mixture in fine, glistening plates which are almost colorless. The yield is 145 g. (97 per cent of the theoretical amount).

(C) *Duroquinone.*—A suspension of 100 g. of the tin compound in a solution of 300 g. of ferric chloride crystals in a mixture of 150 cc. of water and 20 cc. of concentrated hydrochloric acid is allowed to stand overnight at about 30°, and is then filtered. The product is dissolved in 150 cc. of hot 95 per cent ethyl alcohol. The solution is filtered and allowed to stand overnight at 30°. The yield is 40 g. of duroquinone (90 per cent of the theoretical amount) melting at 109–110°.

2. Notes

1. It is better to nitrate the durene in small batches, for a high yield and pure product are obtained only with a minimal contact of the reaction mixture and the nitric acid.

2. Pure durene is absolutely essential for good results. Durene should be recrystallized from methyl alcohol until the melting point is 79–80°.

3. A large excess of nitric acid is undesirable, since it lowers the yield. The concentration of the nitric acid is also of importance, and, to obtain the best results, it should have a specific gravity of 1.5 or more.

4. It is important that the chloroform layer be run into the carbonate solution as quickly as possible, for continued standing in contact with even small amounts of acid leads to the formation of considerable amounts of red, tarry material. This renders the subsequent purification of the nitro compound much more difficult.

5. No mononitrodurene is ever obtained in this process. Either the dinitro compound results, or else unchanged material and oxidation products.

6. A large flask is necessary because the reduction is vigorous and the reaction mixture will boil up and practically fill the flask of the size recommended.

3. Methods of Preparation

Duroquinone has been prepared by the action of alkalies on 2,3-diketopentane [1] or 3,3-dichloro-2-pentanone; [2] from durenol by coupling with diazotized sulfanilic acid, reducing the azo dye, and oxidizing the resulting aminophenol to the quinone; [3] and from durene by the series of reactions used above, [4] which is due originally to Nef. [5] The method of nitration used in preparing dinitrodurene is a modification of a method introduced by Willstätter and Kubli. [6]

EPICHLOROHYDRIN * AND EPIBROMOHYDRIN

$$2XCH_2CHOHCH_2X + Ca(OH)_2 \rightarrow$$
$$2CH_2\text{---}CH\text{---}CH_2X + CaX_2 + 2H_2O$$
$$\underset{O}{\diagdown\diagup}$$

Submitted by Géza Braun.
Checked by W. W. Hartman and G. L. Boomer.

1. Procedure

(A) *Epichlorohydrin.*—In a 5-l. round-bottomed flask, 1350 g. (988 cc., 10.5 moles) of glycerol α,γ-dichlorohydrin (Org. Syn. Coll. Vol. I, **1941,** 292), 840 g. (10 moles) of technical, finely powdered calcium hydroxide (88 per cent), and 840 cc. of water (20°) are shaken vigorously for fifteen minutes (Note 1). The mixture forms a thick paste at the beginning, but the epichlorohydrin soon separates from the calcium salts as a mobile liquid. The flask is fitted with a rubber stopper carrying a wide delivery tube, and the mixture is distilled from a water bath, at first under 40–50 mm. pressure. The pressure is then lowered to 10 mm. and the temperature raised gradually to 95–100° (Note 2).

[1] von Pechmann, Ber. **21,** 1420 (1888).
[2] Faworsky, J. prakt. Chem. (2) **51,** 538 (1895).
[3] Smith, Opie, Wawzonek, and Prichard, J. Org. Chem. **4,** 318 (1939).
[4] Smith and Dobrovolny, J. Am. Chem. Soc. **48,** 1420 (1926).
[5] Nef, Ber. **18,** 2806 (1885); Ann. **237,** 5 (1887).
[6] Willstätter and Kubli, Ber. **42,** 4151 (1909).
* Commercially available; see p. v.

The receiver must be cooled effectively in an ice-salt mixture to $-5°$ or below, to ensure a maximum yield. The distillate is transferred to a separatory funnel, the upper, aqueous, layer returned to the reaction flask, and the distillation repeated. A third distillation in a similar manner gives a small additional amount of epichlorohydrin (Note 3). The lower layers from the successive distillations are combined and distilled through a fractionating column, under reduced pressure. The epichlorohydrin fraction is collected up to $75°/50$ mm., and the residue (about 160–180 cc.), which contains a large percentage of dichlorohydrin, is returned to the original reaction flask, together with 150 cc. of water. This material is distilled once under reduced pressure as described above, and the lower layer of the distillate is combined with the main fraction of epichlorohydrin. The crude product is distilled at ordinary pressure until the temperature of the vapor reaches $115°$; at this point the distillation is stopped and the water layer removed from the distillate. The lower layer of the distillate is dried over anhydrous sodium sulfate and returned to the distilling flask. After a small fore-run, the epichlorohydrin distils at $115–117°$. The yield is 650–700 g. (67–72 per cent of the theoretical amount).

(B) *Epibromohydrin.*—In a 5-l. round-bottomed flask, 2140 g. (1 l., 9.8 moles) of glycerol α,γ-dibromohydrin (p. 308) is suspended in 1.5 l. of water, and 400 g. of technical, powdered calcium hydroxide (88 per cent) is added gradually, with shaking, in the course of about fifteen minutes. A further 400 g. of calcium hydroxide (total, 9.5 moles) is added at once, and the epibromohydrin is distilled at reduced pressure in the manner described for epichlorohydrin (Note 2). The combined lower layers from two such distillations (about 750 cc.) are dried over anhydrous sodium sulfate and fractionated at atmospheric or reduced pressure. The yield of epibromohydrin, b.p. $134–136°$ or $61–62°/50$ mm., is 1130–1200 g. (84–89 per cent of the theoretical amount).

2. Notes

1. The prescribed amount of water should be used; more water causes frothing. The reaction is not exothermic.

2. Epichlorohydrin boils at $30–32°/10$ mm., epibromohydrin at $61–62°/50$ mm. Both these liquids are quite volatile with water vapor under reduced pressure.

3. The volume of the epichlorohydrin layer obtained in the successive distillations is roughly: (1) 500 cc., (2) 200 cc., (3) 20 cc.

3. Methods of Preparation

Epichlorohydrin [1] and epibromohydrin [2] have been prepared by treatment of glycerol dichloro- and dibromohydrins with alkalies in various ways. The procedures described here represent a laboratory application of the Griesheim process.[3]

ERUCIC ACID

$$CH_3(CH_2)_7CH{=}CH(CH_2)_{11}CO_2H$$

(Hydrolysis of Rape-Seed Oil)

Submitted by C. R. NOLLER and R. H. TALBOT.
Checked by H. T. CLARKE and E. R. TAYLOR.

1. Procedure

IN a 5-l. flask fitted with a reflux condenser are placed 2.5 l. of 95 per cent ethyl alcohol and 340 g. (4.5 moles) of commercial (73–75 per cent) potassium hydroxide. The mixture is gently shaken until the hydroxide is dissolved, 1330 g. (1.5 l., approximately 4 equivalents) of rape-seed oil is added with shaking (Note 1), and the mixture is refluxed on a steam bath for twenty-five to thirty hours.

The hot mixture is poured, with stirring, into 15 l. of warm water (50–60°), and is followed by 700 cc. (8.2 moles) of concentrated (36 per cent) hydrochloric acid (Note 2). After standing until the layers are distinct (ten to fifteen minutes), the lower layer is siphoned off as completely as possible and the oil is washed with two 1-l. portions of warm water.

The oil thus obtained, which should amount to 1460–1600 cc., is dissolved in three times its volume of 95 per cent ethyl alcohol and the mixture cooled to −10 to 0°, when the erucic acid crystallizes (Note 3). After six to eight hours at this temperature, the crude erucic acid is collected on a basket centrifuge (Note 4). The mother liquor, on chilling, yields a second crop of erucic acid. The combined product (800–1100 g.) melts either partially or wholly at room temperature, owing to the

[1] Org. Syn. Coll. Vol. I, **1941**, 233.

[2] Berthelot and Luca, Ann. chim. (3) **48**, 306, 311 (1856); Reboul, ibid. (3) **60**, 32 (1860).

[3] Chemische Fabrik Griesheim-Elektron, Ger. pat. 246,242 [Frdl. **10**, 22 (1910–12)]; Braun, J. Am. Chem. Soc. **54**, 1248 (1932).

presence of oleic acid. It is dissolved in an equal volume of alcohol, chilled for six hours at 0°, and centrifuged as before, when it is obtained in well-defined crystalline form. The second crop of this recrystallization resembles that of the first crystallization and must again be recrystallized. The product is finally recrystallized once again from an equal volume of 95 per cent alcohol. The recrystallized acid contains alcohol, which is removed by heating to constant weight on the steam bath under diminished pressure. The yield of acid is 260–360 g. (Note 5); the acid melts at 31–32° (Note 6).

2. Notes

1. If the solution is mixed in this manner, the rape-seed oil emulsifies on being poured into the alkali and refluxing begins more smoothly.

2. If the acid is added before the water, esterification occurs to an appreciable extent, and the yield may fall to less than 200 g.

3. The crystallization is best accomplished by cooling for several hours at 0°. Cooling in an ice-salt mixture is much quicker but the erucic acid so obtained contains more oleic acid.

4. The centrifuge affords an easy way of filtering the erucic acid, since if the acid is not quickly separated from the mother liquor it melts and makes separation impossible. If a centrifuge is not available, the acid must be filtered at 0°, which is inconvenient except during cold weather.

5. The mother liquors from the recrystallizations may be combined, evaporated, and distilled under reduced pressure, taking two equal fractions. The lower (b.p. 200–220°/5 mm.) consists mainly of oleic acid; the higher (b.p. 220–230°/5 mm.) is solid at 20° and yields a further quantity of erucic acid on recrystallization.

6. The acid obtained contains a small percentage of arachidic acid and other higher saturated fatty acids, and has an iodine number of 66.9 (instead of about 75). If not entirely colorless, the product may be distilled under reduced pressure, when it boils at 241–243°/5 mm. or 252–254°/12 mm.; there is practically no loss, only a minute amount of higher-boiling material remaining in the flask.

3. Methods of Preparation

Erucic acid is a constituent of various natural oils but is most conveniently obtained from rape-seed oil. The process described above is essentially that of Reimer and Will.[1] Methods have been developed for

[1] Reimer and Will, Ber. **19**, 3320 (1886); Müller, Rölz, and Wiener, ibid. **67**, 296 (1934).

obtaining pure erucic acid free from saturated acids,[2] but these involve time-consuming procedures of fractional precipitation and crystallization, and necessarily give poor yields. The product obtained above is satisfactory for most purposes.

ETHOXYACETIC ACID AND ETHYL ETHOXYACETATE

(Acetic acid, ethoxy-, and ethyl ester)

$$ClCH_2CO_2H + 2C_2H_5ONa \rightarrow C_2H_5OCH_2CO_2Na + NaCl + C_2H_5OH$$

$$C_2H_5OCH_2CO_2Na + HCl \rightarrow C_2H_5OCH_2CO_2H + NaCl$$

$$C_2H_5OCH_2CO_2H + C_2H_5OH \xrightarrow{\text{(HCl)}} C_2H_5OCH_2CO_2C_2H_5 + H_2O$$

Submitted by REYNOLD C. FUSON and BRUNO H. WOJCIK.
Checked by C. R. NOLLER and J. J. GORDON.

1. Procedure

(A) *Ethoxyacetic Acid.*—In a 2-l. round-bottomed flask, provided with a reflux condenser 70 to 80 cm. long, is placed 1250 cc. of absolute ethyl alcohol (Note 1). Through the condenser tube 69 g. (3 gram atoms) of metallic sodium is added rapidly enough to keep the alcohol refluxing gently. When the sodium has completely dissolved, a solution of 142 g. (1.5 moles) of chloroacetic acid in 180 cc. of absolute alcohol is added slowly in 20-cc. portions to the sodium ethoxide solution (Note 2). After all the acid has been added, the mixture is heated gently for ten minutes. The excess alcohol is removed as completely as possible by distilling from a steam bath and finally by passing steam into the residue. The aqueous solution is cooled, and 140 cc. (1.7 moles) of concentrated hydrochloric acid (sp. gr. 1.19) is added. The sodium chloride is removed by filtration with suction and is washed with two 50-cc. portions of ether. The original filtrate is saturated with dry sodium sulfate (30–35 g.) and is then extracted with the ether which was used for washing the precipitate, together with an additional 100 cc. of ether. The ether is separated from the aqueous layer, and the latter is extracted four times with 100-cc. portions of fresh ether. The ether is removed by distillation from a steam bath, and the residue is distilled under reduced pressure, using a 500-cc. Claisen flask equipped with a 500-cc.

[2] Holde and Wilke, Z. angew. Chem. **35**, 105, 186, 289 (1922); Täufel and Bauschinger, ibid. **41**, 157 (1928).

receiver. The acid boils at 109–111°/17–18 mm. and weighs 115–116 g. (74 per cent of the theoretical amount). By redistilling the low-boiling fraction and collecting the portion boiling at 150–210°, there is obtained an additional 7–10 g. of material which is chiefly ethoxyacetic acid and may be combined with the main fraction for conversion to the ester.

(B) *Ethyl Ethoxyacetate.*—The ethoxyacetic acid, which should amount to about 125 g. (1.2 moles), is placed in a 750-cc. Erlenmeyer flask containing 230 cc. (3.9 moles) of absolute ethyl alcohol. The flask is set in a pan filled with cold water (Note 3), and dry hydrogen chloride is passed into the mixture. After the mixture becomes saturated (Note 4) it is allowed to stand twenty-four hours to ensure the completion of the reaction at room temperature (Note 5). The solution is cooled, and a saturated solution of sodium carbonate is added cautiously and with stirring to avoid excessive foaming. The addition is continued until the mixture is faintly alkaline to litmus; an excess of sodium carbonate lowers the yield and should be avoided. The ester is extracted with four 100-cc. portions of ether; the extract is dried with 25 g. of anhydrous potassium carbonate, and the ether is distilled from a steam bath. The residue is distilled at ordinary pressure. The yield of ester boiling at 153–155° is 110–115 g. (55–58 per cent of the theoretical amount based on the chloroacetic acid, or 69–72 per cent based on the ethoxyacetic acid).

2. Notes

1. Alcohol dried over quicklime is satisfactory.

2. The chloroacetic acid should be added fast enough to keep the solution boiling.

3. Heat is evolved during the reaction, and, unless the flask is kept in cold water, an insufficient amount of hydrogen chloride will be absorbed. This lowers the yield.

4. A considerable quantity of dry hydrogen chloride is required. The gas should be allowed to bubble through the mixture for at least five hours.

5. The yield seems to be limited by the equilibrium between the acid and ester. At least twenty-four hours is required to reach this equilibrium.

3. Methods of Preparation

Ethoxyacetic acid was first prepared by Heintz [1] by the reaction of chloroacetic acid with sodium ethoxide. The procedure described above

[1] Heintz, Ann. Physik **109**, 331 (1860); **111**, 555 (1860); Rothstein, Bull. soc. chim. (4) **51**, 838 (1932).

is essentially that of Sommelet.[2] Ethoxyacetic acid has also been prepared by hydrolysis of ethoxyacetonitrile with concentrated hydrochloric acid [3] and by the action of excess sodium ethoxide on 1,1,1,2-tetrachlorethane and on α,β-dichlorovinyl ethyl ether.[4] A patent reports the synthesis from diethyl ether and carbon dioxide [5] at high pressure.

Ethyl ethoxyacetate has been prepared by the action of ethyl iodide on sodium ethoxyacetate,[6] of sodium ethoxide on ethyl chloroacetate,[7] of alcohol on crude diazoacetic ester,[8] and by the alcoholysis of ethoxyacetonitrile using alcoholic hydrogen chloride.[3, 9]

ETHYL ACETOSUCCINATE

(Succinic acid, acetyl-, diethyl ester)

$$CH_3COCH_2CO_2C_2H_5 + NaOC_2H_5$$
$$\rightarrow CH_3C(ONa){=}CHCO_2C_2H_5 + C_2H_5OH$$

$$CH_3C(ONa){=}CHCO_2C_2H_5 + ClCH_2CO_2C_2H_5$$
$$\rightarrow CH_3COCH(CO_2C_2H_5)CH_2CO_2C_2H_5 + NaCl$$

Submitted by HOMER ADKINS, NEVILLE ISBELL, and BRUNO WOJCIK.
Checked by JOHN R. JOHNSON and H. R. SNYDER.

1. Procedure

IN a 3-l. three-necked, round-bottomed flask, fitted with a mechanical stirrer, reflux condenser, and separatory funnel, is placed 400 cc. of absolute alcohol (Note 1). Through the condenser tube is added slowly 23 g. (1 gram atom) of clean sodium cut into thin slices. The completion of the reaction is hastened by heating the flask on a steam bath. When the sodium has dissolved completely, 143 g. (1.1 moles) of ethyl acetoacetate (Org. Syn. Coll. Vol. I, **1941**, 235) is introduced slowly. After the mechanical stirrer is started, 123 g. (1 mole) of ethyl chloroacetate (Note 2) is added slowly over a period of an hour, and the reac-

[2] Sommelet, Ann. chim. phys. (8) **9**, 489 (1906).

[3] Gauthier, ibid. (8) **16**, 304 (1909).

[4] Geuther and Brockhoff, J. prakt. Chem. (2) **7**, 113 (1873).

[5] Dreyfus, Fr. pat. 671,103 [C. A. **24**, 1867 (1930)].

[6] Heintz, Ann. **129**, 40 (1864).

[7] Henry, Ber. **4**, 706 (1871).

[8] Curtius, J. prakt. Chem. (2) **38**, 424 (1888).

[9] Sommelet, Ann. chim. phys. (8) **9**, 501 (1906).

tion mixture is refluxed for five to six hours. At this point the reaction mixture should no longer give an alkaline reaction with moist litmus.

After cooling, the precipitated sodium chloride is removed by filtering with suction and is washed with two 50-cc. portions of absolute alcohol. The alcohol is removed by distilling through a short column from a steam bath. The residue is filtered and transferred to a round-bottomed flask and is fractionated under reduced pressure through a Widmer column containing an 8-cm. spiral (Note 3). The fraction boiling at 121–124°/5 mm. is collected. The yield is 121–134 g. (56–62 per cent of the theoretical amount) (Note 4).

2. Notes

1. A good grade of absolute alcohol is required. For this purpose ordinary absolute alcohol may be dried by treating with a little sodium, adding a few cubic centimeters of ethyl succinate, and distilling directly into the reaction flask (see Note 1, p. 155, and Org. Syn. Coll. Vol. I, **1941**, 259).

2. Ethyl chloroacetate boiling at 142–145° was used. This ester can be prepared readily by refluxing for six hours a mixture of 200 g. of chloroacetic acid, 120 g. of absolute alcohol, and 25 g. of concentrated sulfuric acid.[1] The product is purified in the conventional way, and the yield is 185 g. (70 per cent of the theoretical amount).

3. It is advantageous to use an electrically heated column for this fractionation. The principal by-product of the reaction is ethyl β-acetotricarballylate[2] (b.p. 190°/16 mm.), formed by further action of ethyl chloroacetate upon the initial product.

4. Ethyl α-acetoglutarate may be prepared in a similar way by using 181 g. (1 mole) of ethyl β-bromopropionate (Org. Syn. Coll. Vol. I, **1941**, 246) instead of ethyl chloroacetate. The product is collected at 132–134°/4 mm. and weighs 120 g. (52 per cent of the theoretical amount).

3. Method of Preparation

Ethyl acetosuccinate has been prepared by the interaction of ethyl sodioacetoacetate and ethyl chloroacetate[1] or bromoacetate.[2] The method given above is a modification[3] of that given by Conrad.[1]

[1] Conrad, Ann. **188**, 218 (1877).

[2] Emery, Ber. **23**, 3755 (1890); Fichter and Pfister, ibid. **37**, 1997 (1904).

[3] Isbell, Wojcik, and Adkins, J. Am. Chem. Soc. **54**, 3685 (1932).

ETHYL ADIPATE

(Adipic acid, diethyl ester)

$$HO_2C(CH_2)_4CO_2H + 2C_2H_5OH \xrightarrow[\text{(Toluene)}]{\text{(H}_2\text{SO}_4)}$$

$$C_2H_5O_2C(CH_2)_4CO_2C_2H_5 + 2H_2O$$

Submitted by V. M. Mićović.
Checked by Reynold C. Fuson and E. A. Cleveland.

1. Procedure

Four hundred thirty-eight grams (3 moles) of adipic acid, 1080 cc. (9 moles) of absolute alcohol, 540 cc. of toluene, and 2.5 cc. of concentrated sulfuric acid (Note 1) are placed in a 3-l. distilling flask. The flask is connected with a downward condenser and heated on an oil bath (Note 2). An azeotropic mixture of alcohol, toluene, and water begins to distil at 75°. Distillation is continued until the thermometer in the neck of the flask rises to 78°, when further heating is suspended.

The distillate is collected in a 2-l. flask containing 450 g. of anhydrous potassium carbonate (Note 3). It is well shaken, filtered through a Büchner funnel, and returned to the distilling flask (Note 4). The flask is again heated until the temperature rises to 78–80°, when distillation is discontinued (Note 5). The residual liquid (Note 6) is emptied into a 1-l. flask, the large flask being rinsed with a little alcohol, and distilled under vacuum. Alcohol and toluene distil first; then the temperature rises and ethyl adipate distils at 138°/20 mm. (Note 7). The yield is 580 to 588 g. (95–97 per cent of the theoretical amount) (Notes 8 and 9).

2. Notes

1. For esterification by this method, three times the theoretical quantity of absolute alcohol is taken. If only twice the theoretical amount is used, esterification is not complete. The quantity of sulfuric acid required is 1 per cent of the weight of adipic acid used. For smaller quantities of organic acid, a few drops of sulfuric acid are sufficient.

2. It is necessary to maintain the temperature of the bath at about 115° until the mixture dissolves and distillation begins. Later a temperature of 100–110° is sufficient.

3. For each mole of adipic acid to be esterified, 150 g. of anhydrous potassium carbonate is required.

4. Filtering through a fluted filter paper directly into a distillation flask is entirely satisfactory when smaller quantities are used.

5. The distillate, which contains alcohol, toluene, and water, can be dried, distilled, and used again for esterification after the addition of the necessary quantity of absolute alcohol; or, by the addition of water, toluene alone may be separated, dried over calcium chloride, and distilled.

6. If the solution is allowed to cool, small crystals of unesterified acid separate. The quantity is negligible.

7. Towards the end of the distillation, the temperature rises several degrees; but distillation should be continued, for when redistilled the ester does not leave a residue.

8. According to the submitter, ethyl esters of the dibasic acids from oxalic through sebacic have been prepared in yields of 94–98 per cent by this same procedure.

9. Ethyl adipate can also be prepared [1] by refluxing 175 g. (1.2 moles) of adipic acid, 175 g. (222 cc.) of ethyl alcohol, 450 cc. of benzene, and 80 g. (43.5 cc.) of concentrated sulfuric acid for five hours on the steam bath. The yield of ester boiling at 136–137/19 mm. is 218 g. (90 per cent of the theoretical amount). (Private communication from P. S. PINKNEY. Checked by LOUIS F. FIESER and T. L. JACOBS.)

3. Methods of Preparation

Ethyl adipate is obtained by boiling adipic acid, alcohol, benzene, and sulfuric acid; [1] from adipic acid, alcohol, and hydrogen chloride; [2] from adipic acid, absolute alcohol, and sulfuric acid; [3] by distilling a mixture of ethyl alcohol, toluene, and adipic acid with the addition of some hydrochloric acid which acts as a catalyst; [4] and by the procedure described above. [5]

[1] Van Rysselberge, Bull. soc. chim. Belg. 35, 312 (1926).

[2] Arppe, J. prakt. Chem. (1) 95, 208 (1865).

[3] Curtius, ibid. (2) 91, 4 (1915); Bouveault and Locquin, Bull. soc. chim. (4) 3, 439 (1908).

[4] Locquin and Elghozy, ibid. (4) 41, 445 (1927).

[5] Mićović, ibid. (5) 4, 1661 (1937).

ETHYL BENZOYLACETATE
(Acetic acid, benzoyl-, ethyl ester)

(A) $2CH_3COCH_2CO_2C_2H_5 + 2C_6H_5COCl + 2Na \rightarrow$

$$2CH_3COCHCO_2C_2H_5 + 2NaCl + H_2$$
$$\overset{|}{COC_6H_5}$$

(B) $CH_3COCHCO_2C_2H_5 + H_2O + NH_3 \rightarrow$
$\overset{|}{COC_6H_5}$

$$C_6H_5COCH_2CO_2C_2H_5 + CH_3CO_2NH_4$$

Submitted by R. L. SHRINER, A. G. SCHMIDT, and L. J. ROLL.
Checked by C. R. NOLLER and I. BERGSTEINSSON.

1. Procedure

(A) *Preparation of Ethyl Benzoylacetoacetate.*—In a 5-l. three-necked flask, fitted with a liquid-sealed mechanical stirrer and a reflux condenser, are placed 3.4 l. of dry benzene (Note 1), 195 g. (1.5 moles) of ethyl acetoacetate (Org. Syn. Coll. Vol. I, **1941**, 235), and 34.5 g. (1.5 gram atoms) of clean sodium. The mixture is heated on a steam cone with stirring and allowed to reflux gently for twenty-four hours. After the suspension of the sodioacetoacetic ester has been cooled slightly, 263 g. (218 cc., 1.9 moles) of benzoyl chloride is added over a period of three hours. The mixture is refluxed with stirring for an additional eight hours. It is then cooled to room temperature, and 375 g. of cracked ice is added. After thorough shaking, the benzene layer, which contains the ethyl benzoylacetoacetate, is separated, washed with 75 cc. of 5 per cent sodium bicarbonate solution, dried with sodium sulfate, and the benzene distilled (Note 2). The residue is distilled *in vacuo* from a 500-cc. Claisen flask with a 50-cm. fractionating side arm. After a small fore-run of benzoyl chloride, the fraction boiling at 142–148°/6 mm., or 177–181°/20 mm., is collected. The yield is 223–263 g. (63–75 per cent of the theoretical amount).

(B) *Hydrolysis of Ethyl Benzoylacetoacetate.*—Thirty-two grams (0.6 mole) of ammonium chloride is dissolved in 150 cc. (8.3 moles) of water in a 500-cc. Erlenmeyer flask, and 10 cc. (0.15 mole) of ammonia (sp. gr. 0.9) is added. The solution is warmed to 42°, 58.5 g. (0.25 mole) of ethyl benzoylacetoacetate at 20° is added quickly, and the mixture is shaken (Note 3). The flask is placed in a water bath at 42° for exactly ten minutes and then cooled rapidly by placing it in an ice bath. The solution is extracted twice with 100-cc. portions of ether, and the ether solution is dried with anhydrous magnesium sulfate. The ether is distilled, and the residue distilled *in vacuo*; the yield is 37.0–37.5 g. (77–78 per cent of the theoretical amount) of ethyl benzoylacetate boiling at 132–137°/4 mm., or 165–169°/20 mm.

2. Notes

1. The benzene was dried by distillation, the first portion of the distillate being discarded.

2. It is essential that these steps in the isolation be carried through as rapidly as possible.

3. Larger amounts gave lower yields. The exact procedure must be followed as to time, temperature, and amounts of reagents, and the procedure must be completed without interruption.

3. Methods of Preparation

Ethyl benzoylacetate has been prepared by the condensation (by means of sodium ethoxide) of ethyl acetate with ethyl benzoate,[1] acetophenone with ethyl carbonate,[2] and acetophenone with ethyl oxalate, with subsequent heating;[3] by treatment of ethyl phenylpropiolate[4] or α-bromocinnamic acid[5] with concentrated sulfuric acid, and of ethyl diazoacetate with benzaldehyde;[6] by the condensation of benzene with the monoethyl ester of malonyl monoacid chloride and aluminum chloride,[7] of benzoyl chloride with the product of the reaction of magnesium and ethyl chloroacetate in ether;[8] by the action of alcohol on benzoylacetimino ethyl ether hydrochloride;[9] and by the hydrolysis of ethyl benzoylacetoacetate.[10]

[1] Claisen and Lowman, Ber. **20**, 651 (1887).

[2] Claisen, ibid. **20**, 655 (1887).

[3] Wislicenus, ibid. **28**, 811 (1895).

[4] Baeyer, ibid. **15**, 2705 (1882).

[5] Michael and Browne, ibid. **19**, 1392 (1886).

[6] Buchner and Curtius, ibid. **18**, 2371 (1885).

[7] Marguery, Bull. soc. chim. (3) **33**, 549 (1905).

[8] Meyer and Tögel, Ann. **347**, 76 (1906).

[9] Haller, Bull. soc. chim. (2) **48**, 23 (1887).

[10] Claisen, Ann. **291**, 71 (1896); Shriner and Schmidt, J. Am. Chem. Soc. **51**, 3636 (1929).

ETHYL BENZOYLDIMETHYLACETATE

(Isobutyric acid, α-benzoyl-, ethyl ester)

$$(CH_3)_2CHCO_2C_2H_5 + NaC(C_6H_5)_3$$
$$\rightarrow Na[(CH_3)_2CCO_2C_2H_5] + HC(C_6H_5)_3$$

$$C_6H_5COCl + Na[(CH_3)_2CCO_2C_2H_5]$$
$$\rightarrow C_6H_5COC(CH_3)_2CO_2C_2H_5 + NaCl$$

Submitted by C. R. HAUSER and W. B. RENFREW, JR.
Checked by LEE IRVIN SMITH and E. C. BALLARD.

1. Procedure

To a freshly prepared solution of 0.187 mole of triphenylmethylsodium (p. 607), contained in the flask into which it was transferred after preparation, is added 21.7 g. (25 cc., 0.187 mole) of pure ethyl isobutyrate (Note 1). The mixture is shaken, and, after it has stood at room temperature for ten minutes, a solution of 26 g. (21.5 cc., 0.186 mole) of pure benzoyl chloride in 50 cc. of dry ether is added, with shaking. The mixture becomes warm, and a white precipitate of sodium chloride separates immediately. After standing at room temperature for several hours the mixture is heated on a water bath and ether is distilled until the volume is reduced to 300–400 cc. A solution of 5 cc. of acetic acid in 300 cc. of water is added, and the mixture is shaken in a separatory funnel until two homogeneous layers separate on standing. The aqueous layer is drawn off and discarded; the ethereal layer is shaken with 10 per cent sodium carbonate solution and dried over calcium chloride or Drierite. The solution is filtered from the drying agent and distilled on a water bath until most of the ether is removed. The residue is cooled in a refrigerator, and the triphenylmethane which crystallizes is filtered with suction and washed with several small portions of dry ether. The filtrate, after removal of ether, is distilled under reduced pressure and all material boiling up to 180° at 15 mm. collected. The distillate is redistilled and the fraction boiling up to 160° at 15 mm. is subjected to a final redistillation. There is obtained 20.5–22.5 g. (50–55 per cent of the theoretical amount) of ethyl benzoyldimethylacetate, b.p. 146–148° at 15 mm. or 133–135° at 9 mm. (Notes 2 and 3).

2. Notes

1. It is essential to use pure reagents. Commercial ethyl isobutyrate may be purified satisfactorily by washing with 10 per cent sodium carbonate solution, drying over Drierite for several days, and fractionating through an effective column; material boiling over a one-degree range (110–111°) is recommended for this preparation. Benzoyl chloride was purified by distillation under reduced pressure, and a fraction collected over less than a two-degree range was used.

2. If the more concentrated solution of triphenylmethylsodium, prepared as described in Note 3, p. 609, is employed, approximately 120 g. of ethyl benzoyldimethylacetate can be prepared in a single experiment.

The bottle containing an ether solution of 0.85 mole of triphenylmethylsodium is immersed to two-thirds of its depth in an ice bath. When the bottle is thoroughly cold, the stopper is removed while a tube delivering a rapid stream of dry nitrogen is held at the mouth of the bottle. The bottle is quickly fitted with a three-holed rubber stopper carrying a mechanical stirrer (with open shaft), a dropping funnel, and a bent tube which extends into the bottle for a distance of 1 cm. below the neck. A slow stream of dry nitrogen passes through the bent tube; the nitrogen supply is also connected to the top of the dropping funnel. One hundred two grams (0.875 mole) of ethyl isobutyrate is added rapidly to the vigorously stirred contents of the bottle. After five minutes, 123 g. (101 cc., 0.875 mole) of benzoyl chloride is added through the dropping funnel at such a rate that the mixture does not reflux too vigorously (the addition takes about five minutes). Stirring is continued for two or three minutes after the addition is complete. The bottle is then removed from the ice bath, stoppered, shaken, and allowed to stand for half an hour. The reaction mixture is transferred to a 2.5-l. separatory funnel and the mercury drained off. The remaining mixture is extracted with 300 cc. of water, any sludge formed being collected separately. The sludge is extracted with a little ether, the extract being added to the ether layer in the funnel. The ether solution is washed with two 100-cc. portions of 10 per cent sodium carbonate solution, and dried by shaking with 50 g. of anhydrous sodium or magnesium sulfate and by standing with 50 g. of Drierite in a stoppered 2-l. Erlenmeyer flask. The solution is filtered into a 2-l. round-bottomed flask and the ether removed by distillation. The residue is transferred to a 1-l. Claisen flask and distilled in vacuum, using a metal or oil bath and collecting all the material boiling below 175° at 15 mm. The distillate is transferred to a 200-cc. round-bottomed flask fitted with a 6-inch Widmer fractionating column and redistilled from a metal bath. The ethyl benzoyldimethylacetate boils at 146–148°/15 mm. and weighs 115–125 g. (61–67 per cent of the theoretical amount based on the quantity of triphenylmethylsodium used). (CHARLES R. HAUSER and BOYD E. HUDSON, JR., private communication.)

3. The submitters report that ethyl isobutyrylisobutyrate (ethyl α,α,γ,γ-tetramethylacetoacetate) may be prepared in a similar manner, using isobutyryl chloride instead of benzoyl chloride. The yield of this ester, b.p. 94.5–95.5°/18 mm., is about 55 per cent of the theoretical amount. Several other examples of the acylation of esters by this procedure are given by Hudson, Jr., and Hauser.[1]

3. Methods of Preparation

Ethyl benzoyldimethylacetate has been prepared by the condensation of ethyl isobutyrate with ethyl benzoate,[2] phenyl benzoate, benzoic anhydride, or benzoyl chloride.[3] The procedure described above [3] is a modification of the method of Scheibler and Stein.[4]

ETHYL α,β-DIBROMO-β-PHENYLPROPIONATE

(Hydrocinnamic acid, α,β-dibromo-, ethyl ester)

$$C_6H_5CH{=}CHCO_2C_2H_5 + Br_2 \rightarrow C_6H_5CHBrCHBrCO_2C_2H_5$$

Submitted by T. W. ABBOTT and DARRELL ALTHOUSEN.
Checked by HENRY GILMAN and G. F. WRIGHT.

1. Procedure

IN a 1-l. round-bottomed flask, provided with a two-holed stopper fitted with a dropping funnel and air-vent, is placed 176.2 g. (1 mole) of ethyl cinnamate (Org. Syn. Coll. Vol. I, **1941**, 252) dissolved in 100 cc. of carbon tetrachloride (Note 1). The flask is placed in ice, and 159.8 g. (51.2 cc., 1 mole) of bromine is added in small quantities with frequent stirring (Note 2).

After standing for one hour, the solution is poured into a large dish and the carbon tetrachloride and unused bromine allowed to evaporate

[1] Hudson, Jr., and Hauser, J. Am. Chem. Soc. **63**, 3159 (1941).
[2] Renfrow, Jr., and Hauser, ibid. **60**, 464 (1938).
[3] Hudson, Jr., Dick, and Hauser, ibid. **60**, 1961 (1938).
[4] Scheibler and Stein, J. prakt. Chem. (2) **139**, 111 (1934).

spontaneously (Note 3). The dibromo ester separates in large crystals which form a solid cake in the bottom of the dish (Note 4). This cake is broken up and spread in a thin layer on a large Büchner funnel and subjected to suction until all traces of bromine have disappeared. The white crystals are then dried by pressing between large filter papers. The yield of crude dibromo ester is 280–285 g. (83–85 per cent of the theoretical amount). It melts at 65–71°.

If pure ester is desired, it may be obtained by recrystallizing from petroleum ether (b.p. 70–90°); the yield is 80–85 g. of ester melting at 74–75° from 100 g. of crude product.

2. Notes

1. Carbon tetrachloride is used instead of ether, which produces a very disagreeable lachrymator.

2. The bromine disappears rapidly at first, but more slowly at the end of the reaction. No hydrogen bromide is evolved. The time of addition is about twenty to twenty-five minutes.

3. This process is rather slow; it may be accelerated by inverting over the dish a large funnel which is connected to a suction pump. In this manner the crystalline cake separates in about two hours.

4. If the reaction is carried out carefully, practically no mother liquor is left. However, if such liquor should remain it will yield, on evaporation, crystals which are impure and must be recrystallized before use.

3. Method of Preparation

Ethyl α,β-dibromo-β-phenylpropionate is prepared by adding bromine to ethyl cinnamate.[1]

[1] Anschütz and Kinnicutt, Ber. **11**, 1220 (1878); Aronstein and Hollemann, ibid. **22**, 1181 (1889); Leighton, Am. Chem. J. **20**, 136 (1898); Sudborough and Thompson, J. Chem. Soc. **83**, 671 (1903).

ETHYL ETHOXALYLPROPIONATE

(Oxalacetic acid, methyl-, diethyl ester)

$$CH_3CH_2CO_2C_2H_5 + \underset{\underset{CO_2C_2H_5}{|}}{CO_2C_2H_5} \xrightarrow{C_2H_5ONa} \underset{\underset{C(ONa)CO_2C_2H_5}{\|}}{CH_3CCO_2C_2H_5}$$

$$\xrightarrow{CH_3CO_2H} \underset{\underset{COCO_2C_2H_5}{|}}{CH_3CHCO_2C_2H_5}$$

Submitted by RICHARD F. B. COX and S. M. McELVAIN.
Checked by REYNOLD C. FUSON and WILLIAM E. ROSS.

1. Procedure

SIXTY-NINE grams (3 gram atoms) of sodium is powdered under xylene in a 3-l. three-necked flask. The mixture is cooled, the xylene is decanted, and the sodium is washed twice with small portions of dry ether. One liter of absolute ether is then added to the powdered sodium. The flask is fitted with a mercury-sealed stirrer, an efficient reflux condenser, and a dropping funnel, the condenser and the funnel being protected from moisture with calcium chloride tubes. One hundred and thirty-eight grams (175 cc., 3 moles) of absolute ethyl alcohol is added drop by drop through the funnel (Note 1). After the alcohol has been added and there is no unchanged sodium (as evidenced by cessation of boiling) the flask is immersed in an ice-water bath, and a mixture of 306 g. (3 moles) of ethyl propionate and 438 g. (3 moles) of ethyl oxalate (Org. Syn. Coll. Vol. I, **1941**, 261) is added slowly through the dropping funnel (Note 2).

After the ester mixture has been added, the stirrer is removed and the condenser set for downward distillation. The ether and the alcohol formed in the reaction are removed by heating on a water bath (Note 3). The residue, which usually solidifies upon cooling, is treated with 600 cc. of cold, 33 per cent acetic acid solution. The mixture is allowed to stand for several hours with occasional shaking in order to decompose the sodium derivative completely, and the product is extracted with four 500-cc. portions of ether. The ether solution is washed with 1 l. of water, with two 500-cc. portions of 10 per cent sodium bicarbonate solution, and finally with 1 l. of water. The ether is then removed by distillation, using a steam bath. The residue is fractionated through an efficient column (Note 4). The portion boiling at 114–116°/10 mm. is collected. The yield is 363–425 g. (60–70 per cent of the theoretical amount).

2. Notes

1. The addition of the alcohol takes from four to six hours, depending on the efficiency of the condenser and stirrer.

2. The addition of the ester mixture should be slow enough so that the ether does not reflux. This addition takes two to three hours.

3. When most of the alcohol has distilled, a yellow scum forms on the surface of the red, viscous liquid. The distillation is stopped at this point. When the solution is cooled, the sodium derivative crystallizes with considerable expansion in volume.

4. No appreciable decomposition of the ethoxalyl ester into ethyl methylmalonate takes place when the distillation is carried out at 10 mm. To prevent overheating, the use of an oil bath and a heated column is recommended.

3. Methods of Preparation

Ethyl ethoxalylpropionate has been prepared by the Claisen condensation of ethyl oxalate with ethyl propionate [1] as above, and by the alkylation of ethyl ethoxalylacetate.[1]

ETHYL ETHYLENETETRACARBOXYLATE

(Ethylenetetracarboxylic acid, tetraethyl ester)

$$2CHBr(CO_2C_2H_5)_2 + Na_2CO_3 \rightarrow$$
$$2NaBr + H_2O + CO_2 + (C_2H_5O_2C)_2C{=}C(CO_2C_2H_5)_2$$

Submitted by B. B. Corson and W. L. Benson.
Checked by C. S. Marvel and H. E. Munro.

1. Procedure

A mixture of 200 g. (1.9 moles) of anhydrous sodium carbonate (Note 1) and 300 g. (1.25 moles) of ethyl bromomalonate (Org. Syn. Coll. Vol. I, **1941**, 245) is heated for three hours in a 1-l. flask (Note 2) immersed in an oil bath at 150–160° (Note 3). After the heating period,

[1] Wislicenus and Arnold, Ann. **246**, 329, 336 (1888).

300 cc. of xylene (Note 4) is added while the contents of the flask are still hot (Note 5). The solid is broken up carefully with a rod, and the mixture is transferred to a 2-l. beaker. The reaction flask is then rinsed with a mixture of 100 cc. of xylene and 100 cc. of water. This washing is poured into the 2-l. beaker, and an additional 600 cc. of water is added. The solid readily dissolves upon stirring. The liquid mixture is transferred to a separatory funnel, shaken well, and allowed to settle (Note 6). The lower aqueous layer is discarded (Note 7). The xylene layer is transferred to a 1-l. distilling flask and distilled until the temperature of the liquid itself is 170°. The liquid residue is then transferred to a 500-cc. Claisen flask and distilled under reduced pressure. The fore-run up to 170°/15 mm. is discarded. The product, which is collected at 170–230°/15 mm. (Notes 8 and 9), solidifies within about fifteen minutes. The yield is 150–160 g. (75–80 per cent of the theoretical amount).

The crude product is dissolved in 75 cc. of 95 per cent alcohol at a temperature of 40°, which is sufficient to ensure complete solution. The alcoholic solution is cooled to 12° (Notes 10 and 11) and filtered. The yield of air-dried, colorless product melting at 52.5–53.5° is 95–110 g. (47–55 per cent of the theoretical amount). By evaporation of the alcohol, distillation of the residue under reduced pressure, and recrystallization of the solidified distillate, the yield may be increased to 110–115 g. (55–57 per cent of the theoretical amount).

2. Notes

1. The sodium carbonate should be fine enough to pass through a 100-mesh sieve.

2. The mixture is heated in an open flask. Owing to the high boiling points of ethyl bromomalonate and ethyl ethylenetetracarboxylate, there is but little loss by volatilization. The yield is lowered by the use of a condenser, presumably because the water formed by the reaction is kept in the reaction mixture with resultant hydrolysis of one or both of the esters.

3. The flask is placed in the cold bath. The three hours are counted from the time the temperature of the bath reaches 150°.

4. Toluene or benzene can be substituted for xylene, but xylene is preferable because of its higher boiling point.

5. If the mixture is allowed to cool, it solidifies and is difficult to remove.

6. A good separation is obtained in ten minutes.

7. Experiment has shown that the amount of product in the aqueous layer is negligible.

8. Ethyl ethylenetetracarboxylate boils at 197°/8 mm.; 203°/13 mm.; 210°/22 mm.; 221°/33 mm.; 234°/48 mm.

9. There is very little residue left in the flask. Distillation should be stopped as soon as dark yellow drops of distillate begin to come over.

10. The crystallization mixture is a thick slush with low heat conductivity, hence a rather long cooling period is necessary to lower the temperature to 12°. A thermometer should be used because it is important that the mixture be cooled to 12°, since the solubility curve begins to rise above 12°. There is no advantage in cooling below 12°.

11. The solubility of ethyl ethylenetetracarboxylate in 100 cc. of 95 per cent ethyl alcohol is as follows:

2.0 g. at 0°	16.0 g. at 30°
2.5 g. at 11°	19.0 g. at 31°
4.0 g. at 16°	28.0 g. at 33°
8.0 g. at 23°	35.0 g. at 34°
9.7 g. at 26°	61.0 g. at 36.5°

3. Methods of Preparation

Ethyl ethylenetetracarboxylate has been prepared from monochloro- and monobromomalonic ester through removal of halogen acid with sodium,[1] sodium ethoxide,[2] potassium acetate,[3] potassium carbonate,[4] sodium urethane,[5] sodium formanilide and sodium acetanilide.[6] It has also been prepared by treating the disodium derivative of ethyl ethane-1,1,2,2-tetracarboxylate with bromine,[7] or iodine;[8] by treating dibromomalonic ester with sodium,[9] or sodium ethoxide,[10] or sodium malonic ester followed by potash;[11] and by treating the disodium derivative of malonic ester with iodine.[12] It is reported that the use of moist benzene as the medium in the reaction between bromomalonic ester and potassium carbonate increases the yield of ethylenetetracarboxylic ester to 80 per cent.[13]

[1] Conrad and Guthzeit, Ber. **16**, 2631 (1883).
[2] Conrad and Guthzeit, Ann. **214**, 76 (1882).
[3] Conrad and Brückner, Ber. **24**, 2998 (1891).
[4] Blank and Samson, ibid. **32**, 860 (1899).
[5] Diels and Heintzel, ibid. **38**, 303 (1905).
[6] Paal and Otten, ibid. **23**, 2591 (1890).
[7] Kötz and Stalmann, J. prakt. Chem. (2) **68**, 163 (1903).
[8] Bischoff and Hausdörfer, Ann. **239**, 130 (1887).
[9] Conrad and Brückner, Ber. **24**, 3004 (1891).
[10] Curtiss, Am. Chem. J. **19**, 699 (1897).
[11] Adickes, J. prakt. Chem. (2) **145**, 236 (1936).
[12] Bischoff and Rach, Ber. **17**, 2781 (1884).
[13] Malachowski and Sienkiewiczowa, ibid. **68**, 33 (1935).

ETHYL HYDROGEN SEBACATE

(Sebacic acid, ethyl acid ester)

$$HO_2C(CH_2)_8CO_2H + C_2H_5OH \xrightarrow{\text{(HCl)}} C_2H_5O_2C(CH_2)_8CO_2H + H_2O$$

Submitted by SHERLOCK SWANN, JR., RENÉ OEHLER, and R. J. BUSWELL.
Checked by LEE IRVIN SMITH and J. W. CLEGG.

1. Procedure

IN a 1-l. modified Claisen flask (Note 1), the side arm of which is corked, is placed a mixture of 202 g. (1 mole) of sebacic acid, 150 g. (0.58 mole) of diethyl sebacate (Note 2), 50 cc. of di-n-butyl ether (Note 3), and 30 g. (25 cc.) of concentrated hydrochloric acid (sp. gr. 1.19). A reflux condenser is connected to the top of the distilling flask.

The flask is heated in a Wood's metal bath at 160–170° until the mixture is completely homogeneous. The temperature of the bath is then lowered to 120–130°, and 60 cc. (1 mole) of 95 per cent ethyl alcohol is added to the solution through the condenser. The mixture is allowed to reflux for two hours. At the end of this period an additional 20-cc. portion of ethyl alcohol is poured into the solution and refluxing is continued for two hours longer.

The Wood's metal bath is allowed to cool to about 75° and the reaction mixture is subjected to distillation under reduced pressure, using a water pump (Note 4). The temperature of the bath is increased slowly and distillation is continued, with a water pump, until the bath reaches a temperature of about 125°. The bath is again cooled to 75–80° and the distillation is continued at lower pressure, using an oil pump.

The first fractions consist of a little alcohol, water, and n-butyl ether (Note 5). The next fraction is ethyl sebacate, b.p. 156–158° at 6 mm. (Note 6). Ethyl hydrogen sebacate is collected at 183–187° at 6 mm. The product melts at 34–36° and weighs 114–124 g. (50–54 per cent of the calculated amount, based on the sebacic acid used). Refractionation of the fore-run (b.p. 175–183°/6 mm.) and after-run (b.p. 187–195°/6 mm.) gives an additional 24–26 g. of pure monoester. The total yield is 138–150 g. (60–65 per cent of the theoretical amount) (Notes 7 and 8).

2. Notes

1. The column of the flask should be at least 35 cm. in length and well insulated. Wrapping the column with 10-mm. asbestos rope is satisfactory.

2. The addition of diester at the beginning of the reaction decreases its formation from the reactants so that the monoester becomes the main product.

Ethyl sebacate is prepared conveniently by refluxing 130 g. (0.65 mole) of sebacic acid with 250 g. of ethyl alcohol and 25 cc. of concentrated sulfuric acid. The yield is about 90 per cent, and the product boils at 156–158° at 6 mm. See, also, Note 8, p. 265.

3. n-Butyl ether is used in preference to other possible compounds because it permits the formation of a homogeneous reaction mixture.

4. At the beginning of the distillation the liquid in the flask foams excessively. It is advisable, therefore, to reduce the pressure gradually and not to use low pressures until the foaming subsides.

5. The n-butyl ether may be recovered in pure condition by a simple distillation after the water has been separated from it.

6. After the ethyl sebacate has distilled, it is well to drain the cooling water from the condenser in order to prevent the monoester from solidifying before reaching the receiver. The recovered ethyl sebacate weighs 150–175 g. and may be used directly in a subsequent preparation.

7. In subsequent runs the distillation residue is allowed to remain in the flask. In this way the yield is increased to 70–77 per cent for runs of one and two moles.

8. The submitters report that ethyl hydrogen adipate, b.p. 155–157°/7 mm., has been prepared in 71–75 per cent yields on a one-mole scale by the same procedure.

3. Methods of Preparation

Ethyl hydrogen sebacate has been prepared by the direct esterification of sebacic acid with ethyl alcohol,[1] by the half-saponification of ethyl sebacate,[2] and by heating equimolecular quantities of sebacic acid and diethyl sebacate for several hours.[3]

[1] Neison, J. Chem. Soc. **29**, 319 (1876).
[2] Walker, ibid. **61**, 713 (1892).
[3] Fourneau and Sabetay, Bull. soc. chim. (4) **43**, 859 (1928); (4) **45**, 834 (1929).

ETHYL N-METHYLCARBAMATE

(Carbamic acid, methyl-, ethyl ester)

$$ClCO_2C_2H_5 + CH_3NH_2 + NaOH \rightarrow CH_3NHCO_2C_2H_5 + NaCl + H_2O$$

Submitted by W. W. HARTMAN and M. R. BRETHEN.
Checked by J. B. CONANT and C. F. BAILEY.

1. Procedure

IN a 2-l. flask, provided with a mechanical stirrer and cooled by an ice-salt mixture, are placed 300 cc. of ether and 186 g. (2 moles) of a 33 per cent aqueous methylamine solution. When the stirred mixture has cooled to 5°, 217 g. (2 moles) of ethyl chloroformate (Note 1) is added without allowing the temperature to rise above 5°. When almost half of the chloroformate has been added a cold solution of 80 g. (2 moles) of pure sodium hydroxide in 120 cc. of water is added gradually along with the rest of the chloroformate at such a rate that the last portions of the two solutions are added simultaneously. Constant mechanical stirring throughout the addition is essential (Note 2). After standing for fifteen minutes, the ether layer is separated and the aqueous solution is extracted with 100 cc. of ether. The combined ether layers are rapidly dried by shaking for a short time with about 8 g. of potassium carbonate in two portions. The ether is then distilled and the residue distilled under reduced pressure, the distillate being collected at 55–60°/12 mm. The yield of colorless oil is 182–185 g. (88–90 per cent of the theoretical amount).

2. Notes

1. Technical ethyl chloroformate (chlorocarbonate) is manufactured by the U. S. Industrial Chemicals Company.

2. The rate of addition is determined by the efficiency with which the heat is removed from the reaction mixture. Five hours was required, using an ice-salt mixture outside.

3. Methods of Preparation

Ethyl N-methylcarbamate has been prepared by adding aqueous methylamine to ethyl chloroformate;[1] and from methylcarbamyl chloride and ethyl alcohol.[2]

[1] Schreiner, J. prakt. Chem. (2) 21, 124 (1880); Pechmann, Ber. 28, 855 (1895); Eistert, Angew. Chem. 54, 124 (1941).

[2] Gattermann, Ann 244, 35 (1888).

ETHYL METHYLMALONATE

(Malonic acid, methyl-, diethyl ester)

(A) (From Ethoxalylpropionic Ester)

$$\begin{array}{c} CH_3CHCO_2C_2H_5 \\ | \\ COCO_2C_2H_5 \end{array} \xrightarrow{\text{(Heat)}} CH_3CH(CO_2C_2H_5)_2 + CO$$

Submitted by RICHARD F. B. Cox and S. M. McELVAIN.
Checked by REYNOLD C. FUSON and WILLIAM E. ROSS.

1. Procedure

THREE HUNDRED FORTY-FIVE grams (1.7 moles) of ethyl ethoxalyl-propionate, b.p. 114–116°/10 mm. (p. 272), is placed in a round-bottomed flask of suitable size carrying a reflux condenser, and a thermometer is suspended from the top of the condenser into the liquid. The ethyl ethoxalylpropionate is then heated until a vigorous evolution of carbon monoxide begins (130–150°). The temperature of the liquid is gradually raised as the gas evolution diminishes, and finally the liquid is refluxed until no more gas comes off. The ethyl methylmalonate is then distilled. It boils at 194–196°/745 mm., and the yield is 288 g. (97 per cent of the theoretical amount).

(B) (From Malonic Ester)

$$CH_3Br + Na[CH(CO_2C_2H_5)_2] \rightarrow CH_3CH(CO_2C_2H_5)_2 + NaBr$$

Submitted by NATHAN WEINER.
Checked by REYNOLD C. FUSON and E. A. CLEVELAND.

1. Procedure

FORTY-SIX grams (2 gram atoms) of sodium, cut into small pieces, is added to 1 l. of absolute ethyl alcohol (Note 1) in a 2-l. three-necked flask equipped with a mechanical stirrer and a reflux condenser with a calcium chloride tube. When all the sodium is in solution, 320 g. (2 moles) of ethyl malonate (Note 2) is added. Then 200 g. (2.1 moles) of methyl bromide (Note 3) is bubbled into the stirred solution through a tube (at least 4 mm. in diameter) in the third neck of the flask which dips to the bottom of the solution.

The addition of the methyl bromide consumes about four hours. The reaction proceeds smoothly and with evolution of heat, occasionally reaching the boiling point of the mixture. During the course of the reaction sodium bromide separates. When the addition of methyl bromide is complete, the solution is pale orange and faintly alkaline. It is boiled for an additional half-hour then neutralized with glacial acetic acid and cooled. The sodium bromide is filtered with suction and washed on the funnel with a little cold alcohol. The major part of the alcohol is then removed by distillation at atmospheric pressure (Note 4). The sodium bromide is dissolved in 600–700 cc. of water containing 10 cc. of concentrated hydrochloric acid. The resulting solution (Note 5) is added to the residue from the distillation and the mixture shaken well. The aqueous lower layer is separated from the ester and extracted twice with ether. The ester and ether extracts are combined and dried by shaking quickly with calcium chloride and filtering immediately. The ether is removed, and the ester is shaken for exactly one minute with a cold solution of 10 g. of sodium hydroxide in 30 cc. of water (Note 6). The alkali is drawn off; the ester is washed with dilute acid and dried as before with calcium chloride. It is then distilled *in vacuo*, and the fraction boiling at 96°/16 mm. is taken. There is almost no fore-run or residue. The yield is 275–290 g. (79–83 per cent of the theoretical amount).

2. Notes

1. The quality of the alcohol is very important, traces of water depressing the yield considerably. Absolute alcohol, as ordinarily supplied, is best dried by distillation from a 2–3 per cent solution of sodium, a fore-run and the last 25 per cent being rejected.

2. Even the better grades of malonic ester have been found to contain sufficient impurities to decrease the yield. However, one distillation, at reduced pressure, of even the technical grade of malonic ester makes it sufficiently pure for this synthesis.

3. Vials of methyl bromide may be conveniently used by cooling in an ice bath, breaking the seal, and attaching to the delivery tube by means of ordinary rubber tubing. The vial is then allowed to come to room temperature, and the evaporation of the liquid keeps it sufficiently cool so that it enters the solution at the desired rate. It is best to dry the gas by passage through a tower of potassium hydroxide pellets.

If methyl bromide is not available, it may be generated as follows: 500 g. of 95 per cent sulfuric acid is added slowly, with shaking, to 400 g. of ordinary methyl alcohol, cooled by an ice bath. Six hundred grams of sodium bromide is suspended in one half of this mixture in a

1-l. round-bottomed flask, fitted with a two-holed stopper holding a dropping funnel and a delivery tube. The evolution of methyl bromide is started by heating the flask with a water bath at 50°, and the remainder of the methyl alcohol-acid mixture is added slowly from the funnel as the volume of the mixture in the flask decreases. As the rate of evolution of the gas falls off the temperature is slowly raised until no more methyl bromide is generated and the contents of the flask have become completely solid. The flask is shaken from time to time during the generation to mix the components more thoroughly. The evolved gas is dried by passage through a tower of potassium hydroxide pellets.

4. The removal of the alcohol at atmospheric pressure is best carried out by returning the filtrate to the original reaction flask, in which the mechanical stirrer is retained but the reflux condenser is replaced by a goose-neck and a condenser for downward distillation. The solution is stirred as the alcohol is distilled, thus eliminating the violent bumping that would otherwise occur at the separation of sodium bromide from the concentrated solution.

5. The use of the sodium bromide in this way serves the multiple purpose of releasing any ester that may be adsorbed on it, of depressing the solubility of the ester in the water, and of making the water layer sufficiently dense for the ester to float on it.

6. The ester is treated in this manner to remove any unchanged ethyl malonate. Michael [1] has shown that this treatment will completely remove unchanged ethyl malonate while hardly attacking ethyl methylmalonate. No ethyl dimethylmalonate is formed when methyl bromide is used as the methylating agent.

3. Methods of Preparation

The most widely used method of preparing ethyl methylmalonate is by the alkylation of ethyl malonate with methyl iodide,[2,1] methyl bromide,[3] or methyl sulfate.[4] A separation of the desired product from traces of unchanged starting material and from ethyl dimethylmalonate cannot be accomplished by distillation as the boiling points of the three esters lie within three and one-half degrees of one another. Michael [1] found that unchanged malonic ester can be removed completely by taking advantage of the greater ease with which it is hydrolyzed by alkali,

[1] Michael, J. prakt. Chem. (2) **72,** 537 (1905).

[2] Züblin, Ber. **12,** 1112 (1879); Conrad and Bischoff, Ann. **204,** 146 (1880); Herzig and Wenzel, Monatsh. **24,** 115 (1903).

[3] Lucas and Young, J. Am. Chem. Soc. **51,** 2536 (1929).

[4] Nef, Ann. **309,** 188 (1899).

while Gane and Ingold [5] obtained a pure product by hydrolysis, crystallization of the methylmalonic acid, and re-esterification. It can be inferred from the results of Salkowski, Jr.,[6] with acetoacetic ester that no disubstitution occurs when methyl bromide is employed as the alkylating agent. Procedure B above, which is based on the work of Michael,[1] has been published.[7]

Pure ethyl methylmalonate can also be prepared by the reduction of methylidenemalonic ester with Raney nickel;[8] from ethyl α-bromopropionate through the nitrile,[9] an expensive process because of the cost of the materials and the poor yields; and by the pyrolysis of ethyl ethoxalylpropionate. This last method, which forms the basis for Procedure A above, was first described by Wislicenus.[10]

ETHYL α-NAPHTHOATE

(1-Naphthoic acid, ethyl ester)

$$C_{10}H_7Br + Mg \rightarrow C_{10}H_7MgBr$$
$$C_{10}H_7MgBr + (C_2H_5)_2CO_3 \rightarrow C_{10}H_7CO_2C_2H_5 + C_2H_5OMgBr$$

Submitted by FRANK C. WHITMORE and D. J. LODER.
Checked by REYNOLD C. FUSON and CHARLES F. WOODWARD.

1. Procedure

A SOLUTION of α-naphthylmagnesium bromide is prepared from 24.3 g. (1 gram atom) of magnesium and 207 g. (1 mole) of α-bromonaphthalene by the procedure given on p. 425 (Note 1), but using a large separatory funnel and a 3-l. three-necked flask. The Grignard reagent is transferred to the separatory funnel, and 177 g. (1.5 moles) of ethyl carbonate (Note 2) and 100 cc. of absolute ether are placed in the flask. Stirring is begun, and the α-naphthylmagnesium bromide is added as rapidly as the refluxing of the solution permits. Stirring is continued for a half hour after the addition is finished, and the reaction mixture is then left to stand overnight.

Hydrolysis is effected by pouring the reaction mixture, with shaking,

[5] Gane and Ingold, J. Chem. Soc. **1926**, 14.

[6] Salkowski, Jr., J. prakt. Chem. (2) **106**, 256 (1923).

[7] Fieser and Novello, J. Am. Chem. Soc. **62**, 1856 (1940).

[8] Wojcik and Adkins, ibid. **56**, 2424 (1934).

[9] Zelinsky, Ber. **21**, 3162 (1888); Steele, J. Am. Chem. Soc. **53**, 286 (1931).

[10] Wislicenus, Ber. **27**, 796 (1894).

into a 5-l. flask containing 1.2–1.5 kg. of cracked ice. The basic magnesium bromide is dissolved by adding gradually 145 cc. of cold 30 per cent sulfuric acid (30 cc. of concentrated acid and 120 cc. of water). The organic layer is separated and the aqueous layer extracted with 100 cc. of ether. The combined ether solutions are concentrated by distilling the solvent from a steam bath until the volume of the residue is about 400 cc. The residue is washed with two 40-cc. portions of 5 per cent sodium carbonate solution (Note 3) and dried with 20 g. of anhydrous calcium chloride. The calcium chloride is removed by filtration and the ether is distilled from a steam bath. The residue is then transferred to a 500-cc. distilling flask and distilled; the fraction boiling from 287° to 307° is collected as crude ethyl α-naphthoate. The crude product is redistilled from a 250-cc. modified Claisen flask. The yield of pure ester, boiling at 143–144.5°/3 mm., is 136–147 g. (68–73 per cent of the theoretical amount) (Note 4).

2. Notes

1. If benzene is not added to the reagent, α-naphthylmagnesium bromide precipitates and a separatory funnel of wide bore must be used to prevent clogging during the addition of the reagent to the ester.

2. Two hundred cubic centimeters of commercial ethyl carbonate is washed with 40 cc. of 10 per cent sodium carbonate solution, then with 40 cc. of saturated calcium chloride, and finally with 60 cc. of water. After standing for two hours over 10 g. of calcium chloride, the ester is distilled and the fraction boiling at 125–126° is collected. Ethyl carbonate should not be allowed to stand over anhydrous calcium chloride for more than a day, for the ester combines with the salt.

3. By acidification of the sodium carbonate solution there is obtained 2–3 g. of α-naphthoic acid.

4. This preparation has been run satisfactorily using five times the amounts of materials specified above.

3. Methods of Preparation

Ethyl α-naphthoate has been prepared by treating α-naphthoyl chloride with absolute ethyl alcohol,[1] by heating a mixture of α-naphthoic acid and ethyl alcohol in the presence of sulfuric acid,[2] and by the procedure described above.[3]

[1] Hofmann, Ber. 1, 42 (1868).
[2] Perkin, J. Chem. Soc. 69, 1178 (1896).
[3] Loder and Whitmore, J. Am. Chem. Soc. 57, 2727 (1935).

ETHYL α-PHENYLACETOACETATE

(α-Toluic acid, α-acetyl-, ethyl ester)

$$CH_3COCH(C_6H_5)CN + C_2H_5OH \xrightarrow{(HCl)} CH_3COCH(C_6H_5)\underset{\underset{NH}{\|}}{C}—OC_2H_5$$

$$CH_3COCH(C_6H_5)\underset{\underset{NH}{\|}}{C}—OC_2H_5 + H_2SO_4 + H_2O \rightarrow$$

$$CH_3COCH(C_6H_5)CO_2C_2H_5 + NH_4HSO_4$$

Submitted by R. H. KIMBALL, GEORGE D. JEFFERSON, and ARTHUR B. PIKE.
Checked by C. R. NOLLER.

1. Procedure

THE apparatus consists of a 1-l. three-necked flask with a mercury-sealed mechanical stirrer and a 6-mm. inlet tube reaching to the bottom of the flask. In the third neck is a cork bearing a low-temperature thermometer and a tube containing phosphorus pentoxide on glass wool, and calcium chloride. The inlet tube is connected to three 20-cm. drying towers, two containing phosphorus pentoxide on glass wool, and one containing calcium chloride. To the last tower is connected a 2-l. Florence flask fitted as a wash bottle with a safety tube and containing 1 l. of concentrated sulfuric acid. The Florence flask is connected to the hydrogen chloride generator described in Org. Syn. Coll. Vol. I, **1941**, 293, in which a single charge of 1.5 l. of concentrated sulfuric acid and 800 cc. of hydrochloric acid is sufficient for this preparation. A pressure equalizer should be provided between the generating flask and the funnel.

After the inlet tube is removed from the reaction flask, 400 cc. of absolute alcohol (Note 1) and 161 g. (1 mole) of dry α-phenylaceto-acetonitrile (m.p. 88.5–89.5°) (p. 487) are added. The neck is temporarily closed by a cork, and the nitrile is dissolved by warming with stirring. The flask is then surrounded by a freezing mixture, and the solution vigorously stirred, so that any nitrile which crystallizes will be finely divided. When the temperature reaches −10°, the inlet tube is inserted and a stream of dry hydrogen chloride passed through, with moderate stirring, at such a rate that the bubbles rising from the 6-mm. tubing in the sulfuric acid wash bottle can just be counted. This is continued for five to eight hours until the mixture is saturated (Note 2).

The ice bath is then removed, stirring continued until all the solid has dissolved (about one hour), and the flask allowed to stand overnight (Note 3).

Most of the excess hydrogen chloride is removed by adding porous tile and evacuating the flask with a water pump for a half hour, while it is surrounded by a water bath maintained at about 40°. Two hundred grams of sodium carbonate is dissolved in 1.2 l. of water in a 5-l. flask, and 2 l. of cracked ice added. Into this solution the reaction mixture is poured in a thin stream with vigorous shaking, and the solution is extracted at once with three 500-cc. portions of ether. The ether extracts are washed countercurrently with four 250-cc. portions of ice-cold 5 per cent sodium chloride solution to remove the alcohol and then combined in a 3-l. flask placed in an ice bath.

A solution of 100 g. of c.p. concentrated sulfuric acid in 700 cc. of water is prepared in a 5-l. flask, 1.5 l. of cracked ice added, and the mixture shaken until ice forms on the outside of the flask. After about half of this solution has been poured into the cold ether solution of the imino ether, using a funnel to remove the excess ice, the mixture is shaken for exactly fifteen seconds (Note 4), allowed to settle, and the layers separated. The remaining acid is added to the ether layer in two portions, the mixture each time being shaken for fifteen seconds, and separated. Since the ether solution, although now free of the imino ether, still contains a small amount of ethyl phenylacetoacetate, it is saved to be combined with the main portion later.

The sulfuric acid solution of the imino ether sulfate quickly turns cloudy because of the separation of ethyl α-phenylacetoacetate. To complete the hydrolysis, the mixture is heated on the steam bath for one-half hour at the temperature at which the ether just boils (about 50°) (Note 5). It is then cooled, the ester layer separated, and the acid extracted once with 250 cc. of ether. The ether solution is washed once with 100 cc. of water which is recombined with the acid. The acid solution is replaced on the steam bath and heated for forty-five minutes after the temperature reaches 80–90°. After the solution is cooled and extracted as before, all ether extracts, including the original from which the imino ether was removed, are combined and washed once with 250 cc. of 5 per cent sodium bicarbonate solution, once with 250 cc. of water, and then dried over 20 g. of anhydrous sodium sulfate (Note 6). The sodium sulfate is removed by filtration and washed with ether, the ether removed from the filtrates, and the residue fractionated *in vacuo* from a 250-cc. Claisen flask having a 25-cm. fractionating side arm. The main fraction boils at 139–143°/12 mm., or 130–134°/5 mm., and weighs 103–167 g. (50–81 per cent of the theoretical amount). By fractional distillation of the fore-run, main fraction, and residue, a product boiling over a one- to two-degree range may be obtained with no change in the yield (Note 7).

2. Notes

1. The alcohol was dried once with lime, and once with sodium, according to Note 1, Org. Syn. Coll. Vol. I, **1941**, 251.

2. More rapid saturation lowers the yield appreciably.

3. A fine precipitate, probably ammonium chloride, settles. The most successful runs showed little or none of this precipitate.

4. Any delay at this point results in hydrolysis of some of the imino ether to the product, which stays in the ether layer.

5. Care should be taken in lifting the flask from the bath, since any mixing may cause the ether to boil out of the flask.

6. Removal of product from time to time during the course of the hydrolysis seems to improve the yield.

7. The liquid ester is an equilibrium mixture, the enol content of which is increased by distillation and falls slowly, on standing, to 30 per cent.[1] The boiling point of 145–147°/11 mm., recorded in the literature, is higher than any noted in the present work.

3. Method of Preparation

Ethyl α-phenylacetoacetate can be prepared by the hydrolysis of α-phenylacetoacetonitrile in absolute alcohol with dry hydrogen chloride.[1] The present method differs in specifying neutralization of the hydrogen chloride with sodium carbonate and hydrolysis of the imino ether in aqueous sulfuric acid, so that the product separates as fast as it forms. This protects the ester from further decomposition, and a considerably increased yield results.

[1] Beckh, Ber. **31,** 3160 (1898); Post and Michalek, J. Am. Chem. Soc. **52,** 4358 (1930).

ETHYL PHENYLCYANOPYRUVATE

(Pyruvic acid, cyanophenyl-, ethyl ester)

$$C_6H_5CH_2CN + \underset{\underset{CO_2C_2H_5}{|}}{CO_2C_2H_5} + C_2H_5ONa \rightarrow$$

$$\underset{\underset{CN}{|}}{C_6H_5C}{=}C(ONa)CO_2C_2H_5 + 2C_2H_5OH$$

$$\underset{\underset{CN}{|}}{C_6H_5C}{=}C(ONa)CO_2C_2H_5 + HCl \rightarrow \underset{\underset{CN}{|}}{C_6H_5CHCOCO_2C_2H_5} + NaCl$$

Submitted by ROGER ADAMS and H. O. CALVERY.
Checked by J. B. CONANT and DORIS BLUMENTHAL.

1. Procedure

IN a 3-l. round-bottomed flask, fitted with a reflux condenser, is placed 650 cc. of absolute alcohol (Note 1), and to it 46 g. (2 gram atoms) of sodium is added as rapidly as possible without loss of material through the condenser. If the sodium does not entirely dissolve, heat is applied. To the hot sodium ethoxide solution 312 g. (2.1 moles) of ethyl oxalate (Org. Syn. Coll. Vol. I, **1941**, 261) is added as rapidly as possible. Then, immediately, 234 g. (2 moles) of benzyl cyanide (Org. Syn. Coll. Vol. I, **1941**, 107) is added, and the reaction mixture is allowed to stand overnight. The solution is transferred to a 3-l. beaker and treated with 250–300 cc. of water (Note 2). It is then warmed to 35° and made strongly acid to litmus with concentrated hydrochloric acid. Mechanical stirring is used during the acidification. On cooling to ordinary temperatures the ester crystallizes. The yield is 360–385 g. of lemon-yellow crystals melting at 126–128°. On recrystallization from 60 per cent alcohol, the ester melts at 130°. The final yield is 300–325 g. (69–75 per cent of the theoretical amount).

2. Notes

1. Absolute alcohol, prepared according to the directions in Org. Syn. Coll. Vol. I, **1941**, 259, is satisfactory.
2. This amount of water dilutes the alcohol so that the ester crystallizes well. If more water is used an oily product separates and then solidifies.

3. Method of Preparation

The only method of preparation mentioned in the literature is that given above.[1]

ETHYL PHENYLMALONATE

(Malonic acid, phenyl-, diethyl ester)

$$C_6H_5CH_2CO_2C_2H_5 + (CO_2C_2H_5)_2 \xrightarrow{(C_2H_5ONa)}$$
$$C_6H_5CH(CO_2C_2H_5)COCO_2C_2H_5 + C_2H_5OH$$

$$C_6H_5CH(CO_2C_2H_5)COCO_2C_2H_5 \xrightarrow{(Heat)} C_6H_5CH(CO_2C_2H_5)_2 + CO$$

Submitted by P. A. LEVENE and G. M. MEYER.
Checked by LOUIS F. FIESER and C. H. FISHER.

1. Procedure

IN a 2-l. three-necked flask, fitted with a mercury-sealed stirrer, reflux condenser, and dropping funnel, is placed 500 cc. of absolute ethyl alcohol (Note 1), and 23 g. (1 gram atom) of cleanly cut sodium is added in portions. When the sodium has dissolved the solution is cooled to 60°, and 146 g. (1 mole) of ethyl oxalate (Note 2) is added in a rapid stream through the funnel with vigorous stirring. This is washed down with a small quantity of absolute alcohol and is followed immediately by the addition of 175 g. (1.06 moles) of ethyl phenylacetate. Stirring is discontinued at once, the reaction flask is lowered from the stirrer, and a 2-l. beaker is made ready. Within four to six minutes after the ethyl phenylacetate has been added crystallization sets in. The contents of the flask are transferred immediately to the beaker at the first sign of crystallization, which is nearly instantaneous.

The nearly solid paste of the sodium derivative is allowed to cool to room temperature and then stirred thoroughly with 800 cc. of dry ether. The solid is collected by suction and washed repeatedly with dry ether. The phenyloxaloacetic ester is liberated from the sodium salt with dilute sulfuric acid (29 cc. of concentrated sulfuric acid in 500 cc. of water). The almost colorless oil is separated, and the aqueous layer is extracted with three 100-cc. portions of ether, which are combined with the oil. The ethereal solution is dried over anhydrous sodium sulfate, and the

[1] Erlenmeyer, Ann. **271**, 173 (1892).

ether is distilled. The residual oil, contained in a modified Claisen flask having a fractionating side arm, is heated under a pressure of about 15 mm. in a bath of Wood's metal. The temperature of the bath is brought gradually to 175° and kept there until the evolution of carbon monoxide is complete. During this process the heating is momentarily discontinued in the event of a temporary increase in pressure. At the end of the reaction (five to six hours) the oil which has distilled is returned to the flask, and the ethyl phenylmalonate is distilled at reduced pressure. The fraction boiling at 158–162°/10 mm. weighs 189–201 g. (80–85 per cent of the theoretical amount).

2. Notes

1. A high grade of absolute alcohol is essential. Ordinary "absolute" alcohol may be treated with about 5 per cent of its weight of sodium and distilled directly into the reaction flask.

2. To ensure absolutely dry and neutral reagents the ethyl oxalate (Org. Syn. Coll. Vol. I, **1941**, 261) and ethyl phenylacetate (Org. Syn. Coll. Vol. I, **1941**, 270) were shaken with anhydrous potassium carbonate and distilled carefully under reduced pressure, after a preliminary heating under atmospheric pressure until their boiling points were reached.

3. Methods of Preparation

The procedure is based upon the standard method of Wislicenus.[1] Ethyl phenylmalonate has also been obtained from benzyl cyanide and ethyl carbonate.[2] Phenylmalonic acid has been prepared by carbonation of the enolate of phenylacetic acid.[3]

[1] Wislicenus, Ber. **27**, 1091 (1894); Ruhemann, J. Chem. Soc. **81**, 1214 (1902); Pickard and Yates, ibid. **95**, 1015 (1909); Forster and Müller, ibid. **97**, 135 (1910); Baker and Ingold, ibid. **1927**, 835; Blum-Bergmann, Ber. **65**, 115 (1932).

[2] Nelson and Cretcher, J. Am. Chem. Soc. **50**, 2760 (1928).

[3] Ivanoff and Spassoff, Bull. soc. chim. (4) **49**, 20 (1931).

N-ETHYL-m-TOLUIDINE

(m-Toluidine, N-ethyl-)

$$m\text{-}CH_3C_6H_4NH_2 + C_2H_5Br \rightarrow m\text{-}CH_3C_6H_4NHC_2H_5 \cdot HBr$$

$$m\text{-}CH_3C_6H_4NHC_2H_5 + HNO_2 \rightarrow m\text{-}CH_3C_6H_4N(NO)C_2H_5 + H_2O$$

$$m\text{-}CH_3C_6H_4N(NO)C_2H_5 + 6[H] \rightarrow$$
$$m\text{-}CH_3C_6H_4NHC_2H_5 + NH_3 + H_2O$$

Submitted by JOHANNES S. BUCK and CLAYTON W. FERRY.
Checked by JOHN R. JOHNSON and P. L. BARRICK.

1. Procedure

IN each of two ordinary 250-cc. (8-oz.) narrow-mouthed bottles are placed 32.1 g. (0.3 mole) of m-toluidine and 33 g. (23 cc., 0.3 mole) of ethyl bromide (Note 1). The bottles are sealed with rubber stoppers wired tightly in place and then allowed to stand for twenty-four hours in a 2-l. beaker filled with water at room temperature (Note 2). The white crystalline mass in each bottle is broken up and the amine is liberated by shaking with 150 cc. of 10 per cent sodium hydroxide solution and 50 cc. of ether. The contents of the two flasks are combined, the lower aqueous layer is separated and discarded, and the ether solution of the amine is washed with 150 cc. of water. When the ether is distilled from a steam bath, the crude amine (90–92 g.) is obtained.

This crude amine is added, with cooling, to a solution of 100 cc. of concentrated hydrochloric acid (sp. gr. 1.18) in 350 cc. of water. The solution of the hydrochloride is cooled in an ice bath, and stirred rapidly, while a solution of 41.5 g. (0.6 mole) of sodium nitrite in 150 cc. of water is added slowly. During this addition the temperature should not be allowed to rise above 12°. After all the nitrite has been added, the mixture is allowed to stand for ten minutes and is then extracted with three 100-cc. portions of ether. The ether is evaporated from the extract by warming gently on a steam bath and blowing a stream of air over the surface. Care must be taken to keep the temperature as low as possible during the evaporation (Note 3).

The crude nitroso compound is added gradually, with continuous shaking, to a solution of 407 g. (1.8 moles) of stannous chloride dihydrate in 420 cc. (4.8 moles) of concentrated hydrochloric acid (sp. gr. 1.18) contained in a 3-l. flask. The reaction is exothermic, and cooling is applied, if necessary, to keep the temperature below 60°. After standing for at least an hour (Note 4), the mixture is made strongly alkaline by the cautious addition of a cold solution of 520 g. (13 moles) of sodium hydroxide in about 800 cc. of water. During the addition of the alkali the mixture is agitated vigorously and cooled in running water.

The resulting milky suspension is distilled with steam until about 2 l. of distillate has collected. The distillate is saturated with sodium chloride and extracted with three 100-cc. portions of benzene. The extract is dried thoroughly overnight with flaked potassium hydroxide and decanted from the spent drying agent. After removal of the solvent by distillation, the amine is distilled under reduced pressure. Practically all the material distils at 111–112°/20 mm., or 115.5–117°/26 mm. The pure amine forms a practically colorless, highly refringent liquid and weighs 51–53 g. (63–66 per cent of the theoretical amount) (Note 5). It develops color rapidly on standing.

2. Notes

1. A pure grade of *m*-toluidine was used. A "practical" grade of ethyl bromide gave satisfactory results.

2. If the initial reaction is allowed to proceed too rapidly, considerable pressure may be developed in the bottles. As a safeguard against explosions it is advisable to enclose the bottles in wire mesh shields.

3. The nitroso compound decomposes on warming or on standing. It should not be stored but treated at once with the reducing agent.

4. The mixture may be allowed to stand for a longer period (overnight) without harm. Frequently a granular precipitate of a tin complex of the amine separates.

5. Other N-alkyl-*m*-toluidines may be prepared by practically the same procedure. The submitters report that *n*-propyl, isopropyl, and *n*-butyl derivatives are obtained readily from *m*-toluidine and the appropriate alkyl iodides (rather than the bromides). With these halides the alkylation is effected by placing the sealed bottle in a beaker of water which is warmed gradually to 70–80° and kept in a warm place until the reaction is completed; usually several days are required.

3. Methods of Preparation

N-Ethyl-*m*-toluidine has been obtained by passing *m*-toluidine and ethyl alcohol over a catalyst at high temperatures,[1] and by the use of

[1] Mailhe and de Godon, Compt. rend. **172**, 1417 (1921); Mailhe, Fr. pat. 23,891 (Chem. Zentr. **1922**, IV, 760).

ethyl p-toluenesulfonate [2] as an alkylating agent. The present method of purification is a modification of a general procedure for secondary amines developed by Diepolder.[3]

ETHYL n-TRIDECYLATE

(n-Tridecanoic acid, ethyl ester)

$$n\text{-}C_{12}H_{25}Br + KCN \longrightarrow n\text{-}C_{12}H_{25}CN + KBr$$

$$n\text{-}C_{12}H_{25}CN + 2H_2O \xrightarrow{\text{(KOH)}} n\text{-}C_{12}H_{25}CO_2H + NH_3$$

$$n\text{-}C_{12}H_{25}CO_2H + C_2H_5OH \xrightarrow{\text{(HCl)}} n\text{-}C_{12}H_{25}CO_2C_2H_5 + H_2O$$

Submitted by JOHN R. RUHOFF.
Checked by W. H. CAROTHERS and W. L. MCEWEN.

1. Procedure

In a 12-l. flask fitted with a reflux condenser and an efficient, liquid-sealed mechanical stirrer (Note 1) are placed 2.5 l. of 95 per cent ethyl alcohol, 872 g. (3.5 moles) of pure n-dodecyl bromide (Org. Syn. Coll. Vol. I, **1941**, 29, and p. 246 above), and 278 g. (3.85 moles) of powdered 90 per cent potassium cyanide (Note 2). The mixture is refluxed for fifteen hours, with stirring, in a water bath (Note 3). At the end of this time another 278-g. portion of potassium cyanide is added and the mixture refluxed, with stirring, for fifteen hours longer.

The flask is allowed to cool, and a solution of 670 g. of 90 per cent potassium hydroxide in 1 l. of water is added (Note 4). The solution is refluxed and stirred for thirty hours. The flask is fitted with a head for steam distillation (Note 5), and steam is passed in until foaming prevents further distillation. When the flask has cooled somewhat, the steam inlet is replaced by a separatory funnel with a stem reaching to the bottom of the flask, and 1.5 l. of concentrated hydrochloric acid (sp. gr. 1.18) or the equivalent amount of more dilute acid is added. The flask is shaken occasionally to prevent stratification of the hydrochloric acid. The dropping funnel is then replaced by the steam inlet, and steam is passed through for about one-half hour. At the end of this time the water layer is siphoned off and discarded (Note 6).

[2] Finzi, Ann. chim. applicata **15**, 41 (1925) [C. A. **19**, 2647 (1925)].
[3] Diepolder, Ber. **32**, 3514 (1899).

The crude acid while still warm and fluid is treated with 815 cc. (14 moles) of 95 per cent ethyl alcohol containing 20–30 g. (3–5 per cent by weight) of anhydrous hydrogen chloride, and the solution is filtered from insoluble impurities. The filtrate is placed in a 3-l. flask fitted with a reflux condenser, 160 g. of anhydrous calcium chloride is added, and the mixture is refluxed for twenty-four hours (Note 7). The lower layer is siphoned off and diluted with twice its volume of water to recover any dissolved ester. The recovered ester is combined with the main portion, 408 cc. of alcohol containing 3–5 per cent of anhydrous hydrogen chloride and 80 g. of calcium chloride are added, and the mixture is refluxed for another twenty-four hours. The lower layer is siphoned off and the dissolved ester recovered as before. The upper layer is washed twice with an equal volume of warm water (30–40°) and transferred to a 2-l. separatory funnel; the volume of the oil is approximately 1 l. About 500 cc. of warm water and a few drops of phenolphthalein are also placed in the funnel, and very dilute aqueous ammonia (concentrated aqueous ammonia diluted with twenty volumes of water) is added, with shaking, until the resulting emulsion has a distinctly pink color (50–100 cc. is required). Alcohol is added in portions of about 100 cc. until the emulsion is broken and the layers begin to separate quite rapidly; usually two or three additions of alcohol suffice. The lower layer is drawn off and set aside (Note 8). The oil is washed three times with warm water; if necessary, alcohol is added to break the emulsion. The ester is dried with anhydrous calcium chloride, filtered, and distilled under reduced pressure. The yield of ethyl tridecylate, b.p. 163–165°/5 mm. (178–180°/20 mm.; 197–198°/60 mm.) is 615–660 g. (73–78 per cent of the theoretical amount). By refractionating the fore- and after-runs, an additional 50–70 g. of product is obtained; 15–30 g. of ester and acid are recovered from the ammoniacal washings and from the calcium chloride used to dry the main portion. The total yield is 685–710 g. (81–84 per cent of the theoretical amount) (Notes 9 and 10).

2. Notes

1. A mercury-sealed stirrer may be used, but a "vaseline seal" is very satisfactory.

2. Potassium cyanide gives better results than sodium cyanide. It is essential that the material be finely powdered.

3. The entire operation should be carried out in a hood. A small wash tub placed on the bench and heated from the side with a Bunsen burner is a convenient water bath.

4. This corresponds to 3.85 moles of potassium hydroxide plus a

quantity sufficient so that, after the hydrolysis is completed, the solution will contain 10 per cent of alkali. When less alkali is used the quantity of unhydrolyzed nitrile is greater.

5. The condenser should lead to a 2-l. suction flask the side arm of which is connected directly to the vent of the hood. The alcohol has the odor of hydrocyanic acid and should be discarded.

6. The crude acid may be used for some purposes, but it contains a small quantity of unhydrolyzed nitrile, some material of very disagreeable odor (possibly isocyanide), and usually a small amount of a blue precipitate (an iron-cyanide complex salt). In the preparation of the ester, the precipitate is removed by filtering the alcohol solution of the acid as its presence will interfere with the removal of the free tridecylic acid after esterification.

7. A few pieces of porous plate should be added to prevent bumping.

8. By addition of acid to the washings a small amount (15–30 g.) of ester and acid may be recovered.

9. For larger runs, the procedure is the same and the percentage yields are somewhat higher. Thus, in runs by the author using twice the above quantities, yields of 88–89 per cent were obtained.

10. Pentadecylic acid and ethyl pentadecylate have been prepared by the same procedure from n-tetradecyl bromide (p. 247) with approximately the same yield.

3. Methods of Preparation

Tridecylic acid has been prepared by the malonic ester synthesis from undecyl iodide,[1] by the oxidation of α-hydroxymyristic acid in acetone with potassium permanganate,[2] by the action of dodecyl bromide on potassium cyanide and subsequent hydrolysis,[3] and from undecyl iodide and ethyl cyanoacetate.[4] The present method is an adaptation of that of Ruzicka, Stoll, and Schinz.[3]

[1] Levene, West, Allen, and van der Scheer, J. Biol. Chem. 23, 73 (1915).

[2] Levene and West, ibid. 18, 465 (1914).

[3] Ruzicka, Stoll, and Schinz, Helv. Chim. Acta 11, 685 (1928).

[4] Robinson, J. Chem. Soc. 125, 230 (1924).

FLUOROBENZENE

(Benzene, fluoro-)

$$C_6H_5NH_2 \cdot HCl + HNO_2 \rightarrow C_6H_5N_2Cl + 2H_2O$$

$$C_6H_5N_2Cl + HBF_4 \rightarrow C_6H_5N_2BF_4 + HCl$$

$$C_6H_5N_2BF_4 \rightarrow C_6H_5F + N_2 + BF_3$$

Submitted by D. T. FLOOD.
Checked by W. W. HARTMAN and J. R. BYERS.

1. Procedure

A MIXTURE of 1350 cc. of water and 1650 cc. (20 moles) of concentrated hydrochloric acid (sp. gr. 1.19) is placed in a 30 by 30 cm. glass jar or a 40-l. crock and stirred mechanically (Note 1) while strongly cooled by an ice-salt mixture. Two thousand and seventy-five grams (16 moles) of aniline hydrochloride (Note 2) and a solution of 1.2 kg. (17 moles) of sodium nitrite in 1.5 l. of water are made ready. When the temperature of the acid has reached 5° or below, about one-third of the aniline hydrochloride is added to it, and diazotization is begun by the slow addition of the nitrite solution, the temperature being held below 7°. Additional aniline hydrochloride is added from time to time in such amounts that an excess of crystals is always present. The entire amount may be added by the time that half of the nitrite has been added. The diazotization is stopped when a positive test for free nitrous acid is obtained with potassium iodide starch paper. This should require nearly all the nitrite solution.

Fluoboric acid is made concurrently with the diazotization by the addition, in small amounts, of 1 kg. (16.2 moles) of boric acid (U.S.P. crystals) to 2150 g. (65 moles) of 60 per cent hydrofluoric acid (Note 3). The addition is carried out in two 3-l. flasks coated with wax (Note 4) which are shaken and kept cold by immersion in iced water. The temperature of the acid should not be allowed to rise above 20–25°.

The ice-cold solution of fluoboric acid is then poured into the diazonium solution, which has been cooled below 0°. The temperature should remain under 10° during the addition, which is carried out fairly rapidly. Powerful stirring is required to agitate the thick magma at this stage (Note 1). After twenty to thirty minutes' stirring, the brown-colored mass is filtered with suction, using two 24-cm. Büchner funnels. The yellowish crystalline solid is washed with about 800 cc. of iced water, with the same volume of methyl alcohol, and with about 900 cc. of commercial ethyl ether. The solid should be sucked as free as possible from liquid after each washing. This washing is important, since it improves the stability of the product (Note 5).

The light-brown, fluffy salt is spread out on absorbent paper overnight in a current of air (a table placed near a hood is effective), and is then placed in a 12-l. flask (Note 6). This is connected by a wide bent tube through a long, wide condenser to three 2-l. Erlenmeyer flasks arranged in series and immersed in ice-salt mixtures. The last flask is fitted with an exit tube leading to a good hood, to carry off the voluminous fumes of boron fluoride, or to an absorption system containing ice and water or soda solution (Note 7). The solid is heated gently at one point near its surface with a small flame until decomposition begins. The flame is then withdrawn and the reaction allowed to continue spontaneously as long as it will. If the reaction becomes too vigorous, it may be necessary to cool the flask by rubbing it with a piece of ice (Note 8). The mixture is heated cautiously from time to time, as may be necessary to keep the reaction going. At the last it should be heated vigorously until no more fumes of boron fluoride are evolved. The last traces of fluorobenzene may be removed from the reaction flask by applying a slight suction to the receiving flask.

The combined distillate is separated from any phenol which may have settled out. It is washed four or five times with 10 per cent sodium hydroxide solution until the washings are almost colorless and then once with water (Note 9). It is dried by shaking with crushed calcium chloride, and then distilled from a 2-l. flask through a short column at a fairly rapid rate. The first runnings may contain a little water and can be further dried. The product is a colorless liquid with an odor resembling that of benzene. The yield of fluorobenzene boiling at 84–85° is 780–870 g. (51–57 per cent of the theoretical amount) (Note 10).

2. Notes

1. A slow-speed, paddle stirrer with several blades is preferable because of the large amount of suspended matter present at several times in the process. A stirrer of wood or metal protected with acid-proof paint is satisfactory, although the paint does not last well. Rubber-covered stirrers may also be used.

2. A technical grade of aniline hydrochloride is used. Technical sodium nitrite, hydrochloric acid, and hydrofluoric acid are also employed.

Aniline (1485 g.) and hydrochloric acid (3 l.) without any added water can be used in place of the aniline hydrochloride, water, and acid. The separation of aniline hydrochloride in a hard cake on the side of the jar, however, leads to difficulty in cooling and stirring.

3. A corresponding amount of 48 per cent or of 52 per cent hydrofluoric acid may also be used. The quantities given represent an excess of fluoboric acid. A larger excess may be taken but does not appreciably influence the yield. Fluoboric acid (40 per cent) is now a commercial product.

Hydrofluoric acid in contact with the skin produces extremely painful burns. It is therefore necessary to use every precaution to protect exposed parts of the body, especially the hands and eyes. Long acid-resisting rubber gloves and rubber goggles should be worn. If one is burned by the acid, the burned surface, which has become white, is held under running water until the natural color returns. Prompt application of a paste made from magnesium oxide and glycerin is said to be helpful in preventing burns from becoming serious [Fredenhagen and Wellmann, Angew. Chem. **45**, 537 (1932)].

4. The fluoboric acid may be more conveniently prepared in a lead jar with mechanical stirring. Such a jar can be readily made from sheet lead by bending to shape and soldering. A piece of iron rod fitted into some narrow lead pipe and having a strip of lead soldered on at the bottom makes an effective stirrer, and a wooden lid covered on the under side with a sheet of lead may serve as a cover. The solder is slowly attacked by the acid and can be protected with an acid-proof grease. The fluoboric acid is siphoned from the jar by means of a rubber tube.

5. The presence of moisture affects the stability of benzenediazonium fluoborate. The moist product, if allowed to stand packed together, may undergo spontaneous decomposition. In damp weather, when spread out on paper, it becomes dark and slowly decomposes. The same result is observed if it is not sufficiently washed.

6. The decomposition may also be carried out in two batches from 5-l. flasks. It is more easily controlled in this way.

7. On account of the large volume of gas evolved, all connections should be made with wide tubing.

8. Normally the decomposition proceeds smoothly under the intermittent heating. If the salt is moist, however, the reaction proceeds more rapidly, and, unless the flask is cooled, it may pass beyond control (see Note 5).

9. Since the density of fluorobenzene is about 1.025, it is important that the right strength of caustic soda solution be used in order to effect clean separation of the two layers. Before washing with water, the caustic soda should be completely removed.

10. Preparations have been successfully carried out with two times and three times the quantities here stated and with almost proportionate yields. A large stoneware filter in addition to the lead pot (Note 4) is then desirable to handle the larger quantities of material.

3. Methods of Preparation

The method for the preparation of fluorobenzene described above is adapted from that of Balz and Schiemann.[1] It is recommended that benzenediazonium fluoborate be precipitated from a solution whose total hydrogen-ion concentration does not exceed 1 mole per liter.[2] The diazonium fluoborate has also been prepared from aniline hydrochloride and nitrosyl borofluoride.[3]

Fluorobenzene has also been prepared in 50 per cent yield from benzene diazopiperidide and hydrofluoric acid,[4] although the reaction is said to proceed violently and cannot be used with quantities greater than 10–15 g. of piperidide. Other methods depend on the formation of benzenediazonium fluoride and its decomposition into fluorobenzene when heated.[5,6,7] The claim of Valentiner and Schwarz [5] that this reaction proceeds in aqueous solution could not be substantiated by the submitter.

[1] Balz and Schiemann, Ber. **60**, 1188 (1927).

[2] E. I. du Pont de Nemours and Company, U. S. pat. 1,916,327 [C. A. **27**, 4539 (1933)].

[3] Voznesenskiĭ and Kurskiĭ, J. Gen. Chem. (U.S.S.R.) **8**, 524 (1938) [C. A. **32**, 8379 (1938)].

[4] Wallach, Ann. **235**, 258 (1886); Wallach and Heusler, ibid. **243**, 219 (1888).

[5] Valentiner and Schwarz, Ger. pat. 96,153 [Frdl. **5**, 910 (1897–1900); Chem. Zentr. **1898**, I, 1224].

[6] Holleman and Beekman, Rec. trav. chim. **23**, 225 (1904).

[7] Swarts, ibid. **27**, 120 (1908).

p-FLUOROBENZOIC ACID

(Benzoic acid, p-fluoro-)

$$C_2H_5O_2CC_6H_4NH_2 + NaNO_2 + 2HCl$$
$$\rightarrow C_2H_5O_2CC_6H_4N_2Cl + NaCl + 2H_2O$$

$$C_2H_5O_2CC_6H_4N_2Cl + HBF_4 \rightarrow C_2H_5O_2CC_6H_4N_2BF_4 + HCl$$

$$C_2H_5O_2CC_6H_4N_2BF_4 \rightarrow C_2H_5O_2CC_6H_4F + N_2 + BF_3$$

$$C_2H_5O_2CC_6H_4F + H_2O \rightarrow FC_6H_4CO_2H + C_2H_5OH$$

Submitted by G. Schiemann and W. Winkelmüller.
Checked by W. W. Hartman, J. R. Byers, and J. B. Dickey.

1. Procedure

In a 5-l. round-bottomed flask are placed 165 g. (1 mole) of ethyl p-aminobenzoate (Note 1), 300 cc. of water, and 204 cc. (2.5 moles) of concentrated hydrochloric acid (sp. gr. 1.19). This mixture is warmed on a steam bath for an hour with occasional shaking. The flask containing the resulting white paste of p-carbethoxyaniline hydrochloride is placed in an ice-salt bath and cooled to 0°. The mixture is stirred mechanically, and a solution of 72.6 g. (1 mole) of 95 per cent sodium nitrite in a minimum quantity of water is run in slowly while the temperature is kept below 7°. The diazotization is complete when a faint positive test for nitrous acid with starch-iodide paper persists for ten minutes.

While the diazotization is in process, 68 g. (1.1 moles) of boric acid is dissolved in 133 g. (4 moles) of 60 per cent hydrofluoric acid (Note 2) in a beaker coated with paraffin-wax. The temperature is kept below 25° during the addition to avoid melting the paraffin-wax, and, after the addition, the solution is chilled in an ice-water bath.

The ice-cold fluoboric acid solution is added rather rapidly, with stirring, to the diazonium solution while the temperature is kept below 10°. A thick paste of p-carbethoxybenzenediazonium fluoborate precipitates; stirring is continued for twenty to thirty minutes. The solid is filtered on an 18.5-cm. Büchner funnel and washed consecutively with 300 cc. of cold water, 300 cc. of commercial methyl alcohol, and 200 cc. of commercial ether; it is sucked as dry as possible between washings. The fluoborate is then dried over concentrated sulfuric acid (sp. gr. 1.84) in a vacuum desiccator (Note 3). The yield of the dried fluoborate is

198–205 g. (75–78 per cent of the theoretical amount); the decomposition point is 93–94°.

The thermal decomposition may be conveniently carried out in a 2-l. distilling flask. A second distilling flask of 1-l. capacity is connected directly to the side arm of the first to serve as a receiver. Attached to the side arm of the receiver is a rubber tube arranged to lead the escaping gases over 2 l. of water in a 5-l. round-bottomed flask. The boron trifluoride dissolves in the water, and the other gases are led into a good hood (Note 4). The p-carbethoxybenzenediazonium fluoborate is placed in the decomposition flask and heated at its upper edge with a Bunsen flame. When the white fumes of boron trifluoride commence to appear, the flame is removed and the decomposition is permitted to proceed spontaneously. The heat is applied as necessary, and finally the flask is strongly heated to complete the decomposition and melt the solid. Some of the ethyl ester, b.p. 105–106°/25 mm., of p-fluorobenzoic acid is collected in the receiver, where it is carried by the gases, but the larger part is left in the decomposition flask. The ester is washed from the decomposition flask and the receiver with ether, and the ether is distilled from a steam bath. The residue is refluxed for one hour on a steam bath with a solution of 56 g. (1 mole) of potassium hydroxide in 80 cc. of 95 per cent ethyl alcohol and 120 cc. of water. The solution is then filtered while still hot. The p-fluorobenzoic acid is precipitated by adding concentrated hydrochloric acid to the hot filtrate until the mixture is acid to Congo paper. After the mixture has cooled, the solid is filtered and allowed to dry. For purification, the p-fluorobenzoic acid is dissolved in a hot solution of 40 g. of potassium carbonate in 400 cc. of water; the solution is treated with Norite and filtered hot. Hydrochloric acid is added with stirring to precipitate the fluorobenzoic acid, which is then cooled, filtered, and dried.

When 85 g. (0.32 mole) of p-carbethoxybenzenediazonium fluoborate is thus decomposed, there is obtained 38–40 g. of p-fluorobenzoic acid (84–89 per cent of the theoretical amount based on the fluoborate; 63–69 per cent based on the ester of p-aminobenzoic acid). The melting point of the purified acid is 186°. The crude acid melts at 183–184°.

2. Notes

1. Ethyl p-aminobenzoate can be prepared by reduction of the corresponding nitro compound as described in Org. Syn. Coll. Vol. I, **1941**, 240, or by esterification of p-aminobenzoic acid: 274 g. (2 moles) of p-aminobenzoic acid is added to 1.4 kg. of ethyl alcohol which has been saturated with dry hydrogen chloride. The reaction mixture is refluxed

for twenty-four hours, poured into water, neutralized with sodium carbonate, and the insoluble ester is separated by filtration. The yield is 80 per cent of the theoretical amount.

2. *Hydrofluoric acid causes extremely painful burns. Exposed parts of the body must be protected when working with this material. Compare Note 3, p. 297.*

For preparing the fluoboric acid a lead jar may be conveniently used instead of the beaker lined with paraffin-wax. By using a lead stirrer of the usual shape, mechanical stirring may be substituted. The stirrer should be thrust through a hole in a lead cover of sufficient size to prevent spattering of the hydrofluoric acid.

Forty per cent fluoboric acid is now available commercially; 200 g. of the commercial product may be used in the procedure above.

3. It is very important that the fluoborate be dry. If this solid is wet, the decomposition is very violent, tar is formed, and the yield is lowered.

4. A simpler apparatus for the decomposition is a 2-l. round-bottomed short-necked flask connected by means of a wide (2-cm.) bent tube to a 1-l. flask containing 500 cc. of water. The gases from the second flask are led to a good hood.

3. Methods of Preparation

p-Fluorobenzoic acid has been prepared by the oxidation of p-fluorotoluene with chromic acid in dilute sulfuric acid at 160°;[1] by the oxidation of p-fluorotoluene with potassium permanganate;[2] by the electrolytic oxidation of p-fluorotoluene;[3] by heating p-carboxybenzenediazonium chloride with fuming hydrofluoric acid;[4] by the oxidation of p-fluorobenzaldehyde;[5] by the oxidation of p,p'-difluorostilbene with potassium permanganate;[6] by the oxidation of 4,4'-difluorobiphenyl with nitric acid;[7] and by the oxidation of 4,4'-difluorobiphenyl with chromic acid.[8] The above method is adapted from that of Balz and Schiemann.[9]

[1] Wallach, Ann. **235**, 263 (1886).

[2] Slothouwer, Rec. trav. chim. **33**, 324 (1914); Holleman, ibid. **25**, 332 (1906); Holleman and Slothouwer, Proc. K. Akad. Wetensch. Amsterdam **19**, 497, 500 (1910) (Chem. Zentr. **1911**, I, 74); Koopal, Rec. trav. chim. **34**, 152 (1915); Meyer and Hub, Monatsh. **31**, 933 (1910).

[3] Fichter and Rosenzweig, Helv. Chim. Acta **16**, 1154 (1933).

[4] Schmitt and Gehren, J. prakt. Chem. (2) **1**, 394 (1870); Paterno and Oliveri, Gazz. chim. ital. **12**, 87 (1882).

[5] Rinkes, Chem. Weekblad **16**, 206 (1919).

[6] Meyer and Hoffmann, Monatsh. **38**, 154 (1917).

[7] v. Hove, Bull. acad. roy. Belg. [5] **12**, 801 (1926) (Chem. Zentr. **1927**, I, 884).

[8] Schiemann and Roselius, Ber. **62**, 1813 (1929).

[9] Balz and Schiemann, ibid. **60**, 1186 (1927).

FUMARIC ACID *

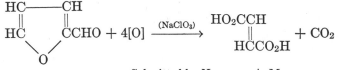

Submitted by NICHOLAS A. MILAS.
Checked by ROGER ADAMS and K. L. AMSTUTZ.

1. Procedure

A 5-l. flat-bottomed flask, equipped with a long (80–90 cm.) wide-bore condenser, a separatory funnel, and a mechanical stirrer (Note 1), is supported 10 cm. above an electric hot plate. Two grams of vanadium pentoxide (Note 2), 450 g. (4.2 moles) of sodium chlorate, and 1 l. of water are placed in the flask; the stirrer is started, the mixture is heated to 70–75° (Note 3), and about 5–10 cc. of 200 g. (2.06 moles) of furfural (Note 4) is added. As soon as a vigorous reaction starts the balance of the furfural is added rapidly enough to maintain it (Note 5). The entire addition requires seventy to eighty minutes. Then the mixture is heated at 70–75°, with stirring, for ten to eleven hours and allowed to stand overnight at room temperature. The crude fumaric acid is filtered with suction and dried in the air. It weighs 155–170 g. (65–72 per cent of the theoretical amount) (Note 6).

More fumaric acid may be obtained from the filtrate by heating it on the water bath with 50 cc. of concentrated hydrochloric acid; the solution usually turns blue at the end of the reaction. The solution is concentrated to about 700 cc. and then cooled with running water. The fumaric acid which separates is collected on a filter and dried in the air. It weighs 10–15 g. and melts at 282–284° in a sealed tube.

The crude product is purified by recrystallization from about 1250 cc. of 1 N hydrochloric acid. This gives 100–110 g. of pure fumaric acid, melting at 282–284° in a sealed tube. An additional amount of the acid may be obtained by concentrating the filtrate to a small volume on a water bath. The total yield of pure fumaric acid is 120–138 g. (50–58 per cent of the theoretical amount).

2. Notes

1. The yields were lower when mechanical stirring was not used.
2. The catalyst is prepared by suspending 20 g. of C.P. ammonium

* Commercially available; see p. v.

metavanadate in 200 cc. of water and adding slowly 30 cc. of concentrated hydrochloric acid (sp. gr. 1.19). The reddish brown semicolloidal precipitate is washed several times with water by decantation and finally suspended in 300 cc. of water and allowed to stand at room temperature for three days. This treatment makes the precipitate granular and easy to filter. The precipitate is collected on a filter using a pump and washed several times with water to free it from hydrochloric acid. It is then dried at 120° for twelve hours, finely powdered, and again dried for twelve hours at 120°.

3. To observe the temperature, a thermometer may be suspended in the flask through the inner tube of the condenser. The mixture may be heated first to 70–75° over a free flame.

4. The furfural used was the technical grade furnished by the Miner Laboratories, Chicago. According to these laboratories the furfural was "about 99 per cent pure." The crude furfural obtained as described in Org. Syn. Coll. Vol. I, **1941**, 280 may be used.

5. The reaction does not seem to start immediately upon the addition of the first few cubic centimeters of furfural. When the vigorous reaction commences, the temperature rises to about 105° and remains there for some time. The yield of fumaric acid seems to depend somewhat upon the rapidity of this stage of the reaction. It is therefore necessary to regulate the addition of furfural so that a vigorous reaction is maintained.

6. The crude fumaric acid is from 74 to 78 per cent pure as found by titration with standard alkali. The only impurity present besides inorganic salts is a little sodium hydrogen maleate which is decomposed by the hydrochloric acid during the purification process.

3. Methods of Preparation

Fumaric acid has been prepared from bromosuccinic acid by heating with water,[1] or dilute hydrobromic acid,[2] and by heating the acid above its melting point.[3] It has also been prepared by heating malic acid,[4] by isomerization of maleic acid;[5] and by the reduction of tartaric acid

[1] Volhard, Ann. **268**, 256 (1892); Müller and Suckert, Ber. **37**, 2598 (1904).

[2] Fittig, Ann. **188**, 90 (1877).

[3] Kekulé, ibid. **130**, 22 (1864); Volhard, ibid. **242**, 158 (1887).

[4] Lassaigne, Ann. chim. phys. (2) **11**, 93 (1819); Pelouze, Ann. **11**, 265 (1834); Wislicenus, ibid. **246**, 91 (1888); Michael, J. prakt. Chem. (2) **46**, 231 (1892); Jungfleisch, Bull. soc. chim. (2) **30**, 147 (1878).

[5] Wislicenus, Ber. **29**, 1080 (abstracts) (1896); Ciamician and Silber, ibid. **36**, 4267 (1903); Skraup, Monatsh. **12**, 107 (1891); Weiss and Downs, J. Am. Chem. Soc. **44**, 1119 (1922); Terry and Eichelberger, ibid. **47**, 1402 (1925).

with phosphorus and iodine.[6] The procedure described is the most satisfactory for laboratory use, and is a slight modification of one described in the literature.[7]

GALLACETOPHENONE

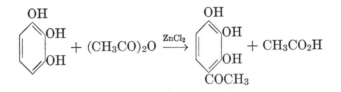

Submitted by I. C. BADHWAR and K. VENKATARAMAN.
Checked by W. W. HARTMAN and L. J. ROLL.

1. Procedure

IN a 250-cc. round-bottomed flask, fitted with a reflux condenser to which is attached a calcium chloride tube, 28 g. (0.21 mole) of freshly fused and finely powdered zinc chloride (Note 1) is dissolved in 38 cc. of glacial acetic acid by heating in an oil bath at 135–140°. Forty grams (0.37 mole) of 95 per cent acetic anhydride is then added to the clear, pale brown liquid, followed by the addition in one lot of 50 g. (0.4 mole) of distilled pyrogallol (Note 2). The mixture is heated at 140–145° (Note 3) for forty-five minutes with frequent and vigorous shaking. The unused acetic anhydride and acetic acid are removed by distilling under reduced pressure. The red-brown cake is broken up by the addition of 300 cc. of water with mechanical stirring for a few minutes. The mixture is cooled in ice water, filtered with suction, and washed with cold water. The crude material, 45–50 g., is crystallized from 500 cc. of boiling water saturated with sulfur dioxide. The yield of straw-colored needles melting at 171–172° is 36–38 g. (54–57 per cent of the theoretical amount). On saturating the mother liquor with salt and cooling to 10°, 4–5 g. of crude material is obtained, which on recrystallization yields 3–4 g. of pure material (Note 4).

[6] Phelps, Ger. pat. 254,420 [Frdl. **11**, 99 (1912–14)]; Org. Chem. Reagents II, Univ. Illinois Bull. **18** (6), 35 (1920).

[7] Milas, J. Am. Chem. Soc. **49**, 2007 (1927); Buluigina, Maslobolno-Zhirovoe Delo **1934**, No. 4, 43 [C. A. **30**, 1743 (1936)].

2. Notes

1. A good quality of zinc chloride must be used, and it is an advantage to fuse it immediately before use.

2. Variations in the proportions of acetic acid, anhydride, and zinc chloride did not result in increased yields.

3. The temperature must be carefully regulated and must not exceed 150°. In this preparation, as well as in the preparation of other ketones by the Nencki reaction, higher temperatures lead to the formation of a highly colored and resinous product which probably contains a little diketone.

4. This method has been used for the preparation of other phenolic ketones such as resacetophenone, 2-acetyl-1-naphthol,[1] 2-phenylacetyl-1-naphthol, and 2-phenylpropionyl-1-naphthol.[2]

3. Methods of Preparation

The method described above is a modification of the process of Nencki and Sieber.[3] Gallacetophenone has also been prepared by treating pyrogallol with acetyl chloride,[4] and from 2-formyl-4-acetylresorcinol by treatment with hydrogen peroxide and alkali.[5]

dl-GLYCERALDEHYDE

$$HOCH_2CHOHCH(OC_2H_5)_2 + H_2O \xrightarrow{(H_2SO_4)}$$
$$HOCH_2CHOHCHO + 2C_2H_5OH$$

Submitted by E. J. Witzemann, Wm. Lloyd Evans, Henry Hass, and E. F. Schroeder.
Checked by Frank C. Whitmore and Harry T. Neher.

1. Procedure

A mixture of 50 g. (0.3 mole) of *dl*-glyceraldehyde acetal (p. 307) and 500 cc. of 0.1 *N* sulfuric acid is allowed to stand for one week at about

[1] Witt and Braun, Ber. **47**, 3227 (1914).
[2] Cheema and Venkataraman, J. Chem. Soc. **1932**, 919.
[3] Nencki and Sieber, J. prakt. Chem. (2) **23**, 151, 538 (1881); Nencki, Ber. **27** 2737 (1894). See also Crabtree and Robinson, J. Chem. Soc. **121**, 1038 (1922).
[4] Einhorn and Hollandt, Ann. **301**, 107 (1898); Fischer, Ber. **42**, 1020 (1909).
[5] Nakazawa, J. Pharm. Soc. Japan **59**, 297 (1939) [C. A. **33**, 8186 (1939)].

20° (Note 1). Thirty cubic centimeters of glacial acetic acid is added; the mixture is neutralized carefully with barium hydroxide solution (Note 2), stirred with 5 g. of decolorizing carbon, and filtered. The filtrate is evaporated at 10 mm. pressure (Note 3). When no more water can be removed, the residue is treated with an equal volume of absolute alcohol to hasten crystallization. The crystals are collected on a filter and dried in a vacuum desiccator over soda-lime and calcium chloride. The yield of product melting at 137–139° is 22 g. (80 per cent of the theoretical amount).

2. Notes

1. During the entire preparation, including the evaporation, the temperature should be kept below 30°. If this precaution is rigidly observed, the glyceraldehyde crystallizes readily.
2. Samples of the filtered solution should give only a very slight opalescence when tested with barium hydroxide and with sulfuric acid.
3. The use of 10 mm. instead of 20 mm. pressure for the evaporation improves the quality of the glyceric aldehyde, making the difference between a syrup which may or may not crystallize and a product which crystallizes even from the concentrated solution.

3. Methods of Preparation

dl-Glyceraldehyde has been obtained by the oxidation of glycerol with nitric acid,[1] with bromine and sodium carbonate,[2] and with hydrogen peroxide in the presence of ferrous salts;[3] by the action of ultraviolet light on glycerol in neutral solution;[4] by the action of sunlight on glycerol in the presence of uranium sulfate;[5] by electrolysis of *dl*-erythronic acid;[6] by the hydrolysis of *dl*-glyceraldehyde acetal;[7] by the oxidation of acrolein;[8] by the oxidation of benzal-

[1] Kiliani, Ber. **54**, 467 (1921); Fischer and Tafel, ibid. **20**, 3385 (1887).
[2] Fischer and Tafel, ibid. **20**, 3385 (1887).
[3] Fenton and Jackson, Chem. News **78**, 187 (1898); J. Chem. Soc. **75**, 5 (1899); Witzemann, J. Am. Chem. Soc. **36**, 2227 (1914).
[4] Bierry, Henri, and Ranc, Compt. rend. **152**, 535 (1911).
[5] Neuberg, Biochem. Z. **13**, 307 (1908).
[6] Neuberg, Scott, and Lachmann, ibid. **24**, 157 (1910).
[7] Wohl, Ber. **31**, 1800, 2395 (1898); Wohl and Neuberg, ibid. **33**, 3100 (1900); Evans and Hass, J. Am. Chem. Soc. **48**, 2706 (1926); Witzemann, ibid. **36**, 1913 (1914); Spoehr and Young, Carnegie Inst. Washington Yearbook, **25**, 177 (1925–1926); Exp. Sta. Record, **57**, 817 (1927) [C. A. **22**, 2368 (1928)].
[8] Neuberg, Biochem. Z. **221**, 492 (1930); **255**, 1 (1932).

1,3-propenediol followed by hydrolysis; [9] and by the alkaline condensation of formaldehyde.[10]

dl-GLYCERALDEHYDE ETHYL ACETAL

$$CH_2{=}CHCH(OC_2H_5)_2 + H_2O + [O] \xrightarrow{(KMnO_4)}$$
$$HOCH_2CHOHCH(OC_2H_5)_2$$

Submitted by E. J. WITZEMANN, WM. LLOYD EVANS, HENRY HASS, and
E. F. SCHROEDER.
Checked by FRANK C. WHITMORE and HARRY T. NEHER.

1. Procedure

IN a 3-l. open flask, equipped with a mechanical stirrer and a thermometer and cooled in an ice bath, is placed a suspension of 65 g. (0.5 mole) of acrolein acetal (p. 17) in 600 cc. of water. The suspension is cooled to 5° (Note 1), and a solution of 80 g. (0.5 mole) of potassium permanganate in 1.5 l. of water is added, with stirring, at the rate of about 25 cc. per minute. The temperature is kept as near 5° as possible during the addition. Soon after the stirring is stopped, the mixture sets to a gel (Note 2). After standing for two hours, the mixture is heated for one hour on the steam bath and then filtered by suction on a 30-cm. Büchner funnel. The residual manganese dioxide is pressed thoroughly and washed with 150 cc. of cold water. The filtrate (about 2.3 l.) is kept cool and treated with 1.2 kg. of freshly dehydrated commercial potassium carbonate. The layers are separated and the water layer is extracted with four 100-cc. portions of ether. The ether extracts are added to the crude acetal layer, and the mixture, which may consist of two layers (Note 3), is dried over 10 g. of potassium carbonate. After removal of the ether, the residue is distilled under reduced pressure. The yield of product boiling at 120–121°/8 mm. is 55 g. (67 per cent of the theoretical amount).

2. Notes

1. The oxidation is very sensitive to changes in temperature. The best results are obtained at 5°; a slight variation causes a marked decrease in the yield.

[9] Fischer, Ahlström, and Richter, Ber. **64**, 611 (1931).
[10] Kuzin, J. Gen. Chem. (U.S.S.R.) **8**, 592 (1938) [C. A. **33**, 1271 (1939)].

2. If the mixture does not set to a gel, the yield is likely to be poor. This is usually due to poor temperature control.

3. Sometimes two layers appear at first, but these disappear when the potassium carbonate is added.

3. Methods of Preparation

dl-Glyceraldehyde acetal has been prepared by heating hydroxy-chloropropionaldehyde acetal with potassium carbonate solution;[1] by treating glyceraldehyde with alcoholic hydrogen chloride;[2] and by oxidation of acrolein acetal with potassium permanganate.[3]

GLYCEROL α,γ-DIBROMOHYDRIN

(2-Propanol, 1,3-dibromo-)

$$3HOCH_2CHOHCH_2OH + 2P + 3Br_2$$
$$\rightarrow 3BrCH_2CHOHCH_2Br + 2P(OH)_3$$

Submitted by GÉZA BRAUN.
Checked by REYNOLD C. FUSON and S. H. BABCOCK.

1. Procedure

IN a 3-l. three-necked, round-bottomed flask, fitted with a powerful glycerol-sealed stirrer, a dropping funnel, and an outlet tube for escaping gases, 1.6 kg. (17.4 moles) of glycerol is thoroughly mixed with 200 g. (6.5 gram atoms) of red phosphorus (Note 1). Nine hundred cubic centimeters (2808 g., 17.5 moles) of bromine (Note 2) is then gradually added, with effective stirring (Note 3), through the dropping funnel in the course of about eight hours. To minimize the escape of bromine, the end of the dropping funnel should reach almost to the bottom of the flask. The by-product gases, consisting mainly of hydrogen bromide and some bromine, are led over concentrated sodium hydroxide solution or to a gas trap. The reaction is exothermic, and the temperature

[1] Wohl, Ber. **31**, 1799 (1898).

[2] Wohl and Neuberg, ibid. **33**, 3103 (1900); Witzemann, J. Am. Chem. Soc. **36**, 2229 (1914).

[3] Wohl, Ber. **31**, 1799 (1898); Evans and Hass, J. Am. Chem. Soc. **48**, 2706 (1926); Witzemann, ibid. **36**, 1912 (1914); Spoehr and Young, Carnegie Inst. Washington Yearbook, **25**, 177 (1925–1926); Expt. Sta. Record, **57**, 817 (1927) [C. A. **22**, 2368 (1928)]; Fischer and Baer, Helv. Chim. Acta **18**, 516 (1935).

quickly rises to 80–100°; then the addition of bromine is so regulated that this temperature is maintained. Toward the end of the period of addition of bromine, the flask is placed in a water bath at 70–75°. After all the bromine has been added, the mixture is allowed to stand overnight and is then warmed on the water bath until all the bromine is consumed (one to two hours).

The reaction mixture is transferred to a 3-l. round-bottomed flask provided with a two-holed rubber stopper carrying a wide delivery tube and a capillary tube. The flask is heated in an oil bath, and distillation is begun under the reduced pressure of a water pump. The receiver is cooled with water. At first a mixture of hydrobromic acid and water passes over; later the dibromohydrin distils. The temperature of the bath is raised as fast as the boiling of the mass permits, and is eventually brought to 180°. The distillation is carefully watched at the end and immediately interrupted at the first sign of decomposition. This is clearly indicated by gas formation, in consequence of which the vacuum cannot be maintained at the previous level. To the straw-yellow distillate a slight excess of solid sodium bicarbonate is added with continuous shaking until effervescence ceases. The inorganic salts are removed by filtration, and the aqueous layer of the filtrate is separated from the crude dibromohydrin. The latter is purified by fractional distillation under reduced pressure from a 2-l. Claisen flask. The distillation is continued until no more water passes over and the inside temperature reaches 100°. Then the dibromohydrin is separated from the water in the distillate, dried with anhydrous sodium sulfate, filtered, and poured back into the distilling flask. By this operation the water is largely removed (Note 4). Then the distillation is continued as before and, after a small fore-run, the dibromohydrin boils at 110–112° under 20 mm. pressure (Note 5). The yield is 2000–2050 g. (52–54 per cent of the theoretical amount) of a colorless product.

The dibromohydrin is a heavy, colorless liquid with a characteristic odor. On standing it gradually becomes yellow. Its specific gravity at 20° is about 2.14.

2. Notes

1. The red phosphorus should be thoroughly mixed with the glycerol before the addition of the bromine. The bromine should not come into contact with the dry phosphorus or a violent reaction will occur.

2. Commercial 98 per cent glycerol and U.S.P. bromine may be used in the preparation.

3. A powerful stirrer is necessary because of the viscous nature of the reaction mixture.

4. The water derives from the chemical interaction of phosphorous acid with the glycerol or with the bromohydrins. Better yields are obtained when the theoretical amount of bromine is used, although on account of this secondary reaction a smaller amount should suffice.

5. The crude dibromohydrin distils without any decomposition at 10–15 mm. pressure if the temperature of the oil bath is not raised over 190°. Above this temperature, formation of acrolein derivatives makes the dibromohydrin lachrymatory.

3. Methods of Preparation

Glycerol α,γ-dibromohydrin has been prepared from glycerol and phosphorus tribromide;[1] from glycerol and bromine;[2] and from glycerol, phosphorus, and bromine.[3, 4, 5]

GLYCINE ETHYL ESTER HYDROCHLORIDE

$$CH_2{=}NCH_2CN + C_2H_5OH + 2HCl + 2H_2O$$
$$\rightarrow HCl \cdot NH_2CH_2CO_2C_2H_5 + CH_2O + NH_4Cl$$

Submitted by C. S. Marvel.
Checked by Louis F. Fieser and S. L. Judkins.

1. Procedure

In a 3-l. round-bottomed flask are placed 500 cc. (400 g., 8.7 moles) of absolute alcohol which has been saturated in the cold with hydrochloric acid gas (Note 1), 870 cc. (680 g., 14.8 moles) of 96 per cent alcohol (Note 2), and 70 g. (1.03 moles) of methyleneaminoacetonitrile (Note 3). This mixture is refluxed on a steam bath for three hours (Note 4). During the refluxing, ammonium chloride separates. After the reaction is complete, the hot alcohol solution is filtered with suction and the filtrate cooled, thus allowing the glycine ester hydrochloride to separate in fine white needles. The product is filtered with suction, sucked as dry as possible on the filter, and then allowed to dry in the air. The yield is about 110 g. The alcohol from the filtrate is distilled

[1] Berthelot and de Luca, Ann. chim. phys. (3) 48, 306 (1856).
[2] Barth, Ann. 124, 349 (1862).
[3] Aschan, Ber. 23, 1826 (1890).
[4] Lespieau, Ann. chim. phys. (7) 11, 236 (1897).
[5] Braun, J. Am. Chem. Soc. 52, 3172 (1930).

(Note 5) until about one-third of its volume is left; it is cooled again, when a second crop of crystals is obtained. The total yield of product, m.p. 142–143°, varies from 125 to 129 g. (87–90 per cent of the theoretical amount). If a very pure product is desired, the material may be recrystallized from absolute alcohol.

2. Notes

1. The 500 cc. of absolute alcohol is cooled in an ice bath and treated with dry hydrogen chloride until 163 g. has been added, an amount sufficient for saturation. The solution should be protected from the moisture of the air with a calcium chloride tube.

2. It is important to use the strengths of alcohol specified in the directions if the best yields are to be obtained, and it is advisable to test the alcohol with a hydrometer just before using. The 870 cc. of 96 per cent alcohol contains just enough water for the hydrolysis. If, therefore, a less concentrated alcohol is used, the glycine ester hydrochloride does not form so readily and does not separate so easily from solution. Experiments using 96 per cent alcohol saturated with hydrochloric acid and 870 cc. of 96 per cent alcohol gave about 8 to 10 g. less product. A more dilute alcohol than 96 per cent gives a much poorer grade and yield of the glycine ester hydrochloride.

3. The crude material as described in Org. Syn. Coll. Vol. I, **1941,** 355, is satisfactory.

4. A cork and not a rubber stopper should be used during the refluxing, as rubber stoppers will cause the product to be colored.

5. It is important that no water get into the alcohol, as glycine ester hydrochloride is quite soluble in water. Concentration of the filtrate on the steam bath should not be done in an open vessel because the solution will take up moisture and the product will not crystallize.

3. Methods of Preparation

Glycine ethyl ester hydrochloride has been prepared by the action of absolute alcohol and hydrogen chloride on glycine;[1] from glycyl chloride and alcohol;[2] by the action of ammonia[3] or hexamethylenetetramine[4]

[1] Curtius and Goebel, J. prakt. Chem. (2) **37,** 159 (1888); Harries and Weiss, Ann. **327,** 365 (1903).

[2] Fischer, Ber. **38,** 2916 (1905).

[3] Hantzsch and Silberrad, ibid. **33,** 70 (1900); Hantzsch and Metcalf, ibid. **29,** 1681 (1896).

[4] Auger, Bull. soc. chim. (3) **21,** 6 (1899); Locquin, ibid. **23,** 662 (1900).

on chloroacetic acid, and subsequent hydrolysis with alcoholic hydro-chloric acid; by the action of hydrogen chloride and alcohol on methyl-eneaminoacetonitrile;[5] and by the reduction of ethyl cyanoformate[6] or the corresponding imido ester hydrochloride.[7]

unsym.-HEPTACHLOROPROPANE

(Propane, 1,1,1,2,2,3,3-heptachloro-)

$$CCl_2{=}CCl_2 + CHCl_3 \xrightarrow{\text{(AlCl}_3)} CCl_3CCl_2CHCl_2$$

Submitted by MARK W. FARLOW.
Checked by FRANK C. WHITMORE and F. W. BREUER.

1. Procedure

IN a 1-l. round-bottomed flask, equipped with a reflux condenser carrying a calcium chloride tube, are placed 166 g. (103 cc., 1 mole) of technical tetrachloroethylene, 300 g. (200 cc., 2.5 moles) of dry chloro-form, and 27 g. (0.2 mole) of anhydrous aluminum chloride. The mix-ture is refluxed gently on the steam bath for fifteen hours (Note 1), cooled to room temperature, and poured into a 1-l. separatory funnel half filled with crushed ice. The organic layer is washed several times with water and dried over calcium chloride or soluble anhydrite. By fractionation at atmospheric pressure through an efficient column, 160–165 g. of chloroform is recovered. Distillation of the unsym.-hepta-chloropropane fraction at diminished pressure gives material boiling at 110–113°/10 mm., or 137–140°/32 mm., and melting at 29–30°. The yield is 250–266 g. (88–93 per cent of the theoretical amount) (Note 2).

2. Notes

1. A small amount of hydrogen chloride is evolved in the initial stages of the reaction.
2. According to Prins[1] the heptachloropropane can be isolated easily by pouring the reaction mixture into water and removing the unreacted

[5] Jay and Curtius, Ber. 27, 60 (1894); Klages, ibid. 36, 1508 (1903).
[6] Ges. für Kohlentechnik M.B. H., Ger. pat. 594,219 [C. A. 28, 3417 (1934)]; 638,577 [C. A. 31, 3066 (1937)]; E. Merck, Ger. pat. 597,305 [C. A. 28, 5078 (1934)].
[7] Ges. für Kohlentechnik M.B. H., Ger. pat. 604,277 [C. A. 29, 812 (1935)].

[1] Prins, Rec. trav. chim. 54, 249 (1935).

materials by steam distillation. The process is stopped when the product begins to distil, and on cooling the residue is obtained as a colorless solid of the correct melting point.

3. Methods of Preparation

The method is essentially that discovered by Böeseken and Prins [2] and studied further by Prins.[1, 3] Pentachloroethane can be used in place of tetrachloroethylene, as it is converted into the unsaturated compound in the presence of aluminum chloride. *unsym.*-Heptachloropropane has been obtained also by the action of phosphorus pentachloride on pentachloroacetone,[4] and by treating dichloroacetyl chloride with aluminum chloride.[5]

HEPTALDOXIME

(Enanthaldehyde oxime)

$$2CH_3(CH_2)_5CHO + 2NH_2OH \cdot HCl + Na_2CO_3 \rightarrow$$
$$2CH_3(CH_2)_5CH{=}NOH + 2NaCl + CO_2 + 3H_2O$$

Submitted by E. W. Bousquet.
Checked by W. H. Carothers and W. L. McEwen.

1. Procedure

In a 5-l. two-necked flask, fitted with a mechanical stirrer, a reflux condenser, a thermometer, and a separatory funnel, are placed an aqueous solution of 348 g. (5 moles) of hydroxylamine hydrochloride (Org. Syn. Coll. Vol. I, **1941**, 318) in 600 cc. of cold water and 460 g. (4 moles) of heptaldehyde (Note 1). Stirring (Note 2) is started, and a solution of 265 g. (2.5 moles) of anhydrous c.p. sodium carbonate in 500 cc. of water is added at such a rate that the temperature of the reaction mixture does not rise above 45°. Stirring is continued at room temperature for an hour after the addition of the sodium carbonate solution is complete.

The oily layer on top of the reaction mixture is separated and washed with two 100-cc. portions of water (Note 3). The washed product is transferred to a 1.5-l. modified Claisen flask and distilled under reduced

[2] Böeseken and Prins, Chem. Zentr. **1911**, I, 466.
[3] Prins, J. prakt. Chem. (2) **89**, 414 (1914).
[4] Fritsch, Ann. **297**, 312 (1897).
[5] Böeseken, Rec. trav. chim. **29**, 109 (1910).

pressure from an oil bath. The first fraction contains a very small amount of water along with a mixture of heptanonitrile and heptaldoxime. The product is collected at 103–107°/6 mm. (temperature of the oil bath, 140–147°) (Note 4). The yield is 420–480 g. (81–93 per cent of the theoretical amount). The product solidifies slowly on cooling and melts at 44–46°. It can be used for reduction to n-heptylamine (p. 318) without further purification.

The product can be purified by recrystallization from 60 per cent ethyl alcohol, using approximately 70 cc. of the solvent to 25 g. of the distilled product. One such recrystallization (Note 5) gives white leaflets melting at 53–55° (Notes 6 and 7). The yield of recrystallized material from a single run is 315–320 g.

2. Notes

1. The heptaldehyde used boiled at 54–59°/16 mm.

2. Since the heptaldehyde and the aqueous solution of hydroxylamine hydrochloride form a heterogeneous mixture, it is necessary to provide rapid, efficient stirring in order to obtain good results.

Ethyl alcohol can be used to provide a homogeneous solution, but the yield seems to be diminished slightly owing to the presence of more high-boiling material.

3. The product is so insoluble in water that an ether extraction is hardly necessary to obtain all the product from the water solution if sufficient time is allowed for the separation of the two layers.

4. The temperature of the oil bath during distillation is important. The first fraction is cut as soon as a constant boiling point is reached, and this constancy of boiling point is obtained sooner if the temperature of the oil bath is regulated to a constant temperature before distillation is started. No more than 50 cc. (of which approximately 10 cc. is water) should come over in the first fraction. If the temperature of the bath is regulated carefully, practically all the product will distil at a constant temperature.

5. The product is dissolved by gentle heating and the solution is then cooled to 0° or below for several hours. The material remaining in the mother liquor (about 30 per cent of the total) may be recovered as impure, oily oxime by evaporation of the alcohol.

6. The melting point given was determined by the capillary-tube method and depended on the rate of heating. The melting point as given in the literature varies from 50° to 58°.

7. Cyclohexanone oxime can be prepared in the same percentage yields by a procedure which differs from the above only in that the

reaction mixture becomes solid before the addition of the sodium carbonate is complete. After all the sodium carbonate has been added, steam is passed in until the oxime is melted, and the mixture is shaken vigorously for fifteen minutes at five-minute intervals. Cyclohexanone oxime boils at 100–105°/10–12 mm. and melts at 87–88°.

3. Method of Preparation

Heptaldoxime has been prepared only by the action of the aldehyde on an aqueous solution of hydroxylamine hydrochloride in the presence of alkali;[1] the method described above is a modification of that given by Westenberger.[1]

n-HEPTANOIC ACID

(Enanthic acid)

$$3C_6H_{13}CHO + 2KMnO_4 + H_2SO_4 \rightarrow$$
$$3C_6H_{13}CO_2H + K_2SO_4 + 2MnO_2 + H_2O$$

Submitted by JOHN R. RUHOFF.
Checked by C. R. NOLLER and M. PATT.

1. Procedure

IN a 5-l. flask, fitted with a mechanical stirrer and cooled in an ice bath, are placed 2.7 l. of water and 350 cc. (644 g.) of concentrated sulfuric acid (sp. gr. 1.84). When the temperature has fallen to 15°, 342 g. (403 cc., 3 moles) of heptaldehyde (Note 1) is added, followed by 340 g. (2.15 moles) of potassium permanganate in 15-g. portions. The permanganate is added at such a rate that the temperature does not rise above 20° (Note 2). When the addition of the permanganate is complete, sulfur dioxide is passed through the solution until it becomes clear (Note 3). The oily layer is separated, washed once with water, and distilled from a modified Claisen flask having a 30-cm. fractionating side arm. The fore-runs are separated from any water and distilled again; this is followed by a redistillation of the high-boiling fractions. The yield of material boiling at 159–161°/100 mm. is 296–305 g. (76–78 per cent of the theoretical amount) (Note 4). This product is sufficiently pure for many purposes; titration indicates a purity of 95–97 per cent.

[1] Westenberger, Ber. **16**, 2992 (1883); Goldschmidt and Zanoli, ibid. **25**, 2593 (1892); Bourgeois and Dambmann, ibid. **26**, 2860 (1893).

For further purification the product is dissolved in a solution of 140 g. (3.5 moles) of sodium hydroxide in 700 cc. of water and steam-distilled from a 2-l. flask until a test portion of the distillate is free of oil. The solution remaining in the flask is cooled to room temperature and acidified with 375 cc. (4.5 moles) of concentrated hydrochloric acid. The heptanoic acid is separated and distilled from a Claisen flask with fractionating side arm. The recovery of acid boiling at 155–157°/80 mm. is 85–90 per cent of the weight of impure material used. Titration indicates it to be 100 per cent pure.

2. Notes

1. Freshly distilled heptaldehyde boiling at 85.5–87.5°/90 mm. was used.

2. The stirring must be vigorous. About two hours is required for the addition of the permanganate.

3. Sulfur dioxide, in the presence of sulfuric acid, reduces the precipitated manganese dioxide to the soluble sulfate; the removal of this large quantity of flocculent material greatly facilitates separation of the heptanoic acid. The addition of sulfur dioxide requires about two hours; an excess is to be avoided. Sodium bisulfite may be used if sulfur dioxide is not available.

4. Occasionally the heptanoic acid has a yellow color which cannot be removed by fractionation.

3. Methods of Preparation

Heptanoic acid has been prepared by the oxidation of heptaldehyde with nitric acid,[1] with potassium permanganate in alkaline aqueous solution [2] or in acetone solution,[3] or with potassium dichromate and sulfuric acid; [4] and by carbonating the reaction product of sodium and 1-chlorohexane.[5]

[1] Tilley, Ann. **67,** 107 (1848); Mehlis, ibid. **185,** 360 (1877)· Krafft, Ber. **15,** 1717 (1882).

[2] Fournier, Bull. soc. chim. (4) **5,** 921 (1909).

[3] Rogers, J. Am. Pharm. Assoc. **12,** 503 (1923) [C. A. **18,** 152 (1924)].

[4] Grimshaw and Schorlemmer, Ann. **170,** 141 (1873).

[5] Morton, LeFevre, and Hechenbleikner, J. Am. Chem. Soc. **58,** 754 (1936).

2-HEPTANOL

$$CH_3CO(CH_2)_4CH_3 \xrightarrow{(Na + C_2H_5OH)} CH_3CH(OH)(CH_2)_4CH_3$$

Submitted by FRANK C. WHITMORE and T. OTTERBACHER.
Checked by HENRY GILMAN and H. J. HARWOOD.

1. Procedure

IN a 3-l. round-bottomed flask, fitted with an efficient Liebig condenser (100 by 1 cm.), 228 g. (2 moles) of methyl n-amyl ketone (Org. Syn. Coll. Vol. I, **1941**, 351) is dissolved in a mixture of 600 cc. of 95 per cent alcohol and 200 cc. of water. One hundred thirty grams (5.6 gram atoms) of sodium in the form of wire is gradually added through the condenser. During the addition of the sodium the flask is cooled with running water (Note 1) so that the reaction does not become unduly violent (Note 2).

When the sodium has dissolved (Note 3), 2 l. of water is added and the mixture is cooled to 15°. The upper oily layer is then separated, washed with 50 cc. of 1 : 1 hydrochloric acid and then with 50 cc. of water, dried over 20 g. of anhydrous sodium sulfate, and distilled with a fractionating column (Note 4). After a small fore-run of low-boiling liquid, the pure heptanol distils at 155–157.5°. The yield is 145–150 g. (62–65 per cent of the theoretical amount).

2. Notes

1. If no cooling is used, condensation products are formed and the yield of heptanol is reduced considerably.

2. The temperature can be held conveniently below 30° by cooling with ice. Such cooling when accompanied by stirring is particularly helpful during the early addition of the sodium (either as wire or in small pieces). With cooling and stirring, very little refluxing takes place and after the addition of about 60 g. of sodium the reaction slows down to such an extent that large amounts of sodium can be added at once without danger of excessive heating.

3. The time required for addition of the sodium may be significantly decreased by the use of mechanical stirring. Although the yield is not increased appreciably by stirring, frothing is prevented and for this reason the sodium may be added more rapidly.

4. The submitters used a Young column with 20 disks 3 cm. apart. The checkers used a Glinsky three-bulbed column.

3. Methods of Preparation

2-Heptanol has been prepared by the action of n-amylmagnesium bromide on acetaldehyde,[1] and by the reduction of methyl n-amyl ketone in alcoholic solution by means of sodium.[2]

n-HEPTYLAMINE

$$CH_3(CH_2)_5CH{=}NOH + 4[H] \xrightarrow{\text{(Na + C}_2\text{H}_5\text{OH)}} CH_3(CH_2)_6NH_2 + H_2O$$

Submitted by W. H. Lycan, S. V. Puntambeker, and C. S. Marvel.
Checked by W. H. Carothers and W. L. McEwen.

1. Procedure

A SOLUTION of 258 g. (2 moles) of heptaldoxime (p. 313) in 4 l. of absolute alcohol (Note 1) is heated to boiling in a 12-l. round-bottomed flask on a steam bath. The flask is equipped with a 150-cm. reflux condenser in which the inner tube is very wide (2.5 cm.). As soon as the alcohol begins to boil, the steam is shut off and the temperature is maintained by introducing strips of sodium through the top of the condenser. The total amount of sodium added is 500 g., and it should be added as rapidly as is possible without loss of alcohol (Note 2). The last 150 g. of sodium melts in the hot mixture and may be added very rapidly without loss of alcohol or amine.

As soon as the sodium has dissolved, the contents of the flask are cooled and diluted with 5 l. of water. The flask is equipped at once with a condenser set for distillation, and the distillate is carried below the surface of a solution of 300 cc. of concentrated hydrochloric acid in 300 cc. of water in a 12-l. flask. The distillation is continued as long as any basic material passes over. When frothing interferes toward the end of the distillation an additional 3 l. of water is added to the distillation flask. The total distillate is 8–9 l.

The alcohol, water, and unreacted oxime are removed by heating the acid distillate on the steam cone under reduced pressure (about 20–30 mm.); the amine hydrochloride crystallizes in the flask. The flask is then cooled and equipped with a reflux condenser through which 1 l. of 40 per cent potassium hydroxide solution is introduced. The hydro-

---

[1] Henry, Rec. trav. chim. **28**, 446 (1909).

[2] Thoms and Mannich, Ber. **36**, 2544 (1903); Pickard and Kenyon, J. Chem. Soc. **99**, 58 (1911).

chloride is washed down from the sides of the flask, and the resulting mixture is cooled and transferred to a separatory funnel. The lower alkaline layer is removed and solid potassium hydroxide is added to the amine in the funnel. The aqueous layer is removed and fresh sticks of potassium hydroxide are added from time to time until no further separation of an aqueous alkaline solution occurs. Twenty-four to thirty hours is required for complete drying. The amine is then decanted through the top of the funnel into a 250-cc. modified Claisen flask and distilled. The *n*-heptylamine is collected at 152–157°. The yield is 140–170 g. (60–73 per cent of the theoretical amount) (Note 3).

2. Notes

1. The yields are poor if the alcohol is not completely dehydrated. A very satisfactory grade of alcohol is obtained by distilling ordinary absolute alcohol from magnesium methoxide (Org. Syn. Coll. Vol. I, **1941**, 249).

2. The best yields are obtained when the reduction is carried out rapidly.

3. Using essentially the same method the following amines have been prepared in 50–60 per cent yields: *n*-butylamine, b. p. 75–80°, from butyraldoxime; *sec.*-butylamine, b. p. 59–65°, from ethyl methyl ketoxime; cyclohexylamine, b. p. 133–135°, from cyclohexanone oxime. Greater care must be observed in drying the butylamines.

3. Methods of Preparation

n-Heptylamine can be prepared by the reduction of heptaldoxime with sodium amalgam and acetic acid,[1] with ammonium amalgam,[2] with sodium and an alcohol,[3] and catalytically;[4] by the reduction of 1-nitroheptane with iron and acetic acid;[5] by the reduction of heptanonitrile with sodium and an alcohol[3, 6] or catalytically;[7] by the reduction of

[1] Goldschmidt, Ber. **20**, 729 (1887).

[2] Takaki and Veda, J. Pharm. Soc. Japan **58**, 276 (1938) [C. A. **32**, 5376 (1938)].

[3] Suter and Moffett, J. Am. Chem. Soc. **56**, 487 (1934).

[4] Mailhe, Compt. rend. **140**, 1692 (1905); Bull. soc. chim. (3) **33**, 963 (1905); Sabatier and Mailhe, Ann. chim. phys. (8) **16**, 102 (1909); Paul, Bull. soc. chim. (5) **4**, 1121 (1937).

[5] Worstall, Am. Chem. J. **21**, 223 (1899).

[6] Forselles and Wahlforss, Ber. **25**, 637 (abstracts) (1892).

[7] Mailhe and de Godon, Bull. soc. chim. (4) **23**, 19 (1918); Mailhe and Bellegarde, ibid. **25**, 591 (1919); Mailhe, Ann. chim. (9) **13**, 203 (1920); Schwoegler and Adkins, J. Am. Chem. Soc. **61**, 3499 (1939).

heptanoamide with sodium and an alcohol [8] or catalytically; [9] by the reduction of heptaldehyde phenylhydrazone with sodium amalgam and acetic acid; [10] and by the catalytic reduction of heptaldehyde and ammonia in alcohol.[11] The formation of a secondary amine, which is a serious limitation in the catalytic reduction of heptanonitrile, can be almost completely suppressed by reducing in the presence of a large amount of ammonia.[11b]

Other methods which also lead to n-heptylamine are the reaction between 1-bromoheptane and ammonia; [12] the Hofmann rearrangement of the amide of caprylic acid; [13] and the Beckmann rearrangement of methyl n-heptyl ketoxime, followed by hydrolysis.[14]

n-HEXADECANE

$$C_{16}H_{33}I + 2[H] \xrightarrow{(Zn + HCl)} C_{16}H_{34} + HI$$

Submitted by P. A. Levene.
Checked by W. W. Hartman, L. A. Smith, and J. B. Dickey.

1. Procedure

In a 2-l. round-bottomed flask, fitted with a liquid-sealed mechanical stirrer, a gas inlet tube, and a tube to carry off hydrogen chloride and acetic acid vapors (Note 1), are placed 915 cc. of glacial acetic acid, 327 g. (5 gram atoms) of zinc dust, and 352 g. (1 mole) of cetyl iodide (m.p. 20–22°) (p. 322). The mixture is saturated with dry hydrogen chloride and then stirred and heated on a steam bath. At the end of every five hours of heating, the mixture is again saturated with hydrogen chloride. After the reaction has proceeded for twenty-five hours, the mixture is allowed to cool, and the layer of hexadecane, which rises to the top of the reaction mixture, is separated. The residue is poured into 3 l. of water and filtered on a Büchner funnel to remove the zinc dust. The zinc dust is washed with 500 cc. of water and then with 250 cc. of ether.

---

[8] Scheuble and Loebl, Monatsh. **25**, 1087 (1904).

[9] Wojcik and Adkins, J. Am. Chem. Soc. **56**, 2419 (1934).

[10] Tafel, Ber. **19**, 1928 (1886).

[11] (a) Mignonac, Compt. rend. **172**, 226 (1921); (b) Schwoegler and Adkins, J. Am. Chem. Soc. **61**, 3499 (1939).

[12] Davis and Elderfield, ibid. **54**, 1503 (1932).

[13] Hofmann, Ber. **15**, 772 (1882); Hoogewerff and Van Dorp, Rec. trav. chim. **6**, 386 (1887).

[14] v. Soden and Henle, Pharm. Ztg. **46**, 1026 (1901) (Chem. Zentr. **1902**, I, 256).

The combined water layers are extracted with two 500-cc. portions of ether. The ether extracts are combined and added to the hexadecane, and the resulting solution is washed with two 250-cc. portions of 20 per cent sodium hydroxide and then with water until free of alkali. The ether solution is dried with 150 g. of anhydrous sodium sulfate, filtered, and distilled from a 500-cc. modified Claisen flask with a fractionating side arm. The yield of n-hexadecane boiling at 156–158°/14 mm. and melting at 16–17° is 192 g. (85 per cent of the theoretical amount).

2. Note

1. If the reaction is run in a hood, an open flask may be used.

3. Methods of Preparation

n-Hexadecane has been prepared by the reduction of cetyl iodide with zinc and hydrochloric acid in alcohol [1] or acetic acid,[2,3] with the zinc-copper couple,[3] and with hydrogen and a palladium catalyst.[3] The hydrocarbon has also been prepared by treating octyl iodide with sodium;[4] by heating mercury dioctyl alone or with zinc dust;[5] by heating palmitic acid with hydriodic acid and red phosphorus;[6] and by reducing 1-hexadecene.[7]

n-Hexadecane has been obtained as a by-product from the preparation of octylmagnesium bromide [8] and from the action of sodium on a mixture of octyl bromide and ethyl bromide,[9] and it is one of the products formed on heating sodium stearate [10] or cetyl ether.[11]

[1] Sorabji, J. Chem. Soc. **47,** 38 (1885).

[2] Levene, West, and van der Scheer, J. Biol. Chem. **20,** 523 (1915).

[3] Carey and Smith, J. Chem. Soc. **1933,** 346.

[4] Zincke, Ann. **152,** 15 (1869).

[5] Eichler, Ber. **12,** 1882 (1879).

[6] Krafft, ibid. **15,** 1701 (1882).

[7] Wibaut and collaborators, Rec. trav. chim. **58,** 360 (1939).

[8] v. Braun, Deutsch, and Schmatloch, Ber. **45,** 1254 (1912).

[9] Lachowicz, Ann. **220,** 180 (1883).

[10] Grün and Wirth, Ber. **53,** 1310 (1920).

[11] Oddo, Gazz. chim. ital. **31** (I) 346 (1901).

n-HEXADECYL IODIDE

(Hexadecane, 1-iodo-)

$$3C_{16}H_{33}OH + PI_3 \rightarrow 3C_{16}H_{33}I + H_3PO_3$$

Submitted by W. W. Hartman, J. R. Byers, and J. B. Dickey.
Checked by W. H. Carothers and W. L. McEwen.

1. Procedure

Two hundred forty-two grams (1 mole) of cetyl alcohol (Note 1), 10 g. (0.32 gram atom) of red phosphorus, and 134 g. (1.06 gram atoms) of resublimed iodine are placed in a 3-l. round-bottomed flask and heated in an oil bath until the alcohol has melted. The flask is then fitted with a reflux condenser and a liquid-sealed mechanical stirrer. With stirring, the mixture is heated at 145–150° (temperature of the oil bath) for five hours. When the reaction mixture has cooled, the cetyl iodide is removed by extracting once with a 250-cc. portion and twice with 125-cc. portions of commercial ether. The combined ether extracts are filtered free of phosphorus and washed with 500 cc. of cold water, 250 cc. of 5 per cent sodium hydroxide solution, and again with 500 cc. of water. The ether solution is dried over anhydrous calcium chloride. After removal of the ether by distilling on a steam bath, the iodide is distilled under reduced pressure. The main fraction, distilling at 220–225°/22 mm. (210–215°/12 mm.), weighs 300 g. (85 per cent of the theoretical amount) and melts at 18–20° (Note 2). Redistillation gives 275 g. (78 per cent of the theoretical amount) boiling at 220–223°/22 mm. (203–205°/9 mm.) and melting at 20–22° (Note 3).

2. Notes

1. Cetyl alcohol prepared according to the directions given on p. 374 and melting at 48–49° is satisfactory. If a poorer grade of cetyl alcohol is used, the yield may be reduced to 70 per cent.

2. This material is probably pure enough for most work. Melting points as high as 25° are recorded in the literature.

3. Traces of iodine come over when the distillation starts and the fore-run is therefore strongly colored. When distillation is started again after being interrupted traces of iodine again appear in the first few drops of the main distillate. A more nearly colorless distillate is obtained if the fractions are cut without interrupting the distillation.

3. Methods of Preparation

The method described is essentially that of Smith.[1] Several other workers have used a similar method.[2] Cetyl iodide has also been prepared by heating cetyl alcohol with yellow phosphorus and iodine in carbon disulfide solution;[3] by repeatedly passing dry hydrogen iodide into the molten alcohol and permitting the reaction mass to stand between additions;[4] and by heating cetyl alcohol or cetyl stearate with 55 per cent hydriodic acid to a temperature of 120° during two hours.[5]

n-HEXALDEHYDE

(Caproaldehyde)

$$C_5H_{11}MgBr + CH(OC_2H_5)_3 \longrightarrow$$
$$C_5H_{11}CH(OC_2H_5)_2 + C_2H_5OMgBr$$

$$C_5H_{11}CH(OC_2H_5)_2 + H_2O \xrightarrow{(H_2SO_4)} C_5H_{11}CHO + 2C_2H_5OH$$

Submitted by G. Bryant Bachman.
Checked by C. R. Noller and W. S. Woon.

1. Procedure

In a 2-l. three-necked, round-bottomed flask, fitted with a liquid-sealed mechanical stirrer, a 250-cc. dropping funnel, and a reflux condenser to which is attached a calcium chloride tube, are placed 30 g. (1.25 gram atoms) of magnesium turnings, 50 cc. of dry ether, and a small crystal of iodine. Stirring is started, and 5 cc. (6 g.) of *n*-amyl bromide is added (Note 1). As soon as the reaction has started, 300 cc. of dry ether is added and then, more slowly, a solution of 183 g. (total, 189 g.; 1.25 moles) of *n*-amyl bromide in 150 cc. of dry ether. If external cooling is provided, all the alkyl halide may be added within half an hour; the solution is refluxed gently for another half-hour to complete the reaction. The heat is removed, the flask cooled to 50°, and 148 g. (1 mole) of ethyl orthoformate (Org. Syn. Coll. Vol. I, **1941**, 258) is added during

---

[1] Smith, J. Chem. Soc. **1932**, 738.

[2] Fridau, Ann. **83**, 9 (1852); Levene, West, and van der Scheer, J. Biol. Chem. **20**, 523 (1915); Delcourt, Bull. soc. chim. Belg. **40**, 284 (1931) [C. A. **25**, 5661 (1931)].

[3] Gascard, Ann. chim. (9) **15**, 372 (1921).

[4] Krafft, Ber. **19**, 2219 (1886).

[5] Guyer, Bieler, and Hardmeier, Helv. Chim. Acta **20**, 1466 (1937).

the course of fifteen to twenty minutes. The mixture is refluxed for six hours (Note 2); at the end of this time the condenser is arranged for distillation and the ether is removed completely with the aid of a steam bath.

The reaction mixture is cooled and treated carefully with 750 cc. of chilled 6 per cent hydrochloric acid. The contents of the flask are kept cool by the occasional addition of ice while the acid is being introduced. As soon as all the solid has dissolved (Note 3), the upper oily layer of hexaldehyde acetal is separated. The acetal is hydrolyzed by distilling it with a solution of 100 g. (55 cc.) of concentrated sulfuric acid in 700 cc. of water. The free aldehyde distils rapidly, and the distillation is complete when a sample of fresh distillate contains 5 per cent or less of immiscible oil. The distillate is collected in a solution of 100 g. (1 mole) of sodium bisulfite in 300 cc. of water. The mixture is shaken vigorously for several minutes; the oily layer remaining undissolved in the bisulfite solution is principally n-amyl alcohol and is discarded. To remove the remainder of the amyl alcohol and other impurities the bisulfite solution is steam-distilled until 200 cc. of distillate has been collected.

The residual aldehyde-bisulfite solution is cooled to 40–50°, a suspension of 80 g. of sodium bicarbonate in 200 cc. of water is added carefully, and the free aldehyde is removed by steam distillation. The upper layer of the distillate is separated, washed with three 50-cc. portions of water (Note 4), dried with 20 g. of anhydrous sodium sulfate, and distilled through a 20-cm. column. The yield of n-hexaldehyde b.p. 126–129°, is 45–50 g. (45–50 per cent of the theoretical amount).

2. Notes

1. The n-amyl bromide was prepared according to the general procedure in Org. Syn. Coll. Vol. I, **1941,** 25, and distilled at 127–129°.

2. Sometimes a white precipitate begins to form immediately, but more often it does not appear until after twenty to thirty minutes of refluxing. If this period of heating is materially decreased, a sudden exothermic reaction occurs when the ether is removed and the yield may be seriously reduced. Longer periods of heating do not increase the yield of hexaldehyde.

3. Solution takes place slowly and is hastened considerably by the use of a mechanical stirrer.

4. The aldehyde dissolved by the wash water may be recovered by steam distillation, but this is scarcely worth while since hexaldehyde is not very soluble in water.

3. Methods of Preparation

Hexaldehyde has been prepared from caproic acid by passing it over zinc dust at 300°,[1] by reaction with amylene at 300° in the presence of thorium oxide,[2] by passing it over manganous oxide at 300–360° with two volumes of formic acid,[3] and by distillation of the calcium salt with calcium formate.[4] It has also been prepared by heating α-hydroxy-heptoic acid or, better, α-acetoxyheptoic acid,[5] and by the procedure described above.[6]

HEXAMETHYLENE GLYCOL

(1,6-Hexanediol)

$$(CH_2)_4{\overset{CO_2C_2H_5}{\underset{CO_2C_2H_5}{\Big\langle}}} + 4H_2 \xrightarrow{(CuCr_2O_4)} (CH_2)_4{\overset{CH_2OH}{\underset{CH_2OH}{\Big\langle}}} + 2C_2H_5OH$$

Submitted by W. A. Lazier, J. W. Hill, and W. J. Amend.
Checked by R. L. Shriner and J. F. Kaplan.

1. Procedure

In a steel reaction vessel (Note 1), capable of withstanding high pressures with an adequate safety factor (Note 2) and having a capacity of 400 cc. or more, are placed 252 g. (1.25 moles) of ethyl adipate (b.p. 144–145°/29 mm.) (p. 264) and 20 g. of copper chromite catalyst, prepared either with or without the addition of barium (p. 142). The reaction vessel is closed, made gas tight, and secured in a suitable agitating device. After connection is made with the hydrogen supply, hydrogen is introduced until a pressure of 2000 to 3000 lb. per sq. in. is reached (Note 2).

Agitation is started, and the reaction system is heated as rapidly as possible to 255°. The temperature is maintained at 255° (Note 3), and hydrogenation is continued until hydrogen absorption is complete (Note 4). The agitation is now stopped, the vessel cooled, and the pressure released. With the aid of four 25-cc. portions of 95 per cent alcohol the

[1] Mailhe, Chem. Ztg. **33**, 243 (1909).
[2] Mailhe, ibid. **34**, 1174 (1910).
[3] Sabatier and Mailhe, Compt. rend. **158**, 986 (1914).
[4] Lieben and Janecek, Ann. **187**, 130 (1877).
[5] Bagard, Bull. soc. chim. (4) **1**, 313 (1907).
[6] Bachman, J. Am. Chem. Soc. **55**, 4281 (1933).

reaction mixture is transferred to a 600-cc. beaker. The catalyst is removed by filtering or centrifuging, and is washed with four more 25-cc. portions of alcohol (Note 5). To the reaction product (Note 6), 50 cc. of 40 per cent sodium hydroxide solution is added, and the alcoholic solution is boiled for two hours under a reflux condenser. The mixture is transferred to a 1-l. distilling flask and the alcohol removed by distilling to a vapor temperature of 95°. The hot residue is transferred to an apparatus for the continuous extraction of liquids (p. 615), using 50 cc. of water to rinse the flask, and the solution is exhaustively extracted with ether (Note 7). The ether is distilled, and, after the removal of water and alcohol, the glycol is distilled under reduced pressure in a 250-cc. Claisen flask. The yield is 125–132 g. (85–90 per cent of the theoretical amount). Hexamethylene glycol boils at 143–144° (bath at 160°) under 4 mm. pressure and melts at 41–42°.

2. Notes

1. Suitable reaction vessels and apparatus for agitation of the reaction mixture are commercially available [1,2] or may be constructed.[3]

2. The pressure of hydrogen to be used is dependent upon the equipment available. Hydrogen in commercial cylinders is sold at a maximum pressure of 2000 lb. per sq. in. Special equipment for compressing hydrogen may be purchased at a reasonable price.[1] The original pressure of hydrogen should not be more than 2000 lb. if the maximum working pressure of the equipment for hydrogenation is 5000 lb. If the working pressure is 10,000 lb. the original pressure in the reaction vessel may be as much as 3000 lb. The full operating pressure is not applied in the beginning since the pressure will rise as the reaction vessel is heated; thus, at 255° the pressure will be 1.8 times as high as it was at 20°. The pressure drops as hydrogenation proceeds; the progress of the reaction may be followed by the change in pressure readings, and completion of the reaction is indicated by the constancy of the pressure readings.

3. The temperature is controlled preferably by an automatic controller operating through a relay which periodically cuts off the supply of electric current.

4. The time (six to twelve hours) required to complete the reaction is

[1] American Instrument Company, Silver Spring, Maryland.

[2] Parr Instrument Company, Moline, Illinois.

[3] Adkins, Ind. Eng. Chem., Anal. Ed. 4, 342 (1932); "Reactions of Hydrogen with Organic Compounds over Copper-Chromium Oxide and Nickel Catalysts," pp. 29–39, University of Wisconsin Press, Madison, Wisconsin, 1937.

a function of the pressure of the hydrogen, activity of the catalyst, and purity of the ethyl adipate. Unless a high pressure of hydrogen is used originally or the reaction vessel is of large capacity (2 l.) it will be necessary to introduce more hydrogen into the reaction vessel; the pressure should never be less than 1500 lb. per sq. in. if the reaction is to run smoothly to completion.

5. The catalyst is most readily removed by centrifuging. If this is not convenient, it may be collected on a sintered glass filter or Büchner funnel.

6. At this point, the amount of ester present may be determined by obtaining the saponification value of the weighed mixture. This procedure is especially desirable when the condition of the apparatus with respect to the possible presence of poisons is unknown, or a new preparation of catalyst is being used. After a batch of catalyst has been tested and the apparatus calibrated so that essentially complete reduction is assured, it is possible to isolate the glycol by fractional distillation at this stage.

According to Burks, Jr., and Adkins [private communication and J. Am. Chem. Soc. **62,** 3300 (1940)] the hydrogenation is reversible and the reaction product always contains esters. In order to obtain hexamethylene glycol free of esters a simpler procedure than that given above is recommended: Thirty grams of crude glycol is dissolved in 50 cc. of water and extracted with four 50-cc. portions of benzene. The water solution is distilled through a modified Widmer column. The recovery of glycol, free of ester, is 93 per cent.

7. The time required for complete extraction varies from twenty-four to fifty hours. It depends on the design of the apparatus and the rate of distillation of the ether. The extraction can be followed by observing the decrease in volume of the aqueous layer containing the glycol. The extraction is complete when the evaporation of a small amount of the supernatant ether on a watch glass leaves no residue. Benzene may be substituted for ether in the extraction.

3. Methods of Preparation

Hexamethylene glycol has been prepared by treating hexamethylene iodide with silver acetate and hydrolyzing the acetate,[4] by hydrolyzing the bromide,[5] by reducing ethyl adipate with sodium and alcohol,[6] and

[4] Hamonet, Bull. soc. chim. (3) **33,** 538 (1905).

[5] Haworth and Perkin, J. Chem. Soc. **65,** 598 (1894).

[6] Bouveault and Blanc, Compt. rend. **137,** 328 (1903); Bull. soc. chim. (3) **31,** 1203 (1904).

by the method here described.[7] The catalytic hydrogenation over copper-chromium oxide of the carbethoxy group to the carbinol group is a very useful and general method for the preparation of mono- and dihydric alcohols.[3, 8]

HIPPURIC ACID

$$ClCH_2CO_2H + 2NH_3 \rightarrow NH_2CH_2CO_2H + NH_4Cl$$

$$NH_2CH_2CO_2H + C_6H_5COCl + 2NaOH \rightarrow$$
$$C_6H_5CONHCH_2CO_2Na + NaCl + 2H_2O$$

$$C_6H_5CONHCH_2CO_2Na + HCl \rightarrow$$
$$C_6H_5CONHCH_2CO_2H + NaCl$$

Submitted by A. W. INGERSOLL and S. H. BABCOCK.
Checked by ROGER ADAMS and B. H. WOJCIK.

1. Procedure

To 3 l. (approximately 45 moles) of concentrated ammonium hydroxide (sp. gr. 0.9) (Note 1) in a 5-l. round-bottomed flask is added, with shaking, a solution of 95 g. (1 mole) of chloroacetic acid (Note 2) in 100 cc. of water. The flask is stoppered and allowed to stand for four days (Note 3). It is then attached to a condenser for distillation, and the solution is concentrated to 600–700 cc. The excess ammonia is recovered during this process by connecting the lower end of the condenser to a wide tube leading to the bottom of a 3-l. bottle containing 1.5 l. of distilled water and connecting this in the same way with a smaller bottle containing a little water. The bottles are cooled by running water. About 2 l. of 20–23 per cent ammonium hydroxide is recovered in the first receiver (Note 1).

The residual solution is then transferred to a 2-l. beaker, a solution of 50 g. (1.25 moles) of sodium hydroxide in 100 cc. of water and a little decolorizing carbon are added, and the mixture is boiled until the odor of ammonia is completely absent (Note 4). The solution is filtered by suction, diluted to 500 cc. with water, and transferred to a 2-l. round-bottomed flask equipped with a mechanical stirrer and cooled by running water. While stirring and cooling below 30°, 150 g. (1.1 moles) of

[7] Lazier, U. S. pat. 2,079,414 [C. A. **31**, 4340 (1937)]; U. S. pat. 2,137,407 [C. A. **33**, 1758 (1939)].

[8] Adkins and Folkers, J. Am. Chem. Soc. **53**, 1095 (1931); **54**, 1145 (1932); Wojcik and Adkins, ibid. **55**, 4939 (1933).

benzoyl chloride (Note 2) and a cold solution of 80 g. (2 moles) of sodium hydroxide in 200 cc. of water are admitted separately from separatory funnels at such rates that the solution is always only slightly alkaline. About an hour is required for adding the reagents, and the mixture is stirred for a half hour longer. It is then poured into 125 cc. of concentrated hydrochloric acid in a 2-l. beaker, and, after cooling, the precipitate is filtered and dried. The weight at this point is 150–160 g. The solid is placed in a beaker with 300 cc. of technical carbon tetrachloride, the beaker is covered with a watch glass, and the mixture is boiled gently for ten minutes. The mixture is then cooled slightly, filtered by gentle suction, and the hippuric acid washed on the filter with 50 cc. of carbon tetrachloride (Note 5). After drying, it weighs 135–140 g. For final purification the acid is dissolved in about 2 l. of boiling water, filtered through a steam-heated funnel and allowed to crystallize without artificial cooling. It then appears in characteristic white needles melting at 186–187°. The yield is 115–122 g. (64–68 per cent of the theoretical amount based on the chloroacetic acid used). Upon concentrating the mother liquor to 200 cc., a further 6–7 g. of slightly brown hippuric acid is obtained.

2. Notes

1. An equivalent amount of recovered ammonium hydroxide (sp. gr. 0.93 or less) was used in a number of runs without reducing the yield.

2. Eastman's "practical" grade reagents were used.

3. The yield was slightly less after two days and no greater after a week.

4. Ammonia must be completely removed to avoid the formation of benzamide.

5. Carbon tetrachloride may be recovered by making the filtrate and washings slightly alkaline with sodium hydroxide, refluxing for a half hour to destroy any benzoyl chloride, and then distilling with steam. Benzoic acid may be recovered by filtering the aqueous residue and acidifying.

3. Methods of Preparation

Hippuric acid has been prepared from the urine of herbivorous animals; [1] by heating benzamide with chloroacetic acid; [2] by heating benzoic anhydride and glycine; [3] by heating benzoic acid and glycine; [4] by heat-

[1] Gregory, Ann. **63**, 125 (1847); Hallowachs, ibid. **106**, 164 (1858); Henneberg, Stohmann, and Rautenberg, ibid. **124**, 200 (1862).

[2] Jazukowitsch, Zeit. für Chem. **1867**, 466.

[3] Curtius, Ber. **17**, 1662 (1884).

[4] Dessaignes, Jahresb. **1857**, 367.

ing benzoyl chloride with silver glycinate suspended in benzene,[5] or with glycine and zinc oxide;[6] and by the action of benzoyl chloride upon an alkaline solution of glycine.[7]

l-HISTIDINE MONOHYDROCHLORIDE

Hydrolysis of
blood proteins →

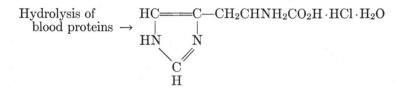

$$HC{=\!\!=\!\!=}C—CH_2CHNH_2CO_2H \cdot HCl \cdot H_2O$$

Submitted by G. L. FOSTER and D. SHEMIN.
Checked by JOHN R. JOHNSON, H. E. CARTER, and P. L. BARRICK.

1. Procedure

IN a large round-bottomed flask are placed 1.4 kg. of dried blood corpuscle paste (Note 1) and 4.5 l. of concentrated hydrochloric acid (sp. gr. 1.18). The flask is warmed on a steam bath until the protein has dissolved, and the mixture is boiled gently under reflux for eighteen to twenty hours. After the hydrolysate has been concentrated under reduced pressure to a thick paste, the distillation is continued with the slow addition of a liter of water, thus eliminating most of the excess hydrochloric acid. The residue is taken up in 8 l. of warm water, cooled, and neutralized to *p*H 4.4–4.6 (methyl orange or bromcresol green) by the addition of concentrated sodium hydroxide solution (Note 2). After the mixture has stood overnight, the precipitated pigment is removed by filtering through a large Büchner funnel fitted with two layers of filter paper and a 2-mm. layer of infusorial earth. The filtrate, which is dull red in color, is decolorized by warming and stirring for ten minutes with 60 g. of Norite.

The pale yellow filtrate and washings from the Norite are diluted to 25 l. with tap water, and a solution of 600 g. of mercuric chloride in 2 l. of hot 95 per cent alcohol is added, with stirring. A concentrated solution of sodium carbonate (corresponding to about 350 g. of anhydrous sodium carbonate) is added slowly, with stirring, until the mixture reaches *p*H 7.0–7.5 (phenol red or litmus). After settling for several

[5] Curtius, J. prakt. Chem. (2) **26**, 145 (1883).
[6] Dessaignes, Ann. **87**, 325 (1853).
[7] Baum, Ber. **19**, 502 (1886); Z. physiol. Chem. **9**, 465 (1885).

hours, preferably overnight, the supernatant liquid is siphoned off, and the crock is refilled with water to the original volume (Note 3). The mixture is allowed to settle, the wash liquid siphoned off, and the precipitate washed twice more in the same fashion. After the third washing, the supernatant liquid is siphoned off and the precipitate collected with suction on a large Büchner funnel fitted with two layers of filter paper and a 2-mm. layer of infusorial earth.

The moist histidine-mercury complex is suspended in 5 l. of water and stirred vigorously while a stream of hydrogen sulfide is introduced. When precipitation of mercuric sulfide is complete, the suspension becomes uniformly black and settles sharply on standing. The filtrate and washings from the mercuric sulfide (Note 4) are concentrated under reduced pressure to a volume of about 1 l. and cleared with 5 g. of Norite. The filtrate and washings from the Norite are concentrated further to a volume of about 250 cc. and mixed with three volumes of 95 per cent alcohol. Crystallization is induced by cooling the solution and scratching the walls of the vessel; the histidine monohydrochloride separates in plates. After the material has remained in an ice chest for three or four days, the crystals are separated by suction filtration. The yield of crude histidine monohydrochloride is 85–90 g. The filtrate, on standing for several weeks in an ice chest, usually deposits an additional 4–5 g. of material.

The crude product is dissolved in five times its weight of water, and, after clearing with a little Norite, the solution is diluted with one and one-half volumes of 95 per cent alcohol. The product separates in well-formed, snow-white crystals, and after standing for several days in an ice chest is collected with suction on a Büchner funnel. The yield of purified histidine monohydrochloride is 75–80 g. (Note 5). The compound melts at 251–252°, with decomposition. The amino acid is not racemized by the procedure employed, and it shows the characteristic optical activity, $[\alpha]_D^{26°} = +8.0°$, in the presence of three moles of hydrochloric acid. The recrystallized product is usually analytically pure and shows the correct Van Slyke amino nitrogen content. Occasionally a second recrystallization is necessary to obtain analytically pure material.

2. Notes

1. Commercial "dried blood corpuscle paste" obtained from Armour and Company, Chicago, was used in this preparation. This paste contains about 15 per cent of moisture and ash, and 200 g. contains about the same amount of crude protein as 1 l. of fresh beef blood (170 g. protein per liter).

If fresh blood is used in this preparation, it is convenient to remove much of the water in the following way. Seven liters of beef blood in a 12-l. round-bottomed flask is treated with 50 cc. of glacial acetic acid and heated on a steam bath, with occasional stirring, until a thick, pasty coagulum results. About 4 l. of water is removed by distillation under reduced pressure, using a steam bath, and the residue is hydrolyzed as described above.

2. About 600 cc. of 50 per cent sodium hydroxide is required for neutralization. An excess of alkali should be avoided.

3. If, as occasionally happens, the histidine-mercury complex settles slowly, the supernatant liquid may be siphoned off and filtered. The small amount of material collected on the filter is then returned to the main portion.

4. The mercuric sulfide may be saved and converted to metallic mercury or mercuric chloride by the usual procedures.

5. About 10 g. of crude histidine monohydrochloride may be recovered from the mother liquor by evaporating under reduced pressure to 50–60 cc. and adding one and one-half volumes of 95 per cent alcohol. On recrystallization, 8–9 g. of pure material is obtained.

3. Methods of Preparation

The preparation of histidine by the hydrolysis of hemoglobin and precipitation with mercuric chloride in alkaline solution was first carried out by Fränkel.[1,2,3] Histidine can also be precipitated as the silver derivative.[4]

[1] Fränkel, Monatsh. **24**, 229 (1903).

[2] Abderhalden, Fleischmann, and Irion, Fermentforschung **10**, 447 (1928) [C. A. **23**, 2994 (1929)].

[3] Gilson, J. Biol. Chem. **124**, 281 (1938).

[4] Vickery and Leavenworth, ibid. **78**, 627 (1928).

HOMOVERATRIC ACID

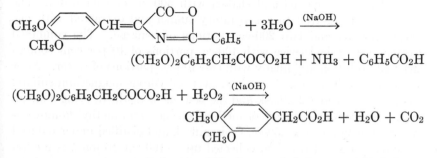

$$(CH_3O)_2C_6H_3CH_2COCO_2H + NH_3 + C_6H_5CO_2H$$

$$(CH_3O)_2C_6H_3CH_2COCO_2H + H_2O_2 \xrightarrow{(NaOH)}$$

Submitted by H. R. SNYDER, J. S. BUCK, and W. S. IDE.
Checked by JOHN R. JOHNSON and P. W. VITTUM.

1. Procedure

(A) *Methyl Homoveratrate.*—In a 3-l. round-bottomed flask are placed 1 l. of 10 per cent sodium hydroxide solution and 200 g. (0.65 mole) of the azlactone of α-benzoylamino-β-(3,4-dimethoxyphenyl)acrylic acid (m.p. 149–150°) (p. 55). The flask is fitted with a reflux condenser and immersed in an oil bath so that the inner level is lower than the oil level of the bath (Note 1). The mixture is refluxed gently for six to seven hours, until the evolution of ammonia is complete. The resulting solution contains the sodium salts of 3,4-dimethoxyphenylpyruvic acid (Note 2) and benzoic acid.

To the above aqueous solution, contained in a 2-l. wide-mouthed Erlenmeyer flask, is added 85 cc. of 40 per cent sodium hydroxide solution. The flask is equipped with a mechanical stirrer and is cooled in an ice-salt mixture. With stirring, 75 cc. of 30 per cent hydrogen peroxide (Merck's "Superoxol") diluted with 75 cc. of water is added at such a rate that the temperature does not rise above 15°. After standing for about ten hours at room temperature (preferably overnight) the solution is acidified by the cautious addition (Note 3) of 450 cc. of concentrated hydrochloric acid (sp. gr. 1.19). The warm acid solution is extracted with one 400-cc. portion and two 200-cc. portions of warm benzene. The combined benzene extracts are dried over anhydrous magnesium sulfate and filtered through a cotton plug into a 3-l. round-bottomed flask.

The benzene is removed by distillation, 1 l. of methyl alcohol (Note 4) containing 15 cc. of concentrated sulfuric acid is added, and the flask is fitted with an efficient reflux condenser provided with a drying tube.

After refluxing gently for five hours the condenser is set downward for distillation and the methyl alcohol is distilled from a steam bath. The residual liquid is cooled and shaken with 500 cc. of cold water. The mixture is transferred to a separatory funnel and extracted with one 400-cc. portion and two 200-cc. portions of benzene. The combined extracts are washed twice with 100-cc. portions of 10 per cent sodium carbonate solution and finally with two 100-cc. portions of water. After the benzene solution has been dried over anhydrous magnesium sulfate, it is transferred to a 2-l. flask and the benzene is distilled, using a steam bath. The residual mixture of methyl benzoate and methyl homoveratrate is transferred to a 250-cc. Claisen flask and distilled under reduced pressure. The first fraction, collected up to 100° at 16 mm., is methyl benzoate (b.p. 87°/16 mm.) and weighs about 75 g. (85 per cent of the theoretical amount). After a small intermediate fraction of 2–3 g., pure methyl homoveratrate is collected at 176–178°/16 mm. or 129–131°/1 mm. The yield is 76–82 g. (56–60 per cent of the theoretical amount based on the azlactone).

(B) *Homoveratric Acid.*—In a 500-cc. round-bottomed flask are placed 250 cc. of 10 per cent sodium hydroxide solution and 76 g. (0.36 mole) of methyl homoveratrate. The flask is fitted with a reflux condenser and the mixture is boiled gently. The saponification proceeds rapidly, and the ester layer disappears after about ten minutes. The mixture is refluxed gently for twenty minutes longer, after which the solution is cooled in an ice bath and then poured slowly, with stirring, into a mixture of 125 cc. of concentrated hydrochloric acid and 350 g. of ice. Crystals of the hydrate of homoveratric acid separate at once. After standing for about thirty minutes the crystalline product is filtered with suction and washed on the filter with two 25-cc. portions of ice water. The crystals are pressed thoroughly on the filter, pulverized, and allowed to stand overnight in a vacuum desiccator containing soda lime (to remove residual hydrochloric acid) and calcium chloride. The yield in the saponification is almost quantitative and amounts to 70 g. (55 per cent of the the theoretical amount based on the original azlactone). This product melts at 96–97° and contains traces of sodium chloride. For purification it is dissolved in 350 cc. of hot benzene and the solution is filtered. To the hot filtrate is added 150 cc. of hot ligroin (b.p. 70–80°), and the solution is covered with a watch glass and allowed to cool slowly. After standing for several hours (preferably overnight) in a cool place, the crystals are filtered with suction and washed with a cold solution of 35 cc. of benzene and 15 cc. of ligroin, followed by 50 cc. of cold petroleum ether. The solvent is removed as completely as possible by pressing on the filter and finally by allowing the product to stand in a vacuum desiccator (Note 5). The purified homoveratric acid weighs 65 g. (51 per cent of the theoretical amount based on the original azlactone) and melts sharply at 98°.

2. Notes

1. The inner level is kept below that of the oil in order to avoid the otherwise uncontrollable bumping of the solution.

2. 3,4-Dimethoxyphenylpyruvic acid can be isolated from this solution in the following way (J. S. Buck and W. S. Ide). The aqueous solution of the sodium salts is saturated with sulfur dioxide, while the temperature is maintained below 40°. The benzoic acid precipitates and is filtered with suction and washed with a small quantity of water. The filtrate and washings are placed in a 3-l. round-bottomed flask provided with a mechanical stirrer and heated to boiling. Concentrated hydrochloric acid is added cautiously, with stirring, until present in excess. The acid must be added carefully since the solution tends to become supersaturated with sulfur dioxide, which is subsequently liberated with violence. A heavy precipitate of 3,4-dimethoxyphenylpyruvic acid separates; after the reaction mixture has cooled, this is filtered with suction, dried, and washed with two 50-cc. portions of ether. The yield of 3,4-dimethoxyphenylpyruvic acid is 110–116 g. (76–80 per cent of the theoretical amount), and the product melts at 181–184°. It can be purified by crystallization from glacial acetic acid.

An alternative procedure for the preparation of homoveratric acid (J. S. Buck and W. S. Ide) consists in isolating the pyruvic acid and subjecting it to the oxidation given in the second paragraph of part (A). This variation obviates the esterification but in the hands of the checkers did not prove so satisfactory as the one described.

3. Large quantities of carbon dioxide are evolved during the addition of the acid.

4. It is unnecessary to use especially dried methyl alcohol. High-grade commercial methanol is quite satisfactory.

5. Since the acid forms a hydrate it is advisable to minimize the exposure of the acid to atmospheric moisture.

3. Methods of Preparation

Homoveratric acid has been prepared by the methylation of homoprotocatechuic acid [1] or homovanillic acid [2] with methyl iodide, and

---

[1] Pictet and Gams, Ber. **42**, 2949 (1909).
[2] Tiemann and Matsmoto, ibid. **11**, 143 (1878).

from veratric aldehyde through the azlactone and 3,4-dimethoxyphenyl-pyruvic acid.[3] Homoveratric acid has also been prepared by hydrolysis of its amide or nitrile: the amide was obtained from veratroyl chloride through the diazoketone,[4] or from veratric aldehyde cyanohydrin through α-chlorohomoveratramide;[5] the nitrile was obtained by catalytic reduction of acyl derivatives of veratric aldehyde cyanohydrin.[6]

The procedure given above is adapted from published directions for the preparation of homoveratric acid[3] and p-methoxyphenylacetic acid.[7]

α-HYDRINDONE

(1-Indanone)

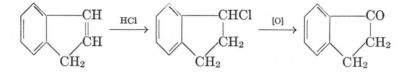

Submitted by R. A. PACAUD and C. F. H. ALLEN.
Checked by LOUIS F. FIESER and W. P. CAMPBELL.

1. Procedure

A 250-cc. flask, fitted with an inlet tube reaching to the bottom, a thermometer, and an exit tube for conducting away unused gas, is immersed in an ice-water bath, and to it is added 80 g. (0.69 mole) of freshly distilled indene (Note 1). While the temperature of the liquid is kept at 5–10°, dry hydrogen chloride is passed in at a moderate rate until 24–27 g. of the gas has been absorbed. The addition takes from eight to ten hours and requires little attention from the operator. The crude α-chlorohydrindene is then transferred to a 250-cc. Claisen flask and distilled at diminished pressure. After a small fore-run (5–10 g.) containing indene, α-chlorohydrindene boiling at 90–103°/15 mm. is collected; the yield is 80–90 g. (Note 2).

In a 500-cc. three-necked flask, fitted with a stirrer, thermometer, and dropping funnel, is placed 100 g. (1 mole) of chromic anhydride dissolved

[3] Haworth, Perkin, and Rankin, J. Chem. Soc. **125**, 1693 (1924).

[4] Arndt and Eistert, Ber. **68**, 205 (1935).

[5] Hahn and Schulz, ibid. **72**, 1302 (1939).

[6] Kindler and Peschke, Arch. Pharm. **271**, 432 (1933); Kindler and Gehlhaar, ibid. **274**, 377 (1936).

[7] Cain, Simonsen, and Smith, J. Chem. Soc. **103**, 1036 (1913).

in 100 cc. of water, and 100 cc. of glacial acetic acid is added. The
α-chlorohydrindene is then admitted through the funnel at such a rate
as to keep the temperature at 35–40°, the flask being cooled externally.
This addition takes about one and one-half hours; after it is complete,
stirring is continued for fifteen minutes before the mixture is poured into
a large beaker and diluted with 300 cc. of water. After the acid present
is neutralized by the addition of solid sodium carbonate (Note 3), the
α-hydrindone is expelled from the mixture by steam distillation, care
being taken to avoid entrainment of froth at the beginning of the process.
When about 2.5 l. of distillate has been collected, the solution again begins to froth. The operation is stopped at this point, since further distillation does not yield an appreciable quantity of product. The α-hydrindone usually solidifies, as soon as the distillate is chilled, to a mixture
of colorless crystals and a yellow solid; the product is collected after
thorough cooling in an ice bath. This moist solid is dissolved in 200 cc.
of benzene, and the solution is subjected to distillation until free from
water. After the benzene is removed by vacuum distillation on the
steam bath, the α-hydrindone is distilled. It boils at 125–126°/17 mm.,
and the distillate is a pale yellow solid, m.p. 39–41°. The yield is 46–
55 g. (50–60 per cent of the theoretical amount, based on the indene
used) (Notes 4 and 5).

2. Notes

1. If technical indene is employed, the fraction boiling at 178–182°
is suitable for the preparation. The checkers used indene, b.p. 181.6–
183.3°, obtained from the Barrett Company, New York City.

2. Most of the material distils at 100–103°/15 mm., but the yield of
α-hydrindone is increased by using the material collected over the wider
range.

3. An excess of sodium carbonate promotes foaming during the distillation and is to be avoided. Since the neutral point is not easily recog-
nized with test paper, the carbonate is added in decreasing amounts
until a fresh portion is no longer decomposed with gas evolution.

4. It is not practicable to dry the steam-distilled product in the air
because α-hydrindone has such a high vapor pressure that the loss is
appreciable. Drying can be accomplished in a vacuum desiccator, but
this takes several days. The distillation specified gives a completely
anhydrous product with the loss of no more than 2–5 g.

5. Crystallization is conveniently accomplished by dissolving the
α-hydrindone in alcohol (1 cc. per gram) at room temperature, adding
water until solid just begins to separate, and cooling to 0°. The substance crystallizes either as long plates or as leaflets.

3. Methods of Preparation

α-Hydrindone has been prepared from indene chlorohydrin,[1] the procedure described above; from indene bromohydrin by heating with dilute sulfuric acid;[2] by cyclization of hydrocinnamoyl chloride with ferric chloride[3] or aluminum chloride;[4] by the cyclization of hydrocinnamic acid with hydrogen fluoride[5] or sulfuric acid;[6] and by the interaction of acrylyl chloride with benzene in the presence of aluminum chloride.[7]

HYDROGEN BROMIDE (ANHYDROUS)

(Hydrobromic acid)

$$H_2 + Br_2 \rightarrow 2HBr$$

Submitted by JOHN R. RUHOFF, ROBERT E. BURNETT, and E. EMMET REID.
Checked by W. H. CAROTHERS and W. L. McEWEN.

1. Procedure

(A) Apparatus.—A 125-cc. distilling flask (B, Fig. 9) (Note 1) is fitted with a two-holed rubber stopper bearing a 50-cc. dropping funnel (Note 2) and an inlet tube about 6 mm. in diameter, both of which reach to the bottom of the flask. The flask is supported in an 800-cc. beaker which serves as a water bath. The side arm of the flask is connected by means of a short piece of rubber tubing to the narrow end (Note 3) of the combustion tube C which is made of Pyrex glass and is 20 mm. in diameter (inside) and 30 cm. in length. It is packed with pieces of porous plate held in place by constrictions as shown in the figure, and it is supported at each end by a small clamp at a sufficient height to give a clearance of 3–4 cm. above a Bunsen burner. The open end of the tube is connected by means of stoppers and a three-way stopcock (Note 4) to the vertical tube D, of the same diameter as C and 60 cm. long. This tube is packed with copper turnings to remove any

[1] Hückel, Sachs, Yantschulewitsch, and Nerdel, Ann. **518**, 172 (1935).

[2] Porter and Suter, J. Am. Chem. Soc. **57**, 2022 (1935).

[3] Wedekind, Ann. **323**, 255 (1902).

[4] Kipping, J. Chem. Soc. **65**, 484 (1894).

[5] Fieser and Hershberg, J. Am. Chem. Soc. **61**, 1278 (1939).

[6] Price and Lewis, ibid. **61**, 2553 (1939).

[7] Moureu, Bull. soc. chim. (3) **9**, 570 (1893); Ann. chim. phys. (7) **2**, 198 (1894); Kohler, Am. Chem. J. **42**, 380 (1909).

uncombined bromine which may escape from the combustion tube. A safety bottle A containing water or some other suitable liquid and having a tube leading to the vent of the hood is placed in the train to provide an outlet for the hydrogen in case an obstruction is formed in the apparatus. It also provides a convenient method for determining the hydrogen pressure (Note 5). Hydrogen is obtained from a cylinder fitted with a reducing valve.

(*B*) *Operation.*—The tube D is disconnected from the combustion tube by turning the stopcock so that the gases pass directly to the hood.

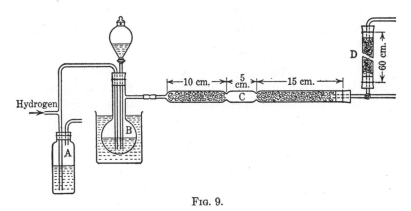

Fig. 9.

Bromine is placed in flask B; the water bath is heated to 38° and maintained at this temperature (Note 6). A slow stream of hydrogen is started through the apparatus, and, when the combustion tube is completely filled with bromine vapors, a low flame is placed under the empty section of the combustion tube. Soon a small yellow flame appears inside the heated portion of the tube, and, when all bromine vapor has been swept from the combustion tube, the latter is connected to the tube D (Note 7). The flame is adjusted to keep the lower part of the tube at a dull red heat; it may be necessary to raise or lower it, according to the rate of operation. The flow of hydrogen is regulated to give the desired output. The apparatus will conveniently produce 300 g. of hydrogen bromide per hour (Note 8).

2. Notes

1. It is convenient to bend the side arm of the distilling flask so that it is perpendicular to the neck.

2. The stopcock of the dropping funnel should be held in place by means of a rubber band.

3. The distilling flask may be connected to the combustion tube by means of a rubber stopper, but it is preferable to seal a piece of 6-mm. tubing to one end and to use a small piece of rubber tubing for the connection. Even with this arrangement the connection must be inspected from time to time to be sure that the tube is not obstructed. It is still more satisfactory to use a ground-glass joint.

4. The three-way stopcock may be dispensed with, but it facilitates the starting operation and provides a quick method for venting the gas in case combination does not take place properly, thus preventing deterioration of the copper turnings.

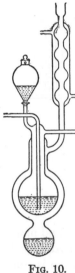

5. When operating to produce about 300 g. of hydrogen bromide per hour the water level is depressed approximately 12 cm.

6. The vapor pressure of bromine should be maintained at half an atmosphere. The vapor pressure of bromine is reported as 324 mm. at 35° and 392 mm. at 40°. If the bath becomes too warm, it should be cooled at once with ice; otherwise more bromine will be evaporated than will combine with the hydrogen present.

A more convenient form of apparatus for the vaporization of bromine is shown in Fig. 10. This is the same in principle as that described in the above procedure, but the bromine container instead of being heated by warm water is heated by the vapor of boiling ethyl bromide (38.4°). Thus no attention to a water bath is required.

FIG. 10.   7. The copper turnings remove any trace of bromine that may be present by converting it to black cupric bromide. If the level of blackened copper rises, bromine is being carried over. The presence of a small quantity of moisture, however, will cause a slight darkening of all the copper.

Alternatively, bromine may be removed by passing the hydrogen bromide through a solution of phenol in carbon tetrachloride (Org. Syn. **20,** 65).

If very dry hydrogen bromide is desired, a small trap surrounded by solid carbon dioxide and placed in the train will condense the water and a small amount of hydrogen bromide. Attention is necessary to prevent clogging of the trap.

8. According to the literature,[1] in the preparation of hydrogen bromide by combination of the elements the chief difficulties are avoiding explosive combination on the one hand and spontaneous extinction of the flame on the other hand. Neither of these difficulties has been en-

[1] Gmelin, "Handbuch der anorganischen Chemie," 8th ed., Part VII, p. 182, Verlag Chemie, Berlin, 1931.

countered in repeated operations extending over many hours with the apparatus described above.

3. Methods of Preparation

The very extensive literature on the preparation of hydrogen bromide is completely and concisely reviewed in Gmelin's "Handbuch." [1] The methods most commonly used involve the hydrolysis of certain bromides, particularly the action of bromine on red phosphorus and water; the action of bromine on hydrocarbons, particularly on tetralin; and the direct combination of the elements. The action of bromine on tetralin is a convenient one for small-scale laboratory operation.[2] The direct combination of the elements has certain advantages of simplicity and cleanliness, especially when considerable amounts of hydrogen bromide are required. Platinum and charcoal have frequently been used as cata-lysts but they can be dispensed with at sufficiently high temperatures.[3]

Detailed directions for preparing hydrogen bromide by the bromination of tetrahydronaphthalene and by the direct combination of hydrogen and bromine are given in Inorganic Syntheses.[4]

p-HYDROXYBENZOIC ACID

(Benzoic acid, *p*-hydroxy-)

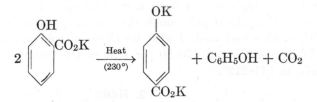

$$2 \quad \underset{(230°)}{\overset{\text{Heat}}{\longrightarrow}} \quad + \; C_6H_5OH + CO_2$$

$$p\text{-}KOC_6H_4CO_2K + 2HCl \longrightarrow p\text{-}HOC_6H_4CO_2H + 2KCl$$

Submitted by C. A. BUEHLER and W. E. CATE.
Checked by JOHN R. JOHNSON and C. P. NICHOLS.

1. Procedure

SIXTY grams (0.43 mole) of potassium carbonate (Note 1) is slowly stirred into a mixture of 100 g. (0.725 mole) of U.S.P. salicylic acid and

[2] For details see Houben, "Die Methoden der organischen Chemie," 3rd ed., Vol. III, p. 1156, Verlag Georg Thieme, Leipzig, 1930.

[3] Ruhoff, Burnett, and Reid, J. Am. Chem. Soc. **56**, 2784 (1934).

[4] Inorganic Syntheses, **I**, 149, McGraw-Hill, New York, 1939.

150 cc. of water contained in a 20-cm. porcelain dish. The solution is evaporated on a steam bath until a thick, pasty residue is obtained. This is broken up into small pieces and dried in an oven at 105–110° for two hours. The solid is then ground as finely as possible, dried for another two hours at 105–110°, and again ground to a fine powder.

The finely powdered mixture of potassium salicylate and carbonate is placed in a 500-cc. round-bottomed flask which is immersed in an oil bath so that only a small portion of the neck protrudes from the bath (Note 2). The bath is heated to 240° (Note 3) and maintained at this temperature for one and one-half hours. During this time the solid in the flask is stirred occasionally with a curved glass rod flattened at the end.

When the reaction is completed (Note 4), the product is transferred as completely as possible, while hot, to a 2-l. flask containing 1 l. of hot water. The reaction flask is rinsed with several portions of the hot solution. The alkaline solution is acidified with concentrated hydrochloric acid (about 75 cc. is required), heated nearly to boiling, treated with 5–6 g. of decolorizing charcoal, and filtered to remove a small quantity of brown resin. The filtrate is cooled under the tap, and the crude brown crystalline product is filtered with suction. The filtrate is concentrated to a volume of approximately 300 cc. and cooled as before. The second crop of crude acid is filtered with suction and combined with the main portion. The total weight of crude p-hydroxybenzoic acid, m.p. 208–211°, is 40–45 g.

The crude acid is dissolved in 300 cc. of hot water, boiled with 4–5 g. of decolorizing charcoal for a few minutes, and the solution filtered. After cooling thoroughly under the tap the purified product is filtered with suction and washed with 10–15 cc. of cold water. The purified acid weighs 35–40 g. (70–80 per cent of the theoretical amount) and melts at 211–212°.

2. Notes

1. An excess of potassium carbonate is used since it prevents the mass from caking during the subsequent heating. Although the original mixture is strongly alkaline a clear solution may not be obtained until the dish is heated.

2. In this way the phenol formed in the reaction is allowed to distil out of the mixture. This operation should be carried out in a hood.

3. The temperature reported is that of the oil bath; the internal temperature is approximately 230°. The temperature should be controlled carefully since pronounced decomposition sets in at higher temperatures.

4. The completeness of the reaction may be determined roughly by treating a small test portion with 3–4 cc. of hot water and acidifying

with concentrated hydrochloric acid. Since p-hydroxybenzoic acid is relatively soluble and salicylic acid only sparingly so, the absence of a precipitate in the warm solution indicates that the reaction is essentially complete.

3. Methods of Preparation

p-Hydroxybenzoic acid has been prepared by heating potassium phenoxide in a stream of carbon dioxide [1] or with carbon tetrachloride,[2] and by heating p-cresol with alkalies and various metallic oxides.[3] The procedure described above is similar to one which appears in the early literature.[4] When the dipotassium salt of salicylic acid is dehydrated by heating in vacuum and is then heated in a carbon dioxide atmosphere, essentially complete conversion to p-hydroxybenzoic acid is reported.[5]

2-HYDROXY-3,5-DIIODOBENZOIC ACID

(Salicylic acid, 3,5-diiodo-)

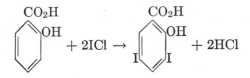

Submitted by G. H. WOOLLETT and W. W. JOHNSON.
Checked by W. W. HARTMAN and E. J. RAHRS.

1. Procedure

TWENTY-FIVE grams (0.18 mole) of salicylic acid (m.p. 159–160°) is dissolved in 225 cc. of glacial acetic acid in a 2-l. beaker provided with a mechanical stirrer (Note 1). To this is added with stirring a solution of 62 g. (0.38 mole) of iodine monochloride (Note 2) in 165 cc. of glacial acetic acid; then 725 cc. of water is added. A yellow precipitate of diiodosalicylic acid appears. The reaction mixture is gradually heated with stirring on a hot plate to 80° and kept at approximately that tempera-

[1] Kolbe, J. prakt. Chem. (2) **10**, 100 (1874); Hartmann, ibid. (2) **16**, 39 (1877); Ost, ibid. (2) **20**, 208 (1879).

[2] Reimer and Tiemann, Ber. **9**, 1285 (1876); Hasse, ibid. **10**, 2186 (1877).

[3] Graebe and Kraft, Ber. **39**, 797 (1906); Friedländer and Löw-Beer, Ger. pat. 170,230 [Frdl. **8**, 158 (1905–07)].

[4] Kolbe, J. prakt. Chem. (2) **11**, 24 (1875); Heyden, Ger. pat. 48,356 [Frdl. **2**, 132 (1887–90)].

[5] Dow Chemical Company, U. S. pat. 1,937,477 [C. A. **28**, 1056 (1934)].

ture for twenty minutes. The entire period of heating should be about forty minutes. Toward the end of the reaction the mixture becomes rather difficult to stir because of the voluminous precipitate. After cooling to room temperature (Note 3), the precipitate is filtered on a Büchner funnel and washed with acetic acid and then with water. When no more water is removed by suction, the solid (75 g.) is dissolved in 100 cc. of warm acetone and filtered by gravity. To the filtrate 400 cc. of water is slowly added with shaking. The fine, flocculent precipitate is filtered by suction, washed with water, and dried. The yield of diiodosalicylic acid melting at 235–236° is 64–64.5 g. (91–92 per cent of the theoretical amount) (Note 4).

2. Notes

1. The amount of glacial acetic acid used may not be sufficient to dissolve the salicylic acid completely. Solution will be completed upon the addition of the iodine chloride solution.

2. Iodine monochloride of sufficient purity for this preparation may be made as follows: Dry chlorine is led in at, or below, the surface of 127 g. (1 mole) of iodine in a 125-cc. distilling flask while the flask is gently shaken. When 34.5 g. (0.97 mole) of chlorine has been introduced, the iodine chloride is distilled in an ordinary distilling apparatus with a filter flask, protected from atmospheric moisture by a calcium chloride tube, as a receiver. The yield of iodine chloride, boiling between 97° and 105°, is 142 g. (87 per cent of the theoretical amount). The product can be preserved in a dry, glass-stoppered bottle. An excess of iodine is essential in this preparation.

3. Free iodine, if present, is removed by the addition of 5 per cent sodium sulfite solution.

4. The checkers found that 4-hydroxy-3,5-diiodobenzoic acid can be made from 4-hydroxybenzoic acid using the above directions with the exception that the product is not recrystallized from acetone, in which it is only slightly soluble. The yield of 4-hydroxy-3,5-diiodobenzoic acid, melting at 278–279° with decomposition, is 59 g. (84 per cent of the theoretical amount).

3. Methods of Preparation

The method given is based on that of Cofman.[1] Diiodosalicylic acid has been prepared by heating salicylic acid with iodine in alcohol;[2] by

[1] Cofman, Gazz. chim. ital. **50** (II) 297 (1920).
[2] Lautemann, Ann. **120,** 300 (1861).

using the same reagents with the addition of mercuric oxide;[3] by treating salicylic acid with iodine in the presence of alkali;[3,4] and by treating salicylic acid with iodine and iodic acid.[5] None of these methods, however, appears to give a good yield or a pure product.

β-HYDROXYETHYL METHYL SULFIDE

(Ethanol, 2-methylmercapto-)

$$(HN{=}\overset{\overset{\displaystyle SCH_3}{|}}{C}{-}NH_2)_2H_2SO_4 + 2NaOH$$

$$\rightarrow H_2N{-}\overset{\overset{\displaystyle NH}{\|}}{C}{-}NHCN + 2CH_3SH + Na_2SO_4 + 2H_2O$$

$$CH_3SH + NaOC_2H_5 \rightarrow CH_3SNa + C_2H_5OH$$

$$CH_3SNa + ClCH_2CH_2OH \rightarrow CH_3SCH_2CH_2OH + NaCl$$

Submitted by WALLACE WINDUS and P. R. SHILDNECK.
Checked by H. T. CLARKE and S. GURIN.

1. Procedure

A 2-l. three-necked flask is fitted with a dropping funnel, a stopcock, and a long condenser. The end of the condenser is connected with the following assembly: a safety trap, consisting of an empty gas-washing bottle; a second wash bottle containing 100 cc. of dilute sulfuric acid (1 volume of concentrated sulfuric acid to 2 volumes of water); a tower, about 30 cm. high, containing calcium chloride; an empty 2-l. flask which acts as a trap; a 2-l. flask containing 80.5 g. (3.5 gram atoms) of clean sodium dissolved in 1.5 l. of absolute alcohol; an empty 1-l. flask and a 1-l. flask containing 500 cc. of a saturated solution of lead acetate (Note 1). The exit tube from this last flask leads to a suction pump.

After 556 g. (2 moles) of methyl isothiourea sulfate (p. 411) is placed in the three-necked flask, a very slow current of air is drawn through the apparatus by means of the suction pump while 800 cc. of cold 5 N sodium hydroxide is added to the methyl isothiourea sulfate through the

[3] Weselsky, ibid. **174**, 103 (1874).
[4] Kekulé, ibid. **131**, 226 (1864).
[5] Liechti, ibid. Spl. **7**, 133, 141 (1870).

dropping funnel. The mixture is warmed very slowly to generate the methyl mercaptan (Note 2). As the reaction nears completion (after about two hours), the mixture is heated strongly for about thirty minutes (Note 3).

The solution of sodium methyl sulfide in absolute alcohol is transferred to a 3-l. three-necked flask, which is placed on a steam bath and fitted with a dropping funnel, a reflux condenser, and a mechanical stirrer. The solution is heated until the alcohol begins to boil. Heating is then discontinued and 302 g. (3.75 moles) of ethylene chlorohydrin (Note 4) is added dropwise with efficient stirring over a period of about two hours (Note 5). The reaction mixture is concentrated by distilling as much of the alcohol as possible on the steam bath. The mixture is then allowed to cool and the sodium chloride removed by filtration. The flask is rinsed and the sodium chloride washed with three 100-cc. portions of 95 per cent alcohol. The combined filtrate and washings are concentrated on the steam bath under reduced pressure until no further distillate passes over. The residue is then transferred to a modified Claisen flask and fractionally distilled under reduced pressure. The yield is 238–265 g. (74–82 per cent of the theoretical amount based on the sodium used) of a product boiling at 68–70°/20 mm. (Note 6).

2. Notes

1. The lead acetate removes any unreacted methyl mercaptan by precipitating it as the lead salt.

2. The rate of heating controls the rate of evolution of the methyl mercaptan. After the rapid evolution of the gas begins, the reaction mixture should be heated very gently. The slight suction aids in obtaining a regular flow of the gas.

3. The evolution of methyl mercaptan is almost complete after one and one-half to two hours. Prolonged vigorous heating increases the amount of ammonia evolved.

4. The ethylene chlorohydrin should be redistilled and the fraction boiling at 126–128° should be used.

5. The reaction is usually complete immediately after the addition of the ethylene chlorohydrin, obviating the necessity for refluxing the mixture. When the reaction is complete the solution is neutral to litmus paper.

6. Quantities of material five times as large as those called for above may be used without decreasing the percentage yield of product. With the larger amounts of material it is more convenient to filter the sodium chloride before concentrating the solution.

3. Methods of Preparation

The methods for preparing methyl mercaptan and β-hydroxyethyl methyl sulfide are essentially those of Arndt [1] and Kirner,[2] respectively.

p-IODOANILINE

(Aniline, *p*-iodo-)

$$C_6H_5NH_2 + I_2 + NaHCO_3 \rightarrow p\text{-}IC_6H_4NH_2 + NaI + H_2O + CO_2$$

Submitted by R. Q. BREWSTER.
Checked by W. H. CAROTHERS and W. L. McEWEN.

1. Procedure

IN a 3-l. beaker are placed 110 g. (1.2 moles) of aniline, 150 g. (1.8 moles) of sodium bicarbonate, and 1 l. of water, and the mixture is cooled to 12–15° by the addition of a small amount of ice. The beaker is then fitted with an efficient mechanical stirrer. The blade of a large porcelain spatula should be inserted into the liquid to overcome the rotary motion and thus obtain better mixing. The stirrer is started and 254 g. (1 mole) of powdered iodine is added in 15–20 g. portions at intervals of two to three minutes so that all the iodine is introduced during one-half hour. Stirring is continued for twenty to thirty minutes. By this time the reaction is complete and the color of the free iodine in the solution has practically disappeared. The crude *p*-iodoaniline, which separates as a dark crystalline mass, is collected on a Büchner funnel, pressed as free from water as possible, and dried in the air. The filtrate may be saved for the recovery of iodine (Note 1).

For the purification of the *p*-iodoaniline, the crude product is placed in a 2-l. flask and 1 l. of gasoline (Note 2) is added. The flask is fitted with an air-cooled reflux condenser and heated in a water bath at a temperature of 75–80° (Note 3). The flask should be shaken frequently, and about fifteen minutes should be allowed for saturation of the solution. The hot gasoline solution is slowly decanted into a beaker set in an ice-salt mixture and stirred constantly. The *p*-iodoaniline crystallizes immediately in practically colorless needles which are filtered and dried in the air (Notes 4 and 5). The filtrate is returned to the flask for

[1] Arndt, Ber. **54**, 2238 (1921).
[2] Kirner, J. Am. Chem. Soc. **50**, 2451 (1928).

use in a second extraction (Note 6). The yield is 165–185 g. (75–84 per cent of the theoretical amount) of a product which melts at 62–63°.

2. Notes

1. The sodium iodide which remains in the aqueous solution may be converted into iodine as follows: To the aqueous filtrate from the p-iodoaniline are added 100 cc. of concentrated sulfuric acid and 200 g. of sodium dichromate in 200 cc. of water. The iodine is allowed to settle, washed three times with water by decantation, collected on a filter, and allowed to dry on a watch glass. The yield of crude iodine is 167–179 g.

2. The gasoline used was a fractionated product (b. p. 70–150°). Ordinary gasoline may be used, but it has the disadvantage that the higher-boiling hydrocarbons are removed very slowly from the p-iodoaniline.

3. If a higher temperature is used a tarry material is sometimes formed and a diminished yield results.

4. If rapid cooling is not obtained, the product often separates as an oil.

5. For drying purposes, a current of warm air from a commercial hair dryer is advantageous.

6. Two extractions are usually sufficient, but, if much undissolved organic material still remains, a third extraction should be made. The first fraction is practically colorless, but the second and third portions are light brown unless a little charcoal is used for decolorizing the solution.

3. Methods of Preparation

p-Iodoaniline has been prepared by the reduction of p-nitroiodobenzene;[1] by the hydrolysis of p-iodoacetanilide formed by the action of iodine monochloride on acetanilide;[2] and by the direct iodination of aniline.[3] The method described here is an adaptation of the procedure used by Wheeler,[4] and by Hann and Berliner[5] for the iodination of the toluidines.

[1] Griess, Zeit. für Chem. 1866, 218; Kekulé, ibid. 687; Körner and Wender, Gazz. chim. ital. 17, 489 (1887); Baeyer, Ber. 38, 2762 (1905); Montagne, ibid. 51, 1490 (1918).

[2] Michael and Norton, ibid. 11, 108 (1878); Chattaway and Constable, J. Chem. Soc. 105, 125 (1914).

[3] Hofmann, Ann. 67, 65 (1848); Bradfield, Orton, and Roberts, J. Chem. Soc. 1928, 783; Militzer, Smith, and Evans, J. Am. Chem. Soc. 63, 436 (1941).

[4] Wheeler and Liddle, Am. Chem. J. 42, 501 (1909); Wheeler, ibid. 44, 128, 500 (1910).

[5] Hann and Berliner, J. Am. Chem. Soc. 47, 1710 (1925).

5-IODOANTHRANILIC ACID

(Anthranilic acid, 5-iodo-)

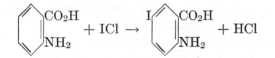

Submitted by V. H. WALLINGFORD and PAUL A. KRUEGER.
Checked by FRANK C. WHITMORE and L. H. SUTHERLAND.

1. Procedure

IN a 3-l. beaker, 110 g. (0.8 mole) of anthranilic acid (Note 1) is dissolved in 1 l. of water and 80 cc. of c.p. concentrated hydrochloric acid (sp. gr. 1.19); the solution is cooled to 20°. In a 2-l. beaker a solution of iodine monochloride in hydrochloric acid is prepared by diluting 140 cc. of c.p. concentrated hydrochloric acid with 500 cc. of cold water, adding just sufficient crushed ice to bring the temperature to 5°, and, during about two minutes, stirring in 131 g. (0.8 mole) of iodine monochloride (p. 197).

The iodine monochloride solution at 5° is stirred rapidly into the anthranilic acid solution at 20°. 5-Iodoanthranilic acid separates almost immediately as a granular, tan to violet precipitate, and the reaction temperature rises to 18–22°. The mixture is stirred for an hour, while warming to room temperature, then filtered on a 13-cm. Büchner funnel. The acid is pressed as dry as possible, washed with three 100-cc. portions of cold water, and then dried at 90–100°. There is obtained 185–189 g. (88–90 per cent of the theoretical amount) of brown to purple acid which melts at 185–190°, with decomposition (Note 2).

5-Iodoanthranilic acid is best purified by recrystallization of its ammonium salt, as follows: To 100 g. of the acid in a 400-cc. beaker is added 200 cc. of hot water, and the acid is dissolved by stirring in concentrated ammonia (sp. gr. 0.9) until solution is complete at 60° and there is a slight excess of ammonia. About 40 cc. of concentrated ammonia is required. Sodium hydrosulfite is added in about 1-g. portions until no further bleaching action is observed, about 5 g. of decolorizing charcoal is added, and, after stirring three minutes, the mixture is filtered with suction on a preheated Büchner funnel into a preheated flask. The filter is washed with 10 cc. of boiling water.

The combined filtrate and washings are transferred to a 400-cc. beaker, allowed to cool slowly without agitation until crystal formation appears complete, then cooled to 5°. The crystalline ammonium salt is then removed by suction filtration on a large Büchner funnel, washed with 15 cc. of ice water, sucked as dry as possible, and spread in a thin layer on an enameled or glass tray and allowed to dry in air at 35–50°. The yield of yellow to violet ammonium salt is 80–89 g., representing a recovery of 76–84 per cent (Notes 3 and 4).

The ammonium salt is dissolved in three parts of hot water, ammonia is added if necessary to effect complete solution at 60°, the solution is again treated with sodium hydrosulfite and 3–4 g. of decolorizing charcoal, filtered hot, and the 5-iodoanthranilic acid is precipitated by adding c.p. hydrochloric acid in 3- to 5-cc. portions, stirring thoroughly after each addition, until the reaction mixture is just faintly acid to Congo red. Ice is then added until the temperature is reduced to 20°, and the precipitated acid is removed by suction filtration, washed freely with cold water, and dried at 100–110° (Note 5). 5-Iodoanthranilic acid is almost quantitatively precipitated from its ammonium salt solution and is obtained as a yellow powder, m.p. 190–195° (decomp.) (Note 6).

2. Notes

1. Commercial "Acid Anthranilic Sublimed" is satisfactory.

2. The decomposition point varies with the method of determination. The values given herein were determined by placing the sample in a capillary tube, inserting the tube in the bath at a temperature 25° below the decomposition point, and heating at the rate of 2–3° per minute.

3. The ammonium salt is not stable under the drying conditions. The dry product is a mixture of 5-iodoanthranilic acid and its ammonium salt. Contact with air may change its color from yellow to violet.

4. Crude 5-iodoanthranilic acid is recovered from the mother liquor by carefully making it just acid to Congo red, using a spot plate, with concentrated hydrochloric acid, cooling to 20°, filtering, washing freely with cold water, and drying at 90–110°. This acid can be added to crude acid in further crystallizations or worked up alone.

5. The drying oven must have good air circulation and uniform temperature control; drying at 120° causes the evolution of considerable amounts of free iodine.

6. This product is sufficiently pure for conversion to m-iodobenzoic acid (p. 353). If greater purity is desired the ammonium salt can be recrystallized from twice its weight of water, between 45° and 5°, with 85–90 per cent recovery. When the salt remains nearly white after drying, the acid precipitated from it melts with decomposition at a temperature above 210°. Four or five recrystallizations may be required.

3. Methods of Preparation

5-Iodoanthranilic acid has been prepared by the reduction of 2-nitro-5-iodobenzoic acid;[1] by treatment of anthranilic acid with iodine in potassium hydroxide solution;[2] by treatment of the anhydride of 5-hydroxymercurianthranilic acid with iodine in aqueous potassium iodide solution;[3] and by iodination of anthranilic acid in glacial acetic acid solution with iodine monochloride,[4] or in dilute acetic acid solution with iodine.[5]

IODOBENZENE

(Benzene, iodo-)

$$C_6H_5NH_2 + NaNO_2 + 2HCl \rightarrow C_6H_5N_2Cl + 2H_2O + NaCl$$

$$C_6H_5N_2Cl + KI \rightarrow C_6H_5I + N_2 + KCl$$

Submitted by H. J. Lucas and E. R. Kennedy.
Checked by John R. Johnson and P. L. Barrick.

1. Procedure

In a 3- or 5-gallon stoneware crock are placed 950 cc. (1130 g., 11.7 moles) of concentrated hydrochloric acid (sp. gr. 1.19), 950 cc. of water, 200 g. (196 cc., 2.15 moles) of aniline, and 2 kg. of ice (Note 1). The mixture is agitated by a mechanical stirrer, and, as soon as the temperature drops below 5°, a chilled solution of 156 g. (2.26 moles) of sodium nitrite in a measured volume (700–1000 cc.) of water is introduced fairly rapidly from a separatory funnel, the stem of which projects below the surface of the reaction mixture. The addition should not be fast enough to cause the temperature to rise to 10° or to cause evolution of oxides of nitrogen. The last 5 per cent of the nitrite solution is added more slowly, and the reaction mixture is tested with starch-iodide paper at intervals until an excess of nitrous acid is indicated.

Stirring is continued for ten minutes, and if necessary the solution is filtered rapidly through a loose cotton plug in a large funnel. An aqueous

[1] Grothe, J. prakt. Chem. (2) 18, 326 (1878).
[2] Wheeler and Johns, Am. Chem. J. 43, 403 (1910); Klemme and Hunter, J. Org. Chem. 5, 230 (1940).
[3] Schoeller and Hueter, Ber. 47, 1938 (1914).
[4] Borsche, Weussmann, and Fritzsche, ibid. 57, 1774 (1924).
[5] Militzer, Smith, and Evans, J. Am. Chem. Soc. 63, 436 (1941).

solution of 358 g. (2.16 moles) of potassium iodide is added and the reaction mixture allowed to stand overnight. The mixture is transferred to a large flask (or two smaller flasks) and heated on a steam bath, using an air-cooled reflux condenser, until no more gas is evolved, then allowed to cool and stand undisturbed until the heavy organic layer has settled thoroughly. A large part of the upper aqueous layer is siphoned off, and discarded (Note 2). The residual aqueous and organic layers are made alkaline by the cautious addition of strong sodium hydroxide solution (100–125 g. of solid technical sodium hydroxide is usually required) and steam-distilled at once. The last one-third of the steam distillate is collected separately and combined with the aqueous layer separated from the earlier portions of the distillate. This mixture is acidified with 5–10 cc. of concentrated sulfuric acid and steam-distilled again. The iodobenzene from this operation is combined with the main portion and dried with 10–15 g. of calcium chloride (Notes 3 and 4). Distillation under reduced pressure gives 327–335 g. (74–76 per cent of the theoretical amount) of iodobenzene, b.p. 77–78°/20 mm. or 63–64°/8 mm. (Note 5).

2. Notes

1. If more ice is used a portion remains unmelted after the diazotization is completed.

2. If a good separation has been made not more than 1–2 g. of iodobenzene is lost with the upper layer.

3. An appreciable amount of iodobenzene is retained by the solid calcium chloride. By treating the spent drying agent with water 8–12 g. of iodobenzene can be recovered.

4. The crude iodobenzene weighs 350–355 g. (80 per cent of the theoretical amount) and is pure enough for many purposes without redistillation.

5. If the distillation is carried too far, the distillate will be colored.

3. Methods of Preparation

The preparation of iodobenzene by iodination of benzene, with iodine and nitric acid, and a survey of preparative methods have been given in an earlier volume.[1] The present procedure, based upon the method of Gattermann,[2] gives a purer product.

[1] Org. Syn. Coll. Vol. I, 1941, 323.

[2] Gattermann-Wieland, "Laboratory Methods of Organic Chemistry," p. 283. Translated from the twenty-fourth German edition by W. McCartney, The Macmillan Company, New York, 1937.

m-IODOBENZOIC ACID

(Benzoic acid, *m*-iodo-)

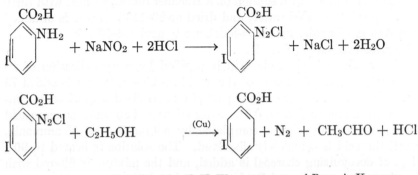

Submitted by V. H. Wallingford and Paul A. Krueger.
Checked by Frank C. Whitmore and L. H. Sutherland.

1. Procedure

In a 1-l. flask 132 g. (0.5 mole) of 5-iodoanthranilic acid (p. 349) and 35 g. (0.5 mole) of sodium nitrite are dissolved in a mixture of 500 cc. of warm water and 60 cc. of 30 per cent sodium hydroxide solution. After cooling to 20° the solution is added from a dropping funnel, over a period of fifteen to twenty minutes, to a well-stirred mixture of 250 cc. of concentrated hydrochloric acid (sp. gr. 1.18) and 250 g. of ice in a 2-l. beaker; more ice is added, as required, to keep the temperature below 20°. The insoluble, yellow diazonium compound separates before completion of the diazotization. After all the solution has been added from the dropping funnel, the reaction mixture is stirred for five minutes and tested for excess nitrous acid with starch-iodide paper. If required, small amounts of solid sodium nitrite are added and the test is repeated at three-minute intervals until a slight excess is definitely established. The diazonium compound is allowed to settle and as much as possible of the supernatant liquor is decanted, leaving a slurry of diazonium compound.

In a 3-l. beaker are placed 750 cc. of 95 per cent ethyl alcohol and 1.5 g. of finely ground copper sulfate, and the mixture is heated on a hot plate or steam bath to 70°. The diazonium slurry is added in about 30-cc. portions to the well-stirred alcohol; the temperature is kept between 60° and 70°, and the nitrogen evolution is allowed to subside considerably between additions. The final traces of diazonium slurry are washed into the alcohol with small amounts of the decanted solution, the remainder of which is then added to the alcohol in 100-cc. portions. The reaction mixture is heated and stirred at 65–70° for half an hour, and then cooled without agitation to 5°. The m-iodobenzoic acid which separates is filtered with suction on a Büchner funnel, washed with three 50-cc. portions of cold water, and dried at 90–110°. There is obtained 107–116 g. of crude, brown m-iodobenzoic acid (86–93 per cent of the theoretical amount) (Note 1).

The crude m-iodobenzoic acid is purified by recrystallization of its ammonium salt. To 100 g. of the acid in a 250-cc. beaker is added 75 cc. of hot water, and the acid is partially neutralized with 24 cc. of concentrated ammonia (sp. gr. 0.9). After stirring at 80° until no more acid dissolves, neutralization is completed by adding 2–5 cc. of ammonia, until the acid is completely dissolved. The solution is heated to 90°, 1 g. of decolorizing charcoal is added, and the mixture is filtered with suction using a preheated Büchner funnel and filter flask. The residue on the filter is washed with 15 cc. of boiling water. The combined washings and filtrate are transferred to a 250-cc. beaker and allowed to cool without agitation to 25–35°, then cooled by any convenient means to 5°. The ammonium m-iodobenzoate crystals are filtered and pressed as dry as possible on a Büchner funnel, spread in a thin layer on a glass or enameled tray, and dried at a temperature not above 60°. There is obtained 86–90 g. of faintly yellow to tan prisms of the ammonium salt, a recovery of 80–84 per cent (Note 2).

The ammonium salt thus obtained is recrystallized until white by dissolving it in an equal weight of water at 80° and cooling to 5°. The recovery in this purification averages 75–85 per cent (Note 3).

m-Iodobenzoic acid melting at 187–188° is obtained by dissolving pure ammonium m-iodobenzoate in four times its weight of hot water, precipitating the acid by acidifying the solution to Congo red with concentrated hydrochloric acid, adding ice to reduce the temperature to 20°, filtering by suction, washing the acid freely with cold water, and drying at 90–110° (Note 4).

2. Notes

1. Evaporation of the mother liquors to incipient turbidity and cooling the concentrate to 5° will produce a few grams of acid, but this is usually very impure and tarry.

2. Ammonium m-iodobenzoate is not stable under these drying conditions. The product at this point contains some free acid.

3. In some runs as many as five recrystallizations failed to furnish white crystals of the ammonium salt, but, in these runs, precipitation

of the acid after three or four crystallizations gave a product having the correct melting point.

Mother liquors from the crystallization are best worked up by acidifying to Congo red with concentrated hydrochloric acid, filtering, washing, drying, and reworking the recovered acid.

4. The total overall yield of pure acid from the crude depends on the amount of crude acid taken, the care exercised to avoid material losses, and the number of recrystallizations required to attain the desired purity. Without recoveries from mother liquors, 100 g. of crude acid, after conversion to the ammonium salt, and three recrystallizations of the latter, gave 40–45 g. of pure acid. Overall recovery of pure acid from 100 g. of the crude gave 50 g. of pure acid melting at 187–188°, 10 g. of less pure acid melting at 184–186°, and about 10 g. of a tarry solid residue.

3. Methods of Preparation

m-Iodobenzoic acid has been prepared by diazotization of *m*-aminobenzoic acid and treatment with potassium iodide in acid solution,[1, 2, 3] and by the action of concentrated nitric acid on a glacial acetic acid solution of iodine and benzoic acid.[4]

p-IODOPHENOL

(Phenol, *p*-iodo-)

$$2 \text{ } p\text{-HOC}_6\text{H}_4\text{NH}_2 + 3\text{H}_2\text{SO}_4 + 2\text{NaNO}_2 \rightarrow$$
$$2 \text{ } p\text{-HOC}_6\text{H}_4\text{N}_2\text{SO}_4\text{H} + \text{Na}_2\text{SO}_4 + 4\text{H}_2\text{O}$$

$$p\text{-HOC}_6\text{H}_4\text{N}_2\text{SO}_4\text{H} + \text{KI} \xrightarrow{\text{(Cu)}} p\text{-HOC}_6\text{H}_4\text{I} + \text{N}_2 + \text{KHSO}_4$$

Submitted by F. B. Dains and Floyd Eberly.
Checked by Reynold C. Fuson and H. H. Hully.

1. Procedure

One hundred nine grams (1 mole) of *p*-aminophenol (Note 1) is dissolved in a mixture of 500 g. of ice, 500 cc. of water, and 65 cc. (120 g., 1.2 moles) of concentrated sulfuric acid (sp. gr. 1.84). To this solution, kept in a freezing mixture at 0°, is added, during the course of an hour

[1] Griess, Ann. **113**, 336 (1860).
[2] Cohen and Raper, J. Chem. Soc. **85**, 1273 (1904).
[3] Cattelain, Bull. soc. chim. (4) **41**, 1546 (1927).
[4] Datta and Chatterjee, J. Am. Chem. Soc. **41**, 294 (1919).

with constant mechanical stirring, a solution of 72 g. (1 mole) of 95 per cent sodium nitrite in 150 cc. of water. The stirring is continued twenty minutes longer, and then 20 cc. (37 g., 0.37 mole) of concentrated sulfuric acid is added.

This solution is poured into an ice-cold solution of 200 g. (1.2 moles) of potassium iodide in 200 cc. of water. After a few minutes, 1 g. of copper bronze (Note 2) is added, with continued stirring, and the solution is warmed slowly on the water bath. The temperature is kept at 75–80° until the evolution of nitrogen ceases; during this process the iodophenol separates as a heavy dark oil. After cooling to room temperature the reaction mixture is extracted three times with 165-cc. portions of chloroform and the combined extracts are washed with dilute thiosulfate solution. The solvent is removed on the water bath and the residue distilled under reduced pressure, the p-iodophenol coming over at 138–140°/5 mm. One crystallization from about 2 l. of ligroin (b.p. 90–110°) gives a colorless product melting sharply at 94°. The yield of recrystallized product is 153–159 g. (69–72 per cent of the theoretical amount).

2. Notes

1. The p-aminophenol used was a commercial product melting at 182–183° with decomposition.

2. Some commercial bronzes used for bronze paints are coated with a film of stearic acid. For chemical work an untreated pure copper bronze should be used.

3. Methods of Preparation

p-Iodophenol was first obtained as a by-product of the action of iodine on salicylic acid in alkaline solution or by heating iodosalicylic acid.[1] It has also been obtained by the action of iodine on phenol in alkaline solution [2] or in the presence of mercuric oxide,[3] or by the action of iodine monochloride.[4] It is best prepared by the diazotization of p-aminophenol and replacement of the diazonium group by iodine [5] although it has also been obtained from p-iodoaniline by diazotization and replacement of the diazonium group by hydroxyl.[6]

[1] Lautemann, Ann. **120**, 299 (1861); Kekulé, ibid. **131**, 221 (1864).

[2] Holleman and Rinkes, Rec. trav. chim. **30**, 96 (1911).

[3] Hlasiwetz and Weselsky, Ber. **2**, 523 (1869).

[4] Schützenberger and Sengenwald, Jahresb. **1862**, 413.

[5] Nölting and Wrzesinski, Ber. **8**, 820 (1875); Nölting and Stricker, ibid. **20**, 3021 (1887); Neumann, Ann. **241**, 74 (1887).

[6] Griess, Zeit. für Chem. **1865**, 427; Holleman and Rinkes, Rec. trav. chim. **30**, 95 (1911).

2-IODOTHIOPHENE

(Thiophene, 2-iodo-)

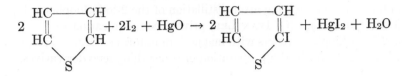

$$2 \begin{array}{c} HC\text{---}CH \\ \| \quad \| \\ HC \quad CH \\ \diagdown \diagup \\ S \end{array} + 2I_2 + HgO \rightarrow 2 \begin{array}{c} HC\text{---}CH \\ \| \quad \| \\ HC \quad CI \\ \diagdown \diagup \\ S \end{array} + HgI_2 + H_2O$$

Submitted by WESLEY MINNIS.
Checked by ROGER ADAMS and H. D. COGAN.

1. Procedure

IN a glass-stoppered, wide-mouthed bottle cooled by ice water are placed 35 g. (0.42 mole) of thiophene (p. 578) and 50 cc. of benzene (Note 1). With constant shaking (Note 2), and cooling when necessary, 75 g. (0.35 mole) of yellow mercuric oxide and 109 g. (0.43 mole) of iodine are added alternately in small amounts during a period of fifteen to twenty minutes. The yellow mercuric oxide changes to crimson mercuric iodide. The mixture is filtered, and the residue is washed with three 25-cc. portions of ether. The ether-benzene filtrate is shaken with a dilute solution of sodium thiosulfate to remove excess iodine and then dried over 5 g. of calcium chloride and filtered. The ether and benzene are removed by distillation on a steam bath (Note 3), and the residue is fractionally distilled under reduced pressure. 2-Iodothiophene distils at 73°/15 mm.; 80–81°/20 mm.; 90–94°/34–38 mm. (Note 4). The yield is 63–66 g. (72–75 per cent of the theoretical amount) (Note 5). If the iodothiophene is still colored by traces of iodine, the color may be removed by shaking with a small amount of mercuric oxide.

2. Notes

1. Ligroin (b.p. 100–120°) may be substituted for benzene.
2. Better yields are obtained when the mixture is vigorously shaken by hand than when mechanical stirring is used. The ordinary stirrer will not keep the mercuric oxide in suspension.
3. Unreacted thiophene can be recovered from the ether-benzene distillate by treating the latter with mercuric oxide and dilute acetic acid, collecting the white precipitate [C$_4$H$_2$S(HgOCOCH$_3$)HgOH] on a Büchner funnel, and decomposing it with concentrated hydrochloric acid.[1]

[1] Dimroth, Ber. **32,** 759 (1899).

In checking this preparation there was not enough unreacted thiophene to be recovered.

4. A small amount of 2,5-diiodothiophene is formed in the reaction. About 4 g. of crystalline diiodothiophene, m.p. 40–41°, can be isolated from the residue remaining after distillation of the 2-iodothiophene (b.p. 65.5–66.5°/9 mm.). (O. IVAN LEE, private communication.)

5. 2-Iodothiophene reacts with magnesium to form a Grignard reagent and is hence useful in the preparation of other thiophene derivatives.

3. Method of Preparation

2-Iodothiophene has been prepared only by the action of iodine and mercuric oxide on thiophene.[2]

ISOBUTYL BROMIDE

(Propane, 1-bromo-2-methyl-)

$$3(CH_3)_2CHCH_2OH + PBr_3 \rightarrow 3(CH_3)_2CHCH_2Br + P(OH)_3$$

Submitted by C. R. NOLLER and R. DINSMORE.
Checked by FRANK C. WHITMORE, D. E. BADERTSCHER, and A. R. LUX.

1. Procedure

IN a 2-l. three-necked flask, fitted with a mechanical stirrer, a thermometer, and a dropping funnel, is placed 518 g. (643 cc., 7 moles) of dry isobutyl alcohol (b.p. 106–108°). The alcohol is cooled to −10° by immersing the flask in an ice-salt bath, and 695 g. (244 cc., 2.56 moles) of phosphorus tribromide (Note 1) is slowly added with stirring at such a rate as to keep the temperature below 0° (about four hours). The cooling bath is removed, and stirring is continued until the mixture reaches room temperature; it is then allowed to stand overnight. The stirrer, funnel, and thermometer are removed, and the flask is fitted with a 30-cm. fractionating column and a condenser. The crude isobutyl bromide is distilled from the reaction mixture under diminished pressure, e.g., at about 50°/200 mm. (Note 2).

The distillate is cooled to about 0° and washed three times with 50-cc. portions of concentrated sulfuric acid cooled to 0°; it is then shaken with

[2] Meyer and Kreis, ibid. **17,** 1558 (1884); Thyssen, J. prakt. Chem. (2) **65,** 5 (1902).

25 g. of anhydrous potassium carbonate until the odor of hydrobromic acid disappears. It is distilled through a 1-m. fractionating column at atmospheric pressure, collecting the portion boiling at 91–93° (88.5–90.5°/728 mm.), or under reduced pressure through a 70 by 2-cm. total reflux, adjustable take-off, adiabatic column (Note 3), b.p. 41–43°/135 mm. The product weighs 525–570 g. (55–60 per cent of the theoretical amount) (Note 4).

2. Notes

1. The phosphorus tribromide boiled at 171–173° (168–170°/725 mm.), and was prepared in 90–95 per cent yield by adding bromine to a stirred suspension of red phosphorus in carbon tetrachloride. A good fractionating column is necessary. Old, red phosphorus containing acids of phosphorus gives a poorer yield.

2. In some runs, the crude bromide was successfully distilled at atmospheric pressures; in others it decomposed violently. With reduced pressure no difficulty was experienced. A water pump with an adjustable leak in the vacuum line was used.

3. The column used with reduced pressure was similar to those described by Whitmore and Lux, J. Am. Chem. Soc. **54**, 3451 (1932). The product, fractionated under reduced pressure using a reflux ratio of 5 : 1, contained less than 1 per cent of tertiary butyl bromide.

4. By similar procedures, the following bromides can be prepared with the yields indicated: sec.-butyl, b.p. 90–93°, 80 per cent; n-propyl, b.p. 70–73°, 95 per cent; isopropyl, b.p. 60–63°, 68 per cent.

In the preparation of these three bromides and isobutyl bromide as well, the phosphorus tribromide procedure described above gives purer products in better yields than the hydrobromic-sulfuric acid method described in Org. Syn. Coll. Vol. I, **1941,** 25. The phosphorus tribromide method is not convenient for the preparation of tertiary butyl bromide; the product is difficult to purify.

3. Methods of Preparation

Isobutyl bromide has been prepared from isobutyl alcohol by the action of bromine and phosphorus,[1] aqueous hydrobromic acid,[2] and gaseous hydrobromic acid;[3] from isobutylene and gaseous hydrogen

[1] Wurtz, Ann. **93**, 114 (1855).

[2] Norris, Am. Chem. J. **38**, 640 (1907).

[3] Fournier, Bull. soc. chim. (3) **35**, 623 (1906); Longinov and Lerman, Khim Farm. Prom. **1933**, 14 [C. A. **27**, 3443 (1933)].

bromide;[4] or hydrogen bromide in glacial acetic acid;[5] and by the rearrangement of tertiary butyl bromide at 210–220°.[6] A number of bromides, including isobutyl bromide, have been prepared by the action of phosphorus tribromide on alcohols.[7] The procedure described above is a modification of one used for preparing cyclopentyl bromide.[8]

ISODURENE

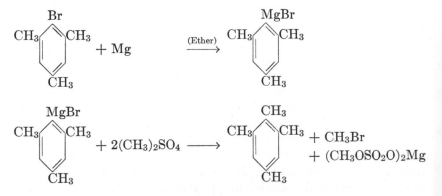

Submitted by LEE IRVIN SMITH.
Checked by ROGER ADAMS and W. W. MOYER.

1. Procedure

A 3-l. three-necked flask fitted with a reflux condenser protected from the air by a calcium chloride tube, a separatory funnel, and a mechanical stirrer is mounted on a steam bath. In the flask are placed 48 g. (2 moles) of magnesium turnings, 150 cc. of anhydrous ether and 100 g. of bromomesitylene (p. 95). The reaction starts slowly, and sometimes it is necessary to add some iodine or use Gilman's catalyst (Note 1). After the reaction starts, it proceeds smoothly and the remaining 298 g. of bromomesitylene (a total of 2 moles) in about 700 cc. of dry ether is added at such a rate that the ether boils briskly. After the last of

[4] Brunel, J. Am. Chem. Soc. 39, 1978 (1917).

[5] Ipatiew and Ogonowsky, Ber. 36, 1988 (1903).

[6] Faworsky, Ann. 354, 343 (1907).

[7] Reynolds and Adkins, J. Am. Chem. Soc. 51, 280 (1929); Tseng and Hou, J. Chinese Chem. Soc. 2, 57 (1934) [C. A. 28, 3711 (1934)].

[8] Adams and Noller, J. Am. Chem. Soc. 48, 1084 (1926).

this ether solution has been added, the mixture is heated on the steam cone until practically all the magnesium has dissolved (Note 2).

The solution of the Grignard reagent is cooled to about 10°, and to it, while stirring vigorously, is added a solution of 600 g. (4.8 moles) of methyl sulfate (Note 3) in about 500 cc. of dry ether. The reaction is very vigorous, and bumping may occur as the result of the separation of insoluble magnesium compounds. The addition of the methyl sulfate requires two to three hours. The reaction mixture is allowed to stand (Note 4) for about twenty-four hours and is then decomposed by adding dilute hydrochloric acid through the separatory funnel. Stirring is started as soon as the mass is fluid enough. When the mixture is decomposed completely, the ether layer is separated and washed three times with water. After the magnesium salts have been removed, the ether layer is evaporated and the residue is added slowly to a solution of 30 g. of sodium in 500 cc. of absolute alcohol (Note 5). The mixture is boiled for about one-half hour. Then the solution is cooled, 150–200 cc. of ether is added, and the alkali and alcohol are removed by washing thoroughly with water (Note 6). Finally the ether solution is dried over calcium chloride, the ether is distilled, and the residue is warmed on a water bath for three to four hours with 25–30 g. of metallic sodium (Note 7). The mixture is filtered, and the filtrate is fractionated carefully under reduced pressure in a modified Claisen flask. The fractions collected are: up to 85°/18 mm.; 85–87°/18 mm.; and residue. The low-boiling fraction weighs about 50 g. and is mainly mesitylene. The second fraction is isodurene and has a melting point of −24.2°. The yield is 140–160 g. (52–60 per cent of the theoretical amount).

2. Notes

1. Gilman's catalyst [1] is prepared readily by heating an alloy of magnesium containing 12.75 per cent of copper with about 20 per cent by weight of iodine in an evacuated flask. Only about 0.25 g. of this catalyst is required to bring about a reaction with a halogen compound. When this catalyst is used, the ordinary magnesium turnings are added as soon as the reaction has started. The reaction can also be started by adding a small amount of an ether solution of any Grignard reagent, such as ethylmagnesium bromide.

2. If Gilman's catalyst is used to start the reaction there will always be an excess of magnesium at the end of the reaction.

---

[1] Gilman, Peterson, and Schulze, Rec. trav. chim. **47**, 19 (1928).

3. The methyl sulfate was distilled carefully under reduced pressure, and a fraction boiling within 1° was used.

Methyl sulfate is extremely toxic, and great care must be taken to avoid breathing the vapors and spilling the liquid on the hands or clothes. Ammonia is a specific antidote for methyl sulfate.

4. The reaction mixture sometimes becomes almost solid; stirring is then useless.

5. Aqueous alkali is not sufficient to remove the excess methyl sulfate. Sometimes the reaction between the excess methyl sulfate and the sodium ethoxide solution is vigorous; therefore the two solutions should be mixed carefully. If a large excess of methyl sulfate is present, more sodium ethoxide may be needed in order to keep the solution alkaline.

6. Emulsions can be broken by acidifying.

7. The treatment with sodium ensures a halogen-free product.

3. Methods of Preparation

Isodurene has been prepared from bromomesitylene, methyl iodide, and sodium;[2] from mesitylene, methyl chloride, and aluminum chloride;[3] from mesitylene, methyl iodide, aluminum chloride, and carbon disulfide;[4] from 1,3,4,5-tetramethylbenzonitrile with hydrogen chloride at 250°;[5] by the action of zinc chloride or iodine on camphor;[6] and in small amounts by the action of concentrated sulfuric acid on acetone.[7] The method of preparation described above has been published.[8]

Methods of preparation in which aluminum chloride is used do not give a pure product; aluminum chloride shifts the methyl groups to give mixtures of the three tetramethylbenzenes from which isodurene cannot be separated efficiently. Compare Note 7, p. 252.

[2] Jannasch, Ber. **8**, 356 (1875); Bielefeldt, Ann. **198**, 380 (1879); Jannasch and Weiler, Ber. **27**, 3442 (1894).

[3] Jacobsen, ibid. **14**, 2629 (1881).

[4] Claus and Foecking, ibid. **20**, 3097 (1887).

[5] Hofmann, ibid. **17**, 1915 (1884).

[6] Armstrong and Miller, ibid. **16**, 2259 (1883).

[7] Orndorff and Young, Am. Chem. J. **15**, 267 (1893).

[8] Smith and MacDougall, J. Am. Chem. Soc. **51**, 3003 (1929).

ISONITROSOPROPIOPHENONE

(1,2-Propanedione, 1-phenyl-, 2-oxime)

$$C_6H_5COCH_2CH_3 + CH_3ONO \xrightarrow{(HCl)} \underset{\underset{NOH}{\|}}{C_6H_5COCCH_3} + CH_3OH$$

Submitted by WALTER H. HARTUNG and FRANK CROSSLEY.
Checked by REYNOLD C. FUSON and R. F. PETERSON.

1. Procedure

A 3-l. three-necked, round-bottomed flask (A, Fig. 11) is provided with a reflux condenser, liquid-sealed mechanical stirrer, and two gas delivery tubes (T_1 and T_2) which ex-
tend as far as possible into the flask. Methyl nitrite is generated in a 2-l. Erlenmeyer flask B which is fitted with a 500-cc. dropping funnel C and connected to A by the tube T_1. Dry hydrogen chloride is introduced through T_2. The apparatus is assembled preferably in a hood with effective draft.

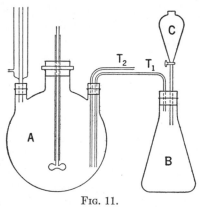

FIG. 11.

In A is placed a solution of 469 g. (3.5 moles) of propiophenone (Note 1) in 2.3 l. of ordinary ethyl ether, and in B a mixture of 290 g. (4 moles)
of 95 per cent sodium nitrite, 180 cc. (142 g., 4.5 moles) of methyl alcohol, and 170 cc. of water. In the dropping funnel C is placed 455 cc. of cold dilute sulfuric acid (prepared by adding one volume of concentrated acid to two volumes of water).

The stirrer is started, and hydrogen chloride is introduced through T_2 at the rate of 6–10 bubbles a second. The acid in C is allowed to drop slowly into B, and the gaseous methyl nitrite (Note 2) is introduced through T_1 into the reaction mixture. The solution in A develops a brown-red color, and after about ten minutes the ether begins to reflux gently (Note 3). The rate of evolution of methyl nitrite is then adjusted so that the ether continues to reflux gently. About four hours is required for addition of the methyl nitrite. Stirring and addition of hydrogen chloride are continued for thirty minutes longer; at the end of this time the solution has ceased boiling and has assumed a clear yellow color.

The reaction mixture is allowed to stand for several hours (preferably overnight) and is then extracted repeatedly with 500-cc. portions of 10 per cent sodium hydroxide solution, until the alkaline extracting medium remains practically colorless when shaken with the ethereal solution (Note 4). Usually five 500-cc. portions of the sodium hydroxide solution are required. The combined alkaline extracts are poured slowly, with stirring, into a mixture of 700–750 cc. of concentrated hydrochloric acid and about 1 kg. of ice. The crystals of isonitrosopropiophenone are filtered with suction and dried. The product weighs 370–390 g. (65–68 per cent of the theoretical amount) and melts at 111–113°. This material can be crystallized from about 550 cc. of toluene and yields 315–335 g. of snow-white crystals, m.p. 112–113° (Note 5).

2. Notes

1. Propiophenone may be prepared in 70–80 per cent yields from benzene and propionyl chloride or propionic anhydride, in the presence of aluminum chloride.

2. For the preparation of small amounts of the isonitrosoketone it is more convenient to employ a higher-boiling alkyl nitrite, such as butyl nitrite (p. 108), which can be added directly to the reaction mixture by substituting a dropping funnel for the tube T_1. Butyl nitrite must be freshly prepared or redistilled shortly before use.

3. The rate of stirring must be kept fairly constant since an abrupt increase in speed may cause the ether to boil at a dangerous rate.

4. The ethereal solution remaining from the alkaline extractions contains unreacted propiophenone which may be recovered by distilling the ether and fractionating the residue. The amount of recovered propiophenone, collected at 210–216°, varies from 80 to 110 g.

5. About 25–30 g. of material may be recovered from the toluene mother liquor by extraction with alkali and reprecipitation with acid.

3. Methods of Preparation

Isonitrosopropiophenone has been prepared from esters of α-benzoylpropionic acid by a process involving saponification, nitrosation, and decarboxylation;[1] from 1-phenyl-1,2-propanedione by the action of hydroxylamine;[2] and from propiophenone by treatment with amyl nitrite,[3] methyl nitrite,[4] or butyl nitrite.[5]

[1] v. Pechmann and Müller, Ber. **21**, 2119 (1888).
[2] Kolb, Ann. **291**, 292 (1896).
[3] Claisen and Manasse, Ber. **22**, 529 (1889).
[4] Slater, J. Chem. Soc. **117**, 590 (1920).
[5] Hartung and Munch, J. Am. Chem. Soc. **51**, 2264 (1929).

ISOPROPYL LACTATE

(Lactic acid, isopropyl ester)

$$CH_3CHOHCO_2H + (CH_3)_2CHOH \xrightarrow{(H_2SO_4)}$$
$$CH_3CHOHCO_2CH(CH_3)_2 + H_2O$$

Submitted by F. A. McDermott.
Checked by C. R. Noller.

1. Procedure

In a 3-l. round-bottomed flask (Note 1) fitted with a 1-meter fractionating column [1] are placed 450 g. (7.5 moles) of anhydrous isopropyl alcohol (Note 2), 212 g. (2 moles) of u.s.p. 85 per cent lactic acid, 1 l. of benzene, and 5 cc. of concentrated sulfuric acid. The flask is placed in an air bath on an electric hot plate (Note 3) and heated until the benzene-isopropyl alcohol-water ternary mixture distils at 66.5°. Distillation is continued slowly (six to seven hours) until the temperature at the head of the column rises to and persists at 71–72° (isopropyl alcohol-benzene binary mixture), and no further separation of water occurs. Ten grams of precipitated calcium carbonate is then added to the mixture, and distillation is continued until the temperature rises to 80° in order to remove most of the benzene and excess isopropyl alcohol (Notes 4 and 5). The contents of the flask are then filtered into a modified Claisen flask and distilled under reduced pressure. Cuts are taken to 60°, 60–75°, 75–80°, and 80–100° at 32 mm. The fraction boiling at 75–80°/32 mm. is isopropyl lactate and weighs 130–160 g. By redistilling the high and low fractions an additional 30–60 g. is obtained, bringing the total yield to 160–180 g. (60–68 per cent of the theoretical amount). The ester may be redistilled at atmospheric pressure (with some loss due to decomposition) at 166–168°.

2. Notes

1. With larger amounts of material it is desirable to employ a two-necked flask; the spare neck is used for introducing the calcium carbonate later in the process.

2. Commercial anhydrous alcohol was used in this preparation. Isopropyl alcohol is very difficult to dry satisfactorily. The water binary mixture, boiling at 80.35°, contains 12.1 per cent of water by weight.

[1] Clarke and Rahrs, Ind. Eng. Chem. **15**, 349 (1923).

The ternary mixture with benzene, boiling at 66.5°, contains 73.8 per cent benzene, 18.7 per cent isopropyl alcohol, and 7.5 per cent water. Hence by adding 120 g. of dry benzene to 100 g. of the isopropyl-water binary mixture, and distilling until the temperature reaches 82°, there will remain 55 to 60 g. of nearly dry isopropyl alcohol.

3. A water or steam bath or oil bath may be used.

4. The temperature of the vapors should not be allowed to rise above 72° before the addition of the calcium carbonate. If too much alcohol is removed before the acid is neutralized, charring and resinification take place with a decrease in the yield of ester.

5. The recovered benzene and excess isopropyl alcohol may be dried by distillation and used in a subsequent run.

3. Methods of Preparation

Isopropyl lactate has been prepared by heating isopropyl alcohol and lactic acid in a sealed tube at 170°,[2] and from silver lactate and isopropyl iodide, together with the isopropyl ester of α-isopropoxypropionic acid.[3] Direct esterification of lactic acid with isopropyl alcohol, using sulfuric acid, has hitherto given less than a 20 per cent yield of impure ester.

ISOPROPYL THIOCYANATE

(Thiocyanic acid, isopropyl ester)

$$(CH_3)_2CHBr + NaSCN \rightarrow (CH_3)_2CHSCN + NaBr$$

Submitted by R. L. Shriner.
Checked by C. R. Noller.

1. Procedure

In a 3-l. round-bottomed flask, fitted with a very efficient mechanical stirrer (Note 1), a reflux condenser, and a 500-cc. separatory funnel, are placed 445 g. (5.5 moles) of sodium thiocyanate (Note 2) and 1250 cc. of 90 per cent ethyl alcohol. The stirrer is started and the mixture is heated to boiling. Then 615 g. (5 moles) of isopropyl bromide (p. 359; Org. Syn. Coll. Vol. I, **1941**, 37) is added slowly during the course of one hour. The mixture is refluxed with stirring for six hours. At the end of this time the precipitated sodium bromide is removed by filtration and

[2] Silva, Bull. soc. chim. (2) **17**, 97 (1872).
[3] Purdie and Lander, J. Chem. Soc. **73**, 298 (1898).

washed with 250 cc. of 95 per cent alcohol. As much of the alcohol as possible is then removed by distillation on the steam bath. To the residue in the flask is added 500 cc. of water, and the upper layer of isopropyl thiocyanate is separated. The aqueous layer is extracted with two 100-cc. portions of ether (Note 3). The ether extracts are added to the crude thiocyanate, and the combined product is dried over anhydrous sodium sulfate (Note 4). The dried material is fractionated twice from a modified Claisen flask with a 25-cm. fractionating column. The following fractions are collected: up to 60°; 60–100°; 100–130°; 130–146°; and 146–151°. The last fraction contains the pure product. The yield is 320–345 g. (63–68 per cent of the theoretical amount). By redistilling the alcohol that was removed on the steam bath through an efficient fractionating column (Note 5) until all the alcohol is removed (Note 6), separating the water, and distilling, there is obtained an additional 55–65 g. of product boiling at 146–151°. The total yield is 385–400 g. (76–79 per cent of the theoretical amount). On redistillation of the combined fractions boiling at 146–151°, practically the entire amount distils at 149–151°.

2. Notes

1. A vigorous mechanical stirrer must be used to prevent the precipitated sodium bromide from settling to the bottom and causing bumping.

2. A technical grade of sodium thiocyanate was used. Potassium thiocyanate does not possess any advantages over the sodium salt.

3. If benzene is used to extract the aqueous layer, three fractionations are necessary to obtain the same yields.

4. The sodium sulfate does not remove the water entirely, and in the subsequent fractionation the water layer should be removed by means of a separatory funnel wherever it appears.

5. An eight-bubbler fractionating column of the type described by Clarke and Rahrs [1] was used.

6. Distillation was continued until water began to appear in the lowest bubbler.

3. Method of Preparation

Isopropyl thiocyanate has been prepared by the action of isopropyl iodide on potassium thiocyanate.[2]

[1] Clarke and Rahrs, Ind. Eng. Chem. **18**, 1092 (1926).
[2] Henry, Ber. **2**, 496 (1869); Gerlich, Ann. **178**, 80 (1875).

ITACONIC ANHYDRIDE AND ITACONIC ACID

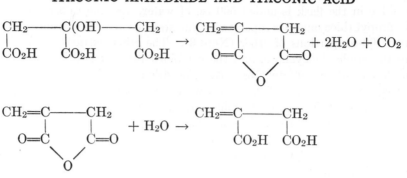

Submitted by R. L. SHRINER, S. G. FORD, and L. J. ROLL.
Checked by C. R. NOLLER.

1. Procedure

(A) *Itaconic Anhydride.*—A 500-cc. Pyrex Kjeldahl flask is fitted with an outlet tube 12 mm. in diameter bent for downward distillation and attached to a 100-cm. water-cooled condenser having an indented Pyrex inner tube (Note 1). Two 250-cc. long-necked distilling flasks, cooled in ice-water baths, are used in series as receivers; the vapors are led to the center of each flask by an adapter and glass tubing.

Two hundred grams (0.95 mole) of U.S.P. citric acid monohydrate is placed in the Kjeldahl flask (Note 2) and heated with a free flame until melted. Then the flask is heated very rapidly with a Meker burner, and the distillation is completed as quickly as possible (ten to twelve minutes). Superheating must be avoided (Note 3). The distillate consists of water and itaconic anhydride, most of which distils at 175–190°. The distillate is immediately poured into a separatory funnel and the lower layer of itaconic anhydride is separated (Note 4). The yield of the anhydride is 40–50 g. (37–47 per cent of the theoretical amount). It is of sufficient purity for use in the preparation of citraconic anhydride (p. 140) (Notes 5 and 6).

(B) *Itaconic Acid.*—Forty grams of itaconic anhydride is refluxed with 100 cc. of water for one hour. The flask is then set aside to cool, and finally placed in an ice bath. The acid crystallizes and is filtered and dried. The yield is 11–18 g. (24–39 per cent of the theoretical amount) of a product which melts at 162–165°. On concentrating the mother liquor to one-third of the original volume, an additional amount of lower-melting product may be obtained (Notes 7 and 8).

2. Notes

1. A Pyrex inner tube is recommended, since the rapid stream of hot vapor often cracks soft glass tubes. A suitable tube may be prepared from a piece of ordinary Pyrex tubing of the proper size by softening it in spots in the blast lamp and applying suction.

2. A clean flask should be used for each run since the presence of residue from a previous run causes excessive foaming during the first part of the decomposition. The flask is cleaned most easily by adding a 25 per cent solution of sodium hydroxide while the residue at the bottom is still molten. Further heating brings about complete solution.

3. Superheating tends to increase the rearrangement to citraconic anhydride. The flask should be heated on all sides over a considerable area, and the distillation should be stopped as soon as the vapors in the reaction flask become yellow.

4. A prompt separation of the anhydride and water layers minimizes hydrolysis of the anhydride. The water layer may be concentrated to give a mixture of itaconic and citraconic acids.

5. The purity of the itaconic anhydride seems to vary greatly with the conditions of the experiment. The crude anhydride always deposits crystals of itaconic acid on standing, probably due to water dissolved or suspended in the anhydride. Some idea of the purity can be obtained by the quantity of itaconic acid that is obtained from it. If the distillation proceeds at exactly the right rate, the anhydride is pure and melts at 67–68°.

6. If larger amounts of itaconic anhydride are desired, it is better to pyrolyze several 200-g. portions of citric acid than a single large portion; the percentage yield of itaconic anhydride usually decreases with larger runs.

7. In some runs no itaconic acid crystallizes. This apparently happens when the distillation of the citric acid has not been carried out rapidly enough and the itaconic anhydride contains a large amount of citraconic anhydride.

8. For preparing considerable amounts of itaconic acid, the following procedure is more convenient than that given above and the yields are much greater.

Nine 120-g. portions of citric acid are distilled rapidly (four to six minutes), using 300-cc. Kjeldahl flasks, and all the distillates are collected in the same receiver. The distillate, which generally does not consist of two layers, is placed in an evaporating dish, 50 cc. of water is added, and the mixture is allowed to stand on a steam bath for three hours. On cooling it sets to a semi-solid mass; this is filtered and washed with 150 cc. of water. The residue consists of 138 g. of perfectly white crystals melting at 165°. By concentrating the filtrate an additional 42 g. of product melting at 157–165° is obtained. The total yield is 26–27 per cent of the theoretical amount. (C. V. WILSON and C. F. H. ALLEN, private communication.)

3. Methods of Preparation

Itaconic anhydride has been made by heating itaconic acid and by distillation of citric acid.[1]

Itaconic acid has been prepared by the distillation of citric acid,[2] of aconitic acid,[3] and of itamalic acid;[4] by heating citric acid with dilute sulfuric acid in a closed tube;[5] by treating aconitic acid with water at 180°;[6] by heating citraconic acid with sodium hydroxide;[7] by heating citraconic anhydride with water at 150°;[8] by heating a concentrated solution of citraconic acid at 120–130° in a sealed tube;[9] and by the action of the fungus *Aspergillus itaconicus* on cane sugar.[10]

A mixture of citraconic and itaconic acids is obtained by flowing a concentrated aqueous solution of citric acid into a heated evacuated vessel, distilling under reduced pressure the mixture of anhydrides formed, and allowing the mixture to react with water.[11]

[1] Anschütz, Ber. **13**, 1541 (1880).
[2] Baup, Ann. **19**, 29 (1836).
[3] Crasso, ibid. **34**, 63 (1840).
[4] Swarts, Zeit. für Chem. **1867**, 649.
[5] Markownikow and von Purgold, ibid. **1867**, 264.
[6] Pebal, Ann. **98**, 94 (1856).
[7] Delisle, ibid. **269**, 86 (1892).
[8] Fittig and Landolt, ibid. **188**, 72 (1877).
[9] Wilm, ibid. **141**, 29 (1867).
[10] Kinoshita, Acta Phytochim. **5**, 271 (1931) [C. A. **26**, 966 (1932)].
[11] Boehringer Sohn A.-G., Brit. pat. 452,460 [C. A. **31**, 1045 (1937)].

2-KETOHEXAMETHYLENIMINE

(Hexamethylenimine, 2-oxo-)

$$(CH_2)_5C=NOH \xrightarrow{(H_2SO_4)} (CH_2)_5\begin{matrix} NH \\ | \\ CO \end{matrix}$$

Submitted by C. S. MARVEL and J. C. ECK.
Checked by LOUIS F. FIESER and C. H. FISHER.

1. Procedure

THE rearrangement of 120 g. (1.06 moles) of pure cyclohexanone oxime (Note 1) is carried out in twelve 10-g. portions according to the procedure given in the first eight lines of (B), p. 77.

The acid solutions of ε-caprolactam from the twelve runs are combined in a 3-l. round-bottomed flask which is fitted with a mechanical stirrer and a separatory funnel and packed in an ice-salt mixture. The solution is cooled to 0° and carefully made faintly alkaline to litmus by the addition of 24 per cent potassium hydroxide solution, added very slowly (five to six hours) with good cooling (Note 2). Usually about 1550 cc. of the alkaline solution is needed. The temperature must be kept below 10° to avoid hydrolysis during this stage of the preparation.

The potassium sulfate which has separated is then removed by filtration and washed with two 100-cc. portions of chloroform. The faintly alkaline aqueous solution is extracted with about five 200-cc. portions (Note 3) of chloroform, and the combined chloroform solutions are washed once with 50 cc. of water to remove any alkali. The chloroform is then distilled and the product fractionated under reduced pressure. The yield of 2-ketohexamethylenimine, boiling at 127–133°/7 mm. and melting at 65–68°, amounts to 71–78 g. (59–65 per cent of the theoretical amount) (Note 4).

2. Notes

1. Pure cyclohexanone oxime, m.p. 86–88° (pp. 76 and 314), must be used since poorer grades char badly when treated with sulfuric acid.

2. Potassium hydroxide gives better results than sodium hydroxide, since the large amount of hydrated sodium sulfate which separates from the solution if sodium hydroxide is used prevents efficient cooling.

3. The extraction is continued until no appreciable amount of product is obtained in the chloroform layer.

4. The boiling point is reported in the literature as 139–140°/12 mm.; the melting point is reported at various temperatures from 65° to 70°.

3. Methods of Preparation

2-Ketohexamethylenimine, ε-caprolactam, has been obtained by heating ε-aminocaproic acid or its ethyl ester [1] and by the rearrangement of cyclohexanone oxime.[2,3,4] The method [4] described above is Ruzicka's [3] modification of Wallach's [2] original directions for the rearrangement of the oxime. This rearrangement can also be run as a continuous process.[5]

LAURYL ALCOHOL *

(Dodecyl alcohol)

$$C_{11}H_{23}CO_2C_2H_5 \xrightarrow{\text{(Na + C}_2\text{H}_5\text{OH)}} C_{11}H_{23}CH_2OH$$

Submitted by S. G. FORD and C. S. MARVEL.
Checked by FRANK C. WHITMORE and D. J. LODER.

1. Procedure

THE central neck of a 5-l. three-necked round-bottomed flask is fitted with a stopper carrying a mercury-sealed mechanical stirrer. One of the side necks is connected by means of a short piece of heavy rubber tubing to a large reflux condenser about 2 m. long, with an inner tube 2.5 cm. in diameter (Note 1). The third neck is fitted with a separatory funnel.

In the flask are placed 70 g. (3 moles) of sodium and 200 cc. of dry toluene (Note 2). The flask is heated in an oil bath until the sodium is melted. The stirrer is then started; when the sodium is finely divided, the oil bath is removed and the mixture allowed to cool. Stirring must be continued during the cooling in order to keep the sodium finely divided. When the mixture has cooled to about 60°, there are added from the

[1] Gabriel and Maass, Ber. **32**, 1271 (1899); von Braun, ibid. **40**, 1840 (1907); Carothers and Berchet, J. Am. Chem. Soc. **52**, 5289 (1930).

[2] Wallach, Ann. **312**, 187 (1900); **343**, 43 (1905).

[3] Ruzicka, Seidel, and Hugoson, Helv. Chim. Acta **4**, 477 (1921).

[4] Eck and Marvel, J. Biol. Chem. **106**, 387 (1934).

[5] E. I. du Pont de Nemours and Company, U. S. pat. 2,234,566 [C. A. **35**, 3650 (1941)]; I. G. Farbenind. A.-G., U. S. pat. 2,249,177 [C. A. **35**, 6599 (1941)].

* Commercially available; see p. v.

separatory funnel, first, a solution of 114 g. (0.5 mole) of ethyl laurate (Note 3) in 150 cc. of absolute alcohol (Note 4), then 500 cc. more of alcohol, as rapidly as is possible (Note 5) without loss of material through the condenser. The time required for the addition of the ester solution and the alcohol is less than five minutes, usually two or three minutes. When the reaction has subsided, the flask is heated on a steam bath until the sodium is completely dissolved (Note 6). The mixture is then steam-distilled to remove the toluene and ethyl alcohol. The contents of the flask are transferred to a separatory funnel while still hot and washed three times with 200-cc. portions of hot water to remove the sodium laurate (Note 7). The lauryl alcohol is extracted with ether from the cooled mixture and the washings. The combined ether extracts are washed with water, sodium carbonate solution, and again with water, and dried over anhydrous magnesium sulfate. The ether is evaporated and the lauryl alcohol distilled under diminished pressure. The yield is 60–70 g. (65–75 per cent of the theoretical amount) of a product boiling at 143–146°/18 mm. or 198–200°/135 mm. (Note 8).

2. Notes

1. The reaction is very vigorous and, unless the condenser has a wide bore, finely divided sodium may be forced out the top and bad fires may result. The inner tube of the condenser may advantageously be made of brass or copper.

2. The toluene is dried by distillation; the first 10 per cent is discarded and the remainder is stored over sodium until used.

3. The ethyl laurate used was prepared by the alcoholysis of cocoanut oil and fractionation of the resulting esters. The material boiled at 127–132°/5 mm. See Org. Syn. **20,** 69, and Organic Chemical Reagents III, Univ. Illinois Bull. **19** (6) 62 (1921).

4. The grade of absolute alcohol used in the reduction is very important. Alcohol dried with magnesium methoxide (Org. Syn. Coll. Vol. I, **1941,** 249) was used in this preparation. Alcohol dried over lime gives very low yields.

5. The best yields are obtained when the reductions are carried out rapidly. If the reaction seems to be about to get out of control, the stirrer is stopped and the mixture is cooled with an ice pack.

6. When several reductions are being made, time is saved by transferring the mixture at this point to another flask, thus having the original apparatus ready for another reduction.

7. Unless the sodium laurate is carefully removed, it causes trouble-some emulsions.

8. Ethyl undecylenate has been reduced to undecylenyl alcohol (b.p. 123–125°/6 mm.) in 70 per cent yields; ethyl myristate to myristyl alcohol (b.p. 170–173°/20 mm.; m.p. 39–39.5°) in 70–80 per cent yields; ethyl palmitate to cetyl alcohol (b.p. 178–182°/12 mm.; m.p. 48.5–49.5°) in 70–78 per cent yields by this same procedure.

3. Methods of Preparation

Lauryl alcohol has been prepared by the reduction of the aldehyde;[1] by the reduction of esters of lauric acid with sodium and absolute alcohol[2] or with sodium, liquid ammonia and absolute alcohol,[3] or catalytically;[4] and by the reduction of lauramide with sodium and amyl alcohol.[5] The method in the above procedure is essentially that described by Levene and Allen.[2]

dl-LYSINE HYDROCHLORIDES

$$C_6H_5CONH(CH_2)_4CHCO_2H \xrightarrow{NH_3} C_6H_5CONH(CH_2)_4CHCO_2H$$
$$\qquad\qquad | \qquad\qquad\qquad\qquad\qquad\qquad\qquad | $$
$$\qquad\qquad Br \qquad\qquad\qquad\qquad\qquad\qquad\qquad NH_2$$

$$\xrightarrow{H_2O + HCl} ClNH_3(CH_2)_4CHCO_2H \xrightarrow{Pyridine} ClNH_3(CH_2)_4CHCO_2H$$
$$\qquad\qquad\qquad\qquad | \qquad\qquad\qquad\qquad\qquad\qquad\qquad | $$
$$\qquad\qquad\qquad\qquad NH_3Cl \qquad\qquad\qquad\qquad\qquad\qquad NH_2$$

Submitted by J. C. Eck and C. S. Marvel.
Checked by C. R. Noller and William Munich.

1. Procedure

(A) *dl-ε-Benzoyllysine.*—A solution of 180 g. (0.57 mole) of ε-benzoyl-amino-α-bromocaproic acid (p. 74) in 2 l. of aqueous ammonia (sp. gr. 0.9) is filtered into a 5-l. flask and allowed to stand for two days. Any

[1] Krafft, Ber. 16, 1718 (1883); Sivkov and Novikova, J. Applied Chem. (U.S.S.R.) 13, 1272 (1940) [C. A. 35, 2110 (1941)].

[2] Bouveault and Blanc, Bull. soc. chim. (3) 31, 674 (1904); Ger. pat. 164,294 [Frdl. 8, 1260 (1905–07)]; Levene and Allen, J. Biol. Chem. 27, 443 (1916); Marvel and Tanenbaum, J. Am. Chem. Soc. 44, 2649 (1922); Adams and Marvel, Org. Chem. Reagents IV, Univ. Illinois Bull. 20 (8) 54 (1922).

[3] Chablay, Compt. rend. 156, 1021 (1913); Ann. chim. (9) 8, 215 (1917).

[4] Adkins and Folkers, J. Am. Chem. Soc. 53, 1095 (1931); Böhme A.-G., Brit. pat. 356,606 [C. A. 26, 5573 (1932)].

[5] Scheuble and Loebl, Monatsh. 25, 348 (1904).

crystals which have formed at the end of this time are filtered, and the filtrate is evaporated on a steam bath at reduced pressure to about 1 l. The crystals are filtered, combined with the first crop, and washed with 100 cc. of alcohol and finally with 100 cc. of ether. The aqueous filtrate is evaporated under reduced pressure to dryness; the residue is washed with two 100-cc. portions of water to remove the ammonium bromide, and then with 50 cc. of alcohol, followed by 50 cc. of ether. The total yield of ε-benzoyllysine, melting at 265–270°, is 100–116 g. (70–81 per cent of the theoretical amount).

(B) *dl-Lysine Dihydrochloride.*—A solution of 100 g. (0.4 mole) of benzoyllysine in a mixture of 600 cc. of hydrochloric acid (sp. gr. 1.18) and 400 cc. of water is boiled under a reflux condenser for ten hours. The mixture is cooled and the benzoic acid removed by filtration. The filtrate is evaporated on a water bath under reduced pressure until a thick syrup remains. The syrup is transferred to a 1.5-l. beaker by means of four volumes (about 400 cc.) of hot absolute alcohol and filtered if necessary. The solution is cooled to 15–20°, and 500 cc. of ether is added slowly with stirring. The precipitate, after filtering and drying, melts at 187–189° and weighs 67–75 g. (76–85 per cent of the theoretical amount); it is analytically pure lysine dihydrochloride (Note 1).

(C) *dl-Lysine Monohydrochloride.*—To a solution of 55 g. (0.25 mole) of lysine dihydrochloride in 1 l. of boiling 95 per cent alcohol (Note 2) is added, with stirring, a solution of 25 g. (0.32 mole) of pyridine in 40 cc. of hot 95 per cent alcohol. The white, crystalline monohydrochloride separates immediately. After cooling overnight in a refrigerator the solid is filtered and washed with two 50-cc. portions of cold absolute alcohol. After drying, the product melts at 260–263° and weighs 42–43 g. (91–94 per cent of the theoretical amount).

For further purification to remove any pyridine hydrochloride, the above product is dissolved in 85 cc. of boiling water, and 650 cc. of boiling 95 per cent alcohol added with stirring. After cooling overnight in the refrigerator the solid is filtered, and washed with one 20-cc. portion of cold absolute alcohol. There is obtained 40–42 g. (95–97 per cent recovery) of monohydrochloride melting at 263–264° (corr.).

2. Notes

1. If a product of lower melting point is obtained, it may be purified by dissolving in 1 l. of hot 95 per cent alcohol, filtering if necessary, cooling, and, without removing any material that may have crystallized, adding slowly with stirring 1.5 l. of ether. If the product separates as an oil it will soon crystallize on standing. The checkers found that one lot

of 75 g. melting at 173–178° when treated in this way gave 67 g. (89 per cent recovery) melting at 187–189°.

2. If the solution is not clear, it should be filtered before the addition of pyridine.

3. Method of Preparation

The above procedure for dl-ϵ-benzoyllysine and dl-lysine hydrochloride is a modification [1] of that published by Braun. [2]

MALONIC ACID

$$ClCH_2CO_2H \xrightarrow{Na_2CO_3} ClCH_2CO_2Na \xrightarrow{NaCN} CNCH_2CO_2Na \rightarrow$$

$$\xrightarrow[H_2O]{NaOH} CH_2(CO_2Na)_2 \xrightarrow{CaCl_2} CH_2(CO_2)_2Ca \xrightarrow{HCl} CH_2(CO_2H)_2$$

Submitted by NATHAN WEINER.
Checked by C. R. NOLLER and M. E. SYNERHOLM.

1. Procedure

In a 5-l. round-bottomed flask, 500 g. (5.3 moles) of chloroacetic acid (Note 1) is dissolved in 700 cc. of water. The solution is warmed to 50°, neutralized with 290 g. (2.7 moles) of anhydrous sodium carbonate, and again cooled to room temperature. Meanwhile, 294 g. (6 moles) of sodium cyanide (97 per cent) is dissolved in 750 cc. of water warmed to 55°; the solution is cooled to room temperature and then added to the sodium chloroacetate solution, with rapid mixing of the two solutions and cooling under the water tap. When the solutions are completely mixed, cooling is stopped and the temperature allowed to rise. When it reaches 95°, the solution is cooled by adding 200 cc. of ice water, and this is repeated, if necessary, until the temperature no longer rises (Note 2). The solution is then heated on the steam bath for one hour to ensure completion of the reaction.

At the end of this time, the solution is cooled to room temperature and 240 g. (6 moles) of solid U.S.P. sodium hydroxide is slowly dissolved in it. When solution is complete, the reaction mixture is again heated on the steam bath under a hood. When the temperature reaches 60–70°, evolution of ammonia begins and becomes more vigorous with rise in

[1] Eck and Marvel, J. Biol. Chem. **106**, 387 (1934).
[2] Braun, Ber. **42**, 844 (1909).

temperature. Most of the ammonia is evolved in forty-five minutes, but the solution is heated for at least three hours, and the last traces of ammonia are removed by bubbling steam through the hot solution for forty-five to sixty minutes more.

A solution of 600 g. of anhydrous calcium chloride in 1.8 l. of water warmed to 40° is added slowly with rapid mixing to the hot sodium malonate solution. A cheese-like precipitate of calcium malonate is formed immediately and becomes coarsely crystalline on standing for twenty-four hours. After the supernatant solution is decanted, the calcium malonate is washed by decantation four or five times with 500-cc. portions of cold water. It is then transferred to a filter, sucked as dry as possible, and dried in the air, or at 45–50°, to constant weight. The yield is 800–900 g.

The dry calcium malonate is placed in a 3-l. round-bottomed flask with sufficient (750–1000 cc.) alcohol-free ether (Note 3) to make a paste which can be stirred. The flask is surrounded by an ice bath, and the well-stirred salt is treated with 1 cc. of 12 N hydrochloric acid for each gram of salt. After the acid has been added slowly through a dropping funnel, the solution is transferred to a continuous extractor (Note 4) and extracted with ether until no more malonic acid is obtained. The product, as obtained from the undried ether solution by concentration, filtration, and drying in the air, melts at 130° or higher and is sufficiently pure for most purposes. The yield is 415–440 g. (75–80 per cent of the theoretical amount).

2. Notes

1. A freshly distilled product boiling over a 3° range was used.

2. If the reaction between the cyanide and the chloroacetate becomes too vigorous, hydrogen cyanide is liberated and partly changed to a brown material, and a corresponding quantity of glycolate is formed. If the temperature of the reaction mixture is allowed to go above 95° spontaneously, the liquid may boil so vigorously and suddenly as to escape from the flask despite the large extra volume provided.

3. Ether is used to avoid unnecessarily increasing the volume of aqueous solution to be extracted. This ether may be used for further extraction. It is necessary to use alcohol-free ether to avoid esterifying the malonic acid during the protracted extraction period.

4. A convenient type of extractor used in this preparation was made as follows by modifying that described by J. Friedrichs:[1] A 20-cm. calcium chloride tower, or other narrow-necked cylinder with a volume of

[1] Friedrichs, Chem. Fabrik **1,** 91 (1928).

about 1.3 l., was used as an extraction chamber. The mantle-tube, conducting the ether vapors to an Allihn condenser, was made of 25-mm. tubing and was about 50 cm. long. The goose-neck to the extraction flask, of 14-mm. tubing, was sealed to the mantle-tube about 8 cm. from the bottom end. The inner tube was of 14-mm. tubing, about 65 cm. long, flanged at the top to a diameter of about 20 mm. A Witt filter plate of the proper diameter may be sealed into the bottom of the tube to make the ether pass up through the water in a stream of fine bubbles, or this can also be accomplished by sealing off the bottom of the tube and piercing it with 3–6 pinholes. The mantle-tube was fitted to the chamber by a properly bored rubber stopper, the condenser to the top of the mantle-tube, and the 500-cc. extraction flask to the goose-neck by charred cork stoppers. With this apparatus 395–400 g. of malonic acid was extracted in seventy-two hours, the ether being changed every twenty-four hours, and the final traces were extracted after an additional twenty-four hours. The extractor as described by Friedrichs is shown in Fig. 12.

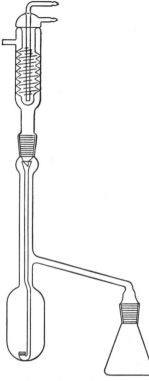

Fig. 12.

3. Methods of Preparation

Malonic acid has been prepared by the hydrolysis of malononitrile with concentrated hydrochloric acid;[2] by the hydration of carbon suboxide;[3] and from an alkali cyanide and ethyl bromoacetate,[4] ethyl chloroacetate,[5] or chloroacetic acid[6] followed by hydrolysis. The preparation using sodium cyanide and chloroacetic acid is the most convenient and economical.

[2] Henry, Compt. rend. 102, 1396 (1886).

[3] Diels and Wolf, Ber. 39, 696 (1906).

[4] Franchimont, ibid. 7, 216 (1874).

[5] Kolbe, Ann. 131, 349 (1864); Müller, ibid. 131, 352 (1864); Petriev, J. Russ. Phys.-Chem. Soc. 10, 64 (1878).

[6] v. Miller, J. prakt. Chem. (2) 19, 326 (1879); Grimaux and Tcherniak, Bull. soc. chim. (2) 31, 338 (1879); Bourgoin, ibid. (2) 33, 574 (1880); Conrad, Ann. 204, 126 (1880).

MALONONITRILE

$$CNCH_2CONH_2 + PCl_5 \rightarrow CH_2(CN)_2 + POCl_3 + 2HCl$$

$$2CNCH_2CONH_2 + POCl_3 \rightarrow 2CH_2(CN)_2 + HPO_3 + 3HCl$$

Submitted by B. B. CORSON, R. W. SCOTT, and C. E. VOSE.
Checked by C. S. MARVEL and J. HARMON.

1. Procedure

ONE HUNDRED FIFTY grams (1.8 moles) of pure cyanoacetamide (Org. Syn. Coll. Vol. I, **1941**, 179) is thoroughly mixed in a large (20-cm.) mortar with 150 g. (0.7 mole) of phosphorus pentachloride (Note 1). This mixture is transferred (Note 2) as quickly as possible to a 1-l. Claisen flask equipped with a 360° thermometer and an air-intake tube (Note 3). The Claisen flask is connected by means of a double-length air condenser to a 250-cc. filter flask which in turn is connected to a water pump (Note 4) through a manometer.

After the system has been evacuated to about 30 mm. of mercury, the Claisen flask is immersed in a boiling water bath. The mixture melts and its color deepens to orange. Boiling commences in about fifteen minutes, before the solid is completely melted, and the pressure rises to about 150 mm. owing to the liberation of hydrogen chloride and phosphorus oxychloride (Note 5). When the evolution of gas has slackened, as indicated by the fall of pressure and the less vigorous boiling of the reaction mixture (thirty to thirty-five minutes), the receiver is changed. The distilling flask is removed from the boiling water bath, wiped dry, and immersed in an oil bath at 140° (Note 6); the fresh receiver is placed in ice water.

The malononitrile begins to distil at 113°/30 mm. (125°/50 mm.). The temperature of the oil bath is slowly raised over a period of twenty-five minutes to 180° (Note 7) and the nitrile collected between 113° and 125°. When the distillation has almost ceased, the oil bath is removed so as to prevent discoloration of the product (Note 8). The yield of crude nitrile is 80–95 g. (67–80 per cent of the theoretical amount) (Note 9). The nitrile may be purified by vacuum distillation (Note 10) with about 10 per cent loss; it is collected between 113° and 120°/30 mm. One distillation yields a water-clear liquid, which quickly freezes to an ice-like solid (Note 11) melting at 28–30°. The product has a faint odor resembling that of acetamide.

2. Notes

1. The weights of the reactants correspond to the ratio of 5 molecules of amide to 2 molecules of pentachloride. The use of larger proportions of phosphorus pentachloride leads to lower yields.[1]
The phosphorus pentachloride should be of good quality—a dry solid.

2. This transfer is greatly facilitated by the use of a 15-cm. (6-inch) glass funnel whose stem has been cut off and the hole enlarged to almost the size of the neck of the Claisen flask. The mixture is emptied into the body of the funnel and pushed through the large hole by means of a glass rod. A gas mask should be worn, for any considerable breathing of the fumes brings on a harsh cough which may last for several days.

3. To avoid plugging, the air-intake tube is not drawn out to a capillary.

4. An oil pump cannot be used because of the corrosive acid fumes.
It is reported that a single water pump is usually unable to remove the evolved gases fast enough, with the result that some material is carried over mechanically with the foam. If two water pumps in parallel are used, this difficulty is overcome. (C. F. H. ALLEN, private communication.)

5. Unless the phosphorus oxychloride is to be recovered, it is advisable to allow it to be carried away by the suction. This is easily accomplished if the receiver is not cooled.

6. If the initial temperature of the oil bath is below 140°, the nitrile distils too slowly (with consequent lowering of yield); if the temperature is above 140°, the nitrile distils too rapidly, some condenses in the rubber tubes leading to the pump and clogs the line.
The flask should be immersed as far as possible in the oil (to within about 100 mm. of the top of the flask). A 1-gal. enamelware coffee-pot makes an excellent oil bath. An ordinary soldered vessel will not stand the temperature.

7. The temperature of the oil bath must not rise much above 180°, else the distillate becomes colored and violent bumping is liable to occur. The distillate should not be allowed to solidify in the condenser. This is most likely to occur during the first part of the distillation. The solid is easily melted by heating with a small flame.

8. The residue in the Claisen flask can be softened by water and broken up with a rod. The small amount remaining is easily removed with concentrated nitric or, preferably, concentrated sulfuric acid.

9. The keynote to success in this preparation is speed. Malononitrile is a sensitive substance and it must be removed from the reaction mix-

[1] Hesse, Am. Chem. J. **18,** 726 (1896).

ture as rapidly as possible. After experience with one or two runs the yields will consistently be better than 67 per cent.

10. Malononitrile can be distilled in small amounts (50 cc. or so) at ordinary pressure; the boiling point is around 220°. However, the longer heating necessary with larger amounts is likely to cause violent decomposition: the liquid darkens, boils spontaneously, and finally spurts from the flask in a cloud of white fumes and burning liquid; the latter partially solidifies to a brittle red solid.

11. Nitrogen determinations indicate a purity of 99.5 per cent. The product remains colorless for at least one year if stored in brown bottles protected from the light. If kept in ordinary bottles and exposed to the light, the nitrile quickly darkens.

12. This work was aided by a grant from the Cyrus M. Warren Fund of the American Academy of Arts and Sciences.

3. Methods of Preparation

Malononitrile has been prepared by the action of phosphorus pentachloride on cyanoacetamide,[1] and by the action of phosphorus pentoxide on malonamide [2] or cyanoacetamide.[3] The preparation using the pentoxide is much less satisfactory than that given above.

MERCURY DI-β-NAPHTHYL

(Mercury, di-2-naphthyl-)

$$2C_{10}H_7N_2Cl \cdot HgCl_2 + 6Cu \rightarrow (C_{10}H_7)_2Hg + 2N_2 + 6CuCl + Hg$$

Submitted by A. N. NESMAJANOW and E. D. KOHN.
Checked by FRANK C. WHITMORE and R. W. BEATTIE.

1. Procedure

IN a 2-l. round-bottomed flask, equipped with a stirrer, are placed 231 g. (0.5 mole) of the addition compound of β-naphthalenediazonium chloride and mercuric chloride (p. 432), 700 cc. of acetone (b.p. 55–57°) and 189 g. (3 moles) of copper powder (Note 1). The mixture is quickly cooled to 20° and stirred for one hour. Seven hundred cubic centimeters of concentrated aqueous ammonia solution (sp. gr. 0.9) is added, mixed

[2] Henry, Compt. rend. **102**, 1394 (1886).
[3] Henry, ibid. **102**, 1395 (1886).

well, and allowed to stand overnight. The supernatant liquid is decanted; the solid is collected on a Büchner funnel and washed successively with 25-cc. portions of water, acetone, and ether. After air-drying, the crude material is recrystallized from xylene, using decolorizing carbon. The crystals thus obtained are slightly yellow (Note 2) and melt at 241.5–243.5°. The yield is 51–55 g. (45–48 per cent of the theoretical amount based on the addition compound used) (Note 3).

2. Notes

1. The same notes apply as in the preparation of β-naphthylmercuric chloride (p. 432).

2. The product from this reaction is never pure white. Colorless mercury di-β-naphthyl can be prepared in good yield from β-naphthylmercuric chloride (p. 432) and sodium iodide, according to the directions given in Org. Syn. Coll. Vol. I, 1941, 231, for mercury di-p-tolyl.[1]

3. Similar results are obtained with other aromatic amines; aniline and p-iodoaniline yield mercury diphenyl and mercury di-p-iodophenyl.

3. Methods of Preparation

Mercury di-β-naphthyl has been prepared by the action of sodium amalgam on β-bromonaphthalene,[2] and by the action of alcoholic sodium iodide on β-naphthylmercuric chloride.[1]

MESACONIC ACID

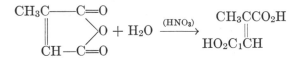

Submitted by R. L. Shriner, S. G. Ford, and L. J. Roll.
Checked by C. R. Noller.

1. Procedure

A mixture of 100 g. (0.89 mole) of citraconic anhydride (Note 1), 100 cc. of water, and 150 cc. of dilute nitric acid (1 part of concentrated

[1] Private communication, Frank C. Whitmore and R. J. Sobatzki.
[2] Chattaway, J. Chem. Soc. 65, 878 (1894); Michaelis, Ber. 27, 251 (1894).

nitric acid to 4 parts of water by volume) is evaporated in a 500-cc. Erlenmeyer flask until the appearance of red fumes (Note 2). The solution is cooled and the mesaconic acid is collected on a filter. The mother liquor is evaporated to 150 cc., cooled, and the crystalline solid which separates is collected on a filter. Further concentration of the mother liquor to 50 cc. yields more product (Note 3). The entire product is recrystallized from 100 cc. of water. The yield of mesaconic acid melting at 203–205° is 50–60 g. (43–52 per cent of the theoretical amount).

2. Notes

1. An equivalent amount of citraconic acid can be used. Directions for preparing citraconic acid and anhydride are given on p. 140.

2. It is necessary to carry the evaporation to the point where red fumes appear in order for the rearrangement to take place. The volume is usually about 250 cc.

3. The concentration of the mother liquor must be carried out in steps in order to obtain an efficient separation of mesaconic acid.

3. Methods of Preparation

Mesaconic acid has been prepared by heating citraconic acid with dilute nitric acid,[1] with hydriodic acid,[2] or with concentrated sodium hydroxide solution;[3] by heating a concentrated water solution of itaconic or citraconic acid at 180–200°;[4] by treating citradibromopyrotartaric acid and mesodibromopyrotartaric acid with potassium iodide and copper at 150°;[5] and by heating citraconic anhydride with nitric acid.[6]

[1] Gottlieb, Ann. **77,** 268 (1851).
[2] Kekulé, Ann. Spl. **2,** 94 (1862).
[3] Delisle, Ann. **269,** 82 (1892).
[4] Swarts, Jahresb. **1873,** 579.
[5] Swarts, Zeit. für Chem. **1868,** 259.
[6] Fittig and Landolt, Ann. **188,** 73 (1877).

dl-METHIONINE

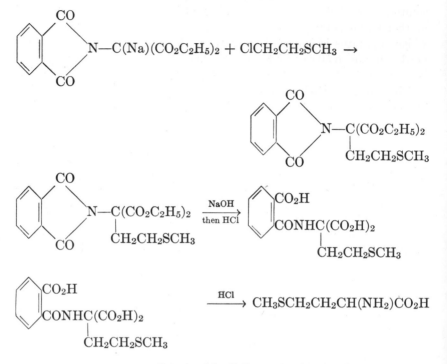

Submitted by G. BARGER and T. E. WEICHSELBAUM.
Checked by H. T. CLARKE and S. GURIN.

1. Procedure

(A) *Ethyl Sodium Phthalimidomalonate.*—To a solution of 9.2 g. (0.4 gram atom) of sodium in 300 cc. of absolute alcohol at 60° is added, with efficient stirring, 126 g. (0.41 mole) of ethyl phthalimidomalonate (Org. Syn. Coll. Vol. I, **1941, 271**). The mixture is rapidly chilled to 0°; the crystalline product is filtered at once by suction and washed successively with two 200-cc. portions of absolute alcohol and two 200-cc. portions of ether. After first drying in a vacuum desiccator and then heating for eight hours under 15 mm. pressure in a flask suspended in an oil bath at 145–155° (Note 1), the product weighs 108–111 g. (82–85 per cent of the theoretical amount).

(B) *Ethyl 1-Methylthiol-3-phthalimidopropane-3,3-dicarboxylate.*—A mixture of 85 g. (0.26 mole) of ethyl sodium phthalimidomalonate and

43 g. (0.39 mole) of β-chloroethyl methyl sulfide (p. 136) is heated in an oil bath at 160–165° in a 1-l. three-necked flask, fitted with a condenser, a thermometer, and a stoppered glass tube for sampling. After one and a half to two hours the mixture is no longer alkaline to litmus. The excess chloroethyl methyl sulfide is distilled under reduced pressure (Note 2), the residual oil is treated with 150 cc. of warm water, and the resulting mixture is transferred to a beaker and chilled. The crystalline material is filtered by suction, washed with 100 cc. of cold water, and recrystallized from the smallest possible quantity of warm absolute alcohol. In this way 75–79 g. (76–80 per cent of the theoretical amount) of a pure product, melting at 66–67°, is obtained.

(*C*) 1-*Methylthiol-3-phthalamidopropane-3,3-dicarboxylic Acid.*—A solution of 25 g. (0.066 mole) of the above ester in 30 cc. of 95 per cent alcohol is heated on the steam bath in a 200-cc. round-bottomed flask, and 70 cc. of 5 N sodium hydroxide is added. The cloudy liquid is heated until a sample gives a clear solution on dilution with water; this occurs after about two hours. The solution is then chilled to 0° and cautiously neutralized to Congo red with 0.2 N hydrochloric acid, whereupon 75 cc. of 5 N hydrochloric acid is slowly added, the temperature being maintained at 0°. The acid separates as colorless crystals. This separation is completed by the slow addition of 60 cc. of concentrated hydrochloric acid (sp. gr. 1.19). The product is filtered by suction and washed free of salt with small quantities of ice-cold water. After drying *in vacuo*, the yield is 21.5–22 g. (95.5–98 per cent of the theoretical amount) of a product melting at 141–143°.

(*D*) *Methionine.*—A suspension of 21.5 g. (0.063 mole) of this tricarboxylic acid in 350 cc. of hot water is heated on the steam bath, and 40 cc. of concentrated hydrochloric acid (sp. gr. 1.19) is added. Carbon dioxide is immediately evolved, and the substance goes into solution. After heating for one and a half hours, 200 cc. more of concentrated hydrochloric acid is added and heating is continued for forty-five minutes longer. The solution, on cooling, deposits phthalic acid; this is filtered and washed with two 50-cc. portions of water (Note 3). The combined filtrate and washings are evaporated to dryness on the steam bath under reduced pressure, and the dry residue is dissolved in 50 cc. of hot water. The resulting solution is treated with 18 cc. of pyridine and poured into 150 cc. of hot absolute alcohol. Methionine rapidly crystallizes; after cooling it is filtered, washed with alcohol, and dried. The first crop weighs about 6.9 g. The mother liquor is evaporated to dryness; the residue is taken up in 15 cc. of hot water and poured into 50 cc. of absolute alcohol, when a further 1.3 g. is obtained. The total 8.2 g. of nearly pure methionine is suspended in 200 cc. of boiling absolute ether, filtered, and dried. In this way, 7.9–8.0 g. of methionine (84–85 per cent of the theoretical amount), melting at 279–280° (corr.), is obtained.

2. Notes

1. The ethyl sodium phthalimidomalonate crystallizes with 1.5 molecules of alcohol, which is removed only on heating above 140° *in vacuo*.

2. About 10–12 g. of a pure product can be recovered by redistilling the distillate.

3. The phthalic acid thus recovered melts at 188° and weighs about 7.8 g. (75 per cent of the theoretical amount). Unless most of the phthalic acid is removed at this point, trouble may be encountered by the separation of pyridine phthalate with the methionine.

3. Methods of Preparation

The first synthesis of methionine, by the Strecker method, gave a very low yield.[1] The procedure given above, based on directions published by the submitters,[2] has the advantage over the process of Windus and Marvel [3] of giving a much higher yield (54–60 per cent as against 13–19 per cent, based on the β-chloroethyl methyl sulfide consumed).

Methionine has also been prepared, without using β-chloroethyl methyl sulfide, from α-benzoylamino-γ-butyrolactone [4] and from α-acetyl-γ-butyrolactone.[5] The first of these syntheses proceeds through ethyl α-benzoylamino-γ-chlorobutyrate, its reaction product with methyl mercaptan, and hydrolysis of that product. The second synthesis proceeds through α-oximino-γ-butyrolactone which is reduced to α-amino-γ-butyrolactone and converted to 3,6-*bis*-(β-hydroxyethyl)-2,5-diketopiperazine. The dihydroxydiketopiperazine is converted to the corresponding dichloroethyl compound, and this, after reaction with methyl mercaptan and subsequent hydrolysis, furnishes methionine.

[1] Barger and Coyne, Biochem. J. **22,** 1417 (1928).

[2] Barger and Weichselbaum, ibid. **25,** 997 (1931).

[3] Windus and Marvel, J. Am. Chem. Soc. **52,** 2575 (1930).

[4] Hill and Robson, Biochem. J. **30,** 248 (1936).

[5] Snyder, Andreen, Cannon, and Peters, J. Am. Chem. Soc. **64,** 2082 (1942).

METHOXYACETONITRILE

(Acetonitrile, methoxy-)

$$NaCN + CH_2O + H_2O \rightarrow HOCH_2CN + NaOH$$

$$HOCH_2CN + (CH_3)_2SO_4 + NaOH$$
$$\rightarrow CH_3OCH_2CN + CH_3NaSO_4 + H_2O$$

Submitted by J. A. SCARROW and C. F. H. ALLEN.
Checked by REYNOLD C. FUSON and CHARLES F. WOODWARD.

1. Procedure

This preparation must be carried out in a hood having good suction. Methyl sulfate has a high vapor pressure in spite of its high boiling point and is very poisonous. Ammonia is a specific antidote and should be kept on hand to destroy any of the ester accidentally spilled. It is advisable to wash the hands in dilute ammonium hydroxide frequently.

In a 1-l. three-necked, round-bottomed flask, fitted with a stirrer, a thermometer for reading low temperatures (Note 1), and a dropping funnel, are placed 98 g. (2 moles) of pulverized sodium cyanide (Note 2) and 200 cc. of water. The stirrer is started, and 60 g. of paraformaldehyde (2 moles) (Note 3) is added in small quantities until the temperature rises to 20–25° and the sodium cyanide has dissolved. The flask is then surrounded by a freezing mixture, and the temperature is kept below 25° during the introduction of the remaining paraformaldehyde.

Two hundred cubic centimeters (270 g., 2.1 moles) of technical methyl sulfate is placed in the dropping funnel, and, when the temperature inside the flask has dropped to 13°, a 20- to 30-cc. portion of the sulfate is added. An exothermic reaction should set in; the ice bath is removed, if necessary, to get the reaction to start (Note 4). When the temperature begins to fall, the remainder of the methyl sulfate is admitted at such a rate as to keep the temperature at 12–15°; this takes at least twenty minutes. When the addition is complete the mixture is stirred an additional forty minutes; during this time the temperature will drop to about 5°. The stirrer is stopped, and the oily, upper layer is separated at once (Note 5). The lower, aqueous layer is returned to the flask and methylated as before with a second 200-cc. portion of methyl sulfate (Note 6).

The oily, upper layer is dried with 10 g. of anhydrous sodium sulfate (Note 7) and is distilled under diminished pressure. For this purpose an efficient fractionating column (Note 8) is used. The portion boiling below 70° at 15 mm. is mainly methoxyacetonitrile and weighs 60–70 g. The upper layer from the second methylation is treated in a similar manner; the distillate weighs 55–60 g. The residue in the distilling flask is methyl sulfate and is used in a subsequent run (Note 9).

The crude fractions are combined and distilled at atmospheric pressure through a good column; about 95 per cent distils at 118–122° as a colorless liquid. The methoxyacetonitrile so prepared weighs 100–110 g. (70–77 per cent of the theoretical amount).

2. Notes

1. A 30- to 35-cm. thermometer reading from −50 to +50° is most convenient since the graduations used are then outside the flask.

2. Potassium cyanide gives poorer results.

3. An equivalent amount of commercial formalin solution can be used with equally good results, provided that allowance is made for the volume of water—the total volume must not exceed 200 cc.

4. The reaction usually starts immediately; occasionally it does not start until the mixture becomes warm.

5. If the mixture is allowed to stand, the temperature rises to 40–50°, a red color develops, and the yield drops to almost nothing.

6. Recovered methyl sulfate can be used without purification.

7. No drying agent is completely satisfactory; sodium sulfate does the least harm.

8. The authors found a Glinsky column most satisfactory. A capillary tube reaching to the bottom of the flask for the admission of air is essential to prevent bumping. In the redistillation a piece of porous plate is sufficient for this purpose.

9. The density of the residual liquid in two runs was found to be 1.31, 1.32. The pure ester has a density of 1.35.

3. Methods of Preparation

Methoxyacetonitrile has been prepared by methylating hydroxyacetonitrile with methyl sulfate;[1] by treating chloromethyl ether with cuprous or mercuric cyanide;[2] and by dehydrating methoxyacetamide with phosphorus pentoxide.[3] The procedure described is a modification of that of Polstorff and Meyer.[1]

[1] Polstorff and Meyer, Ber. **45,** 1911 (1912); Slater and Stephen, J. Chem. Soc. **117,** 312 (1920); Malkin and Robinson, ibid. **127,** 372 (1925).

[2] Gauthier, Ann. chim. phys. (8) **16,** 302, 306 (1909).

[3] Kilpi, Z. physik. Chem. **86,** 671 (1914).

METHYL BENZYL KETONE

(2-Propanone, 1-phenyl-)

(A) (From Phenylacetic and Acetic Acids)

$$C_6H_5CH_2CO_2H + CH_3CO_2H \xrightarrow{(ThO_2)} C_6H_5CH_2COCH_3 + CO_2 + H_2O$$

Submitted by R. M. HERBST and R. H. MANSKE.
Checked by C. R. NOLLER and C. F. LOVE.

1. Procedure

THE reaction is carried out in the apparatus shown in Fig. 13. A–A is a Pyrex combustion tube, 90 cm. long and 2 cm. in diameter, fitted with an inlet chamber B, having a sealed-in side arm C and bearing a

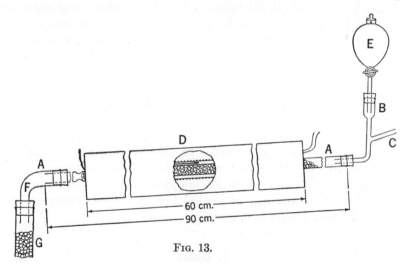

FIG. 13.

separatory funnel E. The center 60 cm. of the tube is filled with thorium oxide catalyst (Note 1) held in place at the lower end by indentations in the combustion tube. The hot junction D of a pyrometer is placed in contact with the glass tube at its center, and the catalyst-filled section of the tube is wrapped with a thin layer of asbestos paper. The tube is supported in an electrically heated cylindrical furnace, 60 cm. in length, which is inclined slightly from the horizontal. The annular space between the tube and the furnace is plugged at the ends with asbestos in order to produce a uniform temperature throughout the tube and to hold the glass tube in place. The lower end of the reaction tube is connected through the adapter F to a vertical glass tube G, 40 cm. long and 2 cm. in diameter, which is filled with glass beads and serves as the condenser. An Erlenmeyer flask or other suitable receiver is placed at the lower end of G.

The furnace is heated to 430–450°, and simultaneously the tube is swept out thoroughly with a stream of carbon dioxide, introduced through the side arm C. The carbon dioxide is passed through a wash bottle of concentrated sulfuric acid to dry it and to estimate the rate of flow. A solution of 136 g. (1 mole) of phenylacetic acid, m.p. 77–79° (Org. Syn. Coll. Vol. I, **1941**, 436), in 120 cc. (120 g., 2 moles) of glacial acetic acid is placed in the separatory funnel E and is introduced into the inlet chamber B at the rate of twelve to fifteen drops per minute. The entire solution should run through in twelve to fifteen hours. Meanwhile a very slow stream of carbon dioxide (one bubble per second) is passed through the tube C, to keep the gases in motion. After all the solution has been added, the funnel is rinsed with 10 cc. of glacial acetic acid, and this is passed through the reaction tube to facilitate removal of the product. The distillate consists of a slightly fluorescent light brown oil and an aqueous layer; both layers are treated with 300 g. of a mixture of ice and water, and rendered alkaline to litmus with a slight excess of 50 per cent sodium hydroxide solution (Note 2).

The oil is separated (Note 3), and the aqueous layer is extracted with two 50-cc. portions of benzene. The extracts are combined with the oil, and the solvent is removed by distillation. Fractionation of the residue under reduced pressure gives 80–95 g. of a methyl benzyl ketone fraction, boiling at 110–120°/21–22 mm., and a residue of dibenzyl ketone (Note 4). The main fraction on redistillation yields 74–87 g. (55–65 per cent of the theoretical amount) of methyl benzyl ketone boiling at 110–115°/21–22 mm. (Notes 5, 6, and 7).

2. Notes

1. The catalyst is prepared as follows: Enough pea-sized pieces of screened pumice to fill the tube is soaked in hot concentrated nitric acid and then washed thoroughly with hot distilled water. In a porcelain dish the pumice is mixed with a solution of 40 g. of thorium nitrate crystals [$Th(NO_3)_4 + 12H_2O$] in 100 cc. of water and is evaporated to dryness, with frequent stirring to ensure uniform deposition of the salt. The impregnated pumice is ignited over a Bunsen burner until decomposition of the nitrate is complete. The pumice carries about 15 g. of thorium oxide.

2. From the alkaline solution about 10–15 per cent of the phenylacetic acid may be recovered by acidification with sulfuric acid. The acid separates as an oil which crystallizes slowly on cooling.

3. Salt may be added to facilitate the separation.

4. When several runs are made, the residues may be combined and distilled under reduced pressure. The fraction boiling at 190–210°/20 mm. amounts to about 19 g. per run and is chiefly dibenzyl ketone.

5. Further purification of the product may be effected by converting the ketone into the bisulfite compound, washing this with ether, decomposing with sodium bicarbonate, and steam-distilling.

6. When several runs are to be made, the catalyst should be regenerated after each run by passing air through the reaction tube for about three hours while the temperature is raised gradually to 550°. The yield in the first run may be low, especially if all the oxides of nitrogen have not been removed from the reaction tube.

7. Using the same procedure the following ketones may be obtained in similar yields: ethyl benzyl ketone from phenylacetic acid and propionic acid, methyl β-phenylethyl ketone from hydrocinnamic acid and acetic acid, and ethyl β-phenylethyl ketone from hydrocinnamic acid and propionic acid.

(B) (From α-Phenylacetoacetonitrile)

$$C_6H_5CH(CN)COCH_3 + H_2O \xrightarrow{(H_2SO_4)} C_6H_5CH_2COCH_3 + NH_3 + CO_2$$

Submitted by PERCY L. JULIAN and JOHN J. OLIVER.
Checked by C. R. NOLLER.

THREE HUNDRED FIFTY cubic centimeters of concentrated sulfuric acid is placed in a 3-l. flask and cooled to −10°. The total first crop of moist α-phenylacetoacetonitrile obtained according to the procedure on p. 487 (corresponding to 188–206 g., or 1.2–1.3 moles of dry product) is added slowly, with shaking, the temperature being kept below 20° (Note 1). After all is added the flask is warmed on the steam bath until solution is complete and then for five minutes longer. The solution is cooled to 0°, 1750 cc. of water added rapidly, and the flask placed on a vigorously boiling water bath and heated for two hours, with occasional shaking. The ketone forms a layer and, after cooling, is separated and the acid layer extracted with 600 cc. of ether. The oil and ether layers are washed successively with 100 cc. of water, the ether combined with the oil and dried over 20 g. of anhydrous sodium sulfate. The sodium sulfate is collected on a filter, washed with ether, and discarded. The ether is removed from the filtrates, and the residue distilled from a modified Claisen flask with a 25-cm. fractionating side arm. The fraction boiling at 109–112°/24 mm. is collected; it weighs 125–150 g. (77–86 per cent of the theoretical amount) (Note 2).

2. Notes

1. If pure dry α-phenylacetoacetonitrile is used, half its weight of water should be added to the sulfuric acid or charring will take place on the steam bath.

2. Usually almost the entire crude product distils in this range with practically no fore-run or residue. Occasionally, however, as much as 30 g. of high-boiling residue, chiefly unchanged nitrile, is obtained. When this happens the yield is correspondingly decreased.

3. Methods of Preparation

Methyl benzyl ketone has been prepared by distilling a mixture of the barium [1] or calcium [2] salts of phenylacetic and acetic acids; by passing the vapors of these acids over a heated thorium oxide catalyst; [3,4] and by heating phenylacetic acid, sodium acetate, and acetic anhydride. [5] Methyl benzyl ketone has also been prepared from phenylacetyl chloride and zinc methyl; [6] from acetyl chloride and cadmium dibenzyl; [7] by rearrangement of α-phenyl-β-methylethylene oxide; [8] by heating α-phenyl-β-methylethylene glycol with dilute sulfuric acid; [9] by heating the addition product of chloroacetone and phenylmagnesium bromide; [9] by treating chloroacetone and benzene with aluminum chloride; [10] and by the hydrolysis of α-phenylacetoacetic ester [11] or phenacylmalonic ester. [12]

[1] Radziszewski, Ber. **3**, 198 (1870).

[2] Young, J. Chem. Soc. **59**, 621 (1891).

[3] Senderens, Ann. chim. phys. (8) **28**, 318 (1913).

[4] Pickard and Kenyon, J. Chem. Soc. **105**, 1124 (1914).

[5] Magidson and Garkusha, J. Gen. Chem. (U.S.S.R.) **11**, 339 (1941) [C. A. **35**, 5868 (1941)].

[6] Popow, Ber. **5**, 500 (1872).

[7] Gilman and Nelson, Rec. trav. chim. **55**, 518 (1936).

[8] Fourneau and Tiffeneau, Compt. rend. **141**, 663 (1905).

[9] Tiffeneau, Ann. chim. phys. (8) **10**, 345, 368 (1907).

[10] Mason and Terry, J. Am. Chem. Soc. **62**, 1622 (1940).

[11] Beckh, Ber. **31**, 3163 (1898).

[12] Metzner, Ann. **298**, 378 (1897).

The procedure in Part (A) above is based on that of Senderens [3] and Pickard and Kenyon.[4] It has been reported that "thoria aërogel" is superior to other forms of thoria catalysts for the preparation of ketones from aliphatic acids or esters.[13]

5-METHYLFURFURAL

(2-Furaldehyde, 5-methyl-)

Levulose (from sucrose) $+ HCl \rightarrow ClCH_2—C_4H_2O—CHO + 4H_2O$

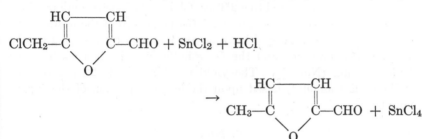

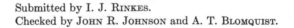

Submitted by I. J. RINKES.
Checked by JOHN R. JOHNSON and A. T. BLOMQUIST.

1. Procedure

IN a 12-l. round-bottomed flask, fitted with a cork bearing a thermometer and a large bent glass tube, is placed 6 l. of 32 per cent hydrochloric acid (sp. gr. 1.163) (Note 1). The acid is heated to 50°, and 1 kg. (2.9 moles) of powdered sugar (Note 2) is dissolved in the liquid with shaking. The dark-colored solution is heated rapidly to 70–72°, kept at this temperature for ten minutes, and poured at once onto 3 kg. of cracked ice in a large earthenware crock (preferably in a hood) (Note 3). When the mixture has cooled to room temperature, 600 g. (2.67 moles) of commercial stannous chloride crystals ($SnCl_2 \cdot 2H_2O$) is added. The reaction mixture is stirred thoroughly for ten minutes and then allowed to stand for twenty-four hours.

The following day the acid liquid is filtered with suction through a large Büchner funnel, to remove large quantities of humus which are produced. The humus on the filter is washed with two 350-cc. portions of water and finally with two 300-cc. portions of benzene. The filtered

[13] Swann, Appel, and Kistler, Ind. Eng. Chem. 26, 388, 1014 (1934).

liquid and aqueous washings have a volume of approximately 10 l. The 5-methylfurfural is removed from the aqueous layer by extraction with benzene, using three 300-cc. portions of the solvent for each 2 l. of liquid (Note 4). The combined benzene extracts (about 5 l.) are divided into two or three portions; each is washed with two 150-cc. portions of 5 per cent sodium carbonate solution and two 100-cc. portions of water. The benzene solution is dried with 100–150 g. of anhydrous magnesium or sodium sulfate, and the benzene is removed by distillation from a 1-l. flask provided with a short fractionating column and a separatory funnel for the continuous introduction of the solution. The distillation is stopped when the temperature of the distilling vapor reaches 85°.

The residue is transferred to a 250-cc. Claisen flask; two small portions of benzene (5–10 cc.) are used to rinse the last drops of the residue into the flask. The last traces of benzene are removed by warming gently under reduced pressure, and the 5-methylfurfural is then collected at 83–85°/15 mm. (Note 5). The yield is 63–70 g. (20–22 per cent of the theoretical amount, based upon the levulose portion of the sugar) (Note 6).

2. Notes

1. Commercial hydrochloric acid gives as satisfactory results as the chemically pure grade. The concentration of the hydrochloric acid should not exceed 32 per cent (sp. gr. 1.163) since stronger acid causes frothing during the heating and gives somewhat lower yields (56–57 g.).

2. Ordinary confectioners' sugar (XXXX sugar) was used. The 3 per cent of starch which it contains has no harmful effect.

3. Lengthening the time of heating or raising the temperature above 72° is definitely harmful.

4. The extractions should be carried out immediately after filtration since further quantities of humus are deposited if the liquid is allowed to stand.

5. 5-Methylfurfural discolors rapidly on standing, and after some time becomes quite black. The usual antioxidants do not retard this alteration perceptibly.

6. 5-Methylfurfural may be prepared by a modification of this method, which is more rapid but gives lower yields.[1] A solution of 800 g. of sucrose in 1 l. of hot water is allowed to flow slowly into a boiling solution of 500 g. of stannous chloride crystals, 2 kg. of sodium chloride, and 4 l. of 12 per cent sulfuric acid in a 12-l. flask. The aldehyde distils as rapidly as it is formed and is steam-distilled from the original distillate

---

[1] Scott and Johnson, J. Am. Chem. Soc. **54,** 2553 (1932).

after being rendered alkaline with sodium carbonate. The product is isolated by benzene extraction of the second distillate and distillation under reduced pressure. The yield is 27–35 g. (10–13 per cent of the theoretical amount).

3. Methods of Preparation

5-Methylfurfural has been prepared by the distillation of rhamnose with dilute mineral acids [2] and by the reduction of 5-bromo- and 5-chloromethylfurfural with stannous chloride.[3] The above procedure, starting from sucrose, has been published by Rinkes.[4]

METHYLHYDRAZINE SULFATE

(Hydrazine, methyl-, sulfate)

$$2C_6H_5CHO + N_2H_4 \cdot H_2SO_4 \xrightarrow{(NH_3)} C_6H_5CH=N-N=CHC_6H_5$$

$$C_6H_5CH=N-N=CHC_6H_5 + (CH_3)_2SO_4 + 3H_2O$$
$$\rightarrow 2C_6H_5CHO + CH_3NHNH_2 \cdot H_2SO_4 + CH_3OH$$

Submitted by H. H. HATT.
Checked by W. W. HARTMAN and W. D. PETERSON.

1. Procedure

(A) *Benzalazine.*—In a 5-l. round-bottomed flask, provided with a stout glass mechanical stirrer, are placed 240 g. (1.85 moles) of powdered hydrazine sulfate (Org. Syn. Coll. Vol. I, **1941**, 309), 1.8 l. of water, and 230 cc. (207 g., 3.4 moles) of 28 per cent aqueous ammonia (sp. gr. 0.90). The mixture is stirred, and, when the hydrazine sulfate has dissolved, 440 cc. (460 g., 4.35 moles) of benzaldehyde (Note 1) is added from a separatory funnel during the course of four to five hours (Note 2). After the mixture has been stirred for a further two hours, the precipitated benzalazine is filtered with suction, washed with water, and pressed thoroughly on a Büchner funnel. The product is dissolved in 800 cc. of boiling 95 per cent ethyl alcohol, and, on cooling, the azine

[2] Maquenne, Ann. chim. phys. (6) **22**, 91 (1891); Runde, Scott, and Johnson, J. Am. Chem. Soc. **52**, 1288 (1930).
[3] Fenton and Gostling, J. Chem. Soc. **79**, 811 (1901).
[4] Rinkes, Rec. trav. chim. **49**, 1123 (1930); **52**, 337 (1933).

separates in yellow needles melting at 92–93°. The yield is 350–360 g. (91–94 per cent of the theoretical amount); an additional 10–15 g. of less pure material can be isolated from the mother liquors. The azine is freed of ethyl alcohol by drying in a vacuum desiccator over calcium chloride.

(B) *Methylhydrazine Sulfate.*—Two hundred grams (0.96 mole) of benzalazine, 350 cc. of dry, thiophene-free benzene, and 100 cc. (133 g., 1.05 moles) of methyl sulfate (Note 3) are mixed in a 3-l. round-bottomed flask, provided with a reflux condenser bearing a calcium chloride tube. The mixture is heated continuously, with occasional shaking, on a water bath to gentle refluxing for five hours. The mixture is cooled, and the solid addition product is decomposed by adding 600 cc. of water and shaking until all the solid material has disappeared. The benzene and benzaldehyde are removed by steam distillation; the residual liquor, after cooling, is treated with 15–20 cc. of benzaldehyde and left overnight. The resin and benzalazine are separated by filtration (Note 4).

The filtrate is evaporated under reduced pressure on a water bath until a semi-crystalline mass remains; this is further desiccated by evaporating twice under reduced pressure with 50-cc. portions of absolute ethyl alcohol. The resulting crystalline cake is crushed with 50 cc. of absolute ethyl alcohol, filtered, and the process repeated. The white, crystalline product is almost pure methylhydrazine sulfate and contains very little hydrazine sulfate. After drying in a vacuum desiccator over calcium chloride, the yield is 105–110 g. (76–80 per cent of the theoretical amount). For purification, the sulfate is dissolved in about 250 cc. of boiling 80 per cent ethyl alcohol, and any undissolved material (chiefly hydrazine sulfate) is filtered. On cooling, the methylhydrazine sulfate separates in white plates, which are filtered with suction and washed with a little absolute alcohol. After drying over calcium chloride, the first fraction, m.p. 141–142°, weighs 70–75 g. (51–54 per cent of the theoretical amount) (Note 5).

2. Notes

1. The benzaldehyde should be freed of benzoic acid by shaking with aqueous sodium carbonate solution.

2. The mixture is stirred vigorously during the reaction, and one or two stout glass rods are clamped in the flask to act as baffles and to break up the lumps of benzalazine.

3. Methyl sulfate is extremely toxic. It should not be spilled; neither should the vapor be inhaled. Ammonia is a specific antidote.

4. Unreacted hydrazine sulfate is removed by conversion to benzala-

zine. The filtrate should not give an appreciable precipitate when mixed with 5 cc. of benzaldehyde and left for four hours.

5. From the mother liquors about 12 g. of less pure material, m.p. 133–136°, can be recovered.

3. Methods of Preparation

The procedure given above is essentially the method of Thiele.[1] Methylhydrazine has also been prepared by reduction and subsequent hydrolysis of nitrosomethylurea,[2] nitromethylurethane,[3] and nitroso-methylamine sulfonic acid;[4] and by methylation of hydrazine hydrate with methyl iodide [5] or diazomethane.[6]

METHYLIMINODIACETIC ACID

(Acetic acid, methyliminodi-)

$$2ClCH_2CO_2H + CH_3NH_2 \xrightarrow{(NaOH)} CH_3N(CH_2CO_2H)_2 + 2HCl$$

Submitted by G. J. BERCHET.
Checked by JOHN R. JOHNSON and P. L. BARRICK.

1. Procedure

IN a 2-l. flask, provided with a mechanical stirrer, separatory funnel, and thermometer, are placed 189 g. (2 moles) of chloroacetic acid and 150 cc. of water. The flask is cooled in ice water, and a cold solution of 160 g. (4 moles) of sodium hydroxide in 500 cc. of water is added, with stirring, at such a rate that the temperature does not exceed 30° (Note 1). After all the alkali has been added, the cooling bath is removed and an aqueous solution (Note 2) containing 31 g. (1 mole) of methyl-amine is added slowly. The reaction is exothermic, and the temperature is kept below 50° by occasional immersion of the flask in ice water. After all the methylamine has been added, the solution is allowed to stand for two hours to complete the reaction.

A solution of 257 g. (1.05 moles) of barium chloride dihydrate in

[1] Thiele, Ann. **376**, 244 (1910).
[2] Brüning, ibid. **253**, 7 (1889).
[3] Backer, Rec. trav. chim. **31**, 193 (1912).
[4] Traube and Brehmer, Ber. **52**, 1286 (1919).
[5] Harries and Haga, ibid. **31**, 56 (1898).
[6] Staudinger and Kupfer, ibid. **45**, 501 (1912).

about 500 cc. of hot water is added to the reaction mixture, with vigorous shaking, and the mixture is heated on a steam bath for one-half hour. A heavy precipitate of the barium salt of the amino acid separates at once. After cooling to room temperature, the barium salt is collected on a suction filter, transferred to a beaker, and washed with two 250-cc. portions of hot water (80°). After drying at 100°, the barium salt weighs 225–230 g. (80–82 per cent of the theoretical amount).

The dry barium salt is placed in a 2-l. flask provided with a mechanical stirrer, 600 cc. of water is added, and the mixture heated to boiling. The calculated quantity of $5 N$ sulfuric acid (Note 3) is introduced gradually from a separatory funnel into the well-stirred mixture over a period of about one hour. The mixture is then centrifuged or filtered with suction (Note 4) through a thin layer of fuller's earth. The barium sulfate precipitate is transferred to a beaker and extracted with two 250-cc. portions of boiling water. The filtrate and washings are transferred to a distilling flask placed in a water bath and concentrated under reduced pressure to a volume of 175–200 cc. (Note 5). The syrupy residue is poured into a large beaker and treated with 500 cc. of absolute methyl alcohol. Crystals of the acid begin to appear at once. The mixture is allowed to stand for three or four hours in an ice bath to complete the precipitation, and the crystalline solid is separated by filtration with suction. After being washed with two 75-cc. portions of methyl alcohol, the product is dried at 100°. The methyliminodiacetic acid forms fine, white crystals, m.p. 215°, and weighs 92–105 g. (63–71 per cent of the theoretical amount).

If an especially pure product is desired, the acid may be reprecipitated by dissolving in an equal weight of warm water and adding three volumes of methyl alcohol. The loss in purification is 4–5 per cent.

2. Notes

1. The temperature must be controlled to avoid formation of glycolic acid. One-half of the alkali is sufficient to neutralize the acid, and the remainder may be added rapidly without danger of raising the temperature.

2. Technical aqueous methylamine solution (28–33 per cent) may be used. The amine content should be determined by titration with standard acid.

3. The barium salt requires 1.416 cc. of $5 N$ sulfuric acid per gram. The acid should be titrated before use.

4. If unchanged barium methyliminodiacetate remains in solution, it peptizes the barium sulfate and the filtration is likely to be trouble-

some. If colloidal barium sulfate is encountered, it is advisable to add a slight excess (less than 1 per cent) of sulfuric acid and continue the heating for twenty minutes longer. Traces of sulfate in the final product may be removed, if necessary, by reprecipitation of the acid with methyl alcohol.

5. The residue should have a syrupy consistency but should be fluid enough to be poured freely from the flask.

3. Methods of Preparation

Methyliminodiacetic acid has hitherto been prepared by the action of methylamine on formaldehyde cyanohydrin with subsequent hydrolysis of the resulting dinitrile.[1] This method was found by the submitter to be much less satisfactory than the procedure given above.

METHYL IODIDE

(Methane, iodo-)

(A) (From Methyl Alcohol)

$$3CH_3OH + PI_3 \rightarrow 3CH_3I + H_3PO_3$$

Submitted by HAROLD S. KING.
Checked by LOUIS F. FIESER.

1. Procedure

IODINE is slowly washed by means of condensed liquid into a mixture of methyl alcohol and phosphorus. The apparatus employed by the submitter, Fig. 14, is rendered more flexible by the use of a three-necked flask, as shown in Fig. 15, and a further improvement is in the provision of a side tube for the addition of iodine to the separatory funnel, A (Note 1). In the following description reference is made to the simpler apparatus, Fig. 14; the changes required for the modified assembly will be obvious.

The cylindrical separatory funnel A has a volume of 1.2 l. and holds 2 kg. (15.75 gram atoms) of iodine crystals (Note 2). A piece of perforated platinum foil B, or a loose plug of glass wool, is placed in the bottom to prevent clogging of the stopcock by iodine crystals or solid

[1] Eschweiler, Ann. **279**, 39 (1894); I. G. Farbenind. A.-G., Fr. pat. 804,497 [C. A. **31**, 3505 (1937)].

impurities. Stopcock C is as large as 5 mm. in bore as a further precaution. For producing up to 4 kg. of methyl iodide, a 5-l. round-bottomed Pyrex flask D is used (Note 3). It is partly submerged in a water bath. Tube E, fitted into the reaction flask by a rubber stopper, is at least 2.5 cm. in internal diameter in order to allow the vapor to rise without interrupting the return of the excess distillate. Tube F is 4 cm. in diameter to allow for preliminary condensation. A space (13 cm.) is left above stopcock G as a reservoir for the distillate. Stopcock G has a bore of 2 or 3 mm. (Note 4). The bottom of the tube from this stopcock is flush with the bottom of the rubber stopper to the iodine container, so that the liquid will flow down the sides of the container instead of forming a channel through the middle of the iodine. Condenser H, 2.5 cm. in diameter and at least 200 cm. long, is attached by a rubber stopper and is well cooled with a strong stream of water (Notes 5 and 6).

In container A is placed 2 kg. of iodine, no crystals being allowed to fall below the platinum foil or glass wool. The iodine is covered with part of 2 l. (50 moles) of methanol (Note 7), the rest of which is added to 200 g. each of red and yellow phosphorus (12.8 gram atoms) in reaction flask D. The apparatus is assembled as illustrated in Fig. 14. The water bath is heated to 70–75°; stopcock G is opened wide and stopcock C part way to allow the solution of iodine to flow slowly into the reaction flask. By the time all the alcohol has been added, enough methyl iodide will have been formed to start the refluxing. Stopcock C is then opened wide and stopcock G partly closed so that only a small stream of methyl iodide flows through the iodine container. The temperature is adjusted so that little of the reflux flows back through tube E. This will require the progressive lowering of the temperature of the water bath as the reaction proceeds (Note 8); a temperature of about 55° is sufficient to promote even boiling after the first portion of iodine has been introduced. If for any reason refluxing becomes too violent, both stopcocks are closed. When the reaction has been brought under control, stopcock G is partly opened and then stopcock C part way, until any liquid in A has run out, after which C is opened wide. When all the iodine has been extracted (about two and one-half hours) both stopcocks are closed, the separatory funnel is lowered with a rotary motion, filled with a second charge of 2 kg. (15.75 gram atoms) of iodine, and raised to its former position. Stopcock C is opened wide and stopcock G part way as before.

After the extraction of this second charge of iodine, the flask is cooled, and a condenser arranged for downward distillation is fitted in place of the special apparatus. To the lower end of the condenser is tightly

attached an adapter dipping under a slush of ice and water. When all
the methyl iodide has distilled (b.p. 40–42.5°) (Note 9), it is separated

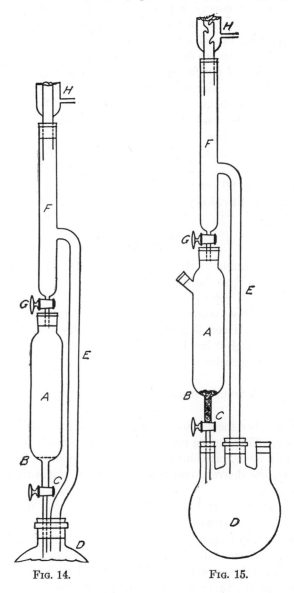

FIG. 14.   FIG. 15.

from the water, dried by shaking with anhydrous calcium chloride,
filtered through glass wool, stored for use in a sealed flask or tubes, and
kept in the dark (Note 10). The yield is 4150–4250 g. (93–95 per cent

of the theoretical amount based on iodine), and the crude product is colorless and sufficiently pure for many purposes. Redistillation gives, with a 95 per cent recovery, a product boiling at 41.0–42.0°. Because of the high vapor pressure of methyl iodide, special precautions should be taken to prevent loss of material. Two different devices that have been found satisfactory are described in Notes 11 and 12.

The discoloration of purified methyl iodide is greatly retarded by the addition of a drop of mercury, and traces of iodine in an old sample may be removed by shaking with mercury (Note 13). A very highly colored product is best washed with very dilute sodium thiosulfate, then with water, before drying.

With minor modifications, this method is suitable for the preparation of ethyl, n-propyl, n-butyl, n-amyl, isobutyl, and isoamyl iodides with yields of more than 90 per cent. A somewhat lower yield is obtained in the preparation of sec.-butyl iodide owing to the formation of hydrogen iodide.

2. Notes

1. The upper opening of the funnel A (Fig. 15) should be the same size as that of tube F, so that the funnel may be used as a receiver in the distillation described below (Note 12).

2. A funnel of this volume and shape is not essential but has been found convenient. Even if a globular separatory funnel is used, the liquid flowing through it has been found to wet the walls.

3. For larger quantities a flask up to 12 l. in capacity may be substituted.

4. The stopcock from a small, broken, separatory funnel serves the purpose.

5. In hot weather there is usually some loss of methyl iodide through the condenser. This can be recovered by a trap made by bending the mouth end of a 50-cc. pipet into a U, which is stoppered into the end of the condenser, the delivery end being just below the surface of a slush of ice and water.

6. In place of the very long condenser, the checker used a modified West condenser with staggered indentures, as shown in Fig. 15. The jacket was 90 cm. in length, the inner tube was 2.2 cm. i.d., the opening at the top was 3.3 cm. i.d. At the top of this condenser was attached a 30-cm. bulb condenser which could be cooled with ice water. With the tap water at 5°, there was no difficulty in effecting complete condensation, and the trap described in Note 5 was not required. In warm weather it will probably be necessary to run the water to each condenser through a large bottle of ice.

7. The methanol need not be absolute. In one experiment it was diluted with 10 per cent of its weight of water. The yield of methyl iodide was identical with that of a check run in which absolute methanol was used.

The refluxing starts more smoothly if the iodine crystals are covered with methanol. The alcoholic solution of iodine forms sufficient methyl iodide to start the refluxing at a lower temperature than would otherwise be necessary.

8. The increased speed of the reaction is largely due to the fact that iodine is more soluble in methyl iodide than in methanol and is therefore more rapidly introduced as the reaction progresses. The reaction is exothermic.

9. It has been pointed out that this distillate is probably the constant-boiling mixture of methyl iodide and methyl alcohol which is composed of 93 per cent by weight of methyl iodide and boils at 39°. The procedure given for recovering the methyl iodide from the water-ice mixture is adequate. (ROBERT A. DINERSTEIN, private communication.)

10. Caution is necessary in disposing of the residues left in the reaction flask because they contain yellow phosphorus. They should be covered at once with water.

11. One end of a coil of 2-mm. glass tubing is sealed to the end of the inner tube of the condenser. The other end of the coil is sealed to a tube leading through the stopper of the collecting vessel, which also carries a Bunsen valve to prevent the entrance of moisture from the air. Both the coil and the collecting vessel are cooled with ice and salt, or with carbon dioxide snow.

12. Another convenient assembly for the distillation is furnished by using the condenser shown in Fig. 15, together with the separatory funnel, A. The methyl iodide is distilled from a 3-l. flask through an inverted U-tube (22 mm.); this brings the vapor down into the condenser, which is also clamped in a vertical position. The separatory funnel serves as the receiver and permits the easy removal of fractions. The side opening is equipped with a small condenser carrying a calcium chloride tube. A simple and adequate method of recording the boiling temperature is to insert in the U-tube in an inverted position a long thermometer with its head imbedded in a small cork, which rests on the bottom of the flask.

13. *Caution.* Methyl iodide in contact with mercury when exposed to the sunlight readily forms methylmercuric iodide, which is exceedingly poisonous.

(B) (*From Methyl Sulfate*)

$$KI + (CH_3)_2SO_4 \rightarrow CH_3I + CH_3SO_4K$$

Submitted by W. W. Hartman.
Checked by John R. Johnson and H. J. Passino.

1. Procedure

A 3-l. three-necked flask is fitted with a thermometer, a separatory funnel, a liquid-sealed mechanical stirrer, and a small fractionating column connected with an efficient condenser set for downward distillation. The condenser leads to a receiving vessel cooled in ice water (Note 1). A solution of 800 g. (4.8 moles) of u.s.p. potassium iodide in 430 cc. of water is placed in the flask, 60 g. of calcium carbonate is added, and the mixture is heated to 60–65°, with stirring.

The temperature is kept at 60–65°, and 630 g. (473 cc., 5 moles) of "practical" methyl sulfate (Note 2) is added gradually through the separatory funnel. The rate of addition is such that the methyl iodide distils briskly; the addition requires about two hours.

After all the methyl sulfate has been added, the temperature is raised to 65–70° for about forty minutes to complete the distillation of the methyl iodide. The product is separated from a small amount of water, dried thoroughly over anhydrous calcium chloride, and decanted into a dry distilling flask. A few crystals of potassium iodide are added, and the material is distilled from a water bath. The yield of methyl iodide, boiling at 41–43°, is 615–640 g. (90–94 per cent of the theoretical amount) (Note 3).

2. Notes

1. It is important to avoid loss of the volatile methyl iodide. The notes to Procedure (A) can be read with profit on this point.
2. Methyl sulfate is extremely toxic. Contact with the liquid or inhaling the vapor should be avoided. Ammonia is a specific antidote.
3. Similar percentage yields are obtained with quantities of material up to ten times as large as those specified in the procedure.

3. Methods of Preparation

Methyl iodide has been prepared by the action of methyl sulfate on an aqueous solution of potassium iodide; [1] by the slow distillation of a

[1] Weinland and Schmid, Ber. **38**, 2327 (1905); Ger. pat. 175,209 [Frdl. **8**, 17 (1905–07)].

mixture of methanol with a large excess of constant-boiling hydriodic acid;[2] by the electrolysis of an aqueous solution of potassium acetate in the presence of iodine or potassium iodide;[3] by the action of an aqueous solution of potassium iodide on methyl p-toluenesulfonate;[4] and by the action of methanol on a solution of phosphorus pentaiodide in methyl iodide.[5] The most generally employed method of preparing methyl iodide has been the interaction of methanol and phosphorus triiodide (or a mixture of iodine and phosphorus, red, yellow, or both).[6] Many modifications of this method have been proposed and certain variants have been suggested for the preparation of the higher alkyl iodides which are applicable to the preparation of methyl iodide.[7]

The procedure described in (A) is a modification of Walker's method.[8] The method of introducing a soluble substance gradually into a reaction flask, described above, is applicable to other reactions, such as the preparation of mercury dialkyls.[9]

The procedure described in (B) can also be employed successfully with iodoform as the iodine-containing starting material, if the iodoform is first heated with alcoholic alkali.[10]

[2] Norris, Am. Chem. J. 38, 639 (1907).

[3] Kaufler and Herzog, Ber. 42, 3860 (1909).

[4] Peacock and Menon, Quart. J. Indian Chem. Soc. 2, 240 (1925); Rodionow, Bull. soc. chim. (4) 39, 323 (1926).

[5] Walker and Johnson, J. Chem. Soc. 87, 1595 (1905).

[6] Dumas and Peligot, Ann. 15, 20 (1835); Ipatiew, J. Russ. Phys.-Chem. Soc. 27, I, 364 (1895) [Ber. 29 (R) 90 (1896)]; J. prakt. Chem. (2) 53, 275 (1896); Crismer, Ber. 17, 652 (1884); Walker and Johnson, J. Chem. Soc. 87, 1592 (1905); Haywood, ibid. 121, 1911 (1922).

[7] Hunt, ibid. 117, 1592 (1920); Bogert and Slocum, J. Am. Chem. Soc. 46, 763 (1924).

[8] Walker, J. Chem. Soc. 61, 717 (1892); Nagai, J. Pharm. Soc. Japan 1916, No. 407, 1 [C. A. 10, 987 (1916)]; Adams and Voorhees, J. Am. Chem. Soc. 41, 796 (1919); Adams, Kamm, and Marvel, Organic Chemical Reagents I, Univ. Illinois Bull. 16 (43) 19 (1919); Reynolds and Adkins, J. Am. Chem. Soc. 51, 280 (1929); King, Proc. Trans. Nova Scotian Inst. Sci. 16 [2] 87 (1924).

[9] Gilman and Brown, J. Am. Chem. Soc. 51, 928 (1929).

[10] Kimball, J. Chem. Education 10, 747 (1933).

METHYL ISOPROPYL CARBINOL

(2-Butanol, 3-methyl-)

$$(CH_3)_2CHBr + Mg \rightarrow (CH_3)_2CHMgBr$$

$$(CH_3)_2CHMgBr + CH_3CHO \rightarrow (CH_3)_2CHCH(OMgBr)CH_3$$

$$(CH_3)_2CHCH(OMgBr)CH_3 + H_2O \rightarrow (CH_3)_2CHCHOHCH_3$$

Submitted by NATHAN L. DRAKE and GILES B. COOKE.
Checked by C. S. MARVEL and B. H. WOJCIK.

1. Procedure

IN a 3-l. three-necked flask, fitted with a mechanical stirrer, a separatory funnel, and a reflux condenser the upper end of which is protected by a calcium chloride tube, are placed 146 g. (6 gram atoms) of dry magnesium turnings (Note 1) and about 250 cc. of dry ether (Note 2).

A solution of 600 g. (4.9 moles) of isopropyl bromide (Note 3) in 300 cc. of dry ether is then added through the separatory funnel. The reaction begins after about 15 cc. of the solution has been added (Note 4). The solution is added at such a rate that the reaction mixture refluxes gently. It is well to arrange to cool the flask with running water if the refluxing becomes too vigorous. The addition of the isopropyl bromide solution should require from three and one-half to four hours. The reaction mixture is refluxed on the water bath for forty minutes after addition of the isopropyl bromide solution is complete.

The flask is then cooled to −5° (Note 5), and a solution of 200 g. (4.5 moles) of acetaldehyde (Note 6) in 250 cc. of dry ether is added at this temperature over a period of one hour.

After addition of the acetaldehyde solution is complete, the product is decomposed by pouring the reaction mixture onto 2 kg. of cracked ice. The excess magnesium may be removed conveniently by decantation at this point. The basic magnesium halide is dissolved by addition of about 1 l. of 15 per cent sulfuric acid. The ether solution is separated, and the aqueous layer is extracted with four 150-cc. portions of ether. The ether solutions are combined, dried over 25 g. of calcined potassium carbonate, filtered, and fractionally distilled, using a short column. The methyl isopropyl carbinol distils at 110–111.5°. The fraction boiling at 37–109° should be dried and refractionated. The total yield is 210–215 g. (53–54 per cent of the theoretical amount).

2. Notes

1. The excess of magnesium is used to increase the yield of Grignard reagent.

2. More ether may be added during the preparation to replace any which may be lost. The ether used should be dried over bright sodium wire.

3. The isopropyl bromide (b.p. 59–60°) was obtained from the Eastman Kodak Company and was used without further treatment. Directions for preparing isopropyl bromide are given in Org. Syn. Coll. Vol. I, **1941,** 37, and on p. 359 above.

4. Heating on the water bath to start the reaction should be unnecessary if all the apparatus and reagents are completely dry.

5. The temperature must not be allowed to rise above −5°.

6. The acetaldehyde may be prepared conveniently by depolymerizing pure, dry paraldehyde with toluenesulfonic acid as a catalyst. Acetaldehyde boiling at 20.5–21° must be used for this preparation. There is no advantage in using gaseous acetaldehyde.

3. Methods of Preparation

Methyl isopropyl carbinol has been prepared by reduction of methyl isopropyl ketone with sodium amalgam,[1] and with sodium;[2] by treatment of chloroacetyl chloride with zinc methyl;[3] by treatment of bromoacetyl bromide with zinc methyl;[4] by treatment of isobutyraldehyde with methylmagnesium bromide;[5] and as a by-product of the reaction between chloroacetone and methylmagnesium iodide.[6]

[1] Münch, Ann. **180,** 339 (1876).

[2] Michael and Zeidler, Ann. **385,** 262 (1911).

[3] Bogomolez, J. Russ. Phys.-Chem. Ges. **13,** 396 (1881) [Ber. **14,** 2066 Ref. (1881)]; Ann. **209,** 86 (1881).

[4] Winogradow, ibid. **191,** 128 (1878).

[5] Henry, Compt. rend. **145,** 22 (1907).

[6] Fourneau and Tiffeneau, ibid. **145,** 438 (1907).

METHYL ISOPROPYL KETONE

(2-Butanone, 3-methyl-)

$$(CH_3)_2C(OH)CH_2CH_3 + Br_2 \rightarrow (CH_3)_2CBrCHBrCH_3 + H_2O$$

$$(CH_3)_2CBrCHBrCH_3 + H_2O \rightarrow (CH_3)_2CHCOCH_3 + 2HBr$$

Submitted by Frank C. Whitmore, W. L. Evers, and H. S. Rothrock.
Checked by W. W. Hartman and L. J. Roll.

1. Procedure

In a 1.5-l. round-bottomed Pyrex flask, fitted with a dropping funnel, an efficient mechanical stirrer, and a thermometer, is placed 176 g. (2 moles) of *tert.*-amyl alcohol (redistilled, b.p. range 0.5°). The flask is surrounded by a water bath, and the temperature of the alcohol is held at 50–60° while 320 g. (103 cc., 2 moles) of bromine is slowly added with stirring during about two hours (Note 1). The stirring is continued for a few minutes until the bromine color has disappeared (Note 2).

To the reaction flask containing the crude trimethylethylene dibromide is added 540 cc. of water, and the flask is fitted with a long reflux condenser and a mercury-sealed stirrer (Note 3). The mixture is refluxed with stirring for three to five hours until hydrolysis is practically complete (Note 4). The reflux condenser is then replaced by a condenser for distillation, and the crude methyl isopropyl ketone is removed from the reaction mixture by direct distillation, with stirring, until the temperature rises (Note 5) and the oil nearly stops coming over (about one and one-half hours), or until the oil coming over is heavier than water. The residue in the flask can be distilled to yield about 380 cc. of constant-boiling hydrobromic acid.

To the distillate, consisting of a yellow oil with a small lower water layer, powdered sodium carbonate (about 10 g.) is added, and the mixture is shaken until the water layer is alkaline and nearly saturated (Note 6). The layers are separated, and the oil is refluxed for about sixteen hours with 20 g. of powdered potassium carbonate and 5 cc. of water (Note 7). The oily layer is again separated and dried over about 6 g. of anhydrous calcium chloride or potassium carbonate (Note 8). It is then placed in a flask containing 2 g. of dry sodium carbonate and distilled through an efficient column. The yield of methyl isopropyl ketone boiling at 92–94° (Note 9) is about 102 g. (59 per cent of the theoretical amount). The product is slightly yellow.

2. Notes

1. The bromine is added at such a rate that only a small amount is present at any time, as shown by an orange-red color. The speed of the reaction depends on the temperature and efficiency of stirring. The temperature of the reaction mixture remains a few degrees above that of the water bath.

After about half the bromine has been added, the mixture becomes cloudy owing to the separation of water, and a somewhat lower temperature (40–45°) suffices to prevent the accumulation of unreacted bromine.

2. The crude trimethylethylene dibromide can be purified by washing with sodium carbonate solution and water, drying, and distilling under reduced pressure; b.p. 49–51°/11 mm.; 60–70 per cent yield. Hydrolysis of the purified dibromide gives a 78 per cent yield of methyl isopropyl ketone. However, the losses in purifying the dibromide result in a slightly decreased over-all yield of ketone so that the purification of the dibromide is not recommended.

3. The quantity of water used is 50 per cent more than the theoretical amount for hydrolyzing the dibromide and converting all the bromine to constant-boiling (47.3 per cent) hydrobromic acid. Less water can be used, but more darkening of the reaction mixture occurs, and the yield of ketone is slightly decreased.

4. Efficient stirring lessens the time required for hydrolysis and avoids loss of material by intermittent sudden boiling. To be sure of complete hydrolysis, refluxing should be continued for at least an hour after the specific gravity of the oily layer becomes less than that of the water layer.

5. The distillate and residue contain a very powerful lachrymator and should be handled under a good hood.

6. The ketone is somewhat soluble in water, and the sodium carbonate serves to salt it out as well as to remove any hydrobromic acid.

7. The crude ketone contains small amounts of bromides, and it darkens even after repeated distillation unless first refluxed over potassium carbonate.

8. It is necessary to dry the ketone thoroughly since water distils with it at about 78°. This fraction can be dried again to recover the ketone.

9. A column 55 cm. high packed with glass tubes 6 mm. in length and diameter was employed with good results. A few grams of trimethylethylene, b.p. 34–36°/735 mm., is always obtained if adequate condensers have been used throughout the preparation. The presence of a little sodium carbonate during the final distillation hinders decomposition.

3. Methods of Preparation

Trimethylethylene dibromide has been obtained by the reaction between bromine and trimethylethylene [1] or tertiary amyl alcohol.[2]

Methyl isopropyl ketone has been obtained by the hydration of isopropylacetylene; [3] from isopropylmagnesium bromide and acetic anhydride; [4] from isopropylmagnesium chloride and ethyl acetate; [5] by the hydrolysis of ethyl dimethylacetoacetate; [6] by passing a mixture of the vapors of isobutyric and acetic acids over a thoria catalyst; [7] and from butane or isobutane, aluminum chloride, and carbon monoxide.[8] Methyl isopropyl ketone can also be obtained by rearrangement of methyl isopropenyl carbinol [9] or trimethylethylene oxide; [10] by heating trimethylethylene glycol with dilute hydrochloric acid; [11] by heating trimethylethylene chlorohydrin (2-methyl-2-hydroxy-3-chlorobutane) with water; [12] and by the hydrolysis of trimethylethylene dibromide.[13]

[1] Michailenko, J. Russ. Phys.-Chem. Soc. **27**, 56 (1895) [Ber. **28** (R) 852 (1895)]; Froebe and Hochstetter, Monatsh. **23**, 1075 (1902).

[2] Ipatiew, J. prakt. Chem. (2) **53**, 266 (1896).

[3] Flavitzki and Krylow, J. Russ. Phys.-Chem. Soc. **10**, 347 (1878); Kutscherow, Ber. **42**, 2761 (1909).

[4] Sweet and Marvel, J. Am. Chem. Soc. **54**, 1188 (1932).

[5] Ivanoff and Spassov, Bull. soc. chim. (5) **2**, 816 (1935).

[6] Schryver, J. Chem. Soc. **63**, 1336 (1893).

[7] Senderens, Compt. rend. **149**, 996 (1909).

[8] Hopff, Nenitzescu, Isacescu, and Cantuniari, Ber. **69**, 2244 (1936).

[9] Kondakow, J. Russ. Phys.-Chem. Soc. **17**, 300 (1885); N. V. de Bataafsche Petroleum Maatschappij, Fr. pat. 777,032 [C. A. **29**, 3686 (1935)].

[10] Ipatiew and Leontowitsch, Ber. **36**, 2018 (1903); Shell Development Company, U. S. pat. 2,106,347 [C. A. **32**, 2542 (1938)].

[11] Bauer, Ann. **115**, 91 (1860).

[12] Krassuski, J. Russ. Phys.-Chem. Soc. **34**, 287 (1902) (Chem. Zentr. **1902**, II, 19).

[13] Ipatiew, J. Russ. Phys.-Chem. Soc. **27**, 359 (1895) [Ber. **29** (R) 90 (1896)]; Froebe and Hochstetter, Monatsh. **23**, 1075 (1902); Colonge, Bull. soc. chim. (5) **3**, 501 (1936).

S-METHYL ISOTHIOUREA SULFATE
(Pseudourea, 2-methyl-2-thio-, sulfate)

$$2(H_2N)_2CS + (CH_3)_2SO_4 \rightarrow (HN{=}\overset{\displaystyle SCH_3}{\underset{|}{C}}{-}NH_2)_2 \cdot H_2SO_4$$

Submitted by P. R. SHILDNECK and WALLACE WINDUS.
Checked by HENRY GILMAN and W. F. SCHULZ.

1. Procedure

In a 2-l. round-bottomed flask are mixed 152 g. (2 moles) of finely divided thiourea and 70 cc. of water. To this is added 138 g. (1.1 moles) of technical methyl sulfate (Note 1). The flask is immediately attached to a long reflux condenser carrying a trap. The reaction is allowed to progress spontaneously (Note 2), with occasional cooling as the reaction becomes more rapid and the flask becomes filled with vapor. After the initial vigorous reaction is completed, the mixture is refluxed for one hour, during which time crystallization takes place (Note 3). The mixture is then allowed to cool, the flask is removed, 200 cc. of 95 per cent ethyl alcohol is added, and the contents of the flask are then filtered with suction. The residue is washed twice with 100-cc. portions of 95 per cent alcohol and allowed to dry in air. The yield is 190 g. of a product which melts with decomposition at 235°. Another crop of crystals weighing 43 g. and melting at 230° can be obtained from the alcoholic filtrate by concentrating it to a paste to which, after cooling, is added 120 cc. of 95 per cent alcohol. The total yield is 220–233 g. (79–84 per cent of the theoretical amount).

2. Notes

1. Technical methyl sulfate, if it has not turned dark brown, need not be distilled before using.

2. If the mixture is cooled too much with ice water, the spontaneous reaction almost ceases and gentle heating is required to start the reaction again. If the mixture is not cooled, the initial vigorous reaction is so violent that material is likely to be lost through the condenser. Since methyl sulfate is poisonous, this must be avoided. Ammonia, which is a specific antidote for methyl sulfate, should be kept at hand.

The checkers observed no spontaneous reaction, and in order to initiate reaction the flask and contents were gently heated with a moving low flame. When reaction sets in, the flame is removed. A container with ice water is kept handy to moderate any unduly vigorous reaction, particularly if the water condenser is less than 125 cm. in length.

3. The completion of the vigorous reaction indicates that half of the thiourea is methylated and that the methyl sulfate has been converted to methylhydrogen sulfate. Vigorous heating is necessary to complete the methylation.

3. Method of Preparation

The procedure described is essentially that of Arndt.[1]

METHYL NITRATE

$$CH_3OH + HONO_2 \xrightarrow{(H_2SO_4)} CH_3ONO_2 + H_2O$$

Submitted by ALVIN P. BLACK and FRANK H. BABERS.
Checked by JOHN R. JOHNSON and H. B. STEVENSON.

1. Procedure

IN a flask cooled in an ice bath are mixed 425 g. (300 cc., 4.7 moles) of C.P. nitrous-free, concentrated nitric acid (sp. gr. 1.42) (Note 1) and 550 g. (300 cc.) of C.P. concentrated sulfuric acid (sp. gr. 1.84). In a second flask, also cooled in an ice bath, 92 g. (50 cc.) of C.P. concentrated sulfuric acid is added to 119 g. (150 cc., 3.7 moles) of pure methyl alcohol (Note 2) while the temperature is maintained below 10°.

One-third of the cold nitric-sulfuric mixture is placed in each of three 500-cc. Erlenmeyer flasks (Note 3), and each portion is treated separately with one-third of the methyl alcohol-sulfuric acid mixture, with constant shaking and thorough mixing (Note 4). The temperature is allowed to rise fairly rapidly to 40° and kept at this point by external cooling. During the addition of the methyl alcohol-sulfuric acid, most of the ester separates as an almost colorless oily layer above the acid. The time required for completion of the reaction is two to three minutes for each flask. The reaction mixtures are allowed to stand in the cold for an additional fifteen minutes but not longer. The lower layer of spent acid is separated promptly and poured at once into a large volume of cold water (about 1 l. for each portion) to avoid decomposition which quickly ensues with copious evolution of nitrous fumes.

[1] Arndt, Ber. 54, 2236 (1921).

The combined ester layers are washed with two 25-cc. portions of ice-cold salt solution (sp. gr. 1.17) (Note 5). A small quantity (8–10 drops) of concentrated sodium hydroxide solution is added to the second wash liquid until it has a faintly alkaline reaction to litmus. The ester is washed free of alkali with ice-cold salt solution and finally washed with two 15-cc. portions of ice water (Note 6). The product is treated with 10–15 g. of anhydrous calcium chloride and allowed to stand with occasional shaking for an hour at 0°. It is then decanted onto a fresh 5-g. portion of the drying agent and after standing for one-half hour is filtered. The crude ester without further purification (Note 7) may be used directly for most synthetic purposes, such as the preparation of phenylnitromethane (p. 512). The yield is 190–230 g. (66–80 per cent of the theoretical amount). *The crude ester should be used promptly and not stored.*

2. Notes

1. Colored specimens of nitric acid may be treated with a small quantity of urea (about 1–2 g. per 100 cc.), but this is unnecessary unless the acid is appreciably colored.

2. Commercial synthetic methanol of high grade was used without further purification. This material is believed to be superior to wood alcohol for this preparation.

3. It should be noted that volume contractions occur in mixing the reagents. The total volume of the mixed acids is about 585 cc. (instead of 600 cc.), and that of the methyl alcohol-sulfuric acid is about 182 cc. (instead of 200 cc.).

4. The treatment of methyl alcohol with a mixture of concentrated nitric and sulfuric acids is not without elements of danger, and *adequate precautionary measures should be taken.* However, the submitters report that more than one hundred preparations were carried out without a single explosion or violent decomposition.

5. This corresponds to a 22 per cent solution of sodium chloride. This particular solution was found to give satisfactory separations and obviate emulsions.

6. Traces of acid remaining in the ester facilitate decomposition, and violent explosions may occur if such specimens are heated.

7. Methyl nitrate may be distilled if adequate precautions are taken. The ester must not be heated suddenly and must not contain any free acid. *Distillation is not recommended,* as the crude ester (after washing and drying) gives as good yields as the distilled product in most synthetic reactions. The loss in distillation is small, and the pure material distils at 64.5–65°. *The residue in the distilling flask must not be superheated.*

3. Methods of Preparation

Methyl nitrate has been prepared by distilling a methyl alcohol-nitric acid mixture to which methyl alcohol-sulfuric acid is added dropwise,[1] and by the use of dilute nitric acid in a procedure otherwise similar to that given above.[2] Neither of these methods is satisfactory, and explosions occur frequently.

METHYL OXALATE

(Oxalic acid, dimethyl ester)

$$(CO_2H)_2 + 2CH_3OH \xrightarrow{(H_2SO_4)} (CO_2CH_3)_2 + 2H_2O$$

Submitted by EVERETT BOWDEN.
Checked by C. S. MARVEL and A. T. KOIDE.

1. Procedure

IN a 500-cc. Pyrex flask, fitted with a cork which loosely carries a glass mechanical stirrer and a separatory funnel, are placed 90 g. (1 mole) of anhydrous oxalic acid (Org. Syn. Coll. Vol. I, **1941**, 421) and 100 cc. (79 g., 2.5 moles) of methanol (Note 1). Then, while the mixture is rapidly stirred, 35 cc. of pure concentrated sulfuric acid (Note 2) is slowly added through the separatory funnel. The mixture is heated, if necessary (Note 3), nearly to boiling, then filtered as rapidly as possible through a 15-cm. filter paper placed in a slightly heated glass funnel, the filtrate being collected in a 500-cc. wide-mouthed Erlenmeyer flask. The first flask is rinsed with 40 cc. of hot methanol, which is poured through the filter paper. After twenty-four hours at 15° (Note 4) the crystals are filtered with suction, sucked as dry as possible, pressed between filter paper, and air-dried for a few minutes. The filtrate, after cooling to about −10°, is filtered rapidly and the product dried as before. A total of 100–115 g. of material, slightly moist with sulfuric acid and melting at 50–52°, is obtained.

For purification, the crude product is dissolved in 100 cc. of redistilled methanol, filtered through a warm funnel, and allowed to crystallize. After several hours the crystals are filtered, and the filtrate is chilled and filtered as before. A total of 80–90 g. (68–76 per cent of the theoretical amount) of methyl oxalate, melting at 52.5–53.5° (Note 5), is obtained.

[1] Lea, Am. J. Sci. (2) **33**, 227 (1862) (Chem. Zentr. **1862**, 602).

[2] Delépine, Bull. soc. chim. (3) **13**, 1044 (1895).

2. Notes

1. The methanol used is the commercial (almost acetone-free) grade known as Columbian Spirits. This material is redistilled for recrystallizing the methyl oxalate.

2. Less sulfuric acid than that employed gives a smaller yield; larger quantities sometimes result in a product that causes difficulty in filtration. It is essential that the acid be added slowly and with vigorous stirring to prevent local superheating and darkening of the solution and product. Some commercial grades of methanol become quite dark in contact with sulfuric acid. The grade of methanol used here does not discolor, the filtrates being only light yellow.

3. Solution of the oxalic acid causes a sharp drop in temperature whereas solution of the sulfuric acid raises the temperature.

4. The major portion of the reaction is complete within a few minutes, but several hours are necessary for complete crystallization.

5. Larger batches give the same percentage yield as the one described. When several batches are to be run, the alcohol from the first recrystallization becomes the starting alcohol for the second batch, etc. This increases the yield somewhat.

3. Methods of Preparation

Methyl oxalate has been prepared by distilling a mixture of oxalic acid, methyl alcohol, and sulfuric acid;[1] by dissolving anhydrous oxalic acid in hot methyl alcohol;[2] by esterifying oxalic acid with methyl alcohol, using anhydrous hydrogen chloride as a catalyst;[3] by methanolysis of ethyl oxalate;[4] by passing vapors of dry methanol through hydrated oxalic acid until the water has been removed;[5] and by a process in which the methyl alcohol-water mixture evolved from hydrated oxalic acid and methanol is dried over potassium carbonate and returned to the reaction flask.[6] The method described in the procedure is simpler than any of these and gives very satisfactory yields.

[1] Dumas and Peligot, Ann. chim. phys. (2) **58**, 44 (1835); Ann. **15**, 32 (1835); Wöhler, ibid. **81**, 376 (1852).

[2] Erlenmeyer, Jahresb. **1874**, 572.

[3] Rising and Stieglitz, J. Am. Chem. Soc. **40**, 726 (1918).

[4] Pfannl, Monatsh. **31**, 316 (1910); Reimer and Downes, J. Am. Chem. Soc. **43**, 950 (1921).

[5] Dutt, J. Chem. Soc. **123**, 2714 (1923).

[6] Kenyon, Org. Syn. Coll. Vol. I, **1941**, 264.

3-METHYLPENTANOIC ACID

(Valeric acid, β-methyl-)

$$CH_3CH_2CH(CH_3)CH(CO_2C_2H_5)_2 + 2KOH \rightarrow$$
$$CH_3CH_2CH(CH_3)CH(CO_2K)_2 + 2C_2H_5OH$$
$$CH_3CH_2CH(CH_3)CH(CO_2K)_2 + H_2SO_4 \rightarrow$$
$$CH_3CH_2CH(CH_3)CH_2CO_2H + CO_2 + K_2SO_4$$

Submitted by E. B. VLIET, C. S. MARVEL, and C. M. HSUEH.
Checked by HENRY GILMAN and R. E. BROWN.

1. Procedure

A SOLUTION of 200 g. (3.6 moles) of potassium hydroxide (Note 1) in 200 cc. of water is placed in a 2-l. round-bottomed flask fitted with a reflux condenser, a mechanical stirrer, and a separatory funnel. The stirrer is started, and to the hot solution 200 g. (0.92 mole) of ethyl sec.-butylmalonate (Note 2) is added slowly. The solution refluxes quietly owing to the heat of saponification. After all the sec.-butyl-malonic ester has been added, the solution is boiled gently for two hours. It is then diluted with 200 cc. of water, and 200 cc. of liquid is distilled from the solution in order to remove all the alcohol formed during the saponification (Note 3).

The residual liquid in the flask is allowed to cool, and a cold solution of 320 g. (3.3 moles) of concentrated sulfuric acid (Note 4) in 450 cc. of water is added through the separatory funnel. This should be done slowly and with stirring in order to prevent foaming. The solution becomes hot and may reflux spontaneously. After all the sulfuric acid has been added the solution is refluxed for about three hours. A layer of organic acid appears and the reflux condenser is replaced by an automatic separator (Note 5). The solution is distilled with the separator attached, and the aqueous portion is returned to the distilling flask. This operation is continued until practically all the organic acid has been driven over; from ten to fifteen hours is required. At the end, about 100 cc. of water is collected in the separator and extracted with ether in order to remove the dissolved acid (Note 6). The ether is distilled; the crude acid is mixed with an equal volume of dry benzene (Note 7) and distilled from a modified Claisen flask with a fractionating side arm. Benzene and water distil first and then 3-methylpentanoic acid distils at 193–196°/743 mm. The yield is 66–69 g. (62–65 per cent of the theoretical amount) (Note 8).

2. Notes

1. Sodium hydroxide is unsatisfactory for this saponification because an organic sodium salt separates, yielding a semi-solid mass.

2. The ethyl *sec.*-butylmalonate was prepared from *sec.*-butyl bromide and malonic ester according to the general method described in Org. Syn. Coll. Vol. I, **1941**, 250. The yield of ester boiling at 124–132°/28 mm. was 80–81 per cent of the theoretical amount. The yields of 3-methylpentanoic acid given in this procedure were obtained with this grade of ester.

3. It is necessary that all the alcohol be eliminated after the saponification. If some is allowed to remain, ethyl 3-methylpentanoate is formed and a considerable amount of low-boiling material is present in the final product.

4. If hydrochloric acid is used, it distils and complicates the purification of the product.

5. The apparatus shown in Org. Syn. Coll. Vol. I, **1941**, 422, Fig. 22, was found to be very effective. Other methods of isolating the acid were not so satisfactory. When extraction with a solvent was used, the yield was about 10 per cent less and the product was contaminated with tarry impurities.

6. The amount of acid recovered by the ether extraction is only 1–2 g.

7. The acid may be dried by other means, but this method was found most convenient.

8. Other acids may be prepared by this general procedure; thus *n*-caproic acid may be obtained from ethyl *n*-butylmalonate in 75 per cent yield.

3. Methods of Preparation

3-Methylpentanoic acid has been prepared by heating *sec.*-butyl-malonic acid,[1] and by the addition of ethylmagnesium bromide to the N-methylanilide of crotonic acid, followed by hydrolysis.[2]

---

[1] Van Romburgh, Rec. trav. chim. **6,** 153 (1887); Kulisch, Monatsh. **14,** 559 (1893); Bentley, J. Chem. Soc. **67,** 267 (1895); Olivier, Rec. trav. chim. **55,** 1032 (1936).

[2] Maxim and Ioanid, Bul. Soc. Chim. România **12,** 28 (1930) [C. A. **25,** 488 (1931)].

α-METHYL-α-PHENYLHYDRAZINE

(Hydrazine, 1-methyl-1-phenyl-)

$$C_6H_5N(NO)CH_3 + 4[H] \xrightarrow{(Zn)} C_6H_5N(NH_2)CH_3 + H_2O$$

Submitted by W. W. HARTMAN and L. J. ROLL.
Checked by LOUIS F. FIESER and J. T. WALKER.

1. Procedure

A MIXTURE of 200 g. (3.1 gram atoms) of zinc dust and 300 cc. of water is placed in a 2-l. three-necked flask equipped with an efficient mechanical stirrer, a thermometer, and a dropping funnel. The suspension is then stirred vigorously while a solution of 100 g. (0.73 mole) of N-nitrosomethylaniline (p. 460) in 200 cc. (210 g., 3.5 moles) of glacial acetic acid is added in a slow stream. The temperature is maintained between 10° and 20° by external cooling or by the addition of finely crushed ice. When all the acid solution has been added (about one and one-half to two hours is required, depending upon the rate of cooling) the mixture is stirred for an hour longer at room temperature and then warmed to 80° on the steam bath. The hot solution is filtered from the unreacted zinc, which is washed with three 100-cc. portions of a warm 5 per cent hydrochloric acid solution. The combined filtrate and washings are cooled and treated with sufficient 40 per cent sodium hydroxide solution to redissolve the zinc hydroxide precipitated. (About 1.2 l. is required.) The oily layer is separated and the aqueous layer extracted with two or three 100-cc. portions of ether. The combined oil and extracts are distilled from a steam bath until the ether is removed; the residue is then distilled under reduced pressure. The yield of colorless (Note 1) α-methyl-α-phenylhydrazine boiling at 106–109°/13 mm. is 46–50 g. (52–56 per cent of the theoretical amount).

2. Note

1. α-Methyl-α-phenylhydrazine darkens on standing.

3. Methods of Preparation

α-Methyl-α-phenylhydrazine has been obtained by reducing nitrosomethylaniline with zinc and acetic acid [1] or electrolytically; [2] by reducing

[1] Fischer, Ann. 190, 152 (1878); 236, 198 (1886).
[2] Wells, Babcock, and France, J. Am. Chem. Soc. 58, 2630 (1936).

N-methylnitroformaldehyde phenylhydrazone with zinc and acetic acid;[3] by heating methylbenzoylphenylhydrazine with concentrated hydrochloric acid;[4] and by methylating phenylhydrazine with sodamide and methyl iodide.[5]

1-METHYL-2-PYRIDONE

[2(1)-Pyridone, 1-methyl-]

$$C_5H_5N + (CH_3)_2SO_4 \rightarrow (C_5H_5NCH_3)SO_4CH_3$$

$$(C_5H_5NCH_3)SO_4CH_3 + 3NaOH + 2K_3Fe(CN)_6 \rightarrow$$

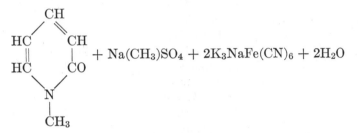

$$+ Na(CH_3)SO_4 + 2K_3NaFe(CN)_6 + 2H_2O$$

Submitted by E. A. PRILL and S. M. MCELVAIN.
Checked by C. S. MARVEL and SIDNEY H. BABCOCK.

1. Procedure

IN a 5-l. round-bottomed flask, fitted with a separatory funnel and a reflux condenser, is placed 145 g. (1.83 moles) of dry pyridine (Note 1); 231 g. (1.83 moles) of methyl sulfate is added dropwise through the separatory funnel. After the addition of the methyl sulfate the flask is heated in a boiling water bath for two hours to complete the reaction.

The flask containing the crude pyridinium salt is removed from the condenser, and the salt is dissolved by adding 400 cc. of water. The flask is fitted with a mechanical stirrer which will stir efficiently any volume of liquid which may be in the flask, and the solution is cooled to 0° in an ice-salt bath. Separate solutions of 1.2 kg. (3.65 moles) of potassium ferricyanide in 2.4 l. of water and of 300 g. (7.5 moles) of sodium hydroxide in 500 cc. of water are prepared and added dropwise from two separatory funnels to the well-stirred solution of the pyri-

[3] Bamberger and Schmidt, Ber. **34**, 591 (1901).
[4] Tafel, ibid. **18**, 1744 (1885).
[5] Grammaticakis, Compt. rend. **210**, 303 (1940).

dinium salt at such a rate that the temperature of the reaction mixture does not rise above 10°. The rate of addition of these two solutions is regulated so that all the sodium hydroxide solution has been introduced into the reaction mixture when one-half of the potassium ferricyanide solution has been added. This usually requires about one hour. The remaining half of the potassium ferricyanide is then added during another hour. The reaction mixture is allowed to stand five hours, during which time it comes to room temperature.

The 1-methyl-2-pyridone is salted out of the reaction mixture by the addition of 400–500 g. of anhydrous sodium carbonate to the well-stirred solution. When no more sodium carbonate dissolves, stirring is discontinued and the yellow or brown oily layer containing most of the desired pyridone, together with some of the unreacted pyridinium salt, water, and inorganic salts, is separated from the aqueous mixture. The aqueous mixture is filtered (Note 2) to remove the excess sodium carbonate and the precipitated potassium or sodium ferrocyanide. The filtrate is divided into three portions, each of which is extracted twice with 200-cc. portions of technical isoamyl alcohol (Note 3). The alcohol used for the second extraction of the first aqueous portion is satisfactory for the first extraction of a second aqueous portion, et cetera, so that a total volume of 800 cc. is used. The isoamyl alcohol extracts are combined and added to the oily layer which was first separated from the reaction mixture. An aqueous layer usually appears and is separated and extracted with another 100-cc. portion of amyl alcohol.

The combined alcohol extract is distilled under reduced pressure from a modified Claisen flask by heating on a water bath. All the solvent is thus recovered for use in a subsequent preparation.

The residue is transferred to a 250-cc. modified Claisen flask and distilled under diminished pressure from an oil bath. There is almost no low-boiling fraction if the alcohol has been carefully removed. The yield of product boiling at 122–124°/11 mm. is 130–140 g. (65–70 per cent of the theoretical amount) (Notes 4 and 5). A small amount of black solid remains in the distillation flask after the pyridone has distilled.

2. Notes

1. A commercial medicinal grade of pyridine was used without further purification.

2. This operation is most satisfactory when a glass-wool plug placed in a glass funnel is used as the filter. The liquid in this funnel should be stirred to prevent the heavy precipitate from settling and clogging the filter.

3. The extraction of the pyridone from the aqueous solution is difficult when benzene or ether is used as a solvent. 1-Methyl-2-pyridone, when dry, is very soluble in ethyl ether, benzene, and most organic solvents. It is, however, practically insoluble in petroleum ether or ligroin. When a mixture of water, benzene, and a little pyridone is shaken together all the pyridone is found in the water layer. When the pyridone is salted out from water by adding a sufficient quantity of potassium carbonate, the oily layer does not dissolve when benzene or ethyl ether is added, but three layers are formed. The pyridone seems to be extracted by ether or benzene only when the aqueous solution is strongly saturated with sodium hydroxide or potassium hydroxide, and then the tendency to form an emulsion is so great that separation of the layers is extremely difficult.

When equal volumes of water, technical isoamyl alcohol, and a little pyridone are shaken together, the pyridone is found to be about equally distributed between the two solvents. Chloroform is also satisfactory for the extraction of pyridone from water.

4. The 1-methyl-2-pyridone when pure is odorless and colorless. It turns dark on standing unless it is kept in a sealed tube.

5. The boiling points at other pressures are given by Fisher and Chur [4] as 250°/740 mm., 130°/14.5 mm., 126°/12.5 mm., and 121°/10 mm.

3. Methods of Preparation

1-Methyl-2-pyridone has been prepared by the methylation of 2-pyridone,[1] by the oxidation of 1-methylpyridinium iodide [2] and of 1-methylpyridinium methyl sulfate [3] with a ferricyanide, and by the electrolytic oxidation of 1-methylpyridinium methyl sulfate.[4]

[1] v. Pechmann and Baltzer, Ber. 24, 3144 (1891).

[2] Decker, J. prakt. Chem. (2) 47, 28 (1893).

[3] Decker and Kaufmann, ibid. (2) 84, 435 (1911); Fargher and Furness, J. Chem. Soc. 107, 690 (1915).

[4] Fisher and Neundlinger, Ber. 46, 2544 (1913); Neundlinger and Chur, J. prakt. Chem. (2) 89, 466 (1914); Fisher and Chur, ibid. (2) 93, 363 (1916).

6-METHYLURACIL

(Uracil, 6-methyl-)

$$NH_2CONH_2 + CH_3COCH_2CO_2C_2H_5 \rightarrow$$
$$CH_3C(NHCONH_2){=}CHCO_2C_2H_5 + H_2O$$
$$CH_3C(NHCONH_2){=}CHCO_2C_2H_5 + NaOH \rightarrow$$
$$CH_3C(NHCONH_2){=}CHCO_2Na + C_2H_5OH$$

$$\underset{\underset{CH_3C{=}CHCO_2Na}{|}}{NHCONH_2} + HCl \rightarrow \underset{\underset{CH_3C{=}CH{-}CO}{|}}{\overset{NH{-}CO{-}NH}{|\qquad|}} + H_2O + NaCl$$

Submitted by JOHN J. DONLEAVY and MEARL A. KISE.
Checked by REYNOLD C. FUSON and WILLIAM E. ROSS.

1. Procedure

EIGHTY grams (1.33 moles) of finely powdered urea is stirred into a mixture of 160 g. (155 cc., 1.23 moles) of ethyl acetoacetate (Note 1), 25 cc. of absolute alcohol (Note 2), and ten drops of concentrated hydrochloric acid in a 5-in. crystallizing dish. The reagents are mixed well, and the dish is covered loosely with a watch glass and placed in a vacuum desiccator over concentrated sulfuric acid. The desiccator is evacuated continuously with a water pump until the mixture has gone to dryness (Note 3), which usually requires from five to seven days (Note 4). The crude β-uraminocrotonic ester when thoroughly dry weighs 200–205 g.

The dry, finely powdered, crude β-uraminocrotonic ester is stirred into a solution of 80 g. (2 moles) of sodium hydroxide in 1.2 l. of water at 95°. The clear solution is then cooled to 65° and carefully acidified, while stirring, by the slow addition of concentrated hydrochloric acid. The 6-methyluracil precipitates almost immediately, and after the mixture is cooled the product is collected on a filter, washed with cold water, alcohol, and ether, and air-dried. The substance is obtained as a colorless powder of a high degree of purity, and the yield is 110–120 g. (71–77 per cent of the theoretical amount). For further purification the pyrimidine may be crystallized from glacial acetic acid. 6-Methyluracil decomposes above 300°.

2. Notes

1. Commercial ethyl acetoacetate can be used with satisfactory results. Directions for preparing this ester are given in Org. Syn. Coll. Vol. I, **1941**, 235.

2. Larger amounts of alcohol increase the period of drying without improving the yield. When no alcohol is used, the condensation proceeds slowly and the yields are low.

3. If the condensation product is used before it is dry, a large amount of carbon dioxide is evolved later in the acidification, indicating incomplete utilization of the ethyl acetoacetate.

4. It is usually advisable to change the sulfuric acid at least daily. Any lumps should be disintegrated occasionally to aid in the drying process.

3. Methods of Preparation

The synthesis of 6-methyluracil from ethyl acetoacetate and urea was described first by Behrend.[1] The substance has been obtained also by the action of lead hydroxide on methylthiouracil in an alkaline medium;[2] by boiling benzal-2-(4-hydroxy-6-methyl)pyrimidylhydrazine with hydrochloric acid;[3] and from urea and diketene.[4]

1,2-NAPHTHALIC ANHYDRIDE

(1,2-Naphthalenedicarboxylic anhydride)

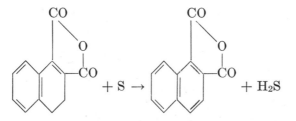

Submitted by E. B. Hershberg and Louis F. Fieser.
Checked by C. R. Noller and S. Kinsman.

1. Procedure

In a 50-cc. Claisen distilling flask with a 50-cc. sealed-on distilling flask as a receiver are placed 20 g. (0.1 mole) of 3,4-dihydro-1,2-naph-

[1] Behrend, Ann. **229,** 5 (1885); Behrend and Roosen, ibid. **251,** 238 (1889). See also Biltz and Heyn, ibid. **413,** 109 (1917).

[2] List, ibid. **236,** 23 (1886).

[3] Thiele and Bihan, ibid. **302,** 308 (1898).

[4] Carbide and Carbon Chemicals Corporation, U. S. pat. 2,138,756 [C. A. **33,** 2152 (1939)]; Standard Oil Development Company, U. S. pat. 2,174,239 [C. A. **34,** 450 (1940)]; Boese, Ind. Eng. Chem. **32,** 16 (1940).

thalic anhydride (p. 194) and 3.2 g. (0.1 gram atom) of sulfur. After the flask is immersed in a bath (Note 1) previously heated to 230–235° and shaken until the globule of sulfur has dissolved (fifteen to twenty minutes), the temperature is raised to 250° for thirty minutes (Note 2). The residue is distilled under reduced pressure (Note 3), and the distillate is crystallized from 150 cc. of benzene to which 50 cc. of ligroin (b.p. 60–80°) has been added at the boiling point. The yield is 15–18 g. (76–91 per cent of the theoretical amount) of light yellow needles melting at 166–167°.

2. Notes

1. A Wood's metal bath or a mixture (m.p. about 150°) of ten parts of potassium nitrate and seven and one-half parts of sodium nitrite may be used.

2. If the heating at 250° is continued until hydrogen sulfide is no longer evolved (about ten hours), the product, after recrystallization, is lighter in color and shrinks less before melting.

3. The material comes over between 210° and 215° at 12–13 mm. with the bath at 260°.

3. Methods of Preparation

1,2-Naphthalic anhydride has been prepared by the hydrolysis of the dinitrile of 1,2-naphthalic acid; [1] by the oxidation of suitably substituted hydrocarbons or ketones; [2] and by the dehydrogenation of the 3,4-dihydro compound with bromine [3] or with sulfur. [4]

[1] Cleve, Ber. **25**, 2475 (1892); Waldmann, J. prakt. Chem. (2) **127**, 197 (1930); Cook, J. Chem. Soc. **1932**, 462.

[2] Freund and Fleischer, Ann. **399**, 186, 210 (1913); Kruber and Schade, Ber. **68**, 11 (1935).

[3] von Auwers and Möller, J. prakt. Chem. (2) **109**, 141 (1925).

[4] Fieser and Hershberg, J. Am. Chem. Soc. **57**, 1853 (1935).

α-NAPHTHOIC ACID

(1-Naphthoic acid)

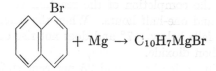

$$C_{10}H_7MgBr + CO_2 \rightarrow C_{10}H_7CO_2MgBr$$

$$2C_{10}H_7CO_2MgBr + H_2SO_4 \rightarrow 2 \underset{\text{}}{\text{(naphthalene)}} CO_2H + MgBr_2 + MgSO_4$$

Submitted by HENRY GILMAN, NINA B. ST. JOHN, and F. SCHULZE.
Checked by C. R. NOLLER.

1. Procedure

IN a 2-l. three-necked flask, fitted with a mechanical stirrer, a reflux condenser, and a separatory funnel, is placed 24.3 g. (1 gram atom) of magnesium turnings (Note 1). The magnesium is covered with 100 cc. of anhydrous ether, and 10 cc. (15 g., 0.07 mole) of α-bromonaphthalene (Note 2) (with a crystal or two of iodine) (Note 3) is added to start the reaction. A warm water bath (45° or higher) is placed under the flask until the reaction starts. The stirrer is started, and a solution of 192 g. (0.93 mole) of α-bromonaphthalene in 500 cc. of anhydrous ether is added to the magnesium at such a rate that the reaction is vigorous but not violent. The addition requires from one and one-half to three hours. The water bath is again placed under the flask, and stirring and refluxing are continued for one-half hour after the addition of the halide is complete. The Grignard reagent which collects as a heavy oil in the bottom of the flask is dissolved by the addition of 533 cc. of dry benzene (Note 4).

The reaction mixture is then cooled by an ice-salt mixture. The separatory funnel is replaced by a two-holed rubber stopper containing a thermometer (bulb immersed in the reaction mixture) and a glass tube drawn out to make a fine capillary (Note 5). When the temperature of the reaction mixture has reached −7°, the condenser is replaced by an entry tube, 10 mm. in diameter and adjusted so that the end is about

50 mm. above the surface of the reaction mixture (Note 6). The reaction mixture is stirred, and dry carbon dioxide is added through this tube (Note 7). The rate of flow of the carbon dioxide is regulated so that the temperature of the reaction mixture does not rise above $-2°$. The time required for the completion of the reaction varies from one and one-fourth to one and one-half hours. When the reaction is complete, the temperature falls below $-7°$ and does not rise on increasing the rate of flow of carbon dioxide.

The flask is placed in an ice bath, and 25 per cent sulfuric acid is added slowly, with stirring, until no further reaction takes place and all the excess magnesium has dissolved (Note 8). The oily layer is separated, and the water layer is extracted with two 100-cc. portions of ether. The combined ether-benzene extracts (Note 9) are shaken with three 100-cc. portions of 25 per cent sodium hydroxide. Each alkaline layer is extracted successively with a 100-cc. portion of ether, and the combined alkaline extracts are then heated to 100° to drive off volatile impurities.

The solution is cooled and acidified strongly with 50 per cent sulfuric acid. The crude α-naphthoic acid is collected on a Büchner funnel, washed until free from sulfate, and dried. The yield of crude material melting at 142–155° is 130–135 g. This is dissolved in 400 cc. of hot toluene, a small amount of Filter-Cel is added, and the solution is filtered through a hot Büchner funnel. The filtrate is cooled in an ice bath, filtered with suction, and the filter cake washed with cold toluene until the filtrate is practically colorless. A light-colored product melting at 159–161° is obtained. The yield is 118–121 g. (68–70 per cent of the theoretical amount) (Note 10).

2. Notes

1. The finer commercial grade of turnings was used. After the reaction had once started, it proceeded smoothly with any grade of commercial turnings.

2. The α-bromonaphthalene (Org. Syn. Coll. Vol. I, **1941**, 121) was purified by distillation under reduced pressure. The fraction boiling at 144–147°/16 mm. was collected.

3. The checker was more successful in starting the reaction within a reasonable length of time when as much as 0.5 g. of iodine was added.

4. It is necessary to dissolve the Grignard reagent in benzene to prevent it from solidifying when the solution is cooled. The benzene should be added to the solution before it cools, as the hardened mass is difficult to redissolve.

5. This is to allow the carbon dioxide which does not react to escape slowly in order that the pressure in the flask does not become too great.

6. The inlet tube is placed at this distance above the surface of the reaction mixture in order to prevent clogging.

7. The carbon dioxide from an ordinary commercial cylinder was dried by passing it through two wash bottles containing sulfuric acid.

8. Cooling the solution externally with ice permits rapid hydrolysis without danger of loss of material through refluxing (cf. Org. Syn. Coll. Vol. I, **1941**, 361).

9. If the ether-benzene solution is not entirely clear, it should be filtered before extraction with sodium hydroxide.

10. A second recrystallization from toluene gives an almost white product melting at 160.5–162°.

3. Methods of Preparation

α-Naphthoic acid has been prepared from α-naphthylmagnesium bromide and carbon dioxide, gaseous [1] or solid; [2] by hydrolysis of ethyl α-naphthoate, prepared in turn from α-naphthylmagnesium bromide; [3] by hydrolysis of α-naphthyl cyanide; [4] and from methyl α-naphthyl ketone and potassium hypochlorite.[2] The acid has been obtained by fusing sodium formate and sodium α-naphthalenesulfonate,[5] and by passing carbamyl chloride into naphthalene and aluminum chloride and hydrolyzing the α-naphthamide thus formed.[6] The methods for preparing α-naphthoic acid have been discussed by Wahl, Goedkoop, and Heberlein.[7]

The procedure described above [8] was developed from the methods of Acree,[1] of Blicke,[1] and of Whitmore and Fox.[1]

[1] Acree, Ber. **37**, 625 (1904); Blicke, J. Am. Chem. Soc. **49**, 2846 (1927); Whitmore and Fox, ibid. **51**, 3363 (1929).

[2] Fieser, Holmes, and Newman, ibid. **58**, 1055 (1936).

[3] Loder and Whitmore, ibid. **57**, 2727 (1935), and p. 282, above.

[4] Merz and Mühlhaüser, Ber. **3**, 712 (1870).

[5] Meyer, ibid. **3**, 364 (1870).

[6] Gattermann and Schmidt, ibid. **20**, 860 (1887).

[7] Wahl, Goedkoop, and Heberlein, Bull. soc. chim. (5) **6**, 533 (1939).

[8] Gilman, St. John, and St. John, Rec. trav. chim. **48**, 594 (1929).

β-NAPHTHOIC ACID

(2-Naphthoic acid)

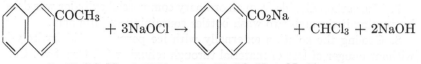

$$C_{10}H_7CO_2Na + HCl \rightarrow C_{10}H_7CO_2H + NaCl$$

Submitted by M. S. Newman and H. L. Holmes.
Checked by W. W. Hartman and F. W. Jones.

1. Procedure

In a 3-l. flask is placed a solution of 184 g. (4.6 moles) of sodium hydroxide in 300–400 cc. of water; sufficient ice is added to make the total volume about 1.5 l. Chlorine is passed into the solution, the temperature being kept below 0° by means of a salt-ice bath, until the solution is neutral to litmus, then a solution of 34 g. of sodium hydroxide in 50 cc. of water is added (Notes 1 and 2). The flask is now supported by a clamp and equipped with a thermometer and an efficient stirrer. The solution is warmed to 55°, and 85 g. (0.5 mole) of methyl β-naphthyl ketone (Note 3) is added. The mixture is vigorously stirred, and, after the exothermic reaction commences, the temperature is kept at 60–70° (Note 4) by frequent cooling in an ice bath until the temperature no longer tends to rise. This requires thirty to forty minutes. The solution is stirred for thirty minutes longer and then the excess hypochlorite is destroyed by adding a solution of 50 g. of sodium bisulfite in 200 cc. of water (Note 5). After cooling to room temperature, the reaction mixture is transferred to a 4-l. beaker and carefully acidified with 200 cc. of concentrated hydrochloric acid. The crude colorless acid is collected on a Büchner funnel, washed with water, and sucked as dry as possible with a rubber dam. After drying, the acid is crystallized (Note 6) from 600 cc. of 95 per cent alcohol, giving 75–76 g. (87–88 per cent of the theoretical amount) of β-naphthoic acid melting at 184–185° (corr.). By distilling 450 cc. of solvent from the mother liquor, an additional 9 g. (10 per cent of the theoretical amount) of acid, m.p. 181–183° (corr.), is obtained (Note 7).

2. Notes

1. It is reported that, in preparing a sodium hypochlorite solution by passing chlorine into sodium hydroxide, it is very difficult to determine

the neutral point because of the instant bleaching of the indicator. If too much chlorine is added, even though the final solution may react alkaline because of the additional sodium hydroxide added later, oxidation of methyl β-naphthyl ketone to β-naphthoic acid does not take place. Consequently the following modified procedure for preparing the hypochlorite solution is recommended.

A solution of 218 g. (5.45 moles) of sodium hydroxide in 300 cc. of water in a 3-l. flask is cooled to room temperature with tap water. Next, 1250 g. of ice is added and chlorine is passed in rapidly until 161 g. (4.5 moles) has been taken up. External cooling is unnecessary. With the amount of ice specified, the solution will be at 0° when the addition of chlorine has been completed. From this point on the procedure is the same as that on p. 428. (Private communication from EDWARD C. STERLING.)

2. The hypochlorite solution also may be prepared conveniently from the calcium hypochlorite sold by the Mathieson Alkali Works under the trade name "HTH" and specified to contain not less than 65 per cent of available calcium hypochlorite.

In a 3-l. round-bottomed flask 250 g. of commercial calcium hypochlorite is dissolved in 1 l. of warm water, and a warm solution of 175 g. of potassium carbonate and 50 g. of potassium hydroxide in 500 cc. of water is added. The flask is stoppered and shaken vigorously until the semi-solid gel which first forms becomes quite fluid. The suspended solid is removed by filtration on a large Büchner funnel, washed with 200 cc. of water, and sucked as dry as possible with the aid of a rubber dam and an efficient suction pump. The filtrate of approximately 1.5 l. is placed in a 3-l. round-bottomed flask and is ready for the addition of methyl β-naphthyl ketone.

Such a solution contains approximately 200 g. (2.3 moles) of potassium hypochlorite. Sodium or potassium hypochlorite may be used, but the calcium salt is not satisfactory because the calcium salt of β-naphthoic acid is sparingly soluble.

3. The Eastman product, m.p. 53–55°, was used.

4. If the mixture is not cooled, the reaction will get out of control because of the rapid evolution of chloroform, and some ketone may steam-distil.

5. It is advisable to test the solution after the addition of the sodium bisulfite with acidified potassium iodide solution to be sure that all the hypochlorite has been destroyed. If hypochlorite is present, the chlorine liberated when the solution is acidified forms a high-melting impurity.

6. The moist acid may be crystallized without drying, but more alcohol must be used to get the product into solution.

7. This method may be used for the preparation of larger quantities, a batch twenty times this size giving a yield of 87 per cent. It may be used also for the preparation of other aromatic acids where suitable ketones are available.

3. Methods of Preparation

β-Naphthoic acid has been prepared principally by the hydrolysis of β-naphthonitrile,[1] the over-all yields from β-naphthylamine, from sodium β-naphthalenesulfonate, and from calcium β-naphthalenesulfonate being given as (approximately) 20 per cent, 21 per cent, and 50 per cent, respectively.[2] The acid has been prepared also by the carbonation of the Grignard reagent from the less accessible β-bromo derivative;[3] by chlorination of β-methylnaphthalene followed by hydrolysis and oxidation;[4] and by the procedure described above.[5] The various methods for preparing β-naphthoic acid have been discussed by Wahl, Goedkoop, and Heberlein.[4]

1,2-NAPHTHOQUINONE

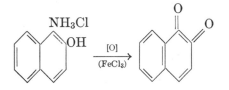

Submitted by Louis F. Fieser.
Checked by C. R. Noller and W. R. White.

1. Procedure

For the best results this preparation must be carried out rapidly. The vessels and reagents required should be made ready in advance. The oxidizing solution is prepared by dissolving 240 g. (0.89 mole) of ferric chloride hexahydrate in a mixture of 90 cc. of concentrated hydrochloric acid and 200 cc. of water with heating, cooling to room

[1] Baeyer and Besemfelder, Ann. **266,** 187 (1891).
[2] Derick and Kamm, J. Am. Chem. Soc. **38,** 408 (1916).
[3] Gilman and St. John, Rec. trav. chim. **48,** 743 (1929).
[4] Wahl, Goedkoop, and Heberlein, Bull. soc. chim. (5) **6,** 533 (1939).
[5] Fieser, Newman, and Holmes, J. Am. Chem. Soc. **58,** 1055 (1936).

temperature by the addition of 200–300 g. of ice, and filtering the solution by suction.

Eighty grams (0.41 mole) of 1,2-aminonaphthol hydrochloride of a high degree of purity (Note 1) is placed in a 5-l. round-bottomed flask and covered with a solution of 5 cc. of hydrochloric acid in 3 l. of water which has been heated to 35°. The material is dissolved quickly by shaking (one to two minutes), and the solution is filtered rapidly by suction from a trace of residue and transferred to a clean 5-l. flask (Note 2). The oxidizing solution is added all at once while rotating the flask vigorously in order to mix the two solutions thoroughly. The quinone separates at once as a voluminous, microcrystalline, yellow precipitate. The product is collected on a Büchner funnel, washed well with water (Note 3), and then for more thorough washing it is transferred to a large beaker, stirred for a few minutes with 2 l. of water (at 30°), and again collected. The filter cake is cut into slices and dried on filter paper at room temperature in an atmosphere free from acid fumes. The yield is 60–61 g. (93–94 per cent of the theoretical amount).

The material is pure golden yellow in color and melts with decomposition at 145–147°, with some softening at about 140°. It contains none of the black, sparingly soluble dinaphthyldiquinhydrone and dissolves without residue in alcohol or benzene. Under the microscope it is seen to consist of a mass of well-formed, fine needles. By crystallization from alcohol or benzene it may be obtained as orange-red needles of good appearance, but the process is unreliable, it involves much loss of material, and the point of decomposition is lowered by 10–20°. It is thus best to use or to preserve the quinone in the form originally obtained; in this condition it is essentially pure and it will keep indefinitely. The material should not be pulverized, for it then becomes highly electrified.

2. Notes

1. The quality of the naphthoquinone is closely dependent upon the purity of the starting material. The crystalline product described on p. 36 may be used without recrystallization and, indeed, without being dried. The excess of oxidizing agent specified is sufficient to allow for a yield of aminonaphthol slightly higher than that given; if the yield is lower the larger excess of ferric chloride does no harm.

2. The solution at this point should be clear, but it may acquire a rather pronounced orange-yellow color when viewed in bulk. It should not, however, turn purple.

3. The yellow color of the wash water is due to a slight solubility of the quinone.

3. Methods of Preparation

The only satisfactory method of preparing 1,2-naphthoquinone (β-naphthoquinone) is by the oxidation of 1,2-aminonaphthol in acid solution, and the chief problem involved is the preparation of this intermediate in suitable yield and purity. This problem and the literature pertaining to it are discussed on p. 38. Most reports of the preparation of the aminonaphthol include some description of its oxidation, but the only particularly helpful comment on the reaction is that ferric chloride is a better oxidizing agent than chromic acid because at a low temperature it does not attack the quinone, even when present in excess.[1]

One other method of preparing β-naphthoquinone is from 1,2-bromonaphthol through the ketonitrobromide.[2] Though the parent quinone itself is such a sensitive compound that the material so obtained decomposes within a few hours, the method is of considerable value for the preparation of certain substituted β-naphthoquinones.[3]

β-NAPHTHYLMERCURIC CHLORIDE

[Naphthalene, 2-(chloromercuri)-]

$$C_{10}H_7NH_2 + NaNO_2 + 2HCl + HgCl_2 \rightarrow$$
$$C_{10}H_7N_2Cl \cdot HgCl_2 + NaCl + 2H_2O$$

$$C_{10}H_7N_2Cl \cdot HgCl_2 + 2Cu \rightarrow C_{10}H_7HgCl + N_2 + 2CuCl$$

Submitted by A. N. Nesmajanow.
Checked by Frank C. Whitmore and R. W. Beattie.

1. Procedure

To 450 cc. of concentrated hydrochloric acid (sp. gr. 1.19) and 500 cc. of water, in a 4-l. (1-gal.) earthenware crock equipped with an efficient stirrer, is added 143 g. (1 mole) of β-naphthylamine. The suspension of the amine hydrochloride is cooled by the addition of 500 g. of cracked ice. When the temperature reaches 5°, solid sodium nitrite (about 69 g.) is added until starch-iodide paper shows an excess. During the diazotization about 600 g. of cracked ice is introduced at such a rate as to keep

[1] Groves, J. Chem. Soc. **45**, 298 (1884).

[2] Fries, Ann. **389**, 315 (1912).

[3] Fries and Schimmelschmidt, ibid. **484**, 245 (1930).

the temperature at 5°. The cold solution of the diazonium salt is filtered to remove a small amount of precipitate and returned to the crock. A solution of 271 g. (1 mole) of mercuric chloride in 300 cc. of concentrated hydrochloric acid is mixed with 300 g. of ice and added slowly to the rapidly stirred solution. A heavy precipitate of the yellow addition compound of β-naphthalenediazonium chloride and mercuric chloride separates. Stirring is continued for a half hour, after which the precipitate is collected on a 20-cm. Büchner funnel, sucked as dry as possible, and washed with two 400-cc. portions of water and two 150-cc. portions of acetone (Note 1). The solid is air-dried at about 20° (Note 2) to constant weight. The yield is 380–390 g. (82–84 per cent of the theoretical amount) (Note 3).

In a 2-l. round-bottomed Pyrex flask equipped with a stirrer are placed 139 g. (0.3 mole) of the addition compound, 700 cc. of acetone (Note 4), and 38 g. (0.6 gram atom) of copper powder (Note 5). The mixture is cooled at once to about 20°, stirred for one hour, and then allowed to stand overnight. The insoluble material is collected on a Büchner funnel (Note 6). The solid mixture is extracted with about 3 l. of boiling commercial xylene and filtered through a hot funnel. On cooling the filtrate, crystals of β-naphthylmercuric chloride separate (Note 7). The yield of product which melts at 266–267° is 52–64 g. (40–49 per cent of the theoretical amount based on the amine used) (Notes 8 and 9).

2. Notes

1. The washing with large amounts of solvents is necessary to remove small amounts of highly colored impurities. The product is practically insoluble in the cold wash liquids.

2. The double salt *explodes violently* if heated. For preparation of the organic mercury compounds the double salt need not be completely dry.

3. It is unsafe to store the addition complexes of mercuric chloride and diazonium chlorides in a closed vessel. The complex salt should be used as soon as possible and, if it is to be kept for any length of time, should be placed in a large open dish or a loosely covered container.

In one of a number of preparations, a batch of about 960 g. of the addition compound from benzenediazonium chloride was air-dried and placed in a large bottle with a screw cap. About four hours later the entire complex decomposed with considerable violence. At the time no heat or strong light was near the bottle. (HENRY GILMAN, private communication.)

4. The acetone employed boiled at 55–57°. The use of methyl or ethyl alcohol decreases the yield slightly.

5. The copper powder is prepared from zinc dust and an excess of copper sulfate solution maintained below 70°.

6. The acetone filtrate contains only a trace of the product.

7. The mother liquor from these crystals may be used again to extract the original crude residue, since the product is practically insoluble in cold xylene.

8. If hydrobromic acid and mercuric bromide are used, the product is β-naphthylmercuric bromide.

9. In a similar way, aniline, the three toluidines, and the aminophenols can be converted into the corresponding chloromercuric compounds in about 40 per cent yields.

3. Methods of Preparation

β-Naphthylmercuric chloride has been prepared by the action of mercuric chloride in amyl alcohol on mercury di-β-naphthyl;[1] from β-naphthalenesulfinic acid and mercuric chloride;[2] from β-naphthylboric acid and mercuric chloride;[3] and by the procedure described above.

m-NITROACETOPHENONE

(Acetophenone, m-nitro-)

$$C_6H_5COCH_3 + HNO_3 \rightarrow m\text{-}NO_2C_6H_4COCH_3 + H_2O$$

Submitted by B. B. CORSON and R. K. HAZEN.
Checked by ROGER ADAMS and W. W. MOYER.

1. Procedure

IN a 1-l. wide-mouthed Erlenmeyer flask, immersed in an ice-salt bath contained in a 2-gal. earthenware crock, is placed 150 cc. of concentrated sulfuric acid. The flask is equipped with an efficient mechanical stirrer, a small dropping funnel, and a thermometer reaching almost to the bottom of the flask (Note 1). The stirrer is started, and, when the sulfuric acid has cooled to ice temperature or below, 60 g. (0.5 mole) of pure acetophenone is slowly dropped in from the dropping funnel at

[1] Michaelis, Ber. **27**, 251 (1894).

[2] Whitmore and Howe, "Organic Compounds of Mercury," by Frank C. Whitmore, The Chemical Catalog Company, New York, 1921, p. 203.

[3] König and Scharrnbeck, J. prakt. Chem. (2) **128**, 168 (1930).

such a rate (about ten minutes for the addition) that the temperature does not rise above 5° (Note 2). After the reaction mixture has cooled, this time to about −7°, the cooled (15–20°) nitrating mixture, consisting of 40 cc. (0.65 mole) of nitric acid (sp. gr. 1.42 at 15.5°) (Note 3) and 60 cc. of concentrated sulfuric acid, is added through the dropping funnel at such a rate (100–120 drops per minute) that the temperature of the reaction mixture remains at 0° or lower (Note 4). After the nitrating acid has been added, stirring is continued for ten minutes longer and the contents of the flask are poured (Note 5), with vigorous manual stirring, into a mixture of 750 g. of cracked ice and 1.5 l. of water. The product separates as a yellow flocculent solid.

After the ice has melted, the product is filtered by suction and the somewhat sticky mass pressed as dry as possible. It is transferred to a mortar and triturated with three successive 300-cc. portions of cold water (to remove acid); it is then stirred to a mush with two successive 25-cc. portions of ice-cold ethyl alcohol (to remove adhering oil); the solid is pressed dry on the suction filter after each of the five washings. The product is pressed on a porous plate and, when fairly dry (about one hour), is dissolved in 100–120 cc. of hot ethyl alcohol (Note 6). The dark solution is filtered quickly through a small suction funnel (Note 7), and the hot filtrate is poured slowly into 1 l. of cold water which is stirred vigorously (Note 8) with a stirring rod during the addition and for several minutes afterward. After standing a few minutes the yellow solid is filtered by suction, washed with 200 cc. of cold water, sucked dry, and pressed out on a porous clay plate.

When the reprecipitated *m*-nitroacetophenone (50–55 g.) is dry it is dissolved in 100 cc. of hot alcohol in a 300-cc. Erlenmeyer flask. The flask is then immersed in an ice bath and shaken vigorously while crystallization is taking place. Because of the great change in solubility between 60° and 50° the agitation of the liquid must be vigorous during this temperature interval, or large clumps of crystals will be formed instead of the purer and more easily dried mush. After the temperature has reached 20° (about one hour) the mixture is filtered by suction (Note 9). The solid (Note 10) is washed with 10 cc. of ice-cold alcohol and pressed dry on a clay plate. The product is light yellow; it softens at 74° and melts at 76–78°. The total yield is 45 g. (55 per cent of the theoretical amount) (Note 11).

2. Notes

1. The only flask found suitable for this preparation is the 1-l. wide-mouthed Erlenmeyer. A large flask is necessary to ensure rapid cooling.

A propeller with long, wide blades agitates the viscous liquid much more efficiently than a stirrer of the centrifugal type. The blades should be as long as allowed by the wide mouth of the flask. The thermometer which is to record the temperature of the reaction mixture should enter at an angle and reach almost to the bottom of the flask, since the amount of liquid is small. The thermometer should have the zero point at least 15 cm. from the bulb in order to facilitate reading the temperature. It is essential that the temperature be watched throughout the experiment. A smaller ice-salt bath than that contained in a 2-gal. earthenware crock is inadequate.

2. The same dropping funnel can be used, without washing, for the addition of the concentrated sulfuric acid, the acetophenone, and the nitrating mixture. With rapid stirring the acetophenone can be easily added in seven minutes without raising the temperature of the reaction mixture above 3°.

3. Nitric acid of lower specific gravity than 1.42 at 15.5° yields an impure product. Ordinary concentrated nitric acid usually has to be strengthened by the addition of fuming nitric acid.

4. Two conditions are necessary for a good nitration: low temperature (0° or below) and rapidity of addition of the nitrating mixture (not longer than forty-five minutes). With efficient cooling the temperature can be held between −5 and 0°. If the temperature should rise once or twice to 3° no harm is done provided that the reaction mixture is quickly cooled back to the correct temperature. In order to avoid local heating the delivery tube of the dropping funnel should be so directed as to deliver the nitrating acid near the site of greatest agitation.

The rate of stirring must be rapid. The optimum speed will be different for different stirrers, depending on the shape and size of the blades. For the stirrer described the speed is 1600 r.p.m. During the addition of acetophenone this high speed is not necessary; in fact, it cannot be maintained on account of excessive splashing. However, during the addition of the nitrating acid the reaction mixture thickens and high-speed stirring becomes possible. The ice-salt mixture must be stirred repeatedly with a stick, and fresh ice and salt must be added from time to time. The temperature of the bath should be around −16°. Enough liquid should be present in the cooling bath so that the lower half of the flask is immersed in brine. If no especial care is exercised, the addition of the nitrating acid requires from two to three hours in order to maintain the temperature at 0° or below. This long exposure to the mixture of sulfuric and nitric acids is as harmful as a rise of temperature; the product is of poor quality and the yield drops from 55 to 15 per cent or less.

By using solid carbon dioxide as the cooling medium and also adding it directly to the reaction mixture, a temperature of −20° can be maintained during the nitration.[1] According to W. W. Hartman, private communication, the use of solid carbon dioxide in this way permits larger runs of *m*-nitroacetophenone to be made.

5. The experiment should not be stopped until the product has been poured into ice water.

6. Although 55° suffices to dissolve the *m*-nitroacetophenone it is desirable to bring the alcoholic solution to a boil to avoid crystallization during filtration.

7. Excessive frothing in the filter flask is easily checked by pinching off the suction tube from time to time.

8. Vigorous stirring and slow pouring are absolutely necessary. In the absence of either, the product will be lumpy rather than flocculent.

9. Since the solubility curve is so flat in the vicinity of room temperature it is not necessary to cool very low before filtering. The difference in solubility between 20° and 10° amounts to only 1 g. per 100 cc. of solution. The solubility of *m*-nitroacetophenone in 96 per cent alcohol is as follows:

TEMPERATURE	WEIGHT IN 10 CC.	TEMPERATURE	WEIGHT IN 10 CC.
8°	0.16	48°	1.88
17	0.25	50	2.38
22	0.31	52	3.08
23	0.34	53.5	4.05
27	0.41	56	5.46
28	0.45	57	8.05
32	0.64	59	11.00
39	0.97	60	15.60
42	1.22		

10. A small additional yield (about 5 g.) may be obtained from the mother liquor by concentrating to 20 cc. and then cooling in ice.

11. This work was aided by a grant from the Cyrus M. Warren Fund of the American Academy of Arts and Sciences.

3. Methods of Preparation

m-Nitroacetophenone has usually been prepared by the nitration of acetophenone;[1,2] it has been made also by the hydrolysis of *m*-nitro-

[1] Barkenbus and Clements, J. Am. Chem. Soc. **56**, 1369 (1934).

[2] Emmerling and Engler, Ber. **3**, 886 (1870); Buchka, ibid. **10**, 1714 (1877); Engler, ibid. **18**, 2238 (1885); v. Kostanecki and Tambor, ibid. **34**, 1691 (1901); Rupe, Braun, and v. Zembruski, ibid. **34**, 3522 (1901); Camps, Arch. Pharm. **240**, 5 (1902).

benzoylacetoacetic ester.[3] The procedure given above has been modified by Morgan and Watson; the modified procedure is reported to
increase the yield of m-nitroacetophenone to 83 per cent of the theoretical
amount.[4]

1-NITRO-2-ACETYLAMINONAPHTHALENE

[Acetamide, N-(1-nitro-2-naphthyl)-]

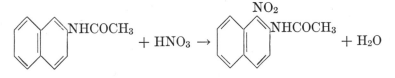

Submitted by W. W. HARTMAN and LLOYD A. SMITH.
Checked by LOUIS F. FIESER and J. T. WALKER.

1. Procedure

IN a 2-l. round-bottomed flask equipped with a mechanical stirrer, a
thermometer, and a dropping funnel, and clamped in a position such
that it may be surrounded by a bath of water and ice when desired, are
placed 300 g. (1.62 moles) of β-acetylaminonaphthalene (m.p. 131–132°)
and 500 cc. of glacial acetic acid. With the mixture at room temperature, the stirrer is set in motion, and 200 g. (143 cc., 2.1 moles) of concentrated nitric acid (sp. gr. 1.4) is added dropwise over a period of forty-
five minutes; the temperature is kept below 40° by occasionally immersing the flask in the cooling bath. When about one-tenth of the nitric
acid has been added, the reaction mixture sets to a mass that is stirred
with difficulty. The addition of nitric acid is stopped at this point.
After three to five minutes the mass becomes fluid, and the addition of
nitric acid is resumed. When about one-fourth of the nitric acid has
been added, all the solid material dissolves. Considerable heat is
evolved at this point, and good cooling is required to keep the temperature from rising above 40°; the mixture must be cooled during the addition of the remainder of the acid. After the addition is complete, stirring is continued for ten minutes longer.

The flask is stoppered and cooled in a bath of ice and water for three
hours; the reaction product should then separate in the form of a yellow,
crystalline paste (Notes 1 and 2). The crystals are collected on a

[3] Gevekoht, Ann. **221**, 334 (1883).
[4] Morgan and Watson, J. Soc. Chem. Ind. **55**, 29T (1936).

19-cm. Büchner funnel and washed, first with 200 cc. of 50 per cent acetic acid, and next with 400 cc. of ordinary ether. This crude, dry product, weighing 270–290 g., is placed in a 3-l. flask and heated under a reflux condenser with 1.7 l. of benzene for twenty minutes. The mixture is then allowed to cool to 40–45° and filtered through a 19-cm. Büchner funnel. The residue is a mixture of sparingly soluble isomers, chiefly 5- and 8-nitro-2-acetylaminonaphthalene. On further cooling of the filtered solution, there is obtained 190–200 g. of 1-nitro-2-acetylaminonaphthalene melting at 117–119°. This material is recrystallized from about 500 cc. of hot 95 per cent ethyl alcohol. Fine yellow crystals melting at 123–124° are obtained. The yield is 175–182 g. (47–49 per cent of the theoretical amount).

2. Notes

1. The material is sometimes slow in crystallizing, and it is then advisable to scratch the walls of the container with a stirring rod. If this fails to induce crystallization, seed may be obtained by diluting a small portion of the solution with water.

2. Crystallization is nearly, but not entirely, complete after this period of time. When the mother liquor was allowed to stand for five days, an additional quantity of material separated. From this material only 3 g. of pure 1-nitro-2-acetylaminonaphthalene was obtained.

3. Method of Preparation

This procedure is based upon previous studies of the nitration of 2-acetylaminonaphthalene.[1]

[1] Jacobson, Ber. **14**, 803 (1881); Liebermann and Jacobson, Ann. **211**, 44 (1882); Friedlaender and Littner, Ber. **48**, 328 (1915); Schiemann and Ley, ibid. **69**, 963 (1936).

NITROBARBITURIC ACID

(Barbituric acid, 5-nitro-)

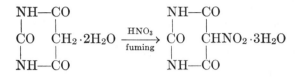

Submitted by W. W. HARTMAN and O. E. SHEPPARD.
Checked by C. S. MARVEL and B. H. WOJCIK.

1. Procedure

IN a 2-l. flask, equipped with a mechanical stirrer and surrounded by an ice bath, is placed 143 cc. of fuming nitric acid (sp. gr. 1.52). Stirring is started, and 100 g. (0.61 mole) of barbituric acid (p. 60) is added over a period of two hours; the temperature is kept below 40° during the addition. The mixture is stirred for one hour after the barbituric acid has been added, and stirring is continued while 430 cc. of water is added and the solution is cooled to 10°. The mixture is filtered, and the residue is washed with cold water and dried on a glass tray at 60–80° (Note 1). The nitrobarbituric acid is dissolved by adding it to 860 cc. of boiling water in a 2-l. flask and heating the mixture on a boiling water bath while steam is blown in until solution is complete (Note 2). After filtration and cooling overnight, the crystals are removed, washed with cold water, and dried in trays in an oven at 90–95° for two to three hours. The product melts with decomposition at 181–183° when heated rapidly. The yield is 139–141 g. (Note 3). On drying the product at 110–115° for two to three hours, the yield is 90–94 g. (85–90 per cent of the theoretical amount) of an anhydrous compound which melts with decomposition at 176°.

2. Notes

1. Unless the product is dried before recrystallization it is difficult to remove all the nitric acid, and the final product will have a strong odor of nitric acid.

2. If a clear yellow solution is not obtained, Norite should be added before filtering.

3. This yield is slightly above the theoretical yield of 139 g., but this is probably due to a greater degree of hydration than is indicated in the formula.

3. Methods of Preparation

Nitrobarbituric acid has been prepared by oxidation of violuric acid,[1] and by treatment of barbituric acid with fuming nitric acid [2] or concentrated nitric acid.[3]

p-NITROBENZALDEHYDE

(Benzaldehyde, *p*-nitro-)

$$(A) \quad p\text{-}O_2NC_6H_4CH_3 + 2(CH_3CO)_2O + 2[O] \xrightarrow{(CrO_3)}$$
$$p\text{-}O_2NC_6H_4CH(OCOCH_3)_2 + 2CH_3CO_2H$$

$$(B) \quad p\text{-}O_2NC_6H_4CH(OCOCH_3)_2 + H_2O \xrightarrow{(H_2SO_4)}$$
$$p\text{-}O_2NC_6H_4CHO + 2CH_3CO_2H$$

Submitted by S. V. LIEBERMAN and RALPH CONNOR.
Checked by JOHN R. JOHNSON and E. A. CLEVELAND.

1. Procedure

(A) *p-Nitrobenzaldiacetate.*—In a 2-l. three-necked, round-bottomed flask equipped with a mechanical stirrer and a thermometer, and surrounded by an ice-salt bath, are placed 570 cc. (600 g.) of glacial acetic acid, 565 cc. (612 g., 6 moles) of acetic anhydride (Note 1), and 50 g. (0.36 mole) of *p*-nitrotoluene (Note 2). To this solution is added slowly, with stirring, 85 cc. (1.5 moles) of concentrated sulfuric acid (Note 3). When the mixture has cooled to 5° 100 g. (1 mole) of chromium trioxide (Note 2) is added in small portions at such a rate that the temperature does not rise above 10° (Note 4), and stirring is continued for ten minutes after the chromium trioxide has been added. The contents of the flask are poured into two 3-l. beakers two-thirds filled with chipped ice, and cold water is added until the total volume is 5–6 l. The solid is separated by suction filtration and washed with cold water until the washings are colorless. The product is suspended in 500 cc. of cold 2 per cent sodium carbonate solution and stirred mechanically. After thorough mixing, the solid is collected on a filter (Note 5) and washed with cold water and finally with 20 cc. of cold alcohol. The product, after drying in a vacuum desiccator, weighs 44–49 g. (48–54 per cent of the theoretical amount), m.p. 120–122° (Note 6).

[1] Ceresole, Ber. **16**, 1134 (1883).
[2] Baeyer, Ann. **130**, 140 (1864).
[3] Fredholm, Z. anal. Chem. **104**, 400 (1936).

The crude material is suitable for hydrolysis, and for other reactions, without further purification. The pure diacetate may be obtained by recrystallizing from 150 cc. of hot alcohol. The hot solution is filtered through a fluted filter to remove insoluble impurities. The yield of pure p-nitrobenzaldiacetate is 43–46 g. (47–50 per cent of the theoretical amount), m.p. 125–126°.

(B) p-Nitrobenzaldehyde.—A mixture of 45 g. (0.18 mole) of crude p-nitrobenzaldiacetate, 100 cc. of water, 100 cc. of alcohol, and 10 cc. of concentrated sulfuric acid is refluxed for thirty minutes and filtered through a fluted filter, and the filtrate is chilled in an ice bath. The crystals are separated by suction filtration, washed with cold water, and dried in a vacuum desiccator. The first crop weighs 22–24 g. (82–89 per cent of the theoretical amount), m.p. 106–106.5°. A second crop amounting to 2–3 g. is obtained by diluting the filtrate with about 300 cc. of water. The total yield is 24–25.5 g. (89–94 per cent of the theoretical amount) (Note 6).

2. Notes

1. The "practical" grade of acetic anhydride (95 per cent) gave yields as high as those obtained with 99–100 per cent acetic anhydride.

2. The p-nitrotoluene used was a "practical" grade, m.p. 50–51°. The chromium trioxide was a U.S.P. grade of 98 per cent purity.

3. If the sulfuric acid is added too rapidly, charring occurs.

4. It is essential that the temperature of the reaction mixture be maintained below 10°. If the oxidant is added so rapidly that this temperature is exceeded, the yield is lowered considerably. With a good ice-salt bath, the time required for the addition is forty-five to sixty minutes.

5. By acidification of the sodium carbonate washings, 7–10 g. of p-nitrobenzoic acid, m.p. 242–243°, is obtained.

6. This procedure appears to be generally applicable to the preparation of those substituted benzaldehydes in which the substituent is not attacked in the oxidation reaction. Thus, the submitters report that p-bromobenzaldehyde may be prepared by the same procedure, substituting 62 g. (0.37 mole) of p-bromotoluene for the p-nitrotoluene and carrying out the oxidation and isolation in the same manner. The yield of crude p-bromobenzaldiacetate is 51–64 g. (48–60 per cent of the theoretical amount), m.p. 90–92°. The pure diacetate is obtained by dissolving the crude product in 150 cc. of hot alcohol, filtering through a fluted filter, and cooling. Filtration gives 39–52 g. of pure material, m.p. 94–95°. A second crop is obtained by diluting the filtrate. The total yield is 47–56 g. (46–54 per cent of the theoretical amount, based

on the bromotoluene). The crude product is hydrolyzed to *p*-bromobenzaldehyde by refluxing 45 g. (0.157 mole) with 100 cc. of water, 150 cc. of alcohol, and 10 cc. of concentrated sulfuric acid, filtering through a fluted filter, cooling, and filtering with suction. A second crop is obtained by diluting the filtrate with about 300 cc. of water. The product weighs 24–28 g. (83–96 per cent of the theoretical amount), m.p. 55–57°.

It is further reported that *p*-cyanotoluene furnishes *p*-cyanobenzaldehyde, the yield being about the same as that in the preparation of *p*-nitrobenzaldehyde. (LEONARD WEISLER, private communication.)

3. Methods of Preparation

p-Nitrobenzaldehyde has been prepared from *p*-nitrotoluene by treatment with isoamyl nitrite in the presence of sodium methoxide [1] or by oxidation with chromyl chloride,[2] cerium dioxide,[3] or chromium trioxide in the presence of acetic anhydride.[4] It can also be prepared by the oxidation of *p*-nitrobenzyl chloride,[5] *p*-nitrobenzyl alcohol,[6] or the esters of *p*-nitrocinnamic acid.[7]

p-NITROBENZYL BROMIDE

(Toluene, α-bromo-*p*-nitro-)

$$p\text{-}NO_2C_6H_4CH_3 + Br_2 \rightarrow p\text{-}NO_2C_6H_4CH_2Br + HBr$$

Submitted by G. H. COLEMAN and G. E. HONEYWELL.
Checked by W. H. CAROTHERS and W. L. McEWEN.

1. Procedure

IN a 1-l. three-necked flask is placed 300 g. (2.2 moles) of technical *p*-nitrotoluene (m.p. 51–52°). The flask is fitted with a liquid-sealed

[1] Angeli and Angelico, Atti accad. Lincei (5) **8**, II, 28 (1899) (Chem. Zentr. **1899**, II, 371).

[2] Richter, Ber. **19**, 1060 (1886); Law and Perkin, J. Chem. Soc. **93**, 1635 (1908).

[3] Farbw. vorm. Meister, Lucius, and Brüning, Ger. pat. 174,238 [Frdl. **8**, 150 (1905–07)].

[4] Thiele and Winter, Ann. **311**, 353 (1900).

[5] Fischer and Greiff, Ber. **13**, 669 (1880); Schmidt, Ger. pat. 15,881 [Frdl. **1**, 60 (1877–87)].

[6] Cohen and Harrison, J. Chem. Soc. **71**, 1057 (1897); Walter, Ger. pat. 118,567 [Frdl. **6**, 131 (1900–02)].

[7] Baeyer, Ger. pat. 15,743 [Frdl. **1**, 60 (1877–87)].

stirrer, reflux condenser, and a separatory funnel arranged so that the stem reaches nearly to the bottom of the flask. The condenser is attached to a gas trap. The flask is heated in an oil bath at 145–150° (Note 1), and 368 g. (118 cc., 2.3 moles) of bromine is added dropwise over a period of two hours (Note 2).

After all the bromine has been added, the heating and stirring are continued for ten minutes. While still liquid, the contents of the flask (Note 3) are poured into a 5-l. round-bottomed flask containing 4 l. of hot ligroin (b.p. 90–100°), and 15 g. of Norite is added. The material is brought into solution by heating on an electric hot plate, and, after boiling for ten minutes, the solution is filtered rapidly with suction (Note 4). After cooling to 20°, the crystals are filtered with suction, pressed thoroughly, and washed with two 50-cc. portions of cold ligroin. The crude product, m.p. 94–97°, weighs 280–315 g. (59–66 per cent of the theoretical amount) and is sufficiently pure for some purposes.

For purification, this material is dissolved in 3–3.5 l. of hot ligroin, boiled with 10–15 g. of decolorizing carbon, and filtered with suction (Note 4). After cooling in an ice bath the crystals are collected on a Büchner funnel, pressed, and washed with two 25-cc. portions of cold ligroin. The pale yellow product melts at 97.5–99° and weighs 250–280 g. (53–59 per cent of the theoretical amount) (Notes 5 and 6).

2. Notes

1. The temperature should be kept within the limits mentioned to obtain the best results.

2. The bromine should be added within two hours even though a small amount may be lost through the condenser.

3. p-Nitrobenzyl bromide and its solution should be handled with caution. If the substance comes in contact with the skin, bathing the affected parts in alcohol will give relief.

4. The inverted filtration method of Bost and Constable (p. 610) is particularly advantageous for filtering hot solutions of nitrobenzyl bromide, since it reduces the fire hazard and the manipulation of the lachrymatory solutions. To avoid clogging, a tube of 8- to 10-mm. bore must be used for connecting the filtering flasks. About 3.5 l. of ligroin may be recovered from the mother liquors.

5. p-Nitrobenzyl bromide has been used as a reagent for the identification of many acids [1] and phenols [2] by conversion into their p-nitrobenzyl esters and ethers.

[1] Reid, J. Am. Chem. Soc. **39**, 126 (1917); Lyons and Reid, ibid. **39**, 1728 (1917).
[2] Reid, ibid. **39**, 304 (1917); Lyman and Reid, ibid. **42**, 615 (1920).

6. It is reported that the bromination procedure described by Brewster [3] is simpler and more convenient than that given above if tungsten lamps are employed as a source of artificial light. The light from two 300-watt tungsten lamps is satisfactory for the bromination of 300 g. of *p*-nitrotoluene. The yield of *p*-nitrobenzyl bromide is 60–70 per cent of the theoretical amount. (LEONARD WEISLER and DONALD PEARLMAN, private communication.)

3. Methods of Preparation

p-Nitrobenzyl bromide has usually been prepared by brominating *p*-nitrotoluene.[1, 3] It has also been prepared by treating *p*-nitrobenzyl alcohol with hydrobromic acid,[4] and by nitrating benzyl bromide.[5]

p-NITRODIPHENYL ETHER

(Ether, *p*-nitrophenyl phenyl,)

$$p\text{-}NO_2C_6H_4Cl + C_6H_5OH + KOH \rightarrow p\text{-}NO_2C_6H_4OC_6H_5 + KCl + H_2O$$

Submitted by RAY Q. BREWSTER and THEODORE GROENING
Checked by W. W. HARTMAN and J. B. DICKEY.

1. Procedure

ONE HUNDRED SIXTY grams (1.7 moles) of a good grade of phenol and 80 g. (1.43 moles) of potassium hydroxide are placed in a 2-l. flask, and the mixture is heated to 130–140° until all the alkali has dissolved. The potassium phenoxide is cooled to 100–110°, and 0.5 g. of copper catalyst (Note 1) and 78.8 g. (0.5 mole) of *p*-nitrochlorobenzene are added. The flask is then fitted with a mechanical stirrer, thermometer, and a reflux condenser. The stirrer is started, and the contents of the flask are warmed with a Bunsen burner to 150–160°, at which temperature a spontaneous reaction begins with ebullition and the separation of potassium chloride. The flame should be removed during this stage of the reaction. Boiling nearly ceases within five to seven minutes, and another 78.8 g. (0.5 mole) of *p*-nitrochlorobenzene is added. The mixture is again heated as before until a second spontaneous reaction begins. This also proceeds for about five minutes without the application of heat. When boiling due to the exothermic reaction has ceased, heat is

[3] Brewster, ibid. **40**, 406 (1918).
[4] Norris, Watt, and Thomas, ibid. **38**, 1077 (1916).
[5] Moureu and Brown, Bull. soc. chim. (4) **29**, 1008 (1921).

applied and a temperature of 150–160° is maintained for an additional thirty minutes. The dark-colored melt is then poured into 1.5 l. of ice water containing 50 g. of sodium hydroxide and stirred well to remove excess phenol. The crude p-nitrodiphenyl ether separates as a dark brown crystalline mass which is allowed to settle. The product is filtered on a Büchner funnel, washed with 2 l. of water, and pressed as free from water as possible. After drying in the air it is distilled from a 500-cc. Claisen flask. The small fraction boiling up to 170°/8 mm., which contains p-nitrochlorobenzene, is discarded. A fraction boiling at 170–188°/8 mm. and weighing 14 g. is collected (Note 2). The main fraction boiling at 188–193°/8 mm. boils at 188–190°/8 mm. on redistillation with no fore-run and practically no residue. p-Nitrodiphenyl ether solidifies on cooling to diamond-shaped crystals melting at 56–58°. The yield is 173–177 g. (80–82 per cent of the theoretical amount) (Note 3).

2. Notes

1. An active copper powder can be prepared from copper sulfate. One hundred grams (0.4 mole) of copper sulfate ($CuSO_4 \cdot 5H_2O$) is dissolved in 350 cc. of hot water in a 1-l. beaker. After cooling to room temperature 35 g. (0.53 gram atom) of zinc dust (more if necessary) is gradually added until the solution is decolorized. The precipitated copper is washed by decantation with water. Dilute hydrochloric acid (5 per cent) is added to the precipitate to remove the excess of the zinc, and agitation is continued until the escape of hydrogen ceases. The copper powder is filtered, washed with water, and kept in a moist condition in a carefully stoppered bottle.

2. This fraction on redistillation yields 4 g. of p-nitrodiphenyl ether.

3. A yield of 84 per cent of o-nitrodiphenyl ether boiling at 183–185°/8 mm. is obtained when o-nitrochlorobenzene is used. For the preparation of m-nitrodiphenyl ether, the method of Ullmann and Sponagel,[1] using m-bromonitrobenzene, seems to be the best, since m-chloronitrobenzene gives large amounts of tarry matter.

3. Methods of Preparation

p-Nitrodiphenyl ether has been prepared by the nitration of diphenyl ether [2] and by heating p-nitrochlorobenzene [3] or p-nitrofluorobenzene [4] with potassium phenoxide and phenol.

[1] Ullmann and Sponagel, Ber. **38**, 2211 (1905).
[2] Suter, J. Am. Chem. Soc. **51**, 2583 (1929).
[3] Haeussermann and Teichmann, Ber. **29**, 1446 (1896).
[4] Rarick, Brewster, and Dains, J. Am. Chem. Soc. **55**. 1289 (1933).

2-NITROFLUORENE AND 2-AMINOFLUORENE

(Fluorene, 2-nitro-, and 2-Fluorenylamine)

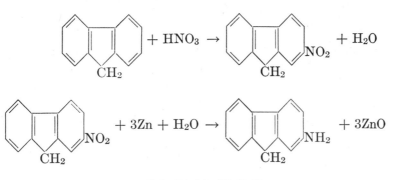

Submitted by W. E. KUHN.
Checked by LOUIS F. FIESER and J. T. WALKER.

1. Procedure

(A) *2-Nitrofluorene.*—Sixty grams (0.36 mole) of fluorene (Note 1) is dissolved in 500 cc. of warm glacial acetic acid in a 1-l. three-necked flask fitted with a thermometer, a mechanical stirrer, and a dropping funnel, and supported in a water bath. The temperature is brought to 50°, and 80 cc. (1.3 moles) of concentrated nitric acid (sp. gr. 1.42) is added with stirring in the course of fifteen minutes. During the addition, the solution becomes slightly yellow, and a small amount of material precipitates. The water bath is slowly brought to 60–65°, when the precipitate dissolves and the color of the solution deepens. Stirring is continued, and heat is applied continuously until the temperature of the mixture reaches 80° (Note 2). After five minutes, the water bath is removed, and the mixture, which now consists of a semi-solid paste of fine, yellow needles, is allowed to cool to room temperature during two hours. The product is collected on a Büchner funnel, sucked as dry as possible, and washed with two 25-cc. portions of cold glacial acetic acid containing 0.5 g. of potassium acetate. It is then washed several times with water and dried. The 2-nitrofluorene so obtained melts at 155–156° and is sufficiently pure for most purposes. The yield is 60 g. (79 per cent of the theoretical amount).

If a purer product is desired, the above material may be crystallized from 200 cc. of glacial acetic acid. The purified product melts at 157° and weighs 56 g.

(B) *2-Aminofluorene.*—In a 2-l. round-bottomed flask, 30 g. (0.14 mole) of dried and powdered 2-nitrofluorene is made into a thin paste with 1 l. of 78 per cent alcohol (820 cc. of 95 per cent alcohol and 180 cc. of water). A solution of 10 g. of calcium chloride in 15 cc. of water, together with 300 g. of zinc dust, is added to the suspension, and the whole is thoroughly mixed. The flask is fitted with an effective reflux condenser, and the mixture is refluxed for two hours.

The sludge of zinc dust and zinc oxide is filtered from the boiling solution and extracted (Note 3) with 50 cc. of boiling 78 per cent alcohol. The combined filtrates are then poured into 2 l. of water, whereupon a white, flocculent precipitate is obtained. This is filtered with suction and recrystallized from 400 cc. of hot 50 per cent alcohol. The purified 2-aminofluorene crystallizes in needles melting at 127.5°. The yield is 20–21 g. (78–82 per cent of the theoretical amount).

2. Notes

1. The fluorene used had a melting point of 113–114° and was obtained from the Gesellschaft für Teerverwertung, Duisberg-Meiderich.

2. The reaction is exothermic at this point, and the temperature of the reaction mixture may be expected to rise ten to fifteen degrees above that of the surrounding bath. If the temperature is allowed to rise above 85°, the 2-nitrofluorene will be highly colored and impure.

3. The sludge of zinc dust and zinc oxide is filtered while hot, through a previously warmed filter. Unless the solution is kept near the boiling point, some of the compound will crystallize during the filtration. If a small amount of zinc dust runs through the filter paper, it is advisable to heat the filtrate to the boiling point and refilter without suction through a folded filter paper in a hot water or steam jacket.

3. Methods of Preparation

The procedures given are essentially those of Diels.[1] 2-Nitrofluorene has also been prepared by passing nitrous vapors into a benzene solution of fluorene.[2]

[1] Diels, Ber. **34,** 1758 (1901).
[2] Monti, Martello, and Valente, Gazz. chim. ital. **66,** 31 (1936).

NITROMESITYLENE

(Mesitylene, 2-nitro-)

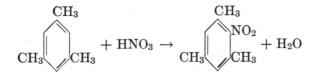

$$\text{(mesitylene)} + HNO_3 \rightarrow \text{(2-nitromesitylene)} + H_2O$$

Submitted by GARFIELD POWELL and F. R. JOHNSON.
Checked by W. W. HARTMAN and J. B. DICKEY.

1. Procedure

IN a 500-cc. three-necked, round-bottomed flask, provided with a mechanical stirrer, a dropping funnel, and a thermometer well, are placed 40 g. (0.333 mole) of mesitylene (Org. Syn. Coll. Vol. I, **1941,** 341) and 60 g. (55.5 cc.) of acetic anhydride (Note 1). The flask is placed in a bath of ice and water, and the reaction mixture is cooled below 10°. A solution of 31.5 g. (20.8 cc., 0.5 mole) of fuming nitric acid (sp. gr. 1.51) in 20 g. (19.1 cc.) of glacial acetic acid and 20 g. (18.5 cc.) of acetic anhydride (Note 2) is added, with stirring, over a period of forty minutes, the temperature being kept between 15° and 20°.

When the addition of the nitric acid solution is complete, the reaction mixture is removed from the ice bath and allowed to stand at room temperature for two hours. The flask is then warmed, with shaking, to 50° on a water bath (Note 3) and maintained at that temperature for ten minutes. The cooled reactants are then poured slowly into 800 cc. of ice water and well stirred. About 40 g. of sodium chloride is added, and the aqueous layer is decanted and extracted with 200–250 cc. of a commercial grade of ether. The ethereal extract is added to the residual nitromesitylene, and this ethereal solution is washed with three or four 30-cc. portions of a 10–15 per cent sodium hydroxide solution until the water extract is distinctly alkaline. The ether is then distilled from a steam bath. The residue, after the addition of about 150 cc. of 10 per cent sodium hydroxide solution, is steam-distilled until a test sample of the distillate is clear and free of oil drops. The steam distillation requires about three hours, during which time 1.5 l. of distillate is collected. The nitromesitylene settles to the bottom of the distillate. The water is decanted through a filter paper, and the residue is dissolved in 30 cc. of ether. The filter paper, if it contains any solid particles, is washed with 10 cc. of ether, and the two fractions are combined and placed in a 150-cc. distilling flask with a water-cooled receiver. After removal of the ether on a steam bath, the yellow residue is distilled with a free flame. All but 0.5 to 1.5 g. of material distils at 243–250°, and the distillate, which weighs 47 g., solidifies on cooling. The yellow crystalline product is purified by dissolving in 25 cc. of commercial methyl alcohol and cooling with stirring in a bath of ice and salt. The crystals are then filtered on a small Büchner funnel and washed twice with 5-cc. portions of cold methyl alcohol. From the wash alcohol, by exactly the same procedure, there is obtained 6–8 g. of crude crystals boiling at 243–249°. This material yields on crystallization 4 g. of pure nitromesitylene. The pure nitromesitylene, which has a light yellow color, melts at 43–44°. The yield is 41–42 g. (74–76 per cent of the theoretical amount).

2. Notes

1. A commercial grade of 85 per cent acetic anhydride is of sufficient purity.

2. The nitric acid is added with care to the acetic acid-acetic anhydride solution, the temperature being kept below 20° by means of a bath of ice and salt. An **explosive** reaction will take place if the nitric acid is added too rapidly.

3. A higher temperature is not advisable.

3. Methods of Preparation

Nitromesitylene has been prepared by the direct nitration of mesitylene with concentrated nitric acid; [1] by the action of benzoyl nitrate on mesitylene in carbon tetrachloride at low temperature; [2] and from dinitromesitylene by reduction of one nitro group and replacement of the resulting amino group by hydrogen. [3]

[1] Fittig and Storer, Ann. **147**, 1 (1868); Biedermann and Ledoux, Ber. **8**, 58 (1875); Schultz, ibid. **17**, 477 (1884); Bamberger and Rising, ibid. **33**, 3625 (1900).

[2] Francis, ibid. **39**, 3801 (1906).

[3] Ladenburg, ibid. **7**, 1135 (1874).

1-NITRO-2-NAPHTHOL

(2-Naphthol, 1-nitro-)

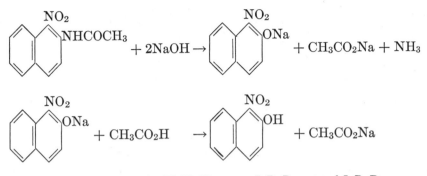

Submitted by W. W. HARTMAN, J. R. BYERS, and J. B. DICKEY.
Checked by LOUIS F. FIESER and J. T. WALKER.

1. Procedure

IN a 3-l. round-bottomed flask, fitted with a reflux condenser, are placed 100 g. (0.435 mole) of 1-nitro-2-acetylaminonaphthalene (p. 438) and a solution of 112 g. (2.8 moles) of sodium hydroxide in 2.7 l. of water (Note 1). The mixture is boiled until ammonia is no longer evolved (six to seven hours). The solution becomes deep red in color. It contains suspended crystals of sodium nitronaphthoxide; these are dissolved by the addition of 1 l. of hot water. The small amount of insoluble material is removed by filtration, washed with hot water until the washings are colorless, and then discarded although it contains a little nitronaphthylamine. The combined washings and filtrate are made acid by adding 500 cc. of glacial acetic acid. The nitronaphthol precipitates as small, bright yellow crystals which are filtered on a 10-cm. Büchner funnel, washed with water, and dried. The yield of material melting at 101–103° is 76–81 g. (92–98 per cent of the theoretical amount).

The product is purified by recrystallization from 500 cc. of methyl alcohol containing 5 cc. of concentrated hydrochloric acid. The first crop of crystals amounts to 60 g. and melts at 103–104°. The mother liquors are concentrated to 150 cc., and a second crop weighing 12–13 g. is collected. The total yield of recrystallized material is 72–73 g. (88–89 per cent of the theoretical amount).

2. Note

1. This concentration of alkali was found to be very satisfactory. The hydrolysis takes place rather rapidly at first when stronger alkali is used, with the precipitation of the sodium salt which forms a thick paste and causes bumping. When more dilute alkali is used, much more time is required to complete the hydrolysis.

3. Methods of Preparation

1-Nitro-2-naphthol has been prepared by the nitration of β-naphthyl ethyl ether;[1,2] by the oxidation of 1-nitroso-2-naphthol;[3] and by the treatment of β-naphthol with diacetylnitric acid.[4] It has also been prepared by the treatment of β-naphthylamine with three moles of sodium nitrite in the presence of an excess of acid;[5] by the decomposition of benzeneazo-β-naphthol with nitric acid;[6] by heating the nitrate of pseudo-cumeneazo-β-naphthol under reduced pressure;[6] by the treatment of 1-bromo-2-naphthol in acetic acid with nitric acid;[7] and by the reaction of nitrogen dioxide with β-naphthol.[7] It has been prepared from 1-nitro-1-bromo-2-ketodihydronaphthalene by treatment with caustic alkali,[8] and by the fusion of α-nitronaphthalene with sodium hydroxide.[9]

The method on which this procedure is based has been described by Andreoni and Biedermann and by others.[10]

[1] Wittkampf, Ber. **17**, 393 (1884).

[2] Gaess, J. prakt. Chem. (2) **43**, 22 (1891).

[3] Stenhouse and Groves, Ann. **189**, 151 (1877); Fierz-David and Ischer, Helv. Chim. Acta **21**, 680 (1938).

[4] Pictet and Krijanowski, Arch. sci. phys. nat. Genève (4) **16**, 191 (1903) (Chem. Zentr. **1903**, II, 1109).

[5] Deninger, J. prakt. Chem. (2) **40**, 300 (1889).

[6] Charrier and Ferreri, Gazz. chim. ital. **44** (I) 176 (1914).

[7] Armstrong and Rossiter, Ber. **24** (R), 720 (1891); J. Chem. Soc. Proc. **1891**, 87, 89.

[8] Fries, Ann. **389**, 317 (1912).

[9] Wohl, Ger. pat. 116,790 [Frdl. **6**, 114 (1900–02)].

[10] Andreoni and Biedermann, Ber. **6**, 342 (1873); Jacobson, ibid. **14**, 806 (1881); Liebermann and Jacobson, Ann. **211**, 46 (1882).

p-NITROPHENYL ISOCYANATE

(Isocyanic acid, *p*-nitrophenyl ester)

$$p\text{-}O_2NC_6H_4NH_2 + COCl_2 \rightarrow p\text{-}O_2NC_6H_4NHCOCl + HCl$$

$$p\text{-}O_2NC_6H_4NHCOCl \rightarrow p\text{-}O_2NC_6H_4NCO + HCl$$

Submitted by R. L. Shriner, W. H. Horne, and R. F. B. Cox.
Checked by W. W. Hartman and J. B. Dickey.

1. Procedure

The apparatus is shown in Fig. 16. Phosgene is introduced at one end of the apparatus, and gentle suction is applied at the other. In the

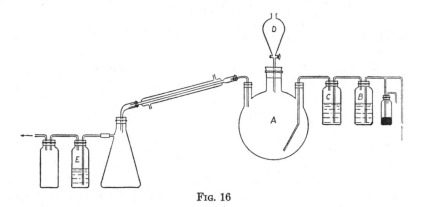

Fig. 16

5-l. flask, A, 500 cc. of dry ethyl acetate (Note 1) is saturated with phosgene at room temperature. The phosgene is purified by bubbling it through cottonseed oil, B, to remove chlorine, and then through concentrated sulfuric acid, C, as shown in the figure (Note 2). A solution of 150 g. (1.09 moles) of *p*-nitroaniline in 1.5 l. of dry ethyl acetate is run in slowly from the separatory funnel, D, over a period of three to four hours. The addition of the *p*-nitroaniline solution must be at such a rate that the precipitate of *p*-nitroaniline hydrochloride that is formed at first is allowed to dissolve and not accumulate (Note 3). During this time a steady stream of phosgene is passed through the solution to ensure an excess (Note 4). Towards the end of the reaction, the solution must be boiled gently with a Bunsen flame to break up the lumps of *p*-nitroaniline hydrochloride which otherwise dissolve very slowly.

After the addition of the last of the p-nitroaniline, the stream of phosgene is continued for five minutes and then shut off. The flame under the flask is then turned up and the ethyl acetate distilled. Care must be taken at the end not to overheat the residue. The brown residue (Note 5) is treated with 800 cc. of hot dry carbon tetrachloride, and the insoluble residue (the disubstituted urea) is removed by filtration.

About two-thirds of the carbon tetrachloride is distilled. The solution is cooled, and the crystals of p-nitrophenyl isocyanate are filtered as quickly as possible in order to avoid prolonged exposure of the compound to the moisture of the air. By concentration of the mother liquor a further crop is obtained. The product is recrystallized from dry carbon tetrachloride and is obtained in the form of light yellow needles melting at 56–57° (Note 6). The yield after one recrystallization is 152–170 g. (85–95 per cent of the theoretical amount) (Note 7).

2. Notes

1. The ethyl acetate, free from ethyl alcohol, is dried with anhydrous magnesium sulfate.

2. This arrangement permits the phosgene reaction to be carried out conveniently and without danger, provided that a good hood and an exhaust fan are available. A slight vacuum is maintained in the system. The excess phosgene is absorbed in 20 per cent sodium hydroxide solution, E.

Although phosgene can be shipped only in steel cylinders in the United States, in some other countries it is available in glass ampoules. If an ampoule of phosgene is used in this preparation, only one-third the weight of p-nitroaniline necessary to react with the phosgene should be taken. In this way an excess of phosgene is assured; compare Note 4. Because of the danger involved, the use of glass ampoules of phosgene is not recommended.

3. The mixture may be warmed if necessary to dissolve the p-nitroaniline hydrochloride.

4. The excess of phosgene retards the following reaction:

$$p\text{-NO}_2\text{C}_6\text{H}_4\text{NHCOCl} + p\text{-NO}_2\text{C}_6\text{H}_4\text{NH}_2$$
$$\rightarrow p\text{-NO}_2\text{C}_6\text{H}_4\text{NHCONHC}_6\text{H}_4\text{NO}_2\text{-}p + \text{HCl}$$

5. The crude residue is a mixture of p-nitroaniline hydrochloride, p-nitrophenyl carbamyl chloride, p-nitrophenyl isocyanate, and p,p'-dinitrodiphenylurea. The p-nitrophenyl carbamyl chloride is converted to p-nitrophenyl isocyanate during recrystallization from the hot carbon tetrachloride.

6. The freshly prepared material melts at 56–57° but after storage soon starts to melt at 54°, particularly if the bottle in which it is stored is opened occasionally. In contact with the moisture of the air, *p,p'*-dinitrodiphenylurea is formed (m.p. 360°). This reaction with water is avoided if the material is sealed in a glass container.

7. *p*-Nitrophenyl isocyanate distils undecomposed at 160–162°/18 mm.

3. Method of Preparation

p-Nitrophenyl isocyanate has been prepared by heating *p*-nitrophenyl carbamyl chloride. The latter has been obtained by the action of phosgene on *p*-nitroaniline in benzene-toluene solutions,[1] and by the action of phosphorus pentachloride on methyl *p*-nitrophenylcarbamate.[2] The preparation given above is based upon directions published by the submitters.[3]

o-NITROPHENYLSULFUR CHLORIDE

(Benzenesulfenyl chloride, *o*-nitro-)

$$NO_2C_6H_4SSC_6H_4NO_2 + Cl_2 \rightarrow 2NO_2C_6H_4SCl$$

Submitted by MAX H. HUBACHER.
Checked by LOUIS F. FIESER and D. J. POTTER.

1. Procedure

IN a 1-l. three-necked flask, fitted with a thermometer, a reflux condenser, and an inlet tube extending to the bottom of the flask and drawn out to a small opening, are placed 600 cc. of dry carbon tetrachloride, 154 g. (0.5 mole) of di-*o*-nitrophenyl disulfide, m.p. 193–195° (Org. Syn. Coll. Vol. I, **1941**, 220), and 0.25 g. of iodine. To the upper end of the condenser is attached a glass tube which dips below the surface of a little carbon tetrachloride contained in a test tube. A current of chlorine, dried with sulfuric acid, is passed into the reaction mixture, the temperature of which is maintained at 50–60°. The rate of flow of the chlorine (about 16–17 g. per hour) is regulated so that little or no gas escapes through the carbon tetrachloride trap. The yellow di-*o*-

---

[1] Vittenet, Bull. soc. chim. (3) **21**, 586 (1899); Van Hoogstraten, Rec. trav. chim. **51**, 418 (1932).

[2] Swartz, Am. Chem. J. **19**, 318 (1897).

[3] Shriner and Cox, J. Am. Chem. Soc. **53**, 1603 (1931); Horne and Shriner, ibid **53**, 3186 (1931).

nitrophenyl disulfide gradually disappears, and after two to two and one-half hours a homogeneous, dark yellow solution is obtained (Note 1). The warm solution is filtered from a small amount of dark residue through a warm Büchner funnel, the flask and filter being rinsed with 30 cc. of warm carbon tetrachloride. The yellow filtrate (Note 2) is cooled to 5° and the product allowed to crystallize. The cake of crystals is broken with a rod, collected on a Büchner funnel, and drained well. The product is dried rapidly (two hours) at 50° and bottled (Note 3). This material melts at 73–74.5° and weighs 126–135 g. (66–71 per cent of the theoretical amount). A further crop is obtained by removing the solvent from the mother liquor by distillation from a water bath (Note 4), the dark, residual oil being poured into an evaporating dish. The last traces of carbon tetrachloride are removed by drying at 50°, during which process the oil crystallizes. This material melts at 67–72° and weighs 48–58 g. It is pure enough for most purposes (Note 5). The total yield is 183–184 g. (96–97 per cent of the theoretical amount) (Notes 6 and 7).

2. Notes

1. An excess of chlorine will do no harm.

2. For many purposes this filtered solution can be used without purification after distilling about 100 cc. of the solvent to remove the excess chlorine. The yield of material in solution is considered as 98 per cent of the theoretical amount.

3. The pure material melts at 74.5–75°. When in contact with moist air, o-nitrophenylsulfur chloride decomposes within a few days, giving off hydrogen chloride. Stored in a brown, glass-stoppered bottle and sealed well, the material can be kept for many months.

4. Since o-nitrophenylsulfur chloride decomposes spontaneously when heated to about 170°, it is advisable to use a water bath when removing the solvent.

5. This material may be crystallized from hot carbon tetrachloride (2 cc. per gram), the filtered solution being cooled to 5°. The recovery of a product melting at 70–73° is 75 per cent.

6. Practically the same method can be used for the preparation of o-nitro-p-chlorophenylsulfur chloride (m.p. 95–97°); the chlorination of the disulfide (m.p. 212–213°) in this case proceeds much more slowly. 2,4-Dinitrophenylsulfur chloride can be prepared by the chlorination of the corresponding disulfide in nitrobenzene suspension at 120–130°. As the chloride is explosive the solvent must be removed by distillation in vacuum at 130°. The crude material melts at 89–92°. After crystallization from carbon tetrachloride it melts at 94–95°.

7. The arylsulfur chlorides are used for the introduction of the ArS- group, their reactions being analogous to those of acid chlorides.

3. Method of Preparation

o-Nitrophenylsulfur chloride has always been prepared by the chlorination of the disulfide.[1]

4-NITROPHTHALIC ACID

(Phthalic acid, 4-nitro-)

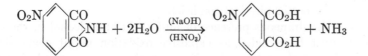

Submitted by E. H. HUNTRESS, E. L. SHLOSS, JR., and P. EHRLICH.
Checked by W. W. HARTMAN and G. W. SAWDEY.

1. Procedure

To a solution of 26.6 g. (0.66 mole) of sodium hydroxide in 240 cc. of water is added 80 g. (0.416 mole) of 4-nitrophthalimide (p. 459) (Note 1). The mixture is heated rapidly to boiling and boiled gently for ten minutes. The solution is made barely acid to litmus with concentrated nitric acid (sp. gr. 1.42); after the neutral point is reached, an additional 70 cc. (100 g., 1.1 moles) of nitric acid is added (Note 2). The solution is again boiled for three minutes, then cooled below room temperature, transferred to a 1-l. separatory funnel, and extracted with two 300-cc. portions of alcohol-free ether (Note 3). Care is taken to ensure thorough mixing before separation of the layers. After the extract is dried over anhydrous sodium sulfate, the ether is distilled until solid begins to separate. The concentrated ether solution is poured into a porcelain dish and the residual solvent allowed to evaporate in a hood (Note 4). The practically white crystals of 4-nitrophthalic acid which separate melt at 163–164° and have a neutralization equivalent of 105.5 (theoretical, 105.5). The yield is 85–87 g. (96–99 per cent of the theoretical quantity).

[1] Zincke, Ber. **44**, 770 (1911); Zincke and Farr, Ann. **391**, 63 (1912).

2. Notes

1. If a large amount of 4-nitrophthalimide is to be hydrolyzed, it will generally be found convenient to carry out a series of small runs of the size given here.

2. The quantity of alkali used at the start is sufficient to neutralize the nitroimide and leave the resulting alkaline solution approximately 1 N. When the solution is neutralized the red color changes to a dirty brown, which turns to a pale yellow on acidification. The 70-cc. portion of concentrated nitric acid suffices to set free all the 4-nitrophthalic acid but avoids the presence of a large excess during the ether extraction.

Nitric acid appears to be preferable to hydrochloric or sulfuric acid. The yields are similar, but the product obtained using nitric acid is pure white and in better physical condition.

3. The simultaneous presence of nitric acid and alcohol in the ether extract must be avoided as explosive oxidation might occur during evaporation of the ether. Furthermore, the presence of alcohol may lead to contamination of the product with traces of the acid ester.

4. Evaporation of the last portion of ether proceeds slowly; the rather soft, fluffy crystals of acid which separate at first gradually become hard and dense. No trouble was experienced from nitric acid remaining with the ether extract, if the ether was free from alcohol and the specified amount of nitric acid was used.

3. Methods of Preparation

4-Nitrophthalic acid has usually been prepared by nitration of phthalic acid [1] or phthalic anhydride,[2,3] followed by separation from the accompanying 3-nitrophthalic acid.[4,5,6] It has also been prepared from 6-nitro-2-naphthol-4-sulfonic acid (obtained from the technical diazoanhydride of 6-nitro-1-amino-2-naphthol-4-sulfonic acid).[7] The present procedure is more convenient than any of the earlier methods.

[1] Miller, Ann. **208**, 223 (1881); Huisinga, Rec. trav. chim. **27**, 261 (1908).

[2] Levy and Stephen, J. Chem. Soc. **1931**, 79.

[3] Culhane and Woodward, Org. Syn. Coll. Vol. I, **1941**, 408.

[4] Bogert and Boroschek, J. Am. Chem. Soc. **23**, 752 (1901).

[5] Cohen, Woodroffe, and Anderson, J. Chem. Soc. **109**, 232 (1916).

[6] Lawrance, J. Am. Chem. Soc. **42**, 1872 (1920).

[7] Ruggli, Knapp, Merz, and Zimmerman, Helv. Chim. Acta **12**, 1043 (1929).

4-NITROPHTHALIMIDE

(Phthalimide, 4-nitro-)

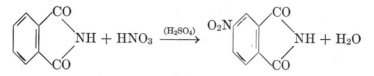

Submitted by E. H. HUNTRESS and R. L. SHRINER.
Checked by FRANK C. WHITMORE and C. P. KRIMMEL.

1. Procedure

Two HUNDRED FORTY cubic centimeters (5.7 moles) of c.p. fuming nitric acid (sp. gr. 1.50) is added to 1.4 l. of concentrated c.p. sulfuric acid (sp. gr. 1.84) in a 3-l. beaker, and the mixture is cooled in an ice bath. As soon as the temperature of the mixed acids reaches 12°, 200 g. (1.36 moles) of commercial phthalimide is stirred in as rapidly as possible while the temperature of the nitrating mixture is kept between 10° and 15°. The reaction mixture is allowed to warm to room temperature in the ice bath as the ice melts, and left overnight.

The clear, pale yellow solution is poured slowly with vigorous stirring onto 4.5 kg. of cracked ice; the temperature of this mixture must not rise above 20°. The crude nitration product is filtered through cloth on a 20-cm. Büchner funnel, using suction, and the mass is pressed as dry as possible. The cake is removed and stirred vigorously with 2 l. of *ice water*. The solid is filtered; the cake is removed and stirred again with 2 l. of *ice water*. This washing is repeated four times. The crude product, after drying in the air, melts at 185–190° and weighs 165–174 g. (63–66 per cent of the theoretical amount). This material is purified by crystallization from 3 to 3.2 l. of 95 per cent ethyl alcohol. This furnishes 136–140 g. (52–53 per cent of the theoretical amount) of 4-nitrophthalimide melting at 198°.

2. Methods of Preparation

4-Nitrophthalimide has been prepared from 4-nitrophthalic acid [1] and by nitrating phthalimide.[2] The procedure described is a modification of the method of Levy and Stephen.[2]

[1] Seidel, Ber. **34**, 4351 (1901); Seidel and Bittner, Monatsh. **23**, 420 (1902); Bogert and Renshaw, J. Am. Chem. Soc. **30**, 1137 (1908).

[2] Levy and Stephen, J. Chem. Soc. **1931**, 79.

N-NITROSOMETHYLANILINE

(Aniline, N-methyl-N-nitroso-)

$$C_6H_5NHCH_3 + NaNO_2 + HCl \rightarrow C_6H_5N(NO)CH_3 + NaCl + H_2O$$

Submitted by W. W. HARTMAN and L. J. ROLL.
Checked by LOUIS F. FIESER and J. T. WALKER.

1. Procedure

A MIXTURE of 107 g. (1 mole) of methylaniline (Note 1), 145 cc. of concentrated hydrochloric acid, and 400 g. of ice is placed in a 3-l. flask equipped with a mechanical stirrer. The mixture is stirred vigorously, and the temperature is maintained at 10° or below by the addition of more ice as required, while a solution of 70 g. (1 mole) of sodium nitrite in 250 cc. of water is added during the course of five or ten minutes. Stirring is then continued for one hour more. The oily layer is separated, and the aqueous portion is extracted with two 100-cc. portions of benzene. The benzene is removed by distillation at ordinary pressure, and the residue is fractionated under reduced pressure. The main fraction of the nitrosomethylaniline distils as a light yellow liquid boiling at 135–137°/13 mm. The yield (Note 1) is 118–127 g. (87–93 per cent of the theoretical amount).

2. Note

1. The yield is dependent upon the quality of the methylaniline used. The higher yield reported was obtained with pure material, b.p. 81–82°/14 mm.

3. Methods of Preparation

N-Nitrosomethylaniline was first prepared by the action of nitrous acid on methylaniline.[1] It has been obtained also by the action of methyl iodide on the sodium salt of benzene diazoic acid followed by reduction;[2] by treating dimethylaniline with tetranitromethane,[3] or with phenylnitrocarbinol;[4] by the acid hydrolysis of nitrosophenylglycine;[5] and by oxidizing dimethyldiphenylhydrazine with nitric oxide.[6]

[1] Hepp, Ber. **10**, 329 (1877).
[2] Bamberger, ibid. **27**, 373 (1894).
[3] Schmidt and Fischer, ibid. **53**, 1538 (1920).
[4] Cohen and Calvert, J. Chem. Soc. **73**, 164 (1898).
[5] Fischer, Ber. **32**, 249 (1899).
[6] Wieland and Fressel, Ann. **392**, 148 (1912).

NITROSOMETHYLUREA

(Urea, α-methyl-α-nitroso-)

I. (From Methylamine Hydrochloride)

$$CH_3NH_3Cl + H_2NCONH_2 \rightarrow CH_3NHCONH_2 + NH_4Cl$$

$$CH_3NHCONH_2 + HNO_2 \rightarrow CH_3N(NO)CONH_2 + H_2O$$

Submitted by F. Arndt.
Checked by C. R. Noller and S. Lieberman.

1. Procedure

In a tared 1-l. flask is placed 200 g. (1.5 moles) of 24 per cent aqueous methylamine solution (Note 1), and concentrated hydrochloric acid is added until the solution is acid to methyl red; about 155 cc. of acid is required. Water is added to bring the total weight to 500 g., 300 g. (5 moles) of urea is added, and the solution is boiled gently under reflux for two and three-quarters hours and then vigorously for one-quarter hour. The solution is cooled to room temperature, 110 g. (1.5 moles) of 95 per cent sodium nitrite is dissolved in it, and the whole is cooled to 0°. A mixture of 600 g. of ice and 100 g. (1 mole) of concentrated sulfuric acid in a 3-l. beaker is surrounded by an efficient freezing mixture, and the cold methylurea-nitrite solution is run in slowly with mechanical stirring at such a rate that the temperature does not rise above 0° (Note 2).

The nitrosomethylurea rises to the surface as a crystalline foamy precipitate which is filtered at once with suction and pressed well on the filter. The crystals are stirred to a paste with about 50 cc. of cold water, sucked as dry as possible (Note 3), and dried in a vacuum desiccator to constant weight. The yield is 105–115 g. (66–72 per cent of the theoretical amount) (Note 4).

2. Notes

1. The methylamine content of the commercial aqueous solution (Röhm and Haas) was determined by titration with standard acid using methyl red as indicator. If the methylamine content is found to be different an equivalent quantity is used.

2. It is convenient to keep the methylurea-nitrite solution in an ice-salt bath and to siphon it into the acid solution, the end of the siphon

dipping below the surface. About one hour is required for the addition.

3. A sample of the moist product should dissolve completely in boiling methyl alcohol. If an appreciable residue remains, which is not usually the case, the washing process is repeated. Each successive washing decreases the yield somewhat.

4. The preparation obtained in this way may be kept indefinitely in a refrigerator. It should not be kept above 20° for more than a few hours. At temperatures in the neighborhood of 30° it may undergo a sudden decomposition without explosion but with the evolution of irritating fumes. It has been reported that the stability is increased by the addition of a few drops of acetic acid.

II. *(From Acetamide)*

$$2CH_3CONH_2 + Br_2 + 2NaOH \longrightarrow$$
$$CH_3NHCONHCOCH_3 + 2NaBr + 2H_2O$$

$$CH_3NHCONHCOCH_3 + H_2O \xrightarrow{\text{(HCl)}} CH_3NHCONH_2 + CH_3CO_2H$$

$$CH_3NHCONH_2 + HONO \longrightarrow CH_3N(NO)CONH_2 + H_2O$$

Submitted by E. D. Amstutz and R. R. Myers.
Checked by C. F. H. Allen and J. Dec.

1. Procedure

(A) *Acetyl Methylurea.*—To a solution (Note 1) of 59 g. (1 mole) of acetamide in 88 g. (0.55 mole) of bromine in a 4-l. beaker is added, dropwise and with hand stirring, a solution of 40 g. (1 mole) of sodium hydroxide in 160 cc. of water. The resulting yellow reaction mixture is heated on a steam bath until effervescence sets in (Note 2), after which heating is continued for an additional two to three minutes. Crystallization of the product from the yellow to red colored solution usually commences immediately (Note 3) and is completed by cooling in an ice bath for one hour. The weight of the white crystalline acetyl methylurea obtained by filtration and air drying is 49–52 g. (84–90 per cent of the theoretical amount) (Notes 4 and 5). It melts at 169–170°.

(B) *Nitrosomethylurea.*—A mixture of 49 g. (0.42 mole) of acetyl methylurea (Note 6) and 50 cc. of concentrated hydrochloric acid is heated, with hand stirring, on a steam bath until it is apparent that no more solid (Note 5) is dissolving. Heating is continued for three or four minutes longer (total time on steam bath—eight to twelve minutes), after which the solution is diluted with an equal volume of water and cooled below 10° in an ice bath. A cold saturated solution of 38 g. (0.55 mole) of sodium nitrite in 55 cc. of water is then run in slowly with stirring. The mixture is allowed to remain in the ice bath for several minutes, after which the nitrosomethylurea is filtered and washed with about 8–10 cc. of ice-cold water. Air drying gives 33–36 g. (76–82 per cent of the theoretical amount) of nitrosomethylurea as pale yellow crystals melting at 123–124°.

2. Notes

1. Gentle heating on a steam bath assists in dissolving the acetamide. Care is necessary, however, to see that only the minimum amount of bromine is lost during the heating.

2. Occasionally this effervescence becomes quite brisk, and for this reason a large container is used.

3. If the solution at this point is perfectly colorless, the product is usually slower in crystallizing, it contains more sodium bromide, and the yield is somewhat lower. For these reasons, the slight excess of bromine used is necessary.

4. When the crude yield is the lower figure given, the remainder of the product can be secured by long cooling of the filtrate in ice.

5. The crude acetyl methylurea contains some sodium bromide, which appears as a white crystalline material insoluble in concentrated hydrochloric acid in Part (B). The sodium bromide dissolves when the solution is diluted and has no effect upon the subsequent treatment with sodium nitrite.

6. The acetyl methylurea prepared in Part (A) may be used without drying.

3. Methods of Preparation

Nitrosomethylurea is always prepared by the nitrosation of methylurea. Methylurea, in turn, can be prepared from (a) methylamine hydrochloride and potassium cyanate;[1] (b) methyl sulfate, ammonia, and potassium cyanate;[1] (c) methylamine hydrochloride and urea;[2] and (d) acetamide, bromine, and alkali.[3] Checked directions for preparing methylurea, and for its subsequent nitrosation, according to (a)

[1] Arndt and Amende, Z. angew. Chem. **43**, 444 (1930); Arndt and Scholz, ibid. **46**, 47 (1933).

[2] Arndt, Loewe, and Avan, Ber. **73**, 606 (1940); Eistert, Angew. Chem. **54**, 124 (1941).

[3] Owen, Ramage, and Simonsen, J. Chem. Soc. **1938**, 1213.

and (b) are described on p. 48 of Volume 15 of Organic Syntheses. The preparation of methylurea, and its subsequent nitrosation, according to (c) and (d) are given in procedures I and II, above. The preparations using potassium cyanate suffer because of the difficulty and the expense of obtaining that reagent. Arndt, Loewe, and Avan [2] have discussed the merits of the different methods of preparing nitrosomethylurea.

NITROSOMETHYLURETHANE

(Carbamic acid, methylnitroso-, ethyl ester)

$$CH_3NHCO_2C_2H_5 + HNO_2 \rightarrow CH_3N(NO)CO_2C_2H_5 + H_2O$$

Submitted by W. W. Hartman and Ross Phillips.
Checked by Louis F. Fieser and J. T. Walker.

1. Procedure

To 206 g. (2 moles) of ethyl N-methylcarbamate (p. 278) and 600 cc. of ordinary ethyl ether in a 5-l. flask is added, along with 200 g. of ice, 650 g. (9 moles) of 96 per cent sodium nitrite (Note 1) dissolved in 1 l. of cold water. The flask is provided with a stopper carrying a thermometer, a tube to lead off evolved nitric oxide, and a separatory funnel with an extension tube reaching to the bottom of the flask. A solution of 1.2 kg. (6.7 moles) of cold 35 per cent nitric acid, prepared by pouring 600 g. (426 cc.) of concentrated acid onto 600 g. of ice, is then cautiously added through the funnel in the course of one and one-half hours. The flask is given an occasional swirl to ensure some mixing, but most of the stirring is done by the evolved gases. Ice is added as required to keep the temperature below 15°. The ether layer first becomes pale red and gradually changes to a blue-green. As soon as the color has changed to green, the ether layer is separated (Note 2), washed twice with cold water, and then with cold potassium carbonate solution until carbon dioxide is no longer evolved. The solution is dried with solid potassium carbonate, and the ether is distilled from a water bath using a 1-l. flask with a 30-cm. column arranged for vacuum distillation. The vacuum is applied as soon as most of the ether has been removed, and the flask is heated gently so that the temperature of the liquid does not exceed 45–50° (Note 3) until the pressure has been reduced below 20 mm. The yield of nitrosomethylurethane boiling at 59–61/10 mm. is 200 g. (76 per cent of the theoretical amount). The density is 1.133 at 20°.

2. Notes

1. A large excess of sodium nitrite is required to give a satisfactory yield. This may be due to reaction according to the following equations:

(1) $$2HNO_2 \rightarrow NO_2 + NO + H_2O$$

(2) $$CH_3NHCO_2C_2H_5 + 2NO_2 \rightarrow CH_3N(NO)CO_2C_2H_5 + HNO_3$$

(1) plus (2)

$$CH_3NHCO_2C_2H_5 + 4HNO_2 \rightarrow$$
$$CH_3N(NO)CO_2C_2H_5 + 2NO + HNO_3 + 2H_2O$$

Nitric oxide (NO) is lost during the reaction. It is not thought advisable to use this by passing in oxygen because of the danger of an explosion, or by passing in air because of the loss of material by evaporation.

2. Nitrosomethylurethane irritates the skin.

3. According to the literature, nitrosomethylurethane explodes when attempts are made to distil it at normal pressure.

3. Methods of Preparation

Nitrosomethylurethane has been prepared by treating ethyl methylcarbamate with sodium nitrite and sulfuric acid,[1] and by passing the gases generated from arsenious oxide and nitric acid into an ethereal solution of ethyl methylcarbamate.[2]

[1] Klobbie, Rec. trav. chim. **9,** 139 (1890).

[2] v. Pechmann, Ber. **28,** 856 (1895); Schmidt, ibid. **36,** 2477 (1903); Brühl, ibid. **36, 3635** (1903).

2-NITROTHIOPHENE

(Thiophene, 2-nitro-)

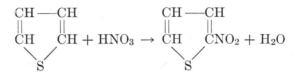

Submitted by V. S. BABASINIAN.
Checked by ROGER ADAMS and A. E. KNAUF.

1. Procedure

EIGHTY-FOUR grams (1 mole) of thiophene (p. 578) is dissolved in 340 cc. of acetic anhydride, and 80 g. (1.2 moles) of fuming nitric acid (sp. gr. 1.51) is dissolved in 600 cc. of glacial acetic acid (Note 1). Each solution is divided into two equal parts. One-half of the nitric acid solution is introduced into a 2-l. three-necked, round-bottomed flask, provided with a thermometer, a motor stirrer, and a separatory funnel. The mixture is cooled to 10°. Then with moderate stirring one-half of the thiophene solution is introduced, drop by drop, and at such a rate as to prevent the heating of the reaction mixture above room temperature. A rapid rise of temperature will occur during the addition of the first fraction of the thiophene solution. In cool weather the temperature is controlled by dipping the nitrating flask into a bath of cold tap water. Cooling to a very low temperature is not necessary, but it is important to avoid superheating the reaction mixture (Note 2). After the addition of the first half of the thiophene, the temperature of the reaction mixture is reduced to 10° and the remainder of the nitric acid solution is rapidly introduced into the flask. Nitration is continued by the gradual addition of thiophene. Throughout the nitration the solution should show a permanent light brown color. The appearance of a pink or dark red color indicates oxidation. The product is allowed to remain at room temperature for two hours. It is then treated with an equal weight of finely crushed ice with rapid shaking. Mononitrothiophene separates in pale yellow crystals. More crystals form if the mixture is allowed to remain in the ice chest for twenty-four hours or longer. The solid is filtered (Note 3) on a Büchner funnel or a Jena glass filter plate at a low temperature, washed thoroughly with ice water, pressed, and dried in a brown desiccator or in the absence of light (Note 4).

The filtrate and the washings contain in solution a small quantity of the product. This is recovered by distillation with steam. The acid distillate consists of snow-white crystals (if it is protected from light) and a solution of the compound. The solid is removed by filtration and washed. The filtrate is made neutral with sodium carbonate and extracted with ether. Upon drying and evaporating, the ethereal layer yields mononitrothiophene contaminated with dinitrothiophene (Note 5).

The total yield is 90–110 g. (70–85 per cent of the theoretical amount) of a product which melts at 44–45°. If this material is steam-distilled and then repeatedly crystallized from petroleum ether (b.p. 20–40°) it is obtained as colorless crystals melting at 45.5° (Notes 6 and 7).

2. Notes

1. The two acids should be mixed gradually, the nitric acid being added to the acetic with shaking. Cooling is often necessary.

2. Success will depend largely upon the proper control of temperature. No trouble may be anticipated if the reaction mixture responds readily to the cooling effect of a cold water bath.

3. Mononitrothiophene is an active poison.[1] The accidental contact of an ethereal solution with the skin has produced painful blisters. In case of accident the compound should be removed from the exposed surface by washing with alcohol.

4. Earlier workers [2] have noted that the compound is sensitive toward light.

5. The yellow color of the nitrothiophene is due to the presence of traces of dinitrothiophene and other impurities.

In order to detect the presence of dinitrothiophene, a few crystals of the solid are dissolved in alcohol and treated with a drop of a weak solution of alcoholic potassium hydroxide. A pink or deep red color will develop at once. An excess of potassium hydroxide will destroy the color.[3]

6. Meyer and Stadler [2] state that nitrothiophene distils without decomposition at 224–225°.

7. Mononitrothiophene has been crystallized from ether, alcohol, benzene, and other solvents. As a rule these solvents fail to yield a snow-white product. It has been found in this work that petroleum ether (b.p. 20–40°) possesses decided advantages in that by prolonged refluxing it extracts mononitrothiophene but does not readily dissolve

[1] Meyer, Ber. **18**, 1772 (1885).
[2] Meyer and Stadler, ibid. **17**, 2648 (1884).
[3] Meyer and Stadler, ibid. **17**, 2780 (1884).

the impurities. With petroleum ether, snow-white crystals have been obtained in needles 10 to 20 cm. in length.

3. Methods of Preparation

Nitrothiophene has been obtained along with dinitrothiophene by drawing a vigorous stream of air charged with thiophene through red fuming nitric acid.[2] It has also been prepared by nitrating thiophene between 0° and 5° with a mixture of acetic anhydride and fuming nitric acid.[4]

OLEYL ALCOHOL

$$C_{17}H_{33}CO_2C_4H_9 + 4Na + 3C_4H_9OH \rightarrow C_{17}H_{33}CH_2OH + 4C_4H_9ONa$$

Submitted by E. E. REID, F. O. COCKERILLE,
J. D. MEYER, W. M. COX, JR., and J. R. RUHOFF.
Checked by C. R. NOLLER and R. BANNEROT.

1. Procedure

A 5-l. round-bottomed flask is fitted with a wide-bore Y adapter and two wide-bore reflux condensers. Three liters of anhydrous butyl alcohol (Note 1) and 507 g. (1.5 moles) of butyl oleate (Note 2) are placed in the flask, and 180 g. (7.8 gram atoms) of clean sodium, cut in approximately 2.5-cm. cubes, is added in one lot and the flask connected to the condensers. The reaction is rather sluggish at first, requiring about one-half hour to reach the boiling point of the butyl alcohol (Note 3), but then becomes quite vigorous. With two condensers no difficulty is encountered in taking care of the reflux, but, if the reaction becomes too vigorous or excessive foaming occurs, wet towels should be placed on the flask until the reaction is again under control (Note 4). Toward the end of the reaction the flask is placed on a heated sand bath and gentle refluxing maintained until all the sodium has reacted. The heating is stopped temporarily, 160 cc. of water is added gradually through the condenser, and the solution is again refluxed gently for one hour (Note 5). At the end of this time the heating is stopped and 1.2 l. of water is added. The flask is well shaken and the mixture allowed to separate into two layers. The lower aqueous layer of sodium hydroxide is siphoned off and discarded (Notes 6 and 11).

[4] Steinkopf and Lützkendorf, Ann. **403**, 27 (1914); Ger. pat. 255,394 [Frdl. **11**, 144 (1912–14)].

About 200 g. of solid sodium chloride (Note 7) is added to the flask, and the butyl alcohol is removed by steam distillation (Note 8). The alcohol layer is separated while still hot (Note 9), transferred to a 1-l. beaker, and heated on a hot plate with stirring until the temperature reaches about 160°. By this time all the water is removed and foaming has stopped (Note 10). The hot liquid is transferred to a 1-l. Claisen flask having a 25-cm. fractionating side arm, and distilled at 3 mm. After a small fore-run of 5-10 g., the main fraction boils at 177–183°/3 mm. and amounts to 330–340 g. (82-84 per cent of the theoretical amount) (Note 11).

2. Notes

1. Commercial butyl alcohol is dried by distillation through a 1-m. column, and the portion boiling at 117.5–118.5° is used. The ordinary commercial alcohol is usually sufficiently pure so that the bulk of the alcohol remaining in the flask after the temperature at the top of the column reaches 117° need not be distilled but may be used directly for the reduction.

2. The butyl oleate was prepared by the alcoholysis of 3 kg. of cold-pressed olive oil by refluxing with 7 l. of butyl alcohol and 150 g. of concentrated sulfuric acid for 20 hours. The mixture was washed three times with 2.5-l. portions of saturated sodium chloride solution. During the last washing methyl orange was added to the solution and enough sodium carbonate was added to neutralize any acid remaining. The excess butyl alcohol was distilled and the residue carefully fractionated from a 2-l. Claisen flask having a 25-cm. fractionating side arm. A total of 3075 g. of distilled esters was obtained, of which 2422 g. (about 70 per cent based on the olive oil used) distilled at 204–208°/3 mm. and had approximately the theoretical iodine number. This product contains a small amount of saturated esters, but it is considerably purer than can be obtained from commercial oleic acid and is satisfactory for most purposes.

3. If one wishes to reduce this time, external heat may be applied until the boiling point is reached. In a run in which the butyl alcohol was heated to boiling before the addition of the sodium, the reaction began more vigorously but the yield was practically unchanged.

4. The reaction mixture should not be cooled below the boiling point of the butyl alcohol as a continuously vigorous reaction is essential for good yields.

5. This saponifies any unreacted ester.

6. The procedure up to this point requires about three and one-half hours and should be continuous.

7. The sodium chloride prevents the formation of an emulsion during the steam distillation.

8. The butyl alcohol is readily recovered and dried with very little loss.

9. The soap, and possibly a high-boiling by-product, causes even the liquid alcohols to set to a jelly on cooling.

10. This procedure largely eliminates the difficulty caused by foaming at the start of the vacuum distillation.

11. The procedure as given is generally applicable for the reduction of esters to alcohols in excellent yields. When preparing the solid normal saturated alcohols, the procedure may be modified, if desired, to permit the recovery of the acid from the unreduced ester. After the alkali is removed the alcohol layer is washed with two successive portions of 20 per cent salt solution which are discarded. Neither the strong alkali nor the salt solution removes an appreciable amount of organic acid. A solution of 50 g. of calcium chloride in 150 cc. of water is added to the butyl alcohol solution, the mixture is steam-distilled until the butyl alcohol is removed, and the flask and contents are allowed to cool. A hole is made in the cake of solid alcohol and the water layer removed. Two liters of toluene is added, and the flask is warmed and shaken until the alcohol dissolves and only fine crystals of the calcium salt of the unreduced acid remain. The solution is cooled to 35° and filtered with suction. The calcium soap is removed from the filter, warmed with about 500 cc. of toluene, cooled, filtered, and washed with a little more toluene. The combined toluene solutions may be concentrated and the alcohol crystallized, or the toluene may be completely distilled and the residue vacuum distilled. The insoluble calcium soap may be decomposed, re-esterified, and used in a subsequent reduction.

3. Methods of Preparation

Oleyl alcohol has been prepared by the reduction of ethyl oleate with hydrogen and a zinc-chromium oxide catalyst [1] or with sodium and absolute ethyl alcohol.[2]

[1] Sauer and Adkins, J. Am. Chem. Soc. **59**, 1 (1937).

[2] Bouveault and Blanc, Bull. soc. chim. (3) **31**, 1210 (1904). See the preparation of lauryl alcohol on p. 372 above.

ORTHANILIC ACID

(Benzenesulfonic acid, o-amino-)

$$o\text{-}O_2NC_6H_4SSC_6H_4NO_2\text{-}o + 4[O] + Cl_2 \xrightarrow{(HNO_3)} 2o\text{-}O_2NC_6H_4SO_2Cl$$

$$o\text{-}O_2NC_6H_4SO_2Cl + H_2O \rightarrow o\text{-}O_2NC_6H_4SO_3H + HCl$$

$$o\text{-}O_2NC_6H_4SO_3H + 6[H] \xrightarrow{(Fe + CH_3CO_2H)} o\text{-}H_2NC_6H_4SO_3H + 2H_2O$$

Submitted by E. Wertheim.
Checked by Reynold C. Fuson and R. S. Schreiber.

1. Procedure

(A) *o-Nitrobenzenesulfonyl Chloride.*—A 3-l. three-necked, round-bottomed flask is fitted with an efficient liquid-sealed stirrer, a reflux condenser, and an inlet tube for introducing chlorine well beneath the surface of the liquid. A glass outlet tube leads from the reflux condenser to the hood. In the flask are placed 200 g. (0.65 mole) of di-*o*-nitrophenyl disulfide (Org. Syn. Coll. Vol. I, **1941**, 220), 1 l. of concentrated hydrochloric acid (sp. gr. 1.18), and 200 cc. of concentrated nitric acid (sp. gr. 1.42). A stream of chlorine is passed into the mixture at the rate of about two bubbles per second, and the solution is warmed on a steam bath to 70°. In about thirty minutes the disulfide melts and the solution becomes orange-red in color. After the disulfide has melted, the heating and addition of chlorine are continued for one hour. The sulfonyl chloride is separated immediately from the supernatant liquid by decantation, washed with two 300-cc. portions of warm water (70°), and allowed to solidify. The water is drained from the solid mass as completely as possible.

The washed chloride is dissolved in 140 cc. of glacial acetic acid at 50–60°, and the solution is quickly filtered by suction. The filtrate is chilled by immersing the flask in cold water and is vigorously stirred in order to cause the sulfonyl chloride to separate in fine crystals. The mixture is now triturated thoroughly with a liter of cold water which is then decanted into a large Büchner funnel. The process is repeated twice. Finally, a liter of cold water is added to the mixture, and then 10 cc. of concentrated ammonium hydroxide (sp. gr. 0.90) is added, with stirring. The crystals are collected at once on the filter, washed with 200 cc. of water, and allowed to dry in the air. The yield is 240 g. (84 per cent of the theoretical amount) of a light yellow product, melting at

64–65°. This material may be used without further purification (and without being dried) for the preparation of orthanilic acid.

(B) *Orthanilic Acid.*—A 3-l. flask, fitted with a reflux condenser and a liquid-sealed stirrer, is placed on a hot plate. In the flask is placed a mixture of 200 g. (0.90 mole) of *o*-nitrobenzenesulfonyl chloride, 100 g. of anhydrous sodium carbonate, and 600 cc. of water. The mixture is heated to boiling and stirred in order to promote the hydrolysis, which is complete within forty-five minutes after the compound has melted. The orange-red solution is filtered, and the filtrate is made just acid to litmus by the addition of acetic acid, about 25 cc. being required. The solution is transferred to a 3-l. three-necked flask which is provided with a reflux condenser and an efficient liquid-sealed stirrer. The solution is heated to boiling on the hot plate, and iron filings (about 20-mesh) are added, with vigorous stirring, at the rate of about 25 g. every fifteen minutes. A total of 350 g. of iron is used. In a few minutes the mixture becomes very deep brown in color and has a tendency to foam. After stirring for four hours, a sample when filtered should yield an almost colorless filtrate; if the filtrate is red or orange, stirring and heating must be continued. When a light-colored filtrate is obtained, 2 g. of decolorizing carbon is added, the hot mixture is filtered by suction, and the residue is washed several times with small amounts of hot water which are added to the main solution. The filtrate is chilled to about 15°, and 95 cc. of concentrated hydrochloric acid is slowly added. The orthanilic acid separates in fine colorless crystals which appear as hexagonal plates under the microscope (Note 1). When the temperature has again fallen to about 15°, the mass is filtered and the precipitate is washed with water and then with ethyl alcohol. If about 20 cc. of concentrated hydrochloric acid is added to the filtrate, an additional deposit of about 1 g. will be obtained after a few hours' standing. The yield is 89 g. (57 per cent of the theoretical amount). The compound is 97–100 per cent pure and for many purposes will not require recrystallization. Material of analytical purity may be obtained by one recrystallization from hot water. The decomposition point is about 325° (bloc Maquenne).

2. Note

1. Solutions of orthanilic acid, when chilled below 13.5°, yield the hydrated form of the acid, which crystallizes as needles (see photographs in the paper by Fierz-David and others).[1]

[1] Fierz-David, Schlittler, and Waldemann, Helv. Chim. Acta 12, 663 (1929).

3. Methods of Preparation

Orthanilic acid was first made by the reduction of nitrobenzenesulfonic acid by ammonium sulfide.[2] This reduction has also been carried out electrolytically, and by the use of iron or zinc.[3] The acid has also been made by the rearrangement of phenylsulfamic acid;[4] by the action of sodium hypobromite upon potassium o-carbaminebenzenesulfonate;[5] by the reduction of the mixed nitrobenzenesulfonic acids followed by separation of the isomers;[6] by the action of methyl alcohol upon o-nitrophenylsulfur chloride;[7] by the action of acid upon diacetyl diphenylsulfamide;[8] by the debromination of p-bromoaniline-o-sulfonic acid;[9] by the reduction of 1,2,6-aminothiophenolsulfonic acid;[10] and by the hydrolysis and reduction of o-nitrobenzenesulfonyl chloride, which was obtained from di-o-nitrophenyl disulfide.[11]

[2] Limpricht, Ann. **177**, 79, 98 (1875).

[3] Wohlfahrt, J. prakt. Chem. (2) **66**, 556 (1902); Goldschmidt and Eckardt, Z. physik. Chem. **56**, 411 (1906); Holleman and Polak, Rec. trav. chim. **29**, 419 (1910); Sharvin, Arbuzov, and Varshavskii, J. Chem. Ind. (Moscow) **6**, 1409 (1929) [C. A. **25**, 501 (1931)].

[4] Bamberger and Hindermann, Ber. **30**, 654 (1897); Bamberger and Kunz, ibid. **30**, 2276 (1897); Bretschneider, J. prakt. Chem. (2) **55**, 286 (1897); Bamberger and Rising, Ber. **34**, 249 (1901); Baumgarten, ibid. **59**, 1976 (1926).

[5] Bradshaw, Am. Chem. J. **35**, 339 (1906).

[6] Bahlmann, Ann. **186**, 307 (1877); Franklin, Am. Chem. J. **20**, 457 (1898); Obermiller, Ger. pat. 281,176 [Frdl. **12**, 125 (1914–16)].

[7] Zincke and Farr, Ann. **391**, 59, 66 (1912).

[8] Wohl and Koch, Ber. **43**, 3301 (1910).

[9] Thomas, Ann. **186**, 128 (1877); Baseler Chem. Fabrik Bindschedler, Ger. pat. 84,141 [Frdl. **4**, 90 (1894–97)]; Kreis, Ann. **286**, 377 (1895); Bradshaw, Am. Chem. J. **35**, 340 (1906); Boyle, J. Chem. Soc. **95**, 1698 (1909); Scott and Cohen, ibid. **121**, 2042 (1922).

[10] Rassow and Döhle, J. prakt. Chem. (2) **93**, 188, 203 (1916).

[11] Elgersma, Rec. trav. chim. **48**, 752 (1929).

PELARGONIC ACID

$$CH_2(CO_2C_2H_5)_2 + C_7H_{15}Br + C_4H_9ONa$$
$$\rightarrow C_7H_{15}CH(CO_2C_2H_5)_2 + C_4H_9OH + NaBr$$

$$C_7H_{15}CH(CO_2C_2H_5)_2 + 2H_2O \xrightarrow{(KOH)} C_7H_{15}CH(CO_2H)_2 + 2C_2H_5OH$$

$$C_7H_{15}CH(CO_2H)_2 \longrightarrow C_7H_{15}CH_2CO_2H + CO_2$$

Submitted by E. Emmet Reid and John R. Ruhoff.
Checked by W. W. Hartman and G. W. Sawdey.

1. Procedure

A 5-l. three-necked flask is fitted with a liquid-sealed mechanical stirrer, reflux condenser, dropping funnel, and thermometer. In the flask is placed 2.5 l. of anhydrous butyl alcohol (Note 1), and 115 g. (5 gram atoms) of clean, bright sodium cut in small pieces is added at one time. Solution of the sodium may be facilitated by stirring, but heating is unnecessary. After the sodium has dissolved completely, the solution is allowed to cool to 70–80°, and then 800 g. (5 moles) of redistilled ethyl malonate (b.p. 135–136°/100 mm.) is added rapidly with stirring. After heating the reaction solution to 80–90°, 913 g. (5.1 moles) of pure heptyl bromide (p. 247, b.p. 179–180°) is added. The bromide should be added rather slowly at first, until precipitation of sodium bromide begins; it may then be added at such a rate that the butyl alcohol refluxes gently. Usually about one hour is required for the introduction of the heptyl bromide. The mixture is refluxed gently until it is neutral to litmus (about one hour).

The entire mixture, including the precipitated sodium bromide, is transferred to a 12-l. flask together with a small amount of water used to rinse the reaction flask. A solution of 775 g. (12.5 moles) of 90 per cent potassium hydroxide in an equal weight of water is added slowly with shaking. The mixture is heated cautiously, with occasional shaking, until refluxing starts (Note 2), and refluxing is continued until saponification is complete (about four or five hours). The flask is fitted at once for steam distillation (Org. Syn. Coll. Vol. I, **1941**, 479), and the mixture is distilled until no more butyl alcohol passes over (Note 3). To the residue 1350 cc. (15.5 moles) of concentrated hydrochloric acid (sp. gr. 1.18) is added carefully, with shaking, and the mixture is refluxed for about one hour (Note 4). After cooling, the water layer is siphoned off and discarded (Note 5).

The oil obtained in the preceding step is transferred to a 3-l. round-bottomed flask and heated under an air-cooled reflux condenser in an oil bath at about 180°. When the evolution of carbon dioxide has ceased (about two hours), the oil is decanted from a small amount of solid material. The solid residue on treatment with 200–300 cc. of concentrated hydrochloric acid gives an additional small quantity of oil which is added to the main portion.

The crude pelargonic acid is distilled in a modified Claisen flask having a fractionating side arm, and the material boiling at 140–142°/12 mm. (188–190°/100 mm.) is collected. The yield is 525–590 g. (66–75 per cent of the theoretical amount). The melting point of the pure acid is 12–12.5° (Note 6).

2. Notes

1. Commercial butyl alcohol was dried over solid potassium carbonate and distilled through a 90-cm. indented column. The portion boiling at 117–118° was used.

2. Two layers are formed at first, but the solution becomes homogeneous as saponification occurs. Boiling chips should be placed in the flask, and heating should be done carefully at first, with occasional shaking, or the reaction may get beyond control.

3. The flask should not be allowed to cool between saponification and distillation. It is advisable to heat the flask to prevent the volume of distillate from becoming too large. Usually about 7 l. of distillate is collected, from which the butyl alcohol can be recovered.

4. When the oily layer ceases to increase, decomposition of the potassium heptylmalonate is complete. A layer of salt sometimes accumulates at the bottom of the flask. Care must be taken in heating to prevent cracking the flask.

5. It is unnecessary to extract the aqueous layer with an organic solvent.

6. n-Caproic acid may be prepared by this method from n-butyl bromide in similar yields (see also p. 417). In this preparation a partial decomposition of the substituted malonic acid is brought about by refluxing the aqueous solution in the 12-l. flask after the addition of the hydrochloric acid. Butylmalonic acid is appreciably soluble in water, and separation of the oily layer does not occur until the malonic acid has been largely decomposed to caproic acid. The time required is about eight to ten hours. It is advisable to heat the acid layer under air reflux as in the preparation of pelargonic acid.

3. Methods of Preparation

Pelargonic acid has been prepared by the oxidation of oleic acid [1] and by hydrolysis of octyl cyanide [2] or heptylacetoacetic ester.[3]

PENTAERYTHRITYL BROMIDE AND IODIDE

[Methane, tetrakis(bromomethyl)- and (iodomethyl)-]

$$3C(CH_2OH)_4 + 4PBr_3 \rightarrow 3C(CH_2Br)_4 + 4H_3PO_3$$

$$C(CH_2Br)_4 + 4NaI \rightarrow C(CH_2I)_4 + 4NaBr$$

Submitted by H. B. Schurink.
Checked by W. H. Carothers and W. L. McEwen.

1. Procedure

(A) *Pentaerythrityl Bromide.*—One hundred twenty-five grams (0.92 mole) of dry pentaerythritol (Org. Syn. Coll. Vol. I, **1941**, 425) is placed in a 500-cc. round-bottomed flask provided with an air-cooled reflux condenser bearing at the upper end a long-stemmed dropping funnel and a bent glass tube. The tube is connected to a suitable trap for absorbing the large quantity of hydrogen bromide which is evolved. The flask is heated on a steam bath, and 500 g. (175 cc., 1.85 moles) of freshly distilled phosphorus tribromide (p. 359) is added cautiously from the dropping funnel. When this addition has been completed the steam bath is replaced by an oil bath and the temperature is raised gradually to 170–180° (Note 1). After heating at this temperature for twenty hours, the orange-red reaction mixture is transferred to a beaker containing 1 l. of cold water and stirred thoroughly to reduce the lumps to a small size. The red, flocculent material is filtered with suction and washed several times with hot water; finally it is washed thoroughly with two 200-cc. portions of cold 95 per cent ethyl alcohol (Note 2). After drying, the material is transferred to a large Soxhlet extractor and extracted exhaustively with 95 per cent alcohol (Note 3). The pentaerythrityl bromide

---

[1] Redtenbacher, Ann. **59**, 52 (1846); Harries and Thieme, ibid. **343**, 355 (1905); Harries and Türk, Ber. **39**, 3737 (1906); Molinari and Soncini, ibid. **39**, 2739 (1906); Molinari and Barozi, ibid. **41**, 2795 (1908); Jegorow, J. prakt. Chem. (2) **86**, 531 (1912).

[2] Zincke and Franchimont, Ann. **164**, 333 (1872).

[3] Jourdan, ibid. **200**, 107 (1880).

separates from the alcohol and after cooling is collected with suction. The yield of crude product melting at 158–160° is 245–270 g. (69–76 per cent of the theoretical amount); this material is sufficiently pure for conversion to the iodide (Note 4). For purification, the crude product may be recrystallized from 95 per cent alcohol, using 30 cc. of solvent per gram; the melting point is raised to 163°, and the recovery is about 85 per cent.

(B) *Pentaerythrityl Iodide.*—In a 500-cc. round-bottomed flask fitted with a reflux condenser, a mixture of 100 g. (0.67 mole) of sodium iodide (dried at 120°), 300 cc. (240 g.) of ethyl methyl ketone (Note 5), and 50 g. (0.13 mole) of crude pentaerythrityl bromide (m.p. 158–160°) is refluxed on a steam bath for forty-eight hours. The condenser is set for distillation, the solvent is distilled, and the residue is washed into a hot Büchner funnel with hot water. The product is washed thoroughly on the funnel with boiling water, pressed well, and transferred to a large Soxhlet extractor. To remove impurities the material is extracted with boiling 95 per cent alcohol until a sample removed from the extraction thimble melts at 233° (Note 6). The product is then removed from the thimble and dried; the yield is 66–73 g. (89–98 per cent of the theoretical amount). The crude product may be recrystallized from hot benzene, using about 18 cc. of solvent per gram. The recovery is about 80 per cent, and the melting point is not changed.

2. Notes

1. The temperature must be raised slowly to avoid formation of spontaneously inflammable hydrides of phosphorus which will ignite and destroy the preparation. A similar result occurs if the phosphorus tribromide is added at 170°.

2. This washing eliminates intermediate bromohydrins.

3. Since the extraction is slow, it is advisable to divide the material into several portions and carry out a number of simultaneous extractions. The extractor used was a modified Soxhlet, arranged so that the extractor tube is heated by the vapor of the solvent (p. 524, Note 3).

4. The crude product has a disagreeable odor, probably due to the presence of a compound of phosphorus. The odor may be eliminated by heating at 120°, followed by several recrystallizations.

5. Acetone may be used as solvent in place of ethyl methyl ketone, but the reaction must then be carried out in a sealed vessel at 95–100° for thirty to thirty-six hours.

6. The extraction requires twelve to sixteen hours, depending upon the rate of refluxing of the alcohol.

3. Methods of Preparation

Pentaerythrityl bromide has been prepared by the action of phosphorus tribromide on pentaerythritol,[1,2] and of an acetic acid solution of hydrobromic acid on pentaerythrityl tetracetate.[3] The iodide has been prepared by the action of red phosphorus and hydriodic acid on pentaerythritol[4] and by treating the bromide with sodium iodide in acetone.[2]

n-PENTANE

$$CH_3(CH_2)_2CHBrCH_3 + Mg \rightarrow CH_3(CH_2)_2CH(MgBr)CH_3$$

$$2CH_3(CH_2)_2CH(MgBr)CH_3 + H_2SO_4 \rightarrow$$
$$2CH_3(CH_2)_3CH_3 + MgBr_2 + MgSO_4$$

Submitted by C. R. NOLLER.
Checked by ROGER ADAMS and L. J. ROLL.

1. Procedure

IN a 5-l. flask, placed on a steam bath and fitted with a mechanical stirrer, a separatory funnel, a thermometer well (Note 1), and a calcium chloride tube, is placed 182 g. (7.5 gram atoms) of magnesium turnings. To this are added a crystal of iodine and 100 cc. of a solution of 1133 g. (7.5 moles) of 2-bromopentane (Note 2) in 750 g. of n-butyl ether (Note 3). The stirrer is started, and the flask is heated with steam until the reaction starts. This may take from fifteen minutes to one hour; the flask must be watched quite closely because the reaction, when once started, is very vigorous and evolves a large amount of heat. As soon as the reaction has started, 750 g. of n-butyl ether is added and then the balance of the solution of 2-bromopentane in n-butyl ether is added at such a rate that the temperature is kept at 50–60°. External cooling is used in order to allow more rapid addition of the 2-bromopentane. After addition is complete (about three hours), the mixture is heated on a steam bath for one hour.

In the meantime, a 12-l. flask containing a solution of 450 cc. of con-

[1] Rave and Tollens, Ann. **276**, 61 (1893); Gustavson, J. prakt. Chem. (2) **54**, 98 (1896).

[2] Backer and Schurink, Rec. trav. chim. **50**, 924 (1931).

[3] Perkin and Simonsen, J. Chem. Soc. **87**, 860 (1905).

[4] Tollens and Wigand, Ann. **265**, 331 (1891).

centrated sulfuric acid in 3 l. of water is placed on a steam bath and fitted with a stirrer, a separatory funnel, and an efficient ice-cooled condenser set for distillation. The stirrer is started, and the solution of the Grignard reagent, prepared above, is added. The acid solution is allowed to become warm but is kept below the boiling point by external cooling. After all the Grignard reagent has been added, the mixture is heated on the steam bath until no more pentane distils. The reaction flask is allowed to cool, and the *n*-butyl ether layer is separated, transferred to a 5-l. flask connected with the condenser, and heated with a free flame until the boiling point of the *n*-butyl ether is reached (Note 4). The combined distillate is separated from a small amount of water, washed twice with 125-cc. portions of cold, concentrated sulfuric acid, and allowed to stand overnight with anhydrous potassium carbonate. After removing the potassium carbonate, the *n*-pentane is fractionated twice through an efficient 100-cm. fractionating column. The yield is 270–290 g. (50–53 per cent of the theoretical amount) of a product which boils at 35.5–36.5°.

2. Notes

1. A closed glass tube containing mercury was used as a thermometer well. It was inserted with the end in the reaction mixture. The thermometer was placed in the well with the bulb in the mercury.
2. The 2-bromopentane used in this preparation was prepared by the action of hydrobromic acid on secondary amyl alcohol (methyl *n*-propyl carbinol) obtained from Stanco Distributors, Inc. Other commercial straight-chain amyl alcohols when converted to the bromides gave a pentane boiling at 33–35°, showing that a considerable quantity of isopentane was present.
3. The use of *n*-butyl ether allows the ready separation of the pentane by distillation. The butyl ether used in this preparation boiled at 142–144°.
4. About 70 per cent of the *n*-butyl ether is recovered readily and may be purified by distillation.

3. Methods of Preparation

n-Pentane has been obtained in many ways, but the reactions which are of preparative interest are few. They include the reduction of 2-bromopentane with zinc and hydrochloric acid;[1] the reduction of 3-iodopentane with zinc and hydriodic acid;[2] the catalytic reduction of

[1] Clarke and Talbot, unpublished results.
[2] Karvonen, Acta Chem. Fennica, **3**, 101 (1930) [C. A. **25**, 2412 (1931)].

pyridine at 50°;[3] the reduction of n-amylene;[4] and the decomposition of sec.-amylmagnesium bromide with ammonium chloride.[5] The use of n-butyl ether in the procedure described above, which eliminates the problem of separating n-pentane and ethyl ether,[5] was suggested by a paper of Marvel, Blomquist, and Vaughn.[6]

PHENACYL BROMIDE

(Acetophenone, α-bromo-)

$$C_6H_5COCH_3 + Br_2 \xrightarrow{(AlCl_3)} C_6H_5COCH_2Br + HBr$$

Submitted by R. M. Cowper and L. H. Davidson.
Checked by Lee Irvin Smith and E. W. Kaiser.

1. Procedure

A solution of 50 g. (0.42 mole) of acetophenone (Org. Syn. Coll. Vol. I, **1941**, 111) in 50 cc. of pure anhydrous ether (Note 1) is placed in a dry three-necked flask fitted with a separatory funnel, mechanical stirrer, and reflux condenser (Note 2). The solution is cooled in an ice bath, 0.5 g. of anhydrous aluminum chloride is introduced (Note 3), and 67 g. (21.5 cc., 0.42 mole) of bromine is added gradually from the separatory funnel, with stirring, at the rate of about 1 cc. per minute. The bromine color disappears rapidly although very little hydrogen bromide is evolved; towards the end of the reaction the solution becomes pink.

After the bromine has been added the ether and dissolved hydrogen bromide are removed at once (Note 4) under reduced pressure with a slight current of air. The phenacyl bromide remains as a solid mass of brownish yellow crystals (Note 5); the color is removed by shaking with a mixture of 10 cc. of water and 10 cc. of petroleum ether. The crystals are filtered with suction and washed several times with fresh portions of the solvent mixture, if necessary, until a white product is obtained (Note 6). The crude phenacyl bromide weighs 74–80 g. (88–96 per cent of the theoretical amount) and melts at 45–48°. This material is sufficiently pure for many purposes. If higher purity is desired the crude product may be recrystallized from 25–30 cc. of methanol, yielding

[3] Skita and Brunner, Ber. **49**, 1598 (1916).

[4] Fettindustrieges. M. B. H., Ger. pat. 329,471 [Frdl. **13**, 178 (1916–21)]; Wibaut and collaborators, Rec. trav. chim. **58**, 337 (1939).

[5] Fischer and Klemm, Z. physik. Chem. A **147**, 275 (1930).

[6] Marvel, Blomquist, and Vaughn, J. Am. Chem. Soc. **50**, 2810 (1928).

54–55 g. (64–66 per cent of the theoretical amount) of white crystals melting at 49–51° (Note 7).

2. Notes

1. Dry carbon tetrachloride may be used as a solvent but is less favorable than dry ether.

2. Quantities up to 200 g. of acetophenone may be brominated in a single operation with equally good yields but this is not generally advisable unless the product is to be used at once, since it becomes discolored on standing.

3. Without aluminum chloride the reaction is slow and incomplete.

4. If the ether and hydrogen bromide are not removed immediately the solution blackens on standing and a lower yield of less pure product results.

5. Phenacyl bromide is a lachrymator and should be manipulated carefully, to avoid contact with the skin and inhalation of the vapor.

6. The water removes yellow color due to residual hydrogen bromide, and the petroleum ether removes unchanged acetophenone or oily by-product. Since the product is quite insoluble in water and only slightly soluble in cold petroleum ether, it may be washed several times with little loss.

7. The checkers observed that all specimens of the product, even after recrystallization, although white at first, became dark and discolored on standing in a vacuum desiccator over calcium chloride.

3. Methods of Preparation

Phenacyl bromide has been prepared by the bromination of acetophenone without a solvent,[1] in carbon disulfide,[2,3] in acetic acid,[3,4,5] and in other organic solvents.[5] The quantitative aspects of the bromination in various solvents have been investigated by Kröhnke.[3] The use of ether in the method described is based on the use of this solvent in the bromination of desoxybenzoin.[6]

Phenacyl bromide is a useful reagent for the identification of organic acids by conversion to crystalline phenacyl esters.[4,7]

[1] Emmerling and Engler, Ber. **4**, 148 (1871).

[2] Hunnius, ibid. **10**, 2007 (1877); Staedel and Kleinschmidt, ibid. **13**, 837 (1880); Staedel, ibid. **16**, 22 (1883).

[3] Kröhnke, ibid. **69**, 921 (1936).

[4] Rather and Reid, J. Am. Chem. Soc. **41**, 77 (1919).

[5] Möhlau, Ber. **15**, 2465 (1882); Lazennec, Bull. soc. chim. (4) **5**, 501 (1909).

[6] Limpricht and Schwanert, Ann. **155**, 68 (1870).

[7] Shriner and Fuson, "The Systematic Identification of Organic Compounds," pp 130 and 132, John Wiley & Sons, New York, 2nd Ed., 1940.

2- AND 3-PHENANTHRENESULFONIC ACIDS

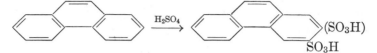

Submitted by Louis F. Fieser.
Checked by W. W. Hartman and G. L. Boomer.

1. Procedure

The sulfonation is carried out in a 2-l. round-bottomed, three-necked flask fitted with a thermometer, a dropping funnel, and a stirrer (Note 1). Five hundred grams (2.8 moles) of pure phenanthrene (Note 2) is melted in the flask, which is clamped in an oil bath heated to 110°. Mechanical stirring is started, and 327 cc. (600 g., 5.8 moles) of concentrated sulfuric acid is run in at such a rate that the internal temperature does not rise above 120° (ten to fifteen minutes). The reaction mixture is stirred and maintained at a temperature of 120–125° for three and one-half hours, when a test portion should give a nearly clear solution in water. The reaction is exothermic, and the bath must be kept at a temperature 5–10° below that of the mixture. Some sulfur dioxide is given off, and the mixture becomes green.

The viscous solution, while still hot, is dissolved in 4 l. of water, and a solution of 400 g. of sodium hydroxide in 600–800 cc. of water is added. After thorough cooling in an ice bath, the precipitated sodium salt is collected on a large funnel, pressed well, and washed thoroughly with 1 l. of half-saturated sodium chloride solution (about 180 g. per liter). The precipitate contains chiefly the sodium 2- and 3-sulfonates. The filtrate contains a mixture of disulfonates and is discarded (Note 3). In order to effect a preliminary concentration of the less-soluble 2-isomer, the sodium salt mixture is dissolved in 7–8 l. of boiling water containing 100 cc. of concentrated hydrochloric acid (Note 4), and the filtered solution is then neutralized with sodium hydroxide and allowed to crystallize. After separating the mother liquor (A), which is saved, the crystals (consisting largely of the sodium 2-sulfonate) are dissolved in about 8 l. of boiling water (Note 5), and 100 g. of solid barium chloride dihydrate is added to the hot solution. The fine white precipitate of barium 2-phenanthrenesulfonate which forms is digested at the boiling point for a short time and then brought onto a 20-cm. Büchner funnel, preheated on the steam bath. The mother liquor (B) is saved.

The precipitated barium 2-sulfonate always retains a certain quantity of the 3-sulfonate, which must be extracted with hot water. The precipitate is digested at the boiling point with 6-l. portions of boiling water until the residual salt is found to be free from isomers. The purity is determined from the melting point of the p-toluidine salt, in the manner described below. Usually two or three such washings are required. The mother liquor (B) from the barium 2-sulfonate is combined with the various washings, and the whole is boiled down to a volume of about 6 l. Sulfuric acid (25–30 cc. of concentrated acid, diluted with water) is then added to precipitate the barium; the filtered solution (Note 6) is concentrated to about 2 l. and neutralized with potassium hydroxide to bring down a mixture of potassium sulfonates (C). The filtrate from the potassium salts may be boiled down further and treated with potassium chloride to ensure complete recovery of the potassium sulfonates, and then discarded. The potassium salt mixture collected in these operations is set aside (C).

The mother liquor (A) from the sodium salt crystallization is boiled down to a volume of about 2–3 l., and 200 g. of potassium chloride is added to the hot solution. The potassium salt mixture which separates on cooling is collected and combined with the potassium salts (C) described in the preceding paragraph; the filtrate is discarded. The combined material is dissolved in the minimum amount of hot water, and the filtered solution is heated to boiling and allowed to cool without disturbance. It deposits a large crop of the potassium 3-sulfonate in a very pure condition (180–210 g.). The mother liquor, containing more of the 3-sulfonate together with some of the 2-sulfonate, is concentrated to a small volume; the product is salted out with potassium chloride and washed free of sulfate ion with potassium chloride solution. A hot aqueous solution of this material is then treated with 10 g. of barium chloride dihydrate, and the precipitated barium 2-sulfonate (5–10 g.) is washed free of isomers. The original filtrate is reserved, but the washings are discarded. From the original filtrate, by evaporation, precipitation of the barium with sulfuric acid, neutralization with potassium hydroxide, and crystallization of the product, there is obtained an additional quantity of the pure potassium 3-sulfonate (50–70 g.).

The yields are: barium 2-phenanthrenesulfonate, 150–200 g. (17–21 per cent); potassium 3-phenanthrenesulfonate, 200–220 g. (24–26 per cent) (Notes 7 and 8).

Identification and Test of Purity.—The method [1] consists in the preparation and examination of a test sample of the p-toluidine salt of the sulfonic acid (Note 9). An aqueous solution of the free sulfonic acid

[1] Fieser, J. Am. Chem. Soc. **51**, 2460, 2471 (1929).

(or of the sodium or potassium salt) is treated with an excess of p-toluidine and hydrochloric acid, enough water is added to bring all the material into solution at the boiling point, and crystallization is allowed to take place. The crystals should be washed well with water. A barium salt should be boiled with dilute sulfuric acid, a little decolorizing carbon added, and the filtered solution treated with p-toluidine. If the amine salt separates in an oily condition, the walls of the vessel should be scratched, for the p-toluidine salt, particularly that of the 3-acid, may remain as an oil for a short time, even when nearly pure. On the other hand, a rather impure acid gives a p-toluidine salt which remains as an oil almost indefinitely. This property characterizes a mixture of isomers nearly as definitely as the depression in the melting point, though this is large. In determining the melting point of an amine salt, the sample may be dried by pressing the material on a filter paper, but the capillary should be placed in the bath when the temperature is below 130°. The preliminary heating gives ample provision for thorough drying, without which the material may melt twenty to thirty degrees below the true melting point.

The p-toluidine salt of 2-phenanthrenesulfonic acid forms flat needles or plates melting at 282° (291° corr.); that of the 3-acid forms thick needles melting at 217° (222° corr.).

2. Notes

1. A convenient form of stirrer is made by bending a glass rod to an angle of 45° about 4 cm. from the end. This type of stirrer can be used to dislodge material adhering to the walls of the flask.

2. A pure grade of phenanthrene should be used. Technical 70 per cent phenanthrene may be purified by the method of Cohen and Cormier.[2] It is important to note that the presence of more than about 2 per cent of anthracene in phenanthrene raises the melting point.

3. It is estimated that no less than twelve isomeric disulfonates are present. No useful products have been obtained from the mixture.

4. The solubility of the sulfonates is increased appreciably by the presence of a mineral acid.

5. A 10-l. enameled pail heated with a ring burner serves as a convenient, if not very durable, vessel for boiling and evaporating the large volumes of solution involved. The contents may be ladled out with a casserole.

6. The addition of a little decolorizing carbon greatly facilitates the coagulation of barium sulfate and aids in retaining it on the filter.

[2] Cohen and Cormier, ibid. **52,** 4363 (1930).

7. The low yield of the monosulfonates is due in large part to the fact that disulfonic acids are invariably formed along with the mono acids.

8. For some reactions, such as alkali fusion, the barium salt of the 2-phenanthrenesulfonic acid is suitable. For conversion into the potassium salt, it is highly advisable to use the barium salt in the moist and finely divided condition in which it is first obtained. Since the reaction is slow at best, several fresh portions of sulfuric acid should be employed for precipitating the barium.

9. This is recommended as a general method of identifying sulfonates. In economy of time and material it is superior to the preparation of the free acid, the acid chloride, the ester, the amide, or the phenol. By this method one can identify quickly a few milligrams of an acid or of any of its metal salts, whether it is in the solid state or in solution.

3. Methods of Preparation

The sulfonation of phenanthrene has been studied by Werner and his students,[3] by Sandqvist,[4] by Fieser,[1] and by Ioffe.[5]

PHENOXTHIN

(Phenoxathiin)

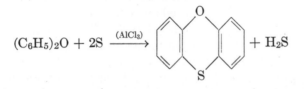

$$(C_6H_5)_2O + 2S \xrightarrow{(AlCl_3)} + H_2S$$

Submitted by C. M. SUTER and CHARLES E. MAXWELL.
Checked by REYNOLD C. FUSON and E. A. CLEVELAND.

1. Procedure

IN a 5-l. flask are placed 1886 g. (11 moles) of phenyl ether (Note 1), 256 g. (8 gram atoms) of sulfur (flowers), and 510 g. (3.8 moles) of anhydrous aluminum chloride. The reactants are mixed well by shaking the flask vigorously; the mixture becomes purple. The flask is fitted with a water-cooled reflux condenser (Note 2) and heated on the steam bath in an efficient hood. The evolution of hydrogen sulfide, vigorous at first,

[3] Werner et al., Ann. **321**, 248 (1902).
[4] Sandqvist, ibid. **392**, 76 (1912).
[5] Ioffe, J. Gen. Chem. (U.S.S.R.) **3**, 448 (1933) [C. A. **28**, 1694 (1934)].

becomes slow after one and one-half hours. After the heating has been continued for a total of four hours, with occasional shaking, the reaction mixture is poured slowly, with stirring, into a 4-l. beaker half filled with ice to which 250 cc. of concentrated hydrochloric acid has been added. More ice is added if necessary. The flask is rinsed with water and the rinsings added to the main product. After the two layers are separated the water layer is discarded and the phenyl ether-phenoxthin layer dried overnight with calcium chloride. This mixture is then distilled at 5 mm. pressure from a 3-l. special Claisen flask having a well-lagged 18-in. column. After removal of the phenyl ether the fraction boiling at 140–160°/5 mm., practically all of which comes over at 150–152°, is collected as phenoxthin (Note 3). The yield is 700 g. (87 per cent of the theoretical amount). This material, which is somewhat colored and has a strong odor, is purified by crystallization from 1.2–1.5 l. of boiling methyl alcohol; the solution should be chilled rapidly and stirred well to prevent the product from separating as an oil. The loss on crystallization is about 3 per cent and the dried material melts at 56–57° (Note 4). A second crystallization gives a product melting about one degree higher.

2. Notes

1. The commercial "diphenyl oxide" is satisfactory.

2. The condenser returns to the flask a small amount of phenyl ether, which would otherwise be carried away by the hydrogen sulfide.

3. The fore-run of phenyl ether, b.p. 98–101°/5 mm., may be used in subsequent runs. Other boiling points are: phenyl ether, 134–137°/15 mm., 259–262°/745 mm.; and phenoxthin, 180–183°/15 mm., 311°/745 mm. Fractionation at 15 mm. gives a lower yield with a larger amount of tarry residue remaining in the flask than fractionation at 5 mm.

4. A pure sample melts at 57.5–58°.

3. Methods of Preparation

Phenoxthin has been obtained by a series of reactions utilizing thio-catechol and 3,5-dinitro-4-chlorobenzoic acid [1] as the starting materials, from phenoxtellurin and sulfur,[2] and by the action of sulfur and aluminum chloride upon phenyl ether.[3]

[1] Mauthner, Ber. **39**, 1340 (1906).

[2] Drew, J. Chem. Soc. **1928**, 519.

[3] Ferrario, Bull. soc. chim. (4) **9**, 536 (1911); Ackermann, Ger. pat. 234,743 [Frdl. **10**, 153 (1910–12)]; Suter, McKenzie, and Maxwell, J. Am. Chem. Soc. **58**, 717 (1936); Bennett, Lesslie, and Turner, J. Chem. Soc. **1937**, 444; Suter and Green, J. Am. Chem. Soc. **59**, 2578 (1937).

α-PHENYLACETOACETONITRILE

(α-Tolunitrile, α-acetyl-)

$$C_6H_5CH_2CN + CH_3CO_2C_2H_5 \xrightarrow{(C_2H_5ONa)} C_6H_5CH(CN)COCH_3 + C_2H_5OH$$

Submitted by Percy L. Julian, John J. Oliver, R. H. Kimball, Arthur B. Pike, and George D. Jefferson.
Checked by C. R. Noller and Martin Synerholm.

1. Procedure

A SOLUTION of sodium ethoxide is prepared from 60 g. (2.6 gram atoms) of clean sodium and 700 cc. of absolute alcohol (Note 1) in a 2-l. round-bottomed flask equipped with a reflux condenser. To the hot solution is added a mixture of 234 g. (2 moles) of pure benzyl cyanide (Note 2) and 264 g. (3 moles) of dry ethyl acetate (Note 3). The mixture is thoroughly shaken, the condenser closed with a calcium chloride tube, and the solution heated on the steam bath for two hours before standing overnight (Note 4). The next morning the mixture is stirred with a wooden rod to break lumps, cooled in a freezing mixture to −10°, and kept at this temperature for two hours. The sodium salt is collected on a 6-in. Büchner funnel and washed four times on the funnel with 250-cc. portions of ether. The filter cake is practically colorless and corresponds to 250–275 g. of dry sodium salt, or 69–76 per cent of the calculated amount. The combined filtrates are placed in the freezing mixture until they can be worked up as indicated below.

The sodium salt still wet with ether is dissolved in 1.3 l. of distilled water at room temperature, the solution cooled to 0°, and the nitrile precipitated by adding slowly, with vigorous shaking, 90 cc. of glacial acetic acid, while the temperature is kept below 10°. The precipitate is separated by suction filtration and washed four times on the funnel with 250-cc. portions of water. The moist cake weighing about 300 g. (Note 5) corresponds to 188–206 g. (59–64 per cent) of dry colorless α-phenylacetoacetonitrile, m.p. 87–89°, which is suitable for most purposes.

If it is desired to recrystallize the crude product, the moist cake is dissolved in 100 cc. of hot methyl alcohol and the solution filtered and cooled, with stirring, to −10°. The crystals are separated by suction filtration and washed once on the filter with 40 cc. of methyl alcohol cooled to −10°. When dry, the product weighs 173–191 g. (54–60 per cent) and melts at 88.5–89.5°.

The filtrates and washings from the separation of the sodium salt are placed in a 5-l. flask and diluted with ice-cold water until the flask is full; the lower layer is removed almost completely by siphoning, most of the ether removed by decantation, and the remainder separated in a separatory funnel. The aqueous layer is extracted twice in a similar manner with 500-cc. portions of ether, and the ether extracts are discarded. The ether remaining in the aqueous layer is removed under diminished pressure by drawing air through the solution for one hour with a suction pump, and the α-phenylacetoacetonitrile is precipitated by adding 60 cc. of glacial acetic acid. If an oil is thrown out, the flask is placed in an ice bath until the precipitate is crystalline. The crystals are separated by suction filtration and washed four times on the funnel with 50-cc. portions of water. When dry the tan-colored product weighs 50–55 g. and melts at 83–86°. It is dissolved in the methyl alcohol mother liquors from the crystallization of the first lot. The solution is boiled with a little Norite, filtered, and cooled to $-10°$. The crystals which form are collected on a filter, washed with 10 cc. of cold methyl alcohol, and dried. There is obtained 43–48 g. of pale straw-colored material, m.p. 87–89°. The product is recrystallized from 25 cc. of pure methyl alcohol and washed with 10 cc. of cold methyl alcohol; there is obtained 37–41 g., m.p. 88.5–89.5°, making a total yield of material of this purity of 210–232 g. (66–73 per cent of the theoretical amount) (Notes 6 and 7).

2. Notes

1. The absolute alcohol may be prepared by drying 95 per cent alcohol twice with lime, or once with lime and once with sodium according to Note 1, Org. Syn. Coll. Vol. I, **1941,** 251, or commercial absolute alcohol may be dried once with lime or sodium just before use.

2. Benzyl cyanide was prepared according to Org. Syn. Coll. Vol. I, **1941,** 107, including the purification with concentrated sulfuric acid.

3. Commercial absolute ethyl acetate was refluxed for one-half hour over phosphorus pentoxide and distilled just before use.

4. If time permits, the procedure may be continued without allowing the mixture to stand overnight. If the drying of the alcohol and ethyl acetate is begun in the morning, however, this is a convenient point at which to interrupt the procedure.

5. If used for the preparation of methyl benzyl ketone (p. 391), the product should not be dried.

6. It does not pay to attempt to recover more pure material by concentration of the mother liquors.

7. The number of steps may be decreased by omitting the isolation

of the sodium salt. If this procedure is followed, the reaction mixture, after standing overnight, is diluted in a 5-l. flask with 2 l. of water and shaken until the sodium salt dissolves. A liter of cracked ice is added and the mixture extracted with one 1-l. and two 500-cc. portions of ether. The extracted aqueous solution is freed of ether as above and precipitated with a solution of 150 cc. of glacial acetic acid in 400 cc. of water, filtered, and washed with water. The product is colored and of lower melting point than that obtained from the purified sodium salt, and must be recrystallized twice from methyl alcohol to reach a melting point of 88.5–89.5°. The total yield of material of this melting point is somewhat less than that given above.

3. Methods of Preparation

α-Phenylacetoacetonitrile has been prepared by the condensation of ethyl acetate with the sodium derivative of benzyl cyanide prepared from benzyl cyanide and sodium amide in ether,[1] and by condensation of ethyl acetate and benzyl cyanide by means of dry[2] or alcoholic[3] sodium ethoxide.

dl-β-PHENYLALANINE

(Alanine, β-phenyl-, *dl*-)

(A) (From the Azlactone of α-Benzoylaminocinnamic Acid)

$$C_6H_5CH{=}C{<}\begin{array}{c}CO{-}O\\|\\N{=}C{-}C_6H_5\end{array} + 2H_2O + 2[H]$$

$$\xrightarrow{\text{(P + HI)}} C_6H_5CH_2CH(NH_2)CO_2H + C_6H_5CO_2H$$

Submitted by H. B. GILLESPIE and H. R. SNYDER.
Checked by W. W. HARTMAN and J. B. DICKEY.

1. Procedure

IN a 1-l. three-necked, round-bottomed flask fitted with a reflux condenser, a mechanical stirrer, and a dropping funnel (Note 1) are placed

[1] Bodroux, Compt. rend. **151**, 234 (1910); Bull. soc. chim. (4) **7**, 848 (1910).
[2] Walther and Schickler, J. prakt. Chem. (2) **55**, 343 (1897).
[3] Beckh, Ber. **31**, 3160 (1898); Post and Michalek, J. Am. Chem. Soc. **52**, 4358 (1930).

25 g. (0.1 mole) of the azlactone of α-benzoylaminocinnamic acid (Notes 2 and 3), 20 g. (0.64 gram atom) of red phosphorus, and 135 g. (125 cc.) of acetic anhydride. During a period of about one hour 195 g. (125 cc., 0.76 mole) of 50 per cent hydriodic acid (sp. gr. 1.56) is added with stirring (Note 4). The mixture is refluxed for three to four hours and, after cooling, is filtered with suction. The unreacted phosphorus is washed on the filter with two 5-cc. portions of glacial acetic acid, and discarded. The filtrate and washings are evaporated to dryness, under reduced pressure, in a 500-cc. Claisen flask heated in a water bath. A 250-cc. distilling flask cooled in ice is used as a receiver, and the distillate is reserved for a second reduction (Note 5).

To the dry residue in the Claisen flask 100 cc. of water is added, and the evaporation to dryness is repeated. The second distillate is discarded. To the residue in the flask 150 cc. of water and 150 cc. of ether are added, and the mixture is shaken until solution is complete. The aqueous layer is separated and extracted three times with 100-cc. portions of ether. The ether extracts are discarded; the water solution is heated on a steam bath with 2–3 g. of Norite and a trace of sodium sulfite until all dissolved ether has been removed. The solution is filtered, and the filtrate is heated to boiling and neutralized to Congo red with 15 per cent ammonia (sp. gr. 0.94). Usually about 25 cc. of ammonia is required. The phenylalanine separates in colorless plates which, when cold, are filtered and washed thoroughly on the filter with two 30-cc. portions of cold water. The yield is 10.5–11 g. (63.6–67 per cent of the theoretical amount) of a product which decomposes at 284–288° (corr.) (Note 6).

2. Notes

1. Clean corks protected by tin foil should be used.

2. This azlactone is prepared readily from benzaldehyde according to the procedure given for the azlactone of α-benzoylamino-β-(3,4-dimethoxyphenyl)-acrylic acid (p. 55). From 53 g. (0.5 mole) of benzaldehyde, 89.5 g. (0.5 mole) of hippuric acid (p. 328), 41 g. of fused sodium acetate, and 153 g. of acetic anhydride there is obtained 78–80 g. (62–64 per cent yield) of an almost pure product melting at 165–166° (corr.). This material is sufficiently pure for use in the preparation of phenylalanine. By crystallization from 150 cc. of benzene a product melting at 167–168° (corr.) is obtained.

3. The reduction may be carried out by the same procedure starting from α-benzoylaminocinnamic acid, and in this way slightly higher yields are obtained. The azlactone may be converted into the free acid in the following way.

In a 12-l. flask fitted with a mechanical stirrer, 62.3 g. (0.25 mole) of the azlactone is suspended in 6 l. of water, and 11 g. (0.275 mole) of sodium hydroxide is added as a 10 per cent solution. The mixture is heated on the steam bath with stirring until solution is complete. This requires three to four hours. The hot solution is filtered and acidified with hydrochloric acid. The α-benzoylaminocinnamic acid separates as white prisms in the hot solution, and when cold it is filtered. The yield is 55.5–64.5 g. (83–96.5 per cent) of almost pure product melting with decomposition over a two-degree range between the limits 224° and 236° (corr.). The crude acid can be recrystallized from alcohol, but its melting point remains unchanged.

4. During the addition the reaction mixture may solidify. If this occurs the stirrer is stopped and one or two portions of about 5 cc. of the hydriodic acid solution are stirred into the cake with a glass rod. The mass then becomes sufficiently fluid to permit use of the mechanical stirrer.

5. For a second run the distillate is placed in a 1-l. flask with 4 cc. of water, 25 g. of the azlactone, and 20 g. of red phosphorus. The mixture is refluxed for three to four hours and treated according to the above procedure. The yield is practically the same as in the first run.

6. The decomposition temperature is extremely variable and depends upon the rate of heating. The temperatures reported here were obtained by immersing the melting-point tube in a bath preheated to 200°, and then heating rapidly.

(B) (*From α-Acetaminocinnamic Acid*)

$$C_6H_5CH{=}CCO_2H \atop \quad\quad | \atop \quad NHCOCH_3 \quad\xrightarrow[\text{(Pt)}]{H_2}\quad C_6H_5CH_2CHCO_2H \atop \quad\quad\quad | \atop \quad\quad NHCOCH_3$$

$$C_6H_5CH_2CHCO_2H \atop \quad\quad | \atop \quad NHCOCH_3 \quad\xrightarrow[\text{(HCl)}]{H_2O}\quad C_6H_5CH_2CH(NH_2)CO_2H$$

Submitted by R. M. HERBST and D. SHEMIN.
Checked by REYNOLD C. FUSON and E. A. CLEVELAND.

1. Procedure

A SOLUTION of 20.5 g. (0.1 mole) of α-acetaminocinnamic acid (p. 1) in 150 cc. of glacial acetic acid (Note 1) is placed in the bottle of a Burgess-Parr reduction apparatus, 0.5 g. of platinum oxide catalyst

(Org. Syn. Coll. Vol. I, **1941**, 463) is added, and the mixture is shaken in an atmosphere of hydrogen under an initial pressure of 40 lb. per sq. in. until the calculated amount of gas is taken up; usually about two hours is required (Notes 2 and 3). When the reduction is complete, the catalyst is removed by suction filtration and washed with a little water. The combined filtrate and washings are evaporated to dryness under diminished pressure on a water bath.

The crystalline residue (Note 4) is taken up in 400 cc. of 1 N hydrochloric acid, transferred to a 1-l. flask fitted with a reflux condenser, and boiled for ten hours (Note 5). The resulting solution is evaporated to dryness under diminished pressure on the water bath; to the residue 100 cc. of water is added slowly through a dropping funnel at about the same rate as that at which it distils, in order to remove the excess hydrochloric acid as completely as possible. The residue is then taken up in 30–40 cc. of boiling water, and the pH of the solution is adjusted until it is basic to Congo red, but still acid to litmus, by careful addition of concentrated ammonia and acetic acid (Note 6). Then two volumes of 95 per cent alcohol is added to aid in the separation of the phenylalanine. The mixture is placed in the refrigerator for a day, after which the product is transferred to a Büchner funnel, and washed first with three 25-cc. portions of ice-cold water and then with alcohol. The yield is 14.5 g. The filtrate is evaporated to dryness under reduced pressure on the water bath, and the residue is extracted with about 70 cc. of ice-cold water in three or four portions. The insoluble material, after washing with 95 per cent alcohol, is added to the main fraction of phenylalanine. The total yield is 16 g. (Note 7).

The combined fractions weighing about 16 g. are dissolved in a minimum amount of boiling water (Note 8), two volumes of 95 per cent alcohol is added, and the flask is placed in a refrigerator overnight to complete crystallization. The phenylalanine is transferred to a Büchner funnel, washed with several small portions of ice-cold water, and finally with alcohol. The yield is 10.5–11 g. By concentrating the filtrate and washings further, 3–3.5 g. of product can be obtained conveniently. The total yield is 14–14.3 g. (85–86 per cent of the theoretical amount) of analytically pure phenylalanine.

2. Notes

1. It may be necessary to warm the mixture in order to dissolve the acetaminocinnamic acid completely in this amount of acetic acid. If so the solution should be allowed to cool to room temperature before it is placed in the reduction apparatus.

2. When the calculated amount of hydrogen is taken up, the catalyst is no longer colloidal and the rate of hydrogen uptake becomes very slow.

3. With freshly prepared and moist catalyst the benzene ring may also be reduced and N-acetylhexahydrophenylalanine formed. When this occurs, the hydrogen uptake continues at a rapid rate even after the amount required for hydrogenation of the side chain has been taken up. After recrystallization from water or dilute alcohol, the hexahydro compound forms needles melting at 178°.

4. Pure N-acetylphenylalanine can be obtained by recrystallizing the residue from hot water or from hot dilute alcohol; it forms colorless needles melting at 150–151°.

5. Hydrolysis with 1 N hydrochloric acid is not complete if less than ten hours is allowed. With higher acid concentrations the hydrolysis can be completed more rapidly.

6. When the solution is made just basic to Congo red, the product separates in an almost solid mass; addition of alcohol facilitates the testing of the pH by disintegrating the mass and also decreases the solubility of the product.

7. This product contains about 2.5 per cent of ammonium chloride; allowing for this the yield of phenylalanine is 94 per cent. Unless absolutely pure phenylalanine is required, the subsequent recrystallization can be omitted.

8. Phenylalanine dissolves rather slowly in boiling water. It is therefore convenient to start with an excess of water and to concentrate the solution over a free flame until crystals begin to separate from the hot solution.

3. Methods of Preparation

dl-Phenylalanine has been prepared by the action of ammonia and hydrogen cyanide on phenylacetaldehyde;[1] by the reduction of the oxime[2] or the phenylhydrazone[3] of phenylpyruvic acid; by the reduction of phenylpyruvic acid in alcoholic-ammoniacal solution;[4] by the reduction of α-aminocinnamic acid or its derivatives;[5] and by the action

[1] Erlenmeyer and Lipp, Ann. **219**, 194 (1883).

[2] Erlenmeyer, ibid. **271**, 169 (1892); Knoop and Hoessli, Ber. **39**, 1479 (1906); Shemin and Herbst, J. Am. Chem. Soc. **60**, 1951 (1938).

[3] Feofilaktov and Vinogradova, Compt. rend. acad. sci. U.R.S.S. **24**, 759 (1939) [C. A. **34**, 1971 (1940)]; J. Gen. Chem. (U.S.S.R.) **10**, 255 (1940) [C. A. **34**, 7283 (1940)].

[4] Knoop and Oesterlin, Z. physiol. Chem. **148**, 311 (1925).

[5] Plöchl, Ber. **17**, 1623 (1884); Erlenmeyer, Ann. **275**, 15 (1893); Bergmann, Stern, and Witte, ibid. **449**, 280 (footnote) (1926); Harington and McCortney, Biochem. J. **21**, 854 (1927); Lamb and Robson, ibid. **25**, 1234 (1931).

of ammonia on α-bromo-β-phenylpyruvic acid [6]—a procedure for which detailed directions are given in Org. Syn. **21**, 99.

PHENYLARSONIC ACID

(Benzenearsonic acid)

$$C_6H_5NH_2 + 2HCl + NaNO_2 \longrightarrow C_6H_5N_2Cl + 2H_2O + NaCl$$

$$C_6H_5N_2Cl + Na_3AsO_3 \xrightarrow{\text{(CuSO}_4\text{)}} C_6H_5AsO_3Na_2 + N_2 + NaCl$$

$$C_6H_5AsO_3Na_2 + 2HCl \longrightarrow C_6H_5AsO_3H_2 + 2NaCl$$

Submitted by R. H. Bullard and J. B. Dickey.
Checked by W. H. Carothers and W. L. McEwen.

1. Procedure

In a 12-l. round-bottomed flask fitted with a mechanical stirrer is placed 1 l. of water. The water is heated to boiling, and 500 g. (4.7 moles) of anhydrous sodium carbonate is added. As soon as the carbonate has dissolved, 250 g. (1.26 moles) of arsenious oxide and 11 g. of crystalline copper sulfate are added with stirring. When all the solids have dissolved the solution is cooled with stirring under a stream of tap water until the temperature falls to 15°.

Concurrently with the preparation of the sodium arsenite solution, a solution of benzenediazonium chloride is prepared. To a well-stirred mixture of 186 g. (2 moles) of technical aniline, 400 cc. (4.8 moles) of concentrated hydrochloric acid (sp. gr. 1.19), 1 l. of water, and enough crushed ice to make a volume of about 3 l., is added slowly a solution of 145 g. (2 moles) of 95 per cent sodium nitrite in 500 cc. of water. This requires about thirty to forty minutes.

The benezenediazonium chloride solution is then added with stirring during a period of one hour to the suspension of sodium arsenite, cooled in an ice and salt bath to 0°. The temperature during the reaction is held below 5° (Note 1). Frothing takes place owing to the escape of nitrogen, but this is easily controlled by the occasional addition of a small quantity of benzene. Stirring is continued for one hour after the addition of the diazonium chloride solution, and the mixture is filtered to remove the solid material which separates. This is washed with 500 cc. of cold water, and the combined liquors are concentrated over a free flame to a volume of about 1.5 l. (Note 2).

[6] Fischer, Ber. **37**, 3064 (1904).

To the hot concentrated solution, which is deep brown in color, concentrated hydrochloric acid is added until no more tarry material separates (Note 3). The tar is filtered and more hydrochloric acid is added until, after filtering, a clear pale yellow solution results. It is important to remove all the tar at this time; otherwise, subsequent recrystallizations will not free the product from color. The phenylarsonic acid is then precipitated by the addition of 250 cc. of concentrated hydrochloric acid (sp. gr. 1.19) (Note 4). When the mixture has cooled (preferably by standing overnight) the phenylarsonic acid is filtered on a Büchner funnel and washed with 200 cc. of cold water. The light yellow crystals are dissolved in 500 cc. of boiling water, 20 g. of Norite is added, the solution filtered hot, and the filtrate allowed to cool. After filtering and drying, the white crystals melt with decomposition at 154–158°, passing into the anhydride, $C_6H_5AsO_2$. The yield is 160–182 g. (39–45 per cent of the theoretical amount) (Note 5).

2. Notes

1. This temperature is advised because it appears to be near the optimum with regard to yield and ease of purifying the product. However, a reaction temperature as high as 15° may be used with good results.

2. The solution is concentrated at atmospheric pressure since frothing occurs if reduced pressure is used.

3. About 100 cc. of hydrochloric acid is required. Care must be exercised not to precipitate any of the phenylarsonic acid.

4. Too large an excess of hydrochloric acid will dissolve some of the product and lower the yield.

5. In occasional runs by this procedure yields as high as 57 per cent have been obtained, but the reasons for this are not known.

3. Methods of Preparation

Phenylarsonic acid has been prepared by oxidizing phenyldichloroarsine [1] or phenyldiiodoarsine [2] with chlorine in water; by the oxidation of phenylarsine with nitric acid or air; [3] by decomposing phenylarsinetetrachloride or phenylarsineoxychloride with water; [4] by heating iodobenzene or bromobenzene with potassium arsenite; [5] by diazotizing p-

[1] Michaelis and Loesner, Ber. **27**, 265 (1894); Rosenheim and Bilecki, ibid. **46**, 551 (1913); Roeder and Blasi, ibid. **47**, 2752 (1914).

[2] Bertheim, ibid. **47**, 274 (1914).

[3] Palmer and Dehn, ibid. **34**, 3599 (1901); Dehn, Am. Chem. J. **33**, 149 (1905).

[4] Michaelis, Ber. **10**, 625 (1877); LaCoste and Michaelis, Ann. **201**, 203 (1880).

[5] Dehn, Am. Chem. J. **33**, 140 (1905); Rosenmund, Ber. **54**, 438 (1921).

arsanilic acid and decomposing in a solution of sodium hydrosulfite and hydrochloric acid;[6] by the action of potassium arsenite on potassium benzeneisodiazo oxide;[7] by the action of benzenediazonium chloride on sodium arsenite in the presence of a copper compound[8] or of magnesium chloride and copper powder;[9] and by the action of a neutral or alkaline mixture containing arsenious oxide, a copper salt, and a reducing agent on benzenediazonium chloride.[10]

The preparation given here in detail is essentially that described by Palmer and Adams.[8] It has been reported that the use of buffers to ensure constant pH increases the yield of phenylarsonic acid in this reaction.[11]

PHENYLBENZOYLDIAZOMETHANE

(Acetophenone, α-diazo-α-phenyl-)

$$\underset{C_6H_5CO}{\overset{C_6H_5}{\diagdown}}C{=}N{-}NH_2 + HgO \rightarrow \underset{C_6H_5CO}{\overset{C_6H_5}{\diagdown}}C{\diagup}\overset{N}{\underset{N}{\diagdown}}\| + Hg + H_2O$$

Submitted by Costin D. Nenitzescu and Eugen Solomonica.
Checked by Louis F. Fieser and Ralph S. Temple.

1. Procedure

THIRTY grams (0.134 mole) of benzil hydrazone (Note 1) is mixed in a mortar with 60 g. (0.28 mole) of yellow mercuric oxide and 15 g. of anhydrous sodium sulfate (Note 2). The mixture is introduced into a 500-cc. glass-stoppered bottle and covered with 200 cc. of absolute ether (Note 3). Four cubic centimeters of a cold, saturated solution of alcoholic potassium hydroxide is added to catalyze the reaction (Note 4), and the mixture is shaken for ten to fifteen minutes. The solution is filtered by gravity through a fine paper, and the residue is washed several times with ether until the liquid is only slightly colored. The combined ethereal extracts are evaporated to dryness at the pressure of the water pump by heating the flask in a water bath to a temperature not

[6] Bertheim, ibid. **41**, 1853 (1908).

[7] Bart, Ger. pat. 250,264 [Frdl. **10**, 1254 (1910–12)].

[8] Chem. Fabrik von Heyden A.-G., Ger. pat. 264,924 [Frdl. **11**, 1030 (1912–14)]; Palmer and Adams, J. Am. Chem. Soc. **44**, 1361 (1922); Norris, J. Ind. Eng. Chem. **11**, 825 (1919); Schmidt, Ann. **421**, 169 (1920).

[9] Bart, Ger. pat. 254,092 [Frdl. **11**, 1030 (1912–14)].

[10] Mouneyrat, Brit. pat. 142,947 [C. A. **14**, 2802 (1920)].

[11] Blas, Génie civil **115**, 448 (1939) [C. A. **34**, 2342 (1940)].

greater than 40° (Note 5). The yellow, crystalline material is dried on a porous plate and recrystallized from anhydrous ether. The yield of azibenzil which melts at about 79° with decomposition is 26–28 g. (87–94 per cent of the theoretical amount) (Note 6).

2. Notes

1. Benzil hydrazone [1] may be prepared as follows: [2] A mixture of 52 g. (0.4 mole) of hydrazine sulfate (Org. Syn. Coll. Vol. I, **1941**, 309), 110 g. (0.8 mole) of sodium acetate, and 250 g. of water is boiled five minutes, cooled to about 50°, and 225 cc. of methyl alcohol added. The precipitated sodium sulfate is filtered and washed with a little alcohol.

A hot solution of 50 g. (0.24 mole) of benzil (Org. Syn. Coll. Vol. I, **1941**, 87) in 75 cc. of methyl alcohol is prepared, and the above solution, heated to 60°, is added. Most of the benzil hydrazone separates immediately, but the yield is increased by refluxing for half an hour. The hydrazone is filtered from the cold solution and washed with a little ether to remove the yellow color. The yield is 50.5 g. (94 per cent of the theoretical amount), melting at 147–151° with decomposition.

Directions for preparing benzil hydrazone from benzil and hydrazine hydrate are given in Org. Syn. **20**, 48.

2. The sodium sulfate absorbs the water formed during the reaction.

3. By using ether, instead of benzene or petroleum ether as specified in the older methods, the evaporation of the solvent after the reaction is facilitated.

4. Without this catalyst the oxidation may require several hours and the results may vary considerably, depending largely upon the quality of the mercuric oxide.

5. The material may explode if the evaporation is carried out at atmospheric pressure on the steam bath.

6. This procedure may be used also for the preparation of diazofluorene.[3]

3. Method of Preparation

Phenylbenzoyldiazomethane, azibenzil, has been prepared by the oxidation of benzil hydrazone with mercuric oxide,[4] using benzene or petroleum ether as the solvent, and without the catalyst here specified.

[1] Curtius and Thun, J. prakt. Chem. (2) **44**, 176 (1891).

[2] Private communication from C. F. H. Allen.

[3] Staudinger and Kupfer, Ber. **44**, 2207 (1911); Staudinger and Gaule, ibid. **49**, 1955 (1916).

[4] Curtius and Thun, J. prakt. Chem. (2) **44**, 182 (1891).

α-PHENYL-β-BENZOYLPROPIONITRILE

(α-Tolunitrile, α-phenacyl-)

$$C_6H_5CH{=}CHCOC_6H_5 + HCN \rightarrow C_6H_5CH(CN)CH_2COC_6H_5$$

Submitted by C. F. H. ALLEN and R. K. KIMBALL.
Checked by J. B. CONANT and HELEN O'BRIEN.

1. Procedure

THE following preparation, through the filtration of the crude solid and washing with water, should be carried out in a hood with a good draft.

Into a 5-l. flask or bottle, set in a water bath and fitted with a stirrer, thermometer, and separatory funnel, are placed 208 g. (1 mole) of benzalacetophenone (Note 1), 3.5 l. of 95 per cent ethyl alcohol (Note 2), and 60 g. (1 mole) of glacial acetic acid (Note 3). The mixture is warmed with stirring to 35°, and a solution of 130 g. (2 moles) of potassium cyanide in 375 cc. of water is added from the separatory funnel over a period of about fifteen minutes. The initial greenish color changes to yellow after all the acetic acid has reacted and the solution has become alkaline (Note 3). Stirring is continued for three hours, the temperature being maintained at 35°. During this time about half of the nitrile crystallizes. The flask is then loosely stoppered and left in a cool place for about fifty hours (conveniently out-of-doors in cold weather, but out of direct sunlight), after which the solid is filtered and washed, first with 500 cc. of cold 50 per cent alcohol, and then with water until free from potassium cyanide (silver nitrate test). The yield of air-dried material melting at 125° is 220–227 g. (93–96 per cent of the theoretical amount). This product is sufficiently pure for most purposes, although it contains traces of a high-melting substance. If a product of higher melting point is desired, it may be recrystallized from 1 l. of 95 per cent alcohol or 375 cc. of acetone. Pure phenylbenzoylpropionitrile melts at 127°.

2. Notes

1. Crude alkali-free, air-dried benzalacetophenone is used (Org. Syn. Coll. Vol. I, **1941**, 78).

2. The preparation may be carried out in a more concentrated solution (1 l. of alcohol) with the same yield, but an inferior product is obtained. If this is done, it is best to stir for fifteen minutes at 50° after

all the cyanide has been added and then cool in tap water. Since the product separates as an oil from a solution of this concentration, it is best to inoculate with a crystal of the nitrile.

3. If the solution becomes too alkaline, the nitrile formed will add to a second molecule of unsaturated ketone so readily that the product will consist almost entirely of a high-melting (284–286°) substance. For this reason it is essential to measure the acetic acid accurately; if too much is used, addition of hydrocyanic acid will not take place.

3. Methods of Preparation

α-Phenyl-β-benzoylpropionitrile has been prepared by the action of sodium or potassium cyanide on β-chlorobenzylacetophenone[1] or benzalacetophenone dibromide;[2] and by the addition of hydrocyanic acid to benzalacetophenone in the presence of sodium or potassium cyanide.[3]

γ-PHENYLBUTYRIC ACID

(Butyric acid, γ-phenyl-)

$$C_6H_5COCH_2CH_2CO_2H + 4[H] \xrightarrow{\text{(Zn + HCl)}} C_6H_5CH_2CH_2CH_2CO_2H + H_2O$$

Submitted by E. L. MARTIN.
Checked by C. R. NOLLER and F. M. McMILLAN.

1. Procedure

AMALGAMATED ZINC is prepared by shaking for five minutes a mixture of 120 g. of mossy zinc, 12 g. of mercuric chloride, 200 cc. of water, and 5–6 cc. of concentrated hydrochloric acid contained in a 1-l. round-bottomed flask. The solution is decanted and the following reagents are added, in the order named, to the zinc: 75 cc. of water, 175 cc. of concentrated hydrochloric acid, 100 cc. of toluene, and 50 g. (0.28 mole) of β-benzoylpropionic acid (p. 81). The flask is fitted with a vertical condenser connected to a gas absorption trap (Note 1), and the reaction

---

[1] Anschütz and Montfort, Ann. **284**, 2 (1895); Rupe and Schneider, Ber. **28**, 960 (1895).

[2] Dodwadmath and Wheeler, Proc. Indian Acad. Sci. **2A**, 438 (1935) [C. A. **30**, 1771 (1936)].

[3] Hann and Lapworth, J. Chem. Soc. **85**, 1358 (1904); Lapworth and Wechsler, ibid. **97**, 41 (1910).

mixture is boiled vigorously for twenty-five to thirty hours (Note 2). Three 50-cc. portions of concentrated hydrochloric acid are added at approximately six-hour intervals during the refluxing period.

After cooling to room temperature the layers are separated. The aqueous layer is diluted with 200 cc. of water and extracted with three 75-cc. portions of ether. The toluene layer and the ether extracts are combined, washed with water, and dried over calcium chloride. The solvents are removed by distillation under reduced pressure on the steam bath, after which the γ-phenylbutyric acid is distilled at 178–181°/19 mm. (148–154°/8–10 mm., 125–130°/3 mm.). The yield of acid, which melts at 46–48° (Note 3), is 38–41 g. (82–89 per cent of the theoretical amount) (Note 4).

2. Notes

1. Considerable hydrogen chloride is driven off during the initial heating, and it might appear that it would be advantageous to use constant-boiling instead of concentrated hydrochloric acid. If this is done, however, the product has a melting point of 40–44° and the yield is somewhat lower.

2. If the refluxing is interrupted for any reason, great care must be exercised to avoid frothing on heating again. The upper part of the flask may be occasionally brushed with a free flame. Once the two layers are well mixed, boiling proceeds smoothly.

3. The recorded melting points vary from 47° to 51°. The acid may be crystallized from hot water (75 cc. per g.) but the recovery is only about 50 per cent. No other suitable solvent or combination of solvents was discovered. Redistillation raises the melting point to 47–48° with only mechanical losses.

4. The procedure described differs from that published in Org. Syn. 15, 64, by the addition of toluene to the reaction mixture. In the presence of toluene the concentration of organic material in the aqueous layer is extremely small and polymolecular reactions take place to a smaller extent than in the original procedure. As a result the yield of pure product is greater.

When the modified procedure is applied to the preparation of higher-melting compounds, for example the γ-naphthylbutyric acids, the layers are separated after cooling to 50–60°, benzene is used for the extraction, and the combined benzene-toluene solution is clarified with Norite while still wet; it is then concentrated somewhat and allowed to cool for crystallization. In preparing methoxylated acids, such as γ-anisyl- or γ-veratrylbutyric acid, some demethylation occurs. When this happens the toluene layer and extracts are mixed with an excess of dilute sodium hydroxide solution and the organic solvents are removed by steam distillation. The alkaline solution is treated at 80° with an excess of methyl sulfate, the solution is clarified with Norite, cooled, and acidified, whereupon the product separates in good condition.

3. Methods of Preparation

Of the several methods by which γ-phenylbutyric acid has been obtained, those of preparative value are the decarboxylation of γ-phenylethylmalonic acid;[1] the carbonation of γ-phenylpropylmagnesium bromide;[2] and the reduction of β-benzoylpropionic acid with amalgamated zinc and hydrochloric acid[3] or of its hydrazone with sodium ethoxide.[4] The use of toluene in the Clemmensen reduction of β-benzoylpropionic acid and almost a score of related compounds has been described by Martin.[5]

o-PHENYLENEDIAMINE

Submitted by E. L. MARTIN.
Checked by W. W. HARTMAN and S. S. FIERKE.

1. Procedure

IN a 1-l. three-necked, round-bottomed flask, fitted with a liquid-sealed mechanical stirrer and reflux condenser, are placed 69 g. (0.5 mole) of *o*-nitroaniline (Org. Syn. Coll. Vol. I, **1941**, 388), 40 cc. of a 20 per cent solution of sodium hydroxide, and 200 cc. of 95 per cent ethanol. The mixture is stirred vigorously and heated on a steam bath until the solution boils gently. The steam is turned off, and 10-g. portions of 130 g. (2 gram atoms) of zinc dust (Note 1) are added frequently enough to keep the solution boiling (Notes 2 and 3). After the addition of zinc dust has been completed the mixture is refluxed with continued

[1] Fischer and Schmitz, Ber. **39**, 2212 (1906).
[2] Grignard, Compt. rend. **138**, 1049 (1904); Rupe and Proske, Ber. **43**, 1233 (1910).
[3] Krollpfeiffer and Schäfer, ibid. **56**, 620 (1923).
[4] Staudinger and Müller, ibid. **56**, 713 (1923).
[5] Martin, J. Am. Chem. Soc. **58**, 1438 (1936).

stirring for one hour; the color of the solution changes from a deep red to nearly colorless. The hot mixture 's filtered by suction, and the zinc residue is returned to the flask and extracted with two 150-cc. portions of hot alcohol. To the combined filtrates is added 2–3 g. of sodium hydrosulfite, and the solution is concentrated under reduced pressure (using a water pump), on a steam bath, to a volume of 125–150 cc. After cooling thoroughly in an ice-salt bath, the faintly yellow crystals are collected, washed once with a small amount of ice water, and dried in a vacuum desiccator. The yield of crude o-phenylenediamine melting at 97–100° is 46–50 g. (85–93 per cent of the theoretical amount). If a purer product is desired, the material is dissolved in 150–175 cc. of hot water containing 1–2 g. of sodium hydrosulfite and treated with decolorizing charcoal. After cooling thoroughly in an ice-salt mixture, the colorless crystals are filtered by suction and washed with 10–15 cc. of ice water. The purified o-phenylenediamine weighs 40–46 g. (74–85 per cent of the theoretical amount) and melts at 99–101° (Notes 4 and 5).

2. Notes

1. The zinc dust should be at least 80 per cent pure, and the amount used should be equivalent to 130 g. of 100 per cent material. A large excess of zinc dust has been used without changing the yield.

2. Great care must be taken not to add too much zinc dust at first as the reaction becomes very vigorous. It is well to have a bath of ice and wet towels at hand in order to control the reaction if it should become too violent.

3. Occasionally the reaction suddenly stops and it is necessary to add an additional 10 cc. of 20 per cent sodium hydroxide solution, which causes the reaction to proceed.

4. The product can also be purified by distillation under reduced pressure in an inert atmosphere, but, unless the material is very nearly pure, considerable decomposition occurs and the distilled product darkens rapidly in contact with air.

5. The free diamine may also be converted into the dihydrochloride, and the salt purified as follows: The crude o-phenylenediamine is dissolved in a mixture of 90–100 cc. of concentrated hydrochloric acid (sp. gr. 1.19) and 50–60 cc. of water containing 2–3 g. of stannous chloride, and the hot solution is treated with decolorizing charcoal. To the hot, colorless filtrate is added 150 cc. of concentrated hydrochloric acid, and the mixture is cooled thoroughly in an ice-salt bath. The colorless crystals are filtered by suction, washed with a small amount of cold concentrated hydrochloric acid, and dried in vacuum over solid sodium hydroxide. The yield of o-phenylenediamine dihydrochloride is 77–81 g.

(85–90 per cent of the theoretical amount based on the weight of o-nitroaniline used).

3. Methods of Preparation

o-Phenylenediamine has been prepared by the reduction of o-nitroaniline by means of tin and hydrochloric acid,[1] stannous chloride and hydrochloric acid,[2] sodium stannite,[3] zinc dust and water,[4] sodium hydrosulfite and sodium hydroxide,[5] and zinc dust and alcoholic alkali;[6] by electrolytic reduction in aqueous alcohol in the presence of sodium acetate;[7] and from o-dichlorobenzene or o-chloroaniline by treatment with aqueous ammonia at 150° under pressure in the presence of copper.[8] The procedure described above is a modification of the method of Hinsberg and König.[6]

o-Phenylenediamine has been proposed as a reagent for the identification of aliphatic acids, by conversion to crystalline 2-alkylbenzimidazoles (p. 65).

α-PHENYLETHYLAMINE

(Benzylamine, α-methyl-)

$$C_6H_5COCH_3 + 2HCO_2NH_4 \rightarrow$$
$$C_6H_5CH(NHCHO)CH_3 + 2H_2O + NH_3 + CO_2$$

$$C_6H_5CH(NHCHO)CH_3 + H_2O + HCl \rightarrow$$
$$C_6H_5CH(NH_3Cl)CH_3 + HCO_2H$$

$$C_6H_5CH(NH_3Cl)CH_3 + NaOH \rightarrow$$
$$C_6H_5CH(NH_2)CH_3 + NaCl + H_2O$$

Submitted by A. W. INGERSOLL.
Checked by REYNOLD C. FUSON and WILLIAM E. ROSS.

1. Procedure

IN a 500-cc. modified Claisen flask are placed 250 g. (4 moles) of ammonium formate (Note 1), 150 g. (1.25 moles) of acetophenone (Note 2),

[1] Zincke and Sintenis, Ber. **6**, 123 (1873); Koerner, Gazz. chim. ital. **4**, 320 (1874); Hübner, Ann. **209**, 361 (1881).

[2] Goldschmidt and Ingebrochtsen, Z. physik. Chem. **48**, 448 (1904); Goldschmidt and Sunde, ibid. **56**, 23 (1906).

[3] Goldschmidt and Eckardt, ibid. **56**, 400 (1906).

[4] Bamberger, Ber. **28**, 250 (1895).

[5] Borsche, Chem. Zentr. **1909**, II, 1550.

[6] Hinsberg and König, Ber. **28**, 2947 (1895).

[7] Rohde, Z. Elektrochem. **7**, 339 (1900).

[8] Soc. pour l'ind. chim. à Bâle, Fr. pat. 788,348 [C. A. **30**, 1395 (1936)].

and a few chips of porous plate. The flask is fitted with a cork carrying a thermometer extending nearly to the bottom, and the side arm is connected to a small condenser set for distillation. On heating the flask with a small flame the mixture first melts to two layers and distillation occurs; at 150–155° it becomes homogeneous and reaction takes place with moderate foaming. The heating is continued, more slowly if necessary, until the temperature reaches 185°. During this process water, acetophenone, and ammonium carbonate distil; about three hours is required and little attention is necessary. At 185° the heating is stopped and the upper layer of acetophenone is separated from the distillate and returned, without drying, to the reaction flask. The mixture is then heated for three hours at 180–185°. The distillate is extracted with 25–30 cc. of benzene to recover acetophenone (Note 3), and the aqueous portion is discarded.

The reaction mixture is cooled and then shaken in a 500-cc. separatory funnel with 150–200 cc. of water to remove ammonium formate and formamide. The crude α-phenylethylformamide is drawn off into the original flask, and the water layer is extracted with two 30-cc. portions of benzene and discarded. The benzene extracts are united with the main portion, and 150 cc. of concentrated hydrochloric acid is added, together with a few pieces of porous plate. The mixture is cautiously heated until the benzene has distilled and then boiled gently for forty to fifty minutes longer. Hydrolysis proceeds rapidly, and the mixture becomes homogeneous except for a small layer of acetophenone and other neutral substances. The mixture is cooled and extracted first with 50 cc. of benzene and then with three or four 25-cc. portions of the solvent. The extracts are saved for the recovery of acetophenone (Note 3).

The aqueous acid solution is transferred to a 1-l. round-bottomed flask provided with a separatory funnel and equipped for steam distillation. A solution of 125 g. of sodium hydroxide in 250 cc. of water is added through the funnel, and the mixture is distilled with steam (Note 4). The first liter of distillate contains most of the amine, but the distillate should be collected until it is only faintly alkaline. A small residue containing di-(α-phenylethyl)-amine and neutral substances remains in the flask and may be discarded.

The distillate is extracted with five 50-cc. portions of benzene, and the benzene solution is dried thoroughly with powdered sodium hydroxide and distilled (Note 5). Most of the amine distils at 184–186°, but the fraction distilling at 180–190° is sufficiently pure for most purposes (Note 6). The yield of this fraction is 80–88 g. By combining the benzene fore-run with the distillation residue, extracting with dilute acid, and recovering the amine as above, an additional 10–12 g. of material can be obtained (Note 7), making the total yield 90–100 g. (60–66 per cent of the theoretical amount based on the acetophenone taken) (Note 8).

2. Notes

1. Ammonium formate may be made in quantity by treating solid ammonium carbonate with a slight excess of commercial 85 per cent formic acid and concentrating the solution, in stages, on a steam bath under reduced pressure. The slightly moist product obtained by suction filtration is suitable for this preparation.

2. Eastman's "practical" acetophenone, m.p. 16–20°, was used. Directions for preparing acetophenone are given in Org. Syn. Coll. Vol. I, **1941**, 111.

3. The benzene solution is washed with dilute alkali, dried, and distilled, the fraction boiling at 198–207° being collected.

4. In the steam distillation it is advisable to heat the distillation flask directly so that the volume remains nearly constant.

5. The amine attacks cork and rubber and absorbs carbon dioxide from the air. It is best distilled in a flask having an in-set side arm and collected in a distilling flask protected by a soda-lime tube.

6. If very pure amine is desired the product described above is dissolved with 1.04 parts of crystalline oxalic acid in 8 parts of hot water. After clarification with Norite, the filtered solution on cooling deposits crystals of the acid oxalate. About 5 g. of the salt remains in each 100 cc. of the mother liquor; most of this can be obtained by evaporation and further crystallization. The amine is liberated from the pure oxalate with potassium hydroxide, distilled with steam, and purified as described above. When a known amount of amine is desired in water solution (as for optical resolution), a weighed amount of the (anhydrous) oxalate is decomposed and the amine is distilled quantitatively with steam.

7. When several runs are to be made the acid solution of the amine may be combined with the next run previous to steam distillation.

8. The method described is rather general. With appropriate modifications for the purification of the amine the method yields α-p-tolylethylamine (72 per cent), α-p-chlorophenylethylamine (65 per cent), α-p-bromophenylethylamine (63 per cent), α-p-xenylethylamine (66 per cent), and α-(β-naphthyl)-ethylamine (84 per cent) from the corresponding ketones.

3. Methods of Preparation

The present procedure was developed from those of Wallach [1] and Freylon,[2] based upon the general method discovered by Leuckart.[3] α-Phenylethylamine also can be prepared satisfactorily by the reduction of acetophenone oxime with sodium and absolute alcohol [4] or sodium amalgam,[5] or with ammonium amalgam,[6] or electrolytically.[7] The amine has been obtained by reducing acetophenone phenylhydrazone with sodium amalgam and acetic acid; [8] from α-phenylethyl bromide and hexamethylenetetramine; [9] by the action of methylmagnesium iodide on hydrobenzamide; [10] and by reducing acetophenone in the presence of ammonia with hydrogen and a nickel catalyst,[11] a method for which detailed directions are given in Volume 23 of this series.

d- AND l-α-PHENYLETHYLAMINE

(Benzylamine, α-methyl-, d- and l-)

Submitted by A. W. INGERSOLL.
Checked by REYNOLD C. FUSON and WILLIAM E. ROSS.

1. Procedure

d-α-Phenylethylamine.—A solution of 100 g. (0.75 mole) (Note 1) of *l*-malic acid in 500 cc. of distilled water is mixed with 120 g. (1 mole) of *dl*-α-phenylethylamine (p. 503); the resulting solution is heated for a short time on the steam bath, filtered into a 1-l. beaker, and allowed to cool slowly. After several hours the crude *d*-α-phenylethylamine-*l*-malate which crystallizes is collected by suction filtration and washed on the filter with 25 cc. of ice water. The filtrate and washings are evaporated on a steam bath to a volume about two-thirds that of the original filtrate, and a second crop of crystals is obtained on cooling (Notes 2 and

[1] Wallach, Ann. **343**, 60 (1905).
[2] Freylon, Ann. chim. phys. (8) **15**, 141 (1908).
[3] Leuckart, Ber. **22**, 1413 (1889).
[4] Mohr, J. prakt. Chem. (2) **71**, 317 (1905).
[5] Kraft, Ber. **23**, 2783 (1890).
[6] Takaki and Ueda, J. Pharm. Soc. Japan **58**, 276 (1938) [C. A. **32**, 5376 (1938)].
[7] Tafel and Pfeffermann, Ber. **35**, 1515 (1902).
[8] Tafel, ibid. **19**, 1929 (1886); **22**, 1856 (1889).
[9] André and Vernier, Compt. rend. **193**, 1192 (1931).
[10] Busch and Leefhelm, J. prakt. Chem. (2) **77**, 5 (1908).
[11] Couturier, Ann. chim. (11) **10**, 610 (1938); Schwoegler and Adkins, J. Am Chem. Soc. **61**, 3499 (1939).

3). By repeating the process it is possible to obtain a third and usually a fourth crop, after which the mother liquor becomes too viscous to permit satisfactory crystallization. The mother liquor is reserved for later use.

The successive crops of crystals are systematically recrystallized as follows, using Norite if necessary. About two-thirds of the first crop is dissolved in about three parts of water and the hot solution allowed to deposit crystals by slow cooling (Note 3). The liquor is filtered or decanted, the remainder of the first crop is dissolved in it, and the process of crystallization is repeated. The remaining crops are then similarly recrystallized in succession from the same liquor, the solution being evaporated to the appropriate volume before each crystallization. The final mother liquor is evaporated in stages, and the viscous residue is united with that from the original crystallization. The various crops are systematically recrystallized from fresh water until pure (Note 4). It is possible to obtain 80–90 g. (63–70 per cent of the theoretical amount) of the pure *d*-base-*l*-acid (anhydrous acid salt).

The pure malate (mol. wt. 255) is decomposed by warming with very slightly more (Note 5) than two equivalents of approximately 2 N sodium hydroxide. The amine is extracted, after cooling, with three or four 25-cc. portions of pure benzene, the solution is dried thoroughly with powdered sodium hydroxide, and the pure amine, b.p. 184–185°, $[\alpha]_D^{25°} + 39.2°$ to $+39.7°$ (without solvent), is obtained by distillation (Note 6). A small amount of the amine distils with the benzene. The yield is 35–40 g. (92–94 per cent of the theoretical amount based on the pure malate).

The mother liquors from the original crystallization and recrystallizations are treated in a similar manner, and the sodium malate solutions are united and reserved for the recovery of *l*-malic acid (Note 7). The recovered amine amounts to 75–80 g. and contains 40–50 per cent excess *l*-amine.

l-α-*Phenylethylamine.*—The recovered amine is converted to the acid tartrate in water solution, using for each gram of amine 1.25 g. of *d*-tartaric acid and 4.0 cc. of water. The solution is boiled with decolorizing carbon, if necessary, filtered, and allowed to cool slowly without disturbance (Note 3). A dense mass of coarse crystals of crude *l*-α-phenylethylamine-*d*-tartrate is deposited. The solution is decanted and evaporated to about two-thirds its original volume, and a second crop of crystals is obtained as before. These operations are repeated once or twice more, after which usually no more coarse crystals, but only a mass of needle-like crystals (mixed salts), can be obtained (Note 8). Meanwhile the first crop is recrystallized from about twice its weight of water

About two-thirds of the weight of crude salt is thus obtained pure; the mother liquor is used as solvent for the second crop, and so forth. The final mother liquor is united with that from the original crystallization and reserved for the recovery of the partially resolved amine. The yield of pure *l*-base-*d*-acid salt (anhydrous acid salt) is 75–100 g., an amount equivalent to somewhat more than the excess of *l*-amine in the mixture taken. The pure salt has a specific rotation of +13.0 to 13.2° in an 8 per cent solution in water.

The pure salt (mol. wt. 271) is dissolved in four parts of water; the amine is liberated with an excess of 20–25 per cent sodium hydroxide solution, extracted with benzene, and purified as described for the *d*-amine. The constants agree closely with those given for the *d*-amine, and the yield is 32–42 g. (94–96 per cent of the theoretical amount based on the pure tartrate, or 53–70 per cent based on the total *l*-amine originally present).

Tne amine recovered in the same way from the mother liquors amounts to 40–50 g. and contains a slight excess of *d*-amine. It may be used conveniently in the next run in place of the *dl*-amine.

2. Notes

1. The amount of malic acid theoretically is sufficient to convert half of the amine to the acid salt and the remaining half to the neutral salt.

2. Small amounts of the volatile amine are lost by hydrolysis during the evaporation and recrystallization of the salt.

3. Fractional crystallization is facilitated by inoculating the warm solution with a crystal of the species expected to separate and allowing crystallization to proceed slowly without disturbance.

4. The purity of *d*-α-phenylethylamine-*l*-malate is not readily determined by its melting point or specific rotation, but rather by its massive crystalline form and solubility. The acid and neutral *l*-base-*l*-acid salts are much more soluble, and usually do not crystallize at all.

5. Excess alkali must be avoided if the malic acid is to be recovered, since it is slowly racemized by heating with concentrated alkali.

6. When the amine is to be used in water solution a weighed amount of the pure salt may be decomposed with alkali, the base distilled quantitatively with steam, and the entire distillate used.

7. The solution containing sodium malate is neutralized with acetic acid, diluted to contain about 5 per cent of sodium malate, and treated at the boiling point with 10 per cent lead acetate solution until lead malate no longer precipitates. The lead malate is collected after cooling and washed by trituration with boiling water. The salt is made into a thin paste with distilled water and decomposed with hydrogen sulfide

(two days). The lead sulfide is filtered and the malic acid solution evaporated to a convenient volume. After titration of an aliquot, the solution may be used instead of pure acid in another run. The recovery is 70–80 per cent.

8. The *l*-base-*d*-acid salt no longer can be obtained pure when the solution contains about equal proportions of the salts of *d*- and *l*-amines. For this reason, also, an initial resolution of the *dl*-amine with *d*-tartaric acid is not feasible.

3. Methods of Preparation

The present method is adapted from that of Lovén.[1] The resolution has been carried out with *d*-α-bromocamphor-π-sulfonic acid (*l*-form);[2,3] with *l*- and *dl*-malic acids (*d*- and *l*-forms);[4] with *l*-quinic acid and *d*-tartaric acid (*d*- and *l*-forms),[5] and with *d*- and *l*-6,6′-dinitrodiphenic acids (*d*- and *l*-forms).[6] Methods employing *d*-benzylmethylacetyl chloride,[7] *d*-oxymethylenecamphor,[8] *l*-quinic acid,[9] and *d*-camphoric anhydride[10] are of theoretical interest only. The *dl*-amine is not resolved by the active camphor-10-sulfonic acids[11] or mandelic acids.[12]

PHENYLGLYOXAL

(Glyoxal, phenyl-)

$$C_6H_5COCH_3 + SeO_2 \rightarrow C_6H_5COCHO + Se + H_2O$$

Submitted by H. A. RILEY and A. R. GRAY.
Checked by LOUIS F. FIESER and C. H. FISHER.

1. Procedure

IN a 1-l. three-necked, round-bottomed flask, fitted with a liquid-sealed stirrer and a reflux condenser, are placed 600 cc. of dioxane

[1] Lovén, J. prakt. Chem. **72**, (2) 307 (1905). See also Lovén, Ber. **29**, 2313 (1896).

[2] Hunter and Kipping, J. Chem. Soc. **83**, 1147 (1903).

[3] Ingold and Wilson, ibid. **1933**, 1502.

[4] Ingersoll, J. Am. Chem. Soc. **47**, 1168 (1925).

[5] André and Vernier, Compt. rend. **193**, 1192 (1931).

[6] Ingersoll and Little, J. Am. Chem. Soc. **56**, 2123 (1934).

[7] Kipping and Salway, J. Chem. Soc. **85**, 444 (1904).

[8] Pope and Read, ibid. **95**, 171 (1909).

[9] Marckwald and Meth, Ber. **38**, 801 (1905).

[10] Freylon, Ann. chim. phys. (8) **15**, 140 (1908).

[11] Pope and Harvey, J. Chem. Soc. **75**, 1110 (1899).

[12] Ingersoll, Babcock, and Burns, J. Am. Chem. Soc. **55**, 411 (1933).

(Note 1), 111 g. (1 mole) of selenium dioxide (Note 2), and 20 cc. of water (Note 3). The mixture is heated to 50–55° and stirred until the solid has gone into solution, 120 g. (1 mole) of acetophenone is added in one lot, and the resulting mixture is refluxed with continued stirring for four hours (Note 4). The hot solution is decanted from the precipitated selenium, and the dioxane and water are removed by distillation through a short column. The phenylglyoxal is distilled at diminished pressure from a 250-cc. Claisen flask (Note 5), and the fraction boiling at 95–97°/25 mm. collected (Note 6). The yield is 93–96 g. (69–72 per cent of the theoretical amount) (Note 7).

The aldehyde sets to a stiff gel on standing, probably as the result of polymerization. It may be recovered without appreciable loss by distillation. Phenylglyoxal may be preserved also in the form of the hydrate, which is conveniently prepared by dissolving the yellow liquid in 3.5–4 volumes of hot water and allowing crystallization to take place (Note 8).

2. Notes

1. Ethyl alcohol (95 per cent) can also be used as solvent. The reaction can be carried out with an excess of acetophenone (2 moles) as solvent, but the results are less satisfactory. The dioxane can be recovered and used in later runs.

2. For the preparation of selenium dioxide, 200 g. (141 cc.) of concentrated nitric acid is heated in a 3-l. beaker on a hot plate under a good hood and 100 g. of selenium is added in portions of 5–10 g. A glass mechanical stirrer to break the foam hastens the process of oxidation. The resulting solution is transferred under the hood to a large evaporating dish and heated on a hot plate at a temperature not exceeding 200° until the selenious acid is completely dehydrated. The crude product is purified by sublimation. A 50-g. portion of the oxide is transferred to a 7-cm. porcelain crucible upon which is placed a 250-cc. filter flask through which a stream of cold water is run from the tap. The crucible is protected with asbestos and heated with a low flame until sublimation is complete (20–30 minutes). When the crucible has cooled, the sublimed selenium dioxide will be found wedged against the condenser. Extreme care should be used when working with selenium dioxide because of its poisonous properties.

An alternative procedure for the oxidation of selenium to the dioxide has been described by Hahn and Schales.[1]

3. Commercial selenious acid (129 g., 1 mole) may be used in place of the mixture of selenium dioxide and water.

[1] Hahn and Schales, Ber. **67**, 1823 (1934).

4. After about two hours the solution becomes clear and little further precipitation of selenium is observable.

5. Several grams of the hydrate may be obtained by adding the forerun to an equal volume of warm water and allowing the product to crystallize.

6. Boiling points reported in the literature are 120°/50 mm. and 142°/125 mm.

7. Phenylacetaldehyde may be used in place of acetophenone. Phenylmethylgloxal is obtained in a similar manner from propiophenone, and many other compounds containing a methylene group adjacent to a carbonyl group may be oxidized by means of selenium dioxide to the corresponding α-ketoaldehyde or α-diketone.[2]

For the oxidation of lepidine and quinaldine to the corresponding quinoline aldehydes, the selenium dioxide should be freshly prepared.[3]

8. The solubility of the hydrate at 20° is given in the literature as one part in about thirty-five parts of water. The melting points recorded range from 73° to 91°; the difference is said to be due to varying degrees of dryness of the samples.[4] The hydrate crystallizes well from water; chloroform, carbon disulfide, alcohol, or ether-ligroin also may be used for the purpose. Phenylglyoxal can be recovered from the hydrate by distillation in vacuum.

3. Methods of Preparation

Phenylglyoxal has been prepared from isonitrosoacetophenone through the bisulfite compound [4, 5] or by treatment with nitrosylsulfuric acid [6] or with nitrous acid.[7] It also has been prepared by the oxidation of benzoylcarbinol with copper acetate,[8] by heating bromophenacyl acetate,[9] and by the oxidation of acetophenone with selenium dioxide.[2]

[2] Riley, Morley, and Friend, J. Chem. Soc. **1932**, 1875; Brit. pat. 354,798 [C. A. **26**, 3804 (1932)]; U. S. pat. 1,955,890 [C. A. **28**, 4067 (1934)].

[3] Kaplan, J. Am. Chem. Soc. **63**, 2654 (1941).

[4] Pinner, Ber. **35**, 4132 (1902); ibid. **38**, 1532 (1905).

[5] v. Pechmann, ibid. **20**, 2904 (1887); Müller and v. Pechmann, ibid. **22**, 2556 (1889); Smedley, J. Chem. Soc. **95**, 218 (1909).

[6] Neuberg and Hofmann, Biochem. Z. **229**, 443 (1930).

[7] Neuberg and Hofmann, ibid. **239**, 495 (1931); Cusmano, Gazz. chim. ital. **68**, 130 (1938).

[8] Nef, Ann. **335**, 271 (1904); Henze, Z. physiol. Chem. **198**, 83 (1931); ibid. **200**, 232 (1931).

[9] Madelung and Oberwegner, Ber. **65**, 935 (1932).

PHENYLNITROMETHANE

(Toluene, α-nitro-)

$$C_6H_5CH_2CN + CH_3ONO_2 + NaOC_2H_5 \rightarrow$$

$$\underset{\underset{CN}{|}}{C_6H_5C}{=}NO_2Na + CH_3OH + C_2H_5OH$$

$$\underset{\underset{CN}{|}}{C_6H_5C}{=}NO_2Na \xrightarrow[H_2O]{NaOH} \underset{\underset{CO_2Na}{|}}{C_6H_5C}{=}NO_2Na$$

$$\xrightarrow{HCl} C_6H_5CH{=}NO_2H + CO_2$$

$$C_6H_5CH{=}NO_2H \rightarrow C_6H_5CH_2NO_2$$

Submitted by ALVIN P. BLACK and FRANK H. BABERS.
Checked by JOHN R. JOHNSON and H. B. STEVENSON.

1. Procedure

(A) *Sodium Phenyl-aci-nitroacetonitrile.*—In a 2-l. round-bottomed flask fitted with an efficient reflux condenser is placed 400 cc. of absolute ethyl alcohol. Through the condenser tube 46 g. (2 gram atoms) of freshly cut metallic sodium is added as rapidly as possible, and the flask is heated in an oil bath after all the sodium has been added. After about one-half hour only a small globule of molten sodium (0.5–1.0 g.) remains and sodium ethoxide begins to precipitate. At this point 100 cc. of absolute alcohol is added and the mixture is cooled to 0°. A second 100-cc. portion of cold absolute alcohol is then poured on top of the solid cake of sodium ethoxide in the flask (Note 1). The reflux condenser is replaced by a stopper carrying a separatory funnel and a calcium chloride tube. An ice-cold mixture of 234 g. (2 moles) of freshly distilled benzyl cyanide (Org. Syn. Coll. Vol. I, **1941,** 107) and 216 g. (180 cc., 2.8 moles) of methyl nitrate (Note 2) is added with constant shaking, at such a rate that the temperature is kept between 4° and 8°. After this addition is completed (about one hour is required), the reaction mixture is allowed to remain at 4–8° and shaken intermittently for one hour. The flask is then provided with a stopper fitted with a Bunsen valve and placed in a freezing mixture for twenty-four hours. The sodium salt of the *aci*-nitro compound which precipitates is filtered with suction on a Büchner funnel, washed thoroughly with dry ether

(Note 3), and air dried. The first crop of material weighs 215–275 g. (58–75 per cent of the theoretical amount). The mother liquor and ether washings are combined and concentrated stepwise to about 150 cc. under reduced pressure. Successive crops of the sodium salt which separate are filtered with suction and washed with dry ether. The total weight of the crude sodium salt is 275–300 g. (75–82 per cent yield). This material is used directly without purification.

(B) *Phenylnitromethane.*—In a 4-l. beaker 300 g. of sodium hydroxide is dissolved in 1.2 l. of water. The beaker is placed in an enameled pan (as a precaution against breakage), and the solution is heated to boiling. Over a period of one hour the air-dried, crude sodium salt of phenyl-nitroacetonitrile (275–300 g.) is added in small portions to the boiling alkali. Boiling is continued until the evolution of ammonia ceases (about three hours); hot water is added from time to time to keep the volume of the solution fairly constant (Note 4). The hot alkaline solution is poured into a shallow porcelain dish and on cooling solidifies to a waxy mass.

The cake of crude sodium salt of phenylnitroacetic acid is broken up with a spoon, transferred to a 4-l. beaker, and stirred with 500 g. of ice. The beaker is placed in a large crock packed with ice-salt mixture, and is provided with a mechanical stirrer. When the solution in the beaker has cooled to $-5°$, concentrated hydrochloric acid is added slowly from a separatory funnel, with vigorous stirring, until the solution is faintly acid to Congo red. During the addition of acid the temperature is not allowed to rise above $-5°$ (Note 5). Usually about 900 cc. of acid is needed and the addition requires about two hours. The cold solution is extracted with one 500-cc. portion of ether, followed by two 250-cc. portions. The combined ether extracts are washed with ice-cold portions of saturated sodium bicarbonate solution until the wash liquid is colorless or only faintly yellow (usually two 100-cc. portions suffice). The ether solution is then washed with 250 cc. of ice water containing two drops of hydrochloric acid, and finally with three 50-cc. portions of ice water. The ether solution is dried over anhydrous sodium sulfate and allowed to stand for three or four days to complete the isomerization of the labile *aci*-form. The solution is filtered and the ether removed at 15–20° under reduced pressure. The residual oil is distilled at 3 mm. pressure in an ordinary Claisen flask, and the phenylnitromethane is obtained as a light yellow oil, b.p. 90–92°/3 mm. (Note 6). The yield is 135–150 g. (50–55 per cent of the theoretical amount, based upon the benzyl cyanide) (Note 7).

2. Notes

1. The supernatant layer of alcohol prevents the reactants dropping directly onto the sodium ethoxide and causing local overheating.

2. The large excess of methyl nitrate increases the yield appreciably. Freshly prepared methyl nitrate (p. 412) was dried and used directly without distillation. It is convenient to use the entire product (usually 210–230 g.) obtained from 120 g. of methyl alcohol.

3. Ordinary ether that has been allowed to stand for several days over anhydrous calcium chloride may be used.

4. The reaction must be watched carefully. Vigorous foaming sometimes occurs, necessitating the addition of small quantities of cold water from a wash bottle.

5. At −5° the *aci*-nitro compound separates as a gray, pasty solid. When the solution becomes acid the colloidal precipitate tends to undergo coagulation.

6. In the distillation of the crude product it is essential to maintain a low pressure and to avoid overheating. If the distillation is pushed too far, decomposition occurs and the distillate discolors rapidly on standing. If the distillation is not carried out carefully violent decomposition may occur which will blow out all connections.

The distilled product decomposes on standing and should be used promptly. Phenylnitromethane should not be stored in a glass-stoppered bottle, as the stopper is likely to become frozen and explosions may occur in attempting to remove it.

7. The procedure outlined can be shortened and simplified considerably by operating at somewhat higher temperatures than those specified. Thus it was found advantageous to add the benzyl cyanide-methyl nitrate solution to the sodium ethoxide with shaking at 5–15°, at which temperature the reaction proceeds smoothly and at a steady rate. The total yield of satisfactory sodium salt is 300–320 g.

The hydrolysis of the nitrile and the acidification operation are best carried out in the same 4-l. beaker, thus avoiding the necessity for making a transfer. In dissolving the sodium hydroxide required for hydrolysis, 1.5 l. of water is used. As soon as no more ammonia is liberated (litmus) the beaker is placed in an ice-salt bath and the mixture stirred vigorously with a mechanically driven glass stirrer. When the temperature has dropped to 30°, 500 g. of ice is added. The acidification with hydrochloric acid (800–850 cc.) is conducted at 0–10° with continued stirring. The reaction mixture is allowed to stand overnight and is extracted the next day. The ethereal solution of the product need be allowed to stand over sodium sulfate for only twenty to twenty-four hours before distillation. The yield is then 153–163 g. of very light-colored product boiling at 92–94°/4 mm. (bath at 135°). Some 10–15 g. of darker material can be distilled from the residue at a slightly higher temperature. (LOUIS F. FIESER and E. BERLINER, private communication.)

3. Methods of Preparation

Phenylnitromethane has been prepared by the nitration of toluene with dilute nitric acid in a sealed tube,[1] by the interaction of benzene-diazonium chloride and nitromethane in alkaline solution,[2] by the action of silver nitrite on benzyl chloride [3] or iodide,[4] and by the condensation of ethyl nitrate with benzyl cyanide and subsequent hydrolysis.[5] The use of methyl nitrate, which can be prepared with less danger and difficulty than ethyl nitrate, is advantageous.

PHENYLPROPIOLIC ACID

(Propiolic acid, phenyl-)

$$C_6H_5CHBrCHBrCO_2C_2H_5 + 3KOH \rightarrow$$
$$C_6H_5C{\equiv}CCO_2K + C_2H_5OH + 2KBr + 2H_2O$$

$$C_6H_5C{\equiv}CCO_2K + H_2SO_4 \rightarrow C_6H_5C{\equiv}CCO_2H + KHSO_4$$

Submitted by T. W. ABBOTT.
Checked by HENRY GILMAN and G. F. WRIGHT.

1. Procedure

A SOLUTION of potassium hydroxide is prepared by dissolving 252.5 g. (4.5 moles) of potassium hydroxide (Note 1) in 1.2 l. of 95 per cent alcohol contained in a 3-l. round-bottomed flask provided with a reflux condenser and heated on a steam bath. To the alkaline solution, cooled to 40–50°, is added 336 g. (1 mole) of crude ethyl α,β-dibromo-β-phenyl-propionate (p. 270). When the initial reaction has subsided, the contents of the flask are refluxed for five hours on the steam bath.

[1] Konowalow, Ber. **28**, 1860 (1895).
[2] Bamberger, Schmidt, and Levinstein, ibid. **33**, 2053 (1900).
[3] Hollemann, Rec. trav. chim. **13**, 405 (1894).
[4] Hantzsch and Schultze, Ber. **29**, 700 (1896).
[5] Wislicenus and Endres, ibid. **35**, 1755 (1902); cf. Gattermann-Wieland, "Laboratory Methods of Organic Chemistry," p. 256. Translated from the twenty-fourth German edition by W. McCartney, The Macmillan Company, New York, 1937.

The reaction mixture is cooled, and the salts which separate are filtered by suction. The filtrate is treated with concentrated hydrochloric acid until neutral to litmus (Note 2), and the salts which precipitate are separated by filtration. The filtrate is then distilled until the vapor reaches 95°. The residue and the precipitated salts, previously separated by filtration, are combined, dissolved in 800 cc. of water, and chilled by the addition of cracked ice to make a volume of 1.8 l. (Note 3). The cooled solution is immersed in an ice-water bath and stirred mechanically while a 20 per cent sulfuric acid solution is added until the solution is strongly acid to litmus. After stirring for twenty minutes the phenylpropiolic acid is filtered by suction and washed with four 30-cc. portions of a 2 per cent sulfuric acid solution.

The acid thus obtained as a light brown, granular product is dissolved in 1 l. of 5 per cent sodium carbonate solution, treated with 20 g. of Norite, and heated on a steam bath for thirty minutes with occasional stirring. The mixture is then filtered and cooled externally, and about 200 g. of cracked ice is added. The solution is stirred mechanically while a 20 per cent solution of sulfuric acid is added slowly. The precipitated acid is filtered by suction, washed first with 50 cc. of a 2 per cent sulfuric acid solution and then with a little water, and air-dried. The yield of acid melting between 115° and 125° is 112–118 g. (77–81 per cent of the theoretical amount).

One hundred grams of the crude acid can be purified by crystallization from 200–300 cc. of carbon tetrachloride, yielding 70 g. of phenylpropiolic acid melting at 135–136°.

2. Notes

1. The best yields are obtained when a 50 per cent excess of potassium hydroxide is used. The concentration of alkali has little or no effect on the yield.

2. The alcohol is best distilled from neutral rather than from alkaline solution.

3. In order to prevent decarboxylation, the temperature should be kept as low as possible. If this precaution is not observed, the yield is lowered and the product is less pure.

3. Methods of Preparation

The procedure described is essentially that of Perkin.[1] Phenylpropiolic acid can also be prepared from ether solutions of β-bromosty-

[1] Perkin, J. Chem. Soc. 45, 172 (1884); Liebermann and Sachse, Ber. 24, 4113 (1891).

rene [2] and β-chlorostyrene [3] with sodium and carbon dioxide; from β-bromostyrene and butyllithium in ether; [4] by the action of alcoholic alkali on α-bromocinnamic acid, [2] β-bromocinnamic acid, [5] or ethyl α-bromocinnamate; [6] and by the action of carbon dioxide on sodium phenylacetylide. [2,7]

The preparation of phenylpropiolic acid by the action of alkali on α,β-dibromocinnamic acid, a more direct synthesis than that involving the ester, has not been much used because of the difficulty of preparing the dibromo acid. It has been reported, however, that α,β-dibromocinnamic acid can be prepared easily and in a 95 per cent yield by the addition of bromine to cinnamic acid in boiling carbon tetrachloride, and that the crude product can be used for the preparation of phenylpropiolic acid. [8] A simplified procedure for the preparation of small amounts of phenylpropiolic acid from α,β-dibromocinnamic acid is described in the same article.

2-PHENYLPYRIDINE

(Pyridine, 2-phenyl-)

$$C_6H_5Br + 2Li = C_6H_5Li + LiBr$$

Submitted by J. C. W. Evans and C. F. H. Allen.
Checked by Reynold C. Fuson, W. E. Ross, and E. A. Cleveland.

1. Procedure

In a 1-l. three-necked flask, fitted with a dropping funnel, a thermometer, mechanical stirrer, and reflux condenser protected from moist-

[2] Glaser, Ann. **154**, 140, 162 (1870).

[3] Erlenmeyer, Ber. **16**, 152 (1883).

[4] Gilman, Langham, and Moore, J. Am. Chem. Soc. **62**, 2328 (1940).

[5] Barisch, J. prakt. Chem. (2) **20**, 180 (1879).

[6] Michael, Ber. **34**, 3647 (1901).

[7] E. I. du Pont de Nemours and Company, U. S. pat. 2,194,363 [C. A. **34**, 4745 (1940)].

[8] Reimer, J. Am. Chem. Soc. **64**, 2510 (1942).

ure (Note 1), the whole being swept with dry nitrogen, are placed 3.5 g. (0.5 gram atom) of lithium, cut into pieces the size of a pea, and 100 cc. of dry ether. The stirrer is started, and about 10 cc. of a mixture of 40 g. of bromobenzene (0.25 mole) in 50 cc. of dry ether is admitted from the dropping funnel; a vigorous reaction usually takes place (Note 2). The remainder of the mixture is added gradually over a half-hour period, when the metal should have largely disappeared (Note 3).

From the dropping funnel is next slowly introduced, with stirring, 40 g. (0.5 mole) of dry pyridine (Note 4) in 100 cc. of dry toluene. The ether is then distilled (Note 5) and the residual suspension stirred at 110° (inside temperature) for eight hours. It is then cooled to about 40°, 35 cc. of water cautiously added, and the liquids filtered if necessary (Note 6). The lower layer is separated and discarded. The toluene layer is dried for an hour with 20 g. of pulverized potassium hydroxide and carefully distilled, using a modified Claisen flask with a fractionating column attached. The material boiling up to 150° is removed at ordinary pressure and the residue distilled *in vacuo*; after two fractional distillations, the yield of 2-phenylpyridine, b.p. 140°/12 mm., is 15.5–19 g. (40–49 per cent of the theoretical amount).

2. Notes

1. The apparatus and reagents must be dried as for the Grignard reaction.

2. Occasionally the reaction will not start without the application of heat; as soon as the reaction begins, however, the source of heat is removed.

3. The yield of phenyllithium is approximately 75 per cent. It can be determined by allowing the phenyllithium to react with an excess of benzophenone and weighing the triphenylcarbinol formed. It is assumed that the carbinol is formed quantitatively.[1]

4. The success of the preparation depends on the dryness of the pyridine. The pyridine was refluxed for eight hours over fresh quicklime and distilled, and then a similar treatment with pulverized potassium hydroxide followed. None of the available barium oxide gave as good results. Merck's medicinal pyridine gave the highest yields.

5. This is easily accomplished by running the water out of the condenser, while heating to 110°.

6. The small particles of unused metal that usually remain hinder the separation into layers.

[1] Gilman, Zoellner, and Selby, J. Am. Chem. Soc. **54**, 1957 (1932).

3. Methods of Preparation

2-Phenylpyridine has been prepared from pyridine and phenyllithium [2] or benzenediazonium chloride.[3] In the procedure described above, replacing ether as a solvent by toluene avoids the necessity of using a sealed tube.[4]

PHENYLPYRUVIC ACID

(Pyruvic acid, phenyl-)

$$C_6H_5CH{=}\underset{\underset{NHCOCH_3}{|}}{C}CO_2H + 2H_2O + HCl \rightarrow$$

$$C_6H_5CH_2COCO_2H + CH_3CO_2H + NH_4Cl$$

Submitted by R. M. HERBST and D. SHEMIN.
Checked by REYNOLD C. FUSON and E. A. CLEVELAND.

1. Procedure

TEN grams (0.05 mole) of α-acetaminocinnamic acid (p. 1) and 200 cc. of 1 N hydrochloric acid (Note 1) are placed in a 500-cc. flask fitted to an upright condenser with a ground-glass joint. Hydrolysis is completed by boiling for three hours. A few droplets of pale green oil may separate from the boiling solution; these are removed by filtration. The crystals of phenylpyruvic acid which separate from the filtrate on cooling (Note 2) are transferred to a Büchner funnel and washed with a little ice-cold water. The combined filtrate and washings are extracted with four 50-cc. portions of ether. The solvent is removed from the ether solution by evaporation at room temperature, finally in a vacuum desiccator (Note 3). The residue is combined with the first crop of crystals and dried in a vacuum desiccator over calcium chloride and potassium hydroxide. The yield is 7.2–7.7 g. (88–94 per cent of the theoretical amount), and the product melts at 150–154° (Notes 4 and 5).

2. Notes

1. Larger quantities of phenylpyruvic acid may be prepared by increasing the amounts of reactants proportionately. However, this is

[2] Ziegler and Zeiser, Ber. **63**, 1847 (1930).
[3] Haworth, Heilbron, and Hey, J. Chem. Soc. **1940**, 352.
[4] Walters and McElvain, J. Am. Chem. Soc. **55**, 4625 (1933).

advisable only when the product is to be used immediately since phenyl-pyruvic acid begins to decompose after standing only a few days.

2. The amount of phenylpyruvic acid which separates from the filtrate is increased if the solution is allowed to stand in the refrigerator several days before filtration. No decomposition was noted when the product was kept suspended in cold, dilute acid.

3. The evaporation may be carried out conveniently at room temperature by passing a stream of dry air or inert gas over the surface of the solution under a glass bell.

4. The melting point varies considerably with the rate of heating.

5. Phenylpyruvic acid may be recrystallized from ethylene chloride, benzene, or chloroform, but losses due to instability of the compound are quite large.

3. Methods of Preparation

Phenylpyruvic acid has been prepared by the hydrolysis of α-benzoyl-aminocinnamic acid with alkalies or acids; [1,2] by the acid hydrolysis of ethyl phenyloxalacetate; [3] by the acid hydrolysis of ethyl phenyl-cyanopyruvate; [2,4] by dehydration of β-phenylglyceric acid with sul-furic acid; [5] and by the alkaline hydrolysis of α-acetaminocinnamic acid. [6]

PHENYL THIENYL KETONE

(Ketone, phenyl 2-thienyl)

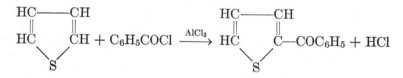

Submitted by WESLEY MINNIS.
Checked by ROGER ADAMS and H. D. COGAN.

1. Procedure

IN a 1-l. three-necked flask, equipped with a mechanical stirrer, a reflux condenser, and a thermometer (with bulb immersed in the liquid),

[1] Plöchl, Ber. **16**, 2817 (1883).

[2] Erlenmeyer, Jr., Ann. **271**, 165, 173 (1892).

[3] Wislicenus, Ber. **20**, 592 (1887).

[4] Erlenmeyer, Jr., and Arbenz, Ann. **333**, 228 (1904); Hemmerlé, Ann. chim. (9) **7**, 229 (1917).

[5] Dieckmann, Ber. **43**, 1034 (1910).

[6] Bergmann and Stern, Ann. **448**, 27 (1926).

are placed 100 g. (0.75 mole) of anhydrous aluminum chloride and 300 g. of carbon disulfide (Note 1). The suspension is cooled to 15–25°, and a solution of 60 g. (0.71 mole) of thiophene (p. 578) and 105 g. (0.75 mole) of benzoyl chloride in 225 g. of carbon disulfide is added through the condenser, with stirring, over a period of three and one-half hours (Note 2). The solution is allowed to warm up to room temperature, and stirring is continued for three more hours; the reaction mixture is then allowed to stand overnight. The mixture is refluxed on the water bath for three and one-half hours, cooled, poured on ice, and extracted with ether. The ether extract is washed successively with sodium carbonate solution and water, and then dried over calcium chloride. The ether is removed by distillation on the water bath, and the residue is distilled under reduced pressure. The yield of product boiling at 200–209°/30–40 mm. is 117–120 g. (88–90 per cent of the theoretical amount). On crystallization from 1 l. of petroleum ether (b.p. 65–110°) there is obtained 110–112 g. of product melting at 52°. Another crystallization from petroleum ether gives a product which melts at 55–56°. The loss on the second crystallization is about 10 per cent.

2. Notes

1. The carbon disulfide was dried over calcium chloride.

2. Thiophene and aluminum chloride react vigorously in carbon disulfide suspension. Subsequent addition of a carbon disulfide solution of benzoyl chloride produces a tar, and a low yield of ketone results.

3. Methods of Preparation

Phenyl thienyl ketone has been prepared by treatment of benzoyl chloride with thienylmercuric chloride;[1] by treatment of thiophene with benzoyl chloride in the presence of thienylmercuric chloride,[2] phosphorus pentoxide,[3] stannic chloride,[4] and aluminum chloride.[5] It has also been prepared from thienylmagnesium iodide and benzonitrile.[6]

[1] Volhard, Ann. **267**, 179 (1892); Steinkopf and Killingstad, ibid. **532**, 288 (1937).

[2] Steinkopf and Bauermeister, ibid. **403**, 70 (1914).

[3] Steinkopf, ibid. **413**, 349 (1917).

[4] Stadnikoff and Rakowsky, Ber. **61**, 269 (1928); Goldfarb, J. Russ. Phys.-Chem. Soc. **62**, 1073 (1930) [C. A. **25**, 2719 (1931)].

[5] Comey, Ber. **17**, 790 (1884).

[6] Thomas and Couderc, Bull. soc. chim. (4) **23**, 289 (1918).

PHLOROACETOPHENONE

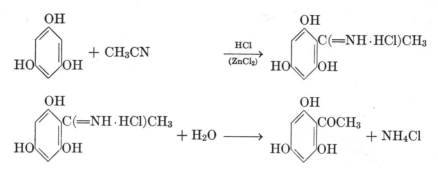

Submitted by K. C. Gulati, S. R. Seth, and K. Venkataraman.
Checked by John R. Johnson and M. T. Bush.

1. Procedure

In a 250-cc. filtering flask, fitted with a calcium chloride tube and a rubber stopper carrying an inverted thistle tube (Note 1) for the introduction of hydrogen chloride, are placed 20 g. (0.16 mole) of well-dried phloroglucinol (Note 2), 13 g. (0.32 mole) of anhydrous acetonitrile, 80 cc. of anhydrous ether, and 4 g. of finely powdered, fused zinc chloride. The flask is cooled in an ice-salt mixture and shaken occasionally while a rapid stream of dry hydrogen chloride is passed through the solution for two hours. The flask is allowed to stand in an ice chest for twenty-four hours, and hydrogen chloride is again passed into the mixture, now pale orange in color, for two hours. The flask is stoppered and allowed to stand in an ice chest for three days.

The bulky orange-yellow precipitate of the ketimine hydrochloride is separated by decanting the ether and washed twice with 20-cc. portions of dry ether. The solid is transferred to a 2-l. round-bottomed flask with 1 l. of hot water. The flask is provided with a reflux condenser, and the yellow solution is boiled vigorously over a wire gauze for two hours. About 3 to 4 g. of Norite is added; the solution is boiled for five minutes longer and filtered with suction while hot. The decolorizing carbon is extracted with two 100-cc. portions of boiling water and this filtrate added to the main portion.

After standing overnight the colorless or pale yellow needles of phloroacetophenone are filtered with suction and dried in an oven at 120° (Note 3). The yield is 20–23.5 g. (74–87 per cent of the theoretical amount) of a product which melts at 217–219° (corr.). This product is quite pure and may be used directly for many purposes. It may be recrystallized from thirty-five times its weight of hot water, with a loss of about 5 per cent. The recrystallized material melts at 218–219° (corr.).

2. Notes

1. A wide-mouthed entry tube for hydrogen chloride is necessary to avoid clogging due to separation of the solid ketimine hydrochloride.

2. All the reagents must be dried carefully. Phloroglucinol (Org. Syn. Coll. Vol. I, **1941**, 455) contains two molecules of water of hydration which is removed by drying overnight at 120°. The acetonitrile and ether used were freshly distilled from phosphorus pentoxide.

3. Phloroacetophenone crystallizes from aqueous solutions with one molecule of water of hydration.[1] The oven-dried crystals take up water readily on exposure to the air. Phloroacetophenone gives a wine-red color with ferric chloride in contrast to the violet color given by phloroglucinol.[2]

3. Methods of Preparation

The Hoesch reaction is the most satisfactory method for preparing phloroacetophenone.[3] The procedure described above is that of Robinson and Venkataraman.[4] Phloroacetophenone has been obtained also by the action of acetyl chloride on phloroglucinol in the presence of aluminum chloride.[2]

PHTHALALDEHYDIC ACID

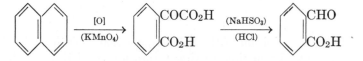

Submitted by J. H. GARDNER and C. A. NAYLOR, JR.
Checked by C. R. NOLLER and CARL LINN.

1. Procedure

IN a 3-l. three-necked flask, provided with a liquid-sealed mechanical stirrer, a reflux condenser, and a dropping funnel, are placed 500 cc. of

[1] Gulati, Seth, and Venkataraman, J. Chem. Soc. **1934**, 1766.
[2] Shriner and Kleiderer, J. Am. Chem. Soc. **51**, 1269 (1929).
[3] Hoesch, Ber. **48**, 1129 (1915).
[4] Robinson and Venkataraman, J. Chem. Soc. **1926**, 2347.

0.5 N sodium hydroxide solution and 32 g. (0.25 mole) of pure naphthalene. The mixture is heated to boiling, and a boiling solution of 212 g. (1.34 moles) of potassium permanganate in 1.5 l. of water is added in small portions during one and one-half hours with vigorous stirring (Note 1). After the last addition, the mixture is boiled for thirty to forty-five minutes to complete the oxidation. Any unchanged permanganate is reduced by the addition of 20 cc. of alcohol, the flask is cooled to solidify the residual naphthalene, and the mixture is filtered.

The filtrate is acidified with 150 cc. (1.8 moles) of concentrated hydrochloric acid (sp. gr. 1.18), evaporated to 500 cc., cooled, and filtered. The filtrate is neutralized with 30 per cent sodium hydroxide solution (150–160 cc.), 50 g. (0.48 mole) of sodium bisulfite is added, and the mixture is evaporated to dryness on a steam bath. The residue is stirred with 100 cc. of concentrated hydrochloric acid and evaporated to dryness on a steam bath. The hydrochloric acid treatment and evaporation are repeated (Note 2).

The residue is extracted thoroughly with benzene in a large Soxhlet extractor (Note 3), and the benzene extract is evaporated to dryness. The crude product is dissolved in 50 cc. of hot water, the solution filtered, and the filtrate cooled in an ice bath with stirring (Note 4). The crystals are filtered with suction and dried in the air. The yield of slightly colored product melting at 94–95° is 15–15.5 g. (40–41 per cent of the theoretical amount).

When this material is recrystallized from 40 cc. of water with 1 g. of decolorizing carbon, and the filtrate cooled to 0°, there is obtained 14–14.5 g. of white crystals melting at 96–96.5°.

2. Notes

1. During the addition of the hot permanganate solution no external heat is applied. From time to time steam is passed through the condenser jacket to return the sublimed naphthalene to the reaction flask.

2. The treatment with hydrochloric acid and the evaporation are repeated in order to decompose completely the phthalaldehydic acid bisulfite compound. Wegscheider and Bondi [1] state that it is necessary to heat the bisulfite compound on the water bath several days with a large excess of hydrochloric acid, but the treatment described above has been found adequate.

3. The material requires two extraction thimbles of the 45 by 125 mm. size, or four of the 30 by 75 mm. size. It is advantageous to use a modi-

[1] Wegscheider and Bondi, Monatsh. **26**, 1055 (1905).

fied Soxhlet extractor of the Clausnitzer type,[2] which allows the vapor of the boiling solvent to surround the extractor tube.

4. It is essential to control carefully the volume of water used in recrystallization. The crude acid contains considerable amounts of phthalic acid; too small a volume of water results in a contaminated product, and too large an amount causes a decided decrease in yield.

3. Methods of Preparation

Phthalaldehydic acid has been prepared by the hydrolysis of 2-bromo- or 2-chlorophthalide,[3] of o-trichloromethylbenzal chloride,[4] of o-cyanobenzal chloride,[5] and of o-dichloromethylbenzoyl chloride;[6] by the ozonization of naphthalene;[7] by the alkaline oxidation of naphthalene,[8] or α-nitronaphthalene[9] followed by the preparation and decomposition of the aniline condensation product[10] or the bisulfite compound[1, 11] of phthalonic acid; and by carbonating the reaction product of o-chlorobenzaldehyde and sodium.[12] Detailed directions for the preparation of phthalaldehydic acid by bromination of phthalide and hydrolysis of the bromination product are given in Volume 23 of this series.

[2] Houben-Weyl, "Die Methoden der organischen Chemie," 3rd. Ed., Vol. I, p. 565, Verlag Georg Thieme, Leipzig, 1925.

[3] Racine, Ann. **239**, 78 (1887); Austin and Bousquet, U. S. pat. 2,047,946 [C. A. **30**, 6011 (1936)].

[4] Coulson and Gautier, Bull. soc. chim. (2) **45**, 507 (1886).

[5] Gabriel and Weise, Ber. **20**, 3197 (1887); Drory, ibid. **24**, 2571 (1891).

[6] Davies, Perkin, and Clayton, J. Chem. Soc. **121**, 2214 (1922).

[7] Seekles, Rec. trav. chim. **42**, 706 (1923).

[8] Tcherniac, Ger. pat. 79,693 [Frdl. **4**, 162 (1894–97)]; Ger. pat. 86,914 [Frdl. **4**, 163 (1894–97)].

[9] Gardner, J. Am. Chem. Soc. **49**, 1831 (1927).

[10] Soc. Chim. des Usines du Rhône, Ger. pat. 97,241 [Frdl. **5**, 139 (1897–1900)]; Fuson, J. Am. Chem. Soc. **48**, 1093 (1926).

[11] Graebe and Trümpy, Ber. **31**, 369 (1898); Sidgwick and Clayton, J. Chem. Soc. **121**, 2263 (1922).

[12] Morton, LeFevre, and Hechenbleikner, J. Am. Chem. Soc. **58**, 754 (1936).

PHTHALIDE

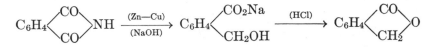

Submitted by J. H. Gardner and C. A. Naylor, Jr.
Checked by H. T. Clarke and D. Blumenthal.

1. Procedure

In a 2-l. round-bottomed flask, 180 g. (2.75 gram atoms) of zinc dust is stirred to a thick paste with a solution of 1 g. of copper sulfate in about 35 cc. of water (Note 1), and 400 g. (327 cc.) of 20 per cent aqueous sodium hydroxide is added. The flask is equipped with a mechanical stirrer, the contents are cooled to 5° by means of an ice bath, and 147 g. (1 mole) of phthalimide (Org. Syn. Coll. Vol. I, **1941,** 457) is added in small portions at such a rate that the temperature does not rise above 8° (about thirty minutes is required). After all the phthalimide has been added, stirring is continued for one-half hour. The mixture is diluted with 400 cc. of water, warmed on a steam bath until evolution of ammonia has ceased (about three hours), and concentrated to a volume of about 400 cc. by distillation under reduced pressure. The material is filtered and the filtrate made acid to Congo red with concentrated hydrochloric acid (about 150 cc. is required). The mixture, in which the phthalide has separated as an oil, is boiled for one hour in order to complete the lactonization of the hydroxymethylbenzoic acid, and transferred while hot to a beaker. On cooling, the oily product solidifies to a hard red-brown cake. After chilling overnight in a refrigerator, when a further quantity of crystalline solid separates from the aqueous layer, the cold mixture is filtered with suction (Note 2). The crude phthalide, which contains a considerable quantity of sodium chloride, is recrystallized in 20-g. portions from 1.5 l. of water; the mother liquor from the first crop is employed for recrystallization of the subsequent portions. Each portion is filtered hot and cooled below 5° before collecting the crystals, which are finally washed with small quantities of ice-cold water (Note 3). The phthalide crystallizes in transparent plates which melt at 72–73°. The yield is 90–95 g. (67–71 per cent of the theoretical amount) (Note 4).

2. Notes

1. In checking these directions complete failure was repeatedly encountered with good commercial grades of zinc dust, and only when the metal was activated with copper sulfate did reduction proceed at all.

2. The mother liquor, on concentrating to less than half its volume, yields no further crystals on chilling.

3. On cooling, a minute amount of a yellow impurity separates with the phthalide at the beginning of crystallization. This impurity apparently cannot be eliminated by boiling with charcoal but is present in too small amount to affect the melting point of the product.

4. Concentration of the final mother liquor to a volume of 500 cc. yields a further small quantity of phthalide, but this operation is scarcely worth while.

3. Methods of Preparation

Phthalide has been prepared by the bromination of *o*-toluic acid followed by hydrolysis;[1] by the reduction of phthalic anhydride,[2] ammonium phthalate,[3] or ethyl acid phthalate;[4] and by the method described in the procedure above.[5]

[1] Hjelt, Ber. **19**, 412 (1886); Salkind and Ssemenow, J. Russ. Phys.-Chem. Soc. **46**, 512 (1914) (Chem. Zentr. **1914**, II, 1271); Davies and Perkin, J. Chem. Soc. **121**, 2207 (1922).

[2] Wislicenus, Ber. **17**, 2181 (1884); Godchot, Bull. soc. chim. (4) **1**, 830 (1907); Sabatier and Kubota, Compt. rend. **172**, 736 (1921); Austin, Bosquet, and Lazier, J. Am. Chem. Soc. **59**, 864 (1937).

[3] Delfino and Somlo, IX Congr. intern. quím. pura applicada **4**, 360 (1934) [C. A. **30**, 2855 (1936)].

[4] E. I. du Pont de Nemours and Company, U. S. pat. 2,114,696 [C. A. **32**, 4607 (1938)].

[5] Reissert, Ber. **46**, 1489 (1913); Kalle and Company, A.-G., Ger. pat. 267,596 [Frdl. **11**, 196 (1912–14)].

SYMMETRICAL AND UNSYMMETRICAL o-PHTHALYL CHLORIDES

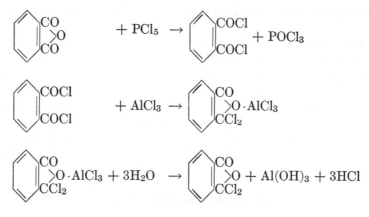

Submitted by ERWIN OTT.
Checked by HENRY GILMAN and F. J. PROCHASKA.

1. Procedure

(A) *Symmetrical o-Phthalyl Chloride.*—A mixture of 148 g. (1 mole) of phthalic anhydride (Note 1) and 220 g. (1.06 moles) of phosphorus pentachloride (Note 2) is placed in a 500-cc. Claisen flask. The flask is equipped with a reflux condenser, the upper end of which is provided with a calcium chloride tube, and the side arm of the flask is closed with a cork. The flask is inclined slightly so that any phosphorus oxychloride which collects in the stoppered side arm will run back into the flask. After heating in an oil bath at 150° for twelve hours, the air condenser and the stopper in the end of the side arm are removed, and the flask is connected to a water-cooled condenser. The temperature is then raised gradually to 250°, during which time most of the phosphorus oxychloride distils into a receiver. The liquid residue is distilled under reduced pressure; at first a small quantity of phosphorus oxychloride distils, and then the *sym.* o-phthalyl chloride distils at 131–133°/9–10 mm. The product thus obtained contains a small amount of phthalic anhydride; it solidifies on cooling in an ice-salt mixture and melts at 11–12° (Note 3). The yield is 187 g. (92 per cent of the theoretical amount).

(B) *Unsymmetrical o-Phthalyl Chloride.*—A mixture of 105 g. of the symmetrical o-phthalyl chloride and 75 g. of finely ground aluminum chloride (Note 4) is heated on a steam bath for eight to ten hours, with exclusion of moisture. The mixture should be stirred frequently until all the powder has dissolved. Upon cooling there is formed a hard mass which is broken into small pieces while still warm. When thoroughly cool, it is triturated with pieces of ice in a mortar, working with small amounts at a time. The white sediment which results is collected on a Büchner funnel and dissolved immediately in about 300 cc. of warm benzene (40–50°). The benzene solution is separated from the small aqueous layer, dried over calcium chloride for eight hours, and filtered. The benzene is distilled under reduced pressure by heating in a water bath at 30–40°. The crystalline residue is extracted in a Soxhlet apparatus with petroleum ether (b.p. 20–40°) until the residue in the thimble consists of practically pure phthalic anhydride (Note 5). The petroleum ether is distilled from the extract, and the crude unsymmetrical phthalyl chloride is purified by fractional crystallization from petroleum ether (b.p. 20–50°). The purified chloride melts at 87–89°. The melting point is not sharp because the unsymmetrical compound begins to revert to the symmetrical isomer. The yield of pure unsymmetrical phthalyl chloride is about 76 g. (72 per cent of the theoretical amount) and is dependent on the quality of the aluminum chloride used.

2. Notes

1. A good grade of sublimed phthalic anhydride should be used (m.p. 128–129°). If this cannot be obtained the ordinary phthalic anhydride can be purified by sublimation.

2. The phosphorus pentachloride should be freed of any phosphorus trichloride or oxychloride present. This may be done by placing the pentachloride in a flask connected to an ice-cooled receiver and heating on a water bath. The pressure in the apparatus is reduced as much as possible by means of a water pump. A calcium chloride tube should be inserted between the pump and the receiver.

3. Pure symmetrical *o*-phthalyl chloride cannot be obtained, even by recrystallization from carbon tetrachloride. A product melting at 16° may be obtained by distilling the unsymmetrical phthalyl chloride at atmospheric pressure.

4. A good quality of aluminum chloride must be used. If this is not available, it may be prepared by heating dried aluminum granules in a current of pure, dry hydrogen chloride. The gas is dried thoroughly by passing it through a tube containing phosphorus pentoxide spread on glass wool. Access of moisture to the aluminum chloride while weighing and pulverizing should be avoided as far as possible.

5. The extraction requires from eight to ten hours.   The residue should be practically pure phthalic anhydride (m.p. 126°).

3. Methods of Preparation

Symmetrical o-phthalyl chloride has been prepared by heating phthalic anhydride with phosphorus pentachloride,[1] thionyl chloride, or benzotrichloride.   With the last two reagents a small amount of zinc chloride is used as a catalyst.[2]

Attempts have been made to prepare pure symmetrical o-phthalyl chloride by repeatedly heating the crude chloride, still containing phthalic acid, with small amounts of phosphorus pentachloride.[3]

Conversion of the symmetrical chloride into the unsymmetrical isomer can also be effected by heating with tin tetrachloride.[4]   The yield, however, is not satisfactory.

[1] Claus and Hoch, Ber. **19,** 1187 (1886).
[2] Kyrides, J. Am. Chem. Soc. **59,** 206 (1937).
[3] Brühl, Ann. **235,** 13 (1886).
[4] Csányi, Ber. **52,** 1792 (1919).

PIMELIC ACID

(A) (From Cyclohexanone)

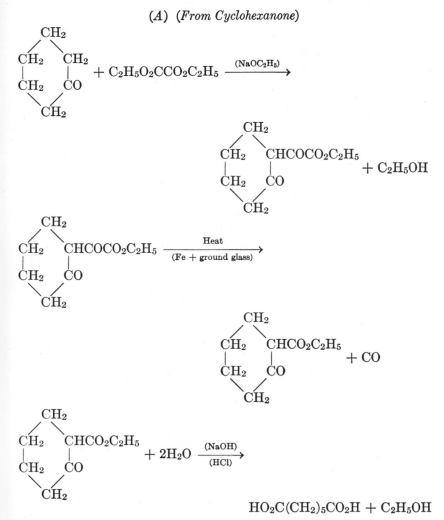

Submitted by H. R. Snyder, L. A. Brooks, and S. H. Shapiro.
Checked by Lee Irvin Smith, R. T. Arnold, and John Moran.

1. Procedure

A solution of sodium ethoxide is prepared by the cautious addition of 46 g. (2 gram atoms) of clean sodium to 600 cc. of anhydrous ethyl alcohol (Note 1) in a 2-l. three-necked flask equipped with a dropping funnel, a mercury-sealed stirrer, and a reflux condenser carrying a calcium chloride tube. The flask is then immersed in an ice-salt bath and the stirrer is started. When the temperature of the solution has reached 10° (Note 2) an ice-cold solution of 196 g. (2 moles) of cyclohexanone (Note 3) in 292 g. (2 moles) of ethyl oxalate (Note 4) is added from the dropping funnel over a period of about fifteen minutes. Vigorous stirring is required to prevent complete solidification of the reaction mixture (Note 5). When the addition is complete, the ice bath is retained for an hour, and then the mixture is stirred at room temperature for about six hours.

The reaction mixture is then decomposed by the careful addition of ice-cold dilute sulfuric acid prepared by adding 56 cc. of concentrated acid (sp. gr. 1.84) to 435 g. of ice. During this neutralization the temperature of the mixture is maintained at about 5–10° by means of an ice-salt bath. The solution, which should now be acid to Congo red paper, is diluted with cold water to a volume of about 4 l. The ethyl 2-keto-cyclohexylglyoxalate separates as a heavy oil and is removed. The aqueous solution is extracted with four 500-cc. portions of benzene. The crude product is combined with the extracts, and the resulting solution is washed with two 200-cc. portions of water. The benzene solution is then allowed to stand in a separatory funnel for a few minutes until it is free from suspended water.

The benzene solution, without drying, is transferred in portions of about 500 cc. to a 1-l. Claisen flask connected to a water-cooled condenser. The flask is heated on the steam bath until the benzene no longer distils. The steam bath is then replaced by an oil bath and the system is gradually evacuated to a pressure of 10–12 mm. while the oil bath is held at about 90°. When all the benzene, unchanged ester, and ketone have distilled (Note 6) the temperature of the oil bath is increased. When the temperature of the distillate reaches 105°/10–12 mm. the receiver is changed. The bath temperature is immediately raised to 175° and all material distilling between 105° and 165° at 10–15 mm. is collected. This requires one-half to one hour. During this time the bath temperature is slowly increased to 200° to obtain the last portions. The yield is 250–265 g. (63–67 per cent of the theoretical amount). The distillate is transferred to a 500–cc. modified Claisen flask, and a trace of iron powder and some finely ground soft glass are added (Note 7). The mixture is distilled at 40 mm. with the bath temperature maintained at 165–175° (Note 8). Carbon monoxide is evolved and the distillate is collected between 125° and 140°. About one and a half to two hours is required for the pyrolysis. The yield of ethyl 2-ketohexahydrobenzoate is 200–210 g. (59–62 per cent of the theoretical amount based on cyclohexanone). This product, whose refractive index at 25° varies from 1.476 to 1.479, is sufficiently pure for use in the next step without redistillation.

In a 2-l. three-necked flask equipped with a dropping funnel, a reflux condenser, and a special stirrer (Note 9) is placed 100 g. (2.5 moles) of sodium hydroxide and 300 cc. of anhydrous methyl alcohol. The stirred mixture is heated for one hour in an oil bath held at 120° in order to effect solution of most of the sodium hydroxide. Stirring and heating are continued while 100 g. (0.59 mole) of ethyl 2-ketohexahydrobenzoate is added over a period of two hours. The resulting mixture is heated for one hour longer with the bath temperature at 120°. It is then diluted with 600 cc. of water and the condenser is arranged for distillation. The methyl alcohol is removed by distillation until a thermometer immersed in the boiling solution reads 98–100°. The residual aqueous solution is vigorously stirred, and exactly 210 cc. of concentrated hydrochloric acid (sp. gr. 1.18) is carefully added drop by drop from a dropping funnel (Note 10). The hot acid solution is treated with 2–4 g. of Darco and is filtered through a heated Büchner funnel. The filtrate is cooled in an ice bath. The pimelic acid is collected on a Büchner funnel and is crystallized from 100 cc. of boiling water for each 45 g. of acid. After drying in the air the acid melts at 103.5–104° and weighs 65–73 g. The mother liquors from the hydrolysis and recrystallization are combined and evaporated to dryness on a steam bath. The resulting solid is extracted with two 500-cc. portions of acetone. The acetone is distilled from a steam bath, and the residual crude pimelic acid is recrystallized from the minimum quantity of benzene, yielding an additional 10–12 g. of pure material. The total yield is 75–83 g. (80–88 per cent of the theoretical amount based on ethyl 2-ketohexahydrobenzoate; 47–54 per cent based on cyclohexanone).

2. Notes

1. Dry sodium methoxide can be used with good results. One hundred eight grams (2 moles) of sodium methoxide (Matheson Alkali Company) and 400 cc. of anhydrous ethyl alcohol are used instead of the sodium ethoxide solution.

2. The dropping funnel may be replaced with a cork carrying a thermometer during this part of the procedure.

3. Commercial cyclohexanone was redistilled through a 20-in. column packed with Carborundum. The fraction boiling at 154°/746 mm. is of sufficient purity.

4. Commercial ethyl oxalate was redistilled from a modified Claisen flask, and the fraction boiling at 83–86°/25 mm. was used.

FIG. 17

5. A precipitate usually appears within an hour or two after addition of the reagents. If the solution is orange in color and no precipitate forms, it may still be worked up with fair results.

6. The last portions of this distillate are tested with ferric chloride. A red or violet color indicates ethyl 2-ketohexahydrobenzoate.

7. More than a milligram of iron, although increasing the rate of pyrolysis, leaves behind a hard residue difficult to remove. Ordinary soft glass is ground in a mortar; from 0.5 to 1 g. is sufficient.

8. If the bath temperature is too high, the unpyrolyzed ester will distil. A high refractive index indicates the presence of unchanged ester.

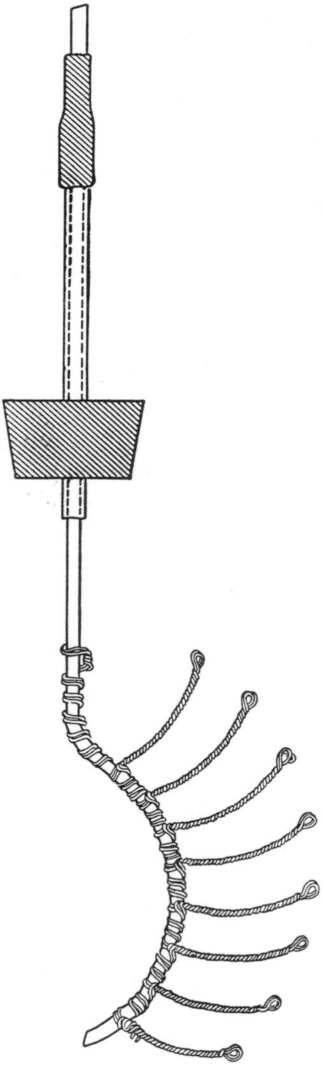

9. The stirrer (Fig. 17) is constructed by wrapping a double strand of copper magnet wire around a semicircular glass stirring rod. At intervals a loop of the wire is twisted to make a projection which extends about 2' in. from the glass rod. The stirrer is introduced into the center neck of the three-necked flask. The projections are arranged so that they touch the sides of the flask when the stirrer is in use. This adjustment is made by a steel or glass rod introduced through one of the side necks. The scraping of the sides of the flask by the copper wire prevents the deposition of the salt on the walls, with consequent moderation of bumping. With an ordinary stirrer the yield in this step is at least 10 per cent less.

10. The solution becomes almost clear when sufficient acid has been added.

(B) (From Salicylic Acid)

$$o\text{-}HOC_6H_4CO_2H + H_2O + 4[H] \xrightarrow{\text{(Na + } iso\text{-}C_5H_{11}OH)} HO_2C(CH_2)_5CO_2H$$

Submitted by ADOLF MÜLLER.
Checked by REYNOLD C. FUSON, J. R. LITTLE, and WESLEY FUGATE.

1. Procedure

FOUR HUNDRED cubic centimeters of freshly distilled isoamyl alcohol (b.p. 128–132°) is heated to 90–100° on an oil bath in a 5-l. two-necked flask fitted with a dropping funnel and reflux condenser. Two hundred forty grams (10.4 gram atoms) of clean sodium is then added, and the temperature of the oil bath is raised rapidly until the alcohol is brought to vigorous boiling (Note 1). A solution of 100 g. (0.73 mole) of salicylic acid in 2 l. of isoamyl alcohol is allowed to flow into the flask at the rate of 100 cc. every four minutes, so that the entire amount is added over a period of eighty minutes. The dropping funnel is then rinsed with 100 cc. of isoamyl alcohol. The solution in the flask is clear at first, then becomes cloudy with the addition of the salicylic acid. The temperature of the oil bath is regulated so that the alcohol refluxes rapidly throughout the course of the experiment. The sodium goes completely into solution in seven to eight hours.

The flask is then allowed to cool to 100°, and 800 cc. of hot water is added with vigorous shaking (Note 2). The hot mixture is transferred to a 5-l. separatory funnel, the flask being rinsed with 200 cc. of hot water (Note 3). The mixture is shaken well and the layers separated. The isoamyl alcohol layer is extracted with four or five 200-cc. portions of nearly boiling water (Note 4). The combined aqueous extracts are then steam-distilled in order to remove any isoamyl alcohol from the aqueous solution of the sodium salt. About 500 cc. of distillate is collected (Note 5).

The flask is allowed to cool, and 920 cc. of hydrochloric acid (sp. gr. 1.19) is added. The unchanged salicylic acid is then steam-distilled from the mixture. The flask is strongly heated, so that its contents are concentrated and, towards the end of the distillation, sodium chloride begins to precipitate. The removal of salicylic acid is practically complete when about 10–12 l. of distillate has come over, although the distillate still gives a ferric chloride test for salicylic acid. The solution in the flask is allowed to cool overnight and is finally chilled in an ice bath.

The crystalline precipitate, which is a mixture of sodium chloride and pimelic acid, is then collected on a suction filter without washing. The product is dried in an evaporating dish on the water bath for several hours, with frequent stirring, whereby the brown mass is partially melted. The mass, after cooling, is transferred to a 500-cc. Soxhlet thimble. The Soxhlet apparatus is mounted over a hot plate and the solid extracted with about 650 cc. of benzene. Complete extraction is indicated when the liquid in the siphon is no longer turbid (Note 6). The extraction is complete when evaporation of a small quantity of the solution in the extractor leaves no residue (two to three hours). After the extraction is complete, the benzene solution is concentrated to about 300 cc. and the pimelic acid allowed to crystallize. The crystals are collected on a filter, washed with cold benzene, and dried in the air. The product melts at 104–105° and weighs 50–58 g. (43–50 per cent of the theoretical amount).

2. Notes

1. The yield of pimelic acid is materially reduced if the alcohol is not refluxing rapidly at this point.

2. The water must be added slowly and with thorough shaking at first, since traces of unchanged sodium may still be present.

3. Unless hot water (85–90°) is used for the extractions troublesome emulsions are likely to form. If such emulsions do form they may be broken by passing steam into the solution while in the separatory funnel.

4. A large portion of the isoamyl alcohol may be recovered for use in subsequent preparations. The moist alcohol from the water extractions is directly distilled, and a fraction boiling at 128–135° collected. The alcohol layer in the fore-run may be separated from the water layer and redistilled with the moist alcohol from the following run.

5. Ethyl pimelate may be prepared by proceeding as follows: The combined aqueous extracts, instead of being subjected to steam distillation, are evaporated on a steam bath until the thick crystalline magma which forms has only a faint odor of isoamyl alcohol. The magma is dissolved in 600 cc. of water, and 800 cc. of concentrated hydrochloric acid is added. The mixture is cooled to room temperature and filtered, and 170 cc. of concentrated hydrochloric acid is added to the filtrate. The acid mixture is extracted with three 400-cc. portions of ether—a continuous extractor such as that shown in Org. Syn. Coll. Vol. I, **1941,** 277, can be used to advantage—and the ether is removed from the extract by distillation. Upon cooling the residue in an ice bath, crystals separate; they are cooled thoroughly, collected on a Büchner funnel, and pressed on a porous plate. The filtrate, after the addition of a few pieces of porous plate, is left overnight in a vacuum desiccator over concentrated sulfuric acid. The crystals which separate are collected on a

filter, pressed on a porous plate, and added to the rest of the product. The total weight is 60–70 g.

The crude material, which is a mixture of pimelic and salicylic acids, is esterified by boiling for four hours with 520 cc. of absolute ethyl alcohol and 6 cc. of concentrated sulfuric acid. Two-thirds of the alcohol is then removed by distillation. To the residue, 600 cc. of water and 400 cc. of ether are added, the mixture is shaken, and the aqueous layer is drawn off. The ether solution is shaken with two 200-cc. portions of 2 N sodium hydroxide solution to remove ethyl salicylate, and then with water until the disappearance of an alkaline reaction. The ether is evaporated and the residue is distilled under reduced pressure. Ethyl pimelate boils at 153–156°/24 mm.; 148–152°/22 mm. The yield is 54–60 g. (35–38 per cent of the theoretical amount based upon the salicylic acid).

In this procedure it is important to remove the brown sludge formed by partial acidification of the alkaline solution of the sodium salts. If all the hydrochloric acid is added at once, ether extraction of the entire precipitate is tedious because of the slow separation of the aqueous and ether layers. It is also important to wash the ether solution of ethyl pimelate and ethyl salicylate carefully with sodium hydroxide solution in order to avoid loss of product by hydrolysis. (Private communication by ADOLF MÜLLER and ERICH RÖLZ. Checked by W. H. CAROTHERS and W. L. McEWEN.)

6. During some runs the pimelic acid begins to crystallize in the siphon. This difficulty may be minimized by fashioning an asbestos jacket around the extraction chamber. Occasionally it is necessary to pass steam over the siphon.

3. Methods of Preparation

Pimelic acid has been obtained as a by-product of the reaction between trimethylene bromide and sodium cyanoacetic ester;[1] by the action of carbon dioxide upon pentamethylene-1,5-dimagnesium bromide;[2] by hydrolysis of pentamethylene cyanide;[3] by the action of sodium and amyl alcohol upon salicylic acid, guaiacol carboxylic acid,[4] or anthranilic acid;[5] and from 2-cyanocyclohexanone.[6]

[1] Carpenter and Perkin, J. Chem. Soc. **75**, 933 (1899).

[2] Grignard and Vignon, Compt. rend. **144**, 1359 (1907).

[3] Hamonet, ibid. **139**, 60 (1904); Bull. soc. chim. (3) **33**, 532 (1905); v. Braun, Ber. **37**, 3591 (1904).

[4] Einhorn and Lumsden, Ann. **286**, 259, 266 (1895); Walker and Lumsden, J. Chem. Soc. **79**, 1198 (1901).

[5] Einhorn and Meyenberg, Ber. **27**, 2467 (1894).

[6] Meyer, Helv. Chim. Acta **16**, 1293 (1933).

The preparation of pimelic acid from cyclohexanone, described in Part (*A*) above, and from salicylic acid, described in Part (*B*) above,[7] are new procedures in Organic Syntheses. The older directions for preparing ethyl pimelate [8] are given as Note 5 on p. 536.

PIPERONYLIC ACID

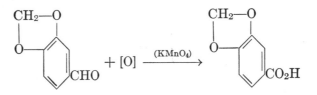

Submitted by R. L. Shriner and E. C. Kleiderer.
Checked by H. T. Clarke and S. Graff.

1. Procedure

Sixty grams (0.4 mole) of piperonal (Note 1) and 1.5 l. of water are placed in a 5-l. flask fitted with an efficient mechanical stirrer. The flask is placed on a steam bath, heated to 70–80°, and the stirrer started (Note 2). A solution of 90 g. (0.56 mole) of potassium permanganate in 1.8 l. of water is allowed to flow into the emulsion of piperonal and water over a period of forty to forty-five minutes (Note 3). The stirring and heating are continued for an hour longer, at the end of which time the permanganate is reduced. Enough 10 per cent potassium hydroxide solution is added to make the solution alkaline. The mixture is filtered while hot, and the manganese dioxide is washed with three 200-cc. portions of hot water. The combined filtrate and washings are cooled. At this point any unreacted piperonal that separates must be filtered (Note 4). The solution is now acidified with hydrochloric acid, the acid being added until no further precipitate forms. The resulting piperonylic acid is filtered, washed with cold water until free of chlorides, and dried. The yield is 60–64 g. (90–96 per cent of the theoretical amount) of a colorless product melting at 224–225° (corr.). For most purposes the crude material is pure enough, but it may be crystallized from ten times its weight of 95 per cent ethyl alcohol, yielding 52–56 g. (78–84 per cent of the theoretical amount) of needles melting at 227–228° (corr.).

[7] Müller, Monatsh. **65**, 18 (1935).
[8] Müller and Rölz, ibid. **48**, 734 (1927); Org. Syn. **11**, 42.

2. Notes

1. The commercial grade of piperonal, m.p. 36°, may be used without purification.
2. The stirring must be sufficiently vigorous to emulsify the molten piperonal thoroughly with the water.
3. Addition of the piperonal to the permanganate lowers the yield to 50 per cent.
4. If the temperature of the reaction mixture is kept between 70° and 80° and the permanganate is added at the rate stated, the piperonal will be entirely oxidized.

3. Methods of Preparation

Piperonylic acid has been made by the oxidation of piperic acid,[1] piperonal,[1] safrole,[2] and isosafrole [2] with potassium permanganate. It has also been prepared by the action of methylene iodide on protocatechuic acid [3] in the presence of alkali.

POTASSIUM ANTHRAQUINONE-α-SULFONATE

(1-Anthraquinonesulfonic acid, potassium salt)

$$C_6H_4\diagup\!\!\!\diagdown_{CO}^{CO}\!\!\!\diagdown\!\!\!\diagup C_6H_4 + H_2S_2O_7 \xrightarrow{(HgSO_4)}$$

$$C_6H_4\diagup\!\!\!\diagdown_{CO}^{CO}\!\!\!\diagdown\!\!\!\diagup C_6H_3SO_3H(\alpha) + H_2SO_4$$

$$C_6H_4\diagup\!\!\!\diagdown_{CO}^{CO}\!\!\!\diagdown\!\!\!\diagup C_6H_3SO_3H(\alpha) + KCl \rightarrow$$

$$C_6H_4\diagup\!\!\!\diagdown_{CO}^{CO}\!\!\!\diagdown\!\!\!\diagup C_6H_3SO_3K(\alpha) + HCl$$

Submitted by W. J. Scott and C. F. H. Allen.
Checked by Louis F. Fieser and E. B. Hershberg.

1. Procedure

A 500-cc. three-necked flask fitted with a mechanical stirrer (Note 1) and a thermometer is half immersed in an oil bath mounted in a hood.

[1] Fittig and Mielck, Ann. **152**, 40 (1869).
[2] Ciamician and Silber, Ber. **23**, 1160 (1890).
[3] Fittig and Remsen, Ann. **168**, 94 (1873).

In the flask are placed 120 g. of 19–22 per cent oleum and 1 g. of yellow mercuric oxide, the bath is warmed to 100°, and 100 g. (0.48 mole) of anthraquinone (Note 2) is added through a powder funnel attached by a piece of wide rubber tubing. The mixture is stirred vigorously and heated at 147–152° for forty-five to sixty minutes, after which the bath is removed and the flask is lowered from the stirrer and replaced by a 2-l. beaker containing 1 l. of hot water. While the water is being stirred, the hot acid solution is poured cautiously down the inner wall of the beaker; the mixture is then boiled for five minutes longer with stirring. After the unchanged anthraquinone (53–59 g.) (Note 3) is collected by suction on a 20-cm. Büchner funnel provided with a cotton filter cloth, it is washed with 200 cc. of hot water. The light brown filtrate, together with the wash water, is heated to 90°, and a solution of 32 g. of potassium chloride in 250 cc. of water is added. After cooling to room temperature (Note 4), the potassium salt, which crystallizes in the form of pale yellow leaflets, is collected on a large Büchner funnel (filter paper) and washed with 200 cc. of cold water. The yield of product, dried at 100° *in vacuo*, is 57–55 g. (Note 5) (77–86 per cent of the theoretical amount based on the anthraquinone converted) (Notes 6 and 7).

2. Notes

1. It is necessary to provide an efficient stirrer driven by a powerful motor. The yields reported were obtained using a Hershberg Chromel wire stirrer (p. 117); with other stirrers the yields were about 10 per cent lower.

2. The anthraquinone employed melted at 284.5–285.5° (corr.). It may be prepared conveniently by oxidation of anthracene (p. 554).

3. The recovered material contains mercury and other impurities and melts at 265–275°. When this is used as such in a second run, there is considerably more disulfonation than with pure anthraquinone.

4. Under the conditions specified very little disulfonation occurs, so that the crystallizing mixture may be allowed to cool to 25° without danger of contamination of the product with disulfonates. When recovered anthraquinone is employed, it is advisable to collect the product when the mixture has cooled to 60°, for the disulfonates present are then retained in the mother liquor. The solubilities of the 1,5- and 1,8-disulfonates increase more rapidly with increasing temperature than does the solubility of the α-monosulfonate.

5. The 57-g. yield refers to the experiment in which 53 g. of starting material was recovered; the percentage yield improves with increase in the amount of anthraquinone recovered.

6. The purity of the product may be checked by converting a sample to α chloroanthraquinone (p. 128) and taking the melting point; that from the above salt melted at 158–160° (corr.).

7. The conditions adopted in this procedure favor the production of the α-monosulfonate in a state of high purity at the expense of a high conversion of anthraquinone. A better conversion can be achieved by conducting the sulfonation at a higher temperature, or by using more oleum, but in either case there is a considerable increase in the amount of disulfonic acids formed. The extent of β-sulfonation is not influenced greatly by the temperature but is dependent chiefly on the amount of mercuric salt present in the solution. The amount specified corresponds approximately to the limit of solubility of the salt in the acid employed, and very little of the β-acid is formed. As the potassium β-sulfonate is more soluble than the α-salt, traces of this isomer are easily eliminated by crystallization.

3. Method of Preparation

The only practical method for the preparation of anthraquinone-α-sulfonates is based upon the discovery [1] that in the presence of a small amount of mercuric salt anthraquinone is sulfonated chiefly in the α-rather than in the β-position. Detailed procedures are described by Fierz-David,[2] by Lauer,[3] and by Groggins;[4] the above directions are based largely upon the observations of Lauer.[3]

PROPIONALDEHYDE

$$3CH_3CH_2CH_2OH + K_2Cr_2O_7 + 4H_2SO_4 \rightarrow$$
$$3CH_3CH_2CHO + K_2SO_4 + Cr_2(SO_4)_3 + 7H_2O$$

Submitted by CHARLES D. HURD and R. N. MEINERT.
Checked by W. L. McEWEN and W. H. CAROTHERS.

1. Procedure

ONE HUNDRED grams (125 cc., 1.7 moles) of n-propyl alcohol, b.p. 96–96.6°, is placed in a 2-l. three-necked, round-bottomed flask fitted with a mercury-sealed stirrer (Note 1), a dropping funnel, and a 60-cm.

[1] Iljinsky, Ber. **36**, 4194 (1903); Schmidt, ibid. **37**, 66 (1904).
[2] Fierz-David, Helv. Chim. Acta **10**, 197 (1927).
[3] Lauer, J. prakt. Chem. (2) **130**, 185 (1931); ibid. (2) **135**, 164 (1932).
[4] Groggins, "Unit Processes in Organic Synthesis," pp. 268–269, McGraw-Hill Book Company, New York, 1935.

ulb condenser (Note 2) set at an angle of 45°. Water at 60° is kept
circulating through this condenser. A condenser set for downward distillation is connected to the top of the first condenser. Cold water
circulates through the second condenser. By means of an adapter, the
lower end of the second condenser is fitted to a receiver which is cooled
in ice water.

The alcohol in the flask is heated to boiling, stirred, and a mixture
of g. (0.56 mole) of potassium dichromate, 120 cc. of concentrated
acid (2.2 moles), and 1 l. of water is added through the dropping
funnel. The addition takes about thirty minutes, and during this time
the contents of the flask are kept vigorously boiling. After all the oxidizing mixture has been added, the contents of the flask are boiled for
fifteen minutes to distil the last of the aldehyde. The propionaldehyde
which collects in the receiver is dried with 5 g. of anhydrous sodium
sulfate and fractionally distilled. The yield of propionaldehyde boiling
at 48–55°, and having a refractive index of 1.364 (Note 3), is 44–47 g.
(45–49 per cent of the calculated amount).

2. Notes

1. The yield of propionaldehyde depends largely upon the efficiency
of the stirrer.

2. The purpose of the first condenser is to condense and return to the
flask any propyl alcohol which escapes.

3. The recorded value for the index of refraction ($N_D^{20°}$) is 1.3636.

3. Methods of Preparation

Propionaldehyde has been prepared by passing propyl alcohol over
finely powdered reduced copper;[1] by passing a mixture of propyl alcohol
and air over a hot platinum spiral[2] or a silver catalyst containing a small
amount of samarium oxide;[3] by passing a mixture of steam and propylene oxide over silica gel at 300°;[4] by adding propyl alcohol to a dichromate oxidizing mixture[5] or adding a dichromate oxidizing mixture to
propyl alcohol;[6] by heating propylene glycol to 500°;[7] by heating a

[1] Sabatier and Senderens, Compt. rend. 136, 923 (1903).

[2] Trillat, Bull. soc. chim. (3) 29, 38 (1903).

[3] Day and Eisner, J. Phys. Chem. 36, 1912 (1932).

[4] I. G. Farbenind. A.-G., Ger. pat. 618,972 [C. A. 30, 1066 (1936)].

[5] Lieben and Zeisel, Monatsh. 4, 14 (1883); Lagerev, J. Gen. Chem. (U.S.S.R.) 5,
515 (1935) [C. A. 29, 6887 (1935)].

[6] Hurd, Meinert, and Spence, J. Am. Chem. Soc. 52, 1140 (1930).

[7] Nef, Ann. 335, 203 (1904).

mixture of calcium formate and calcium propionate; [8] by the action of ethylmagnesium iodide on amyl formate; [9] by catalytic hydrogenation of acrolein; [10] by electrolysis of calcium chloride or dilute sulfuric acid solutions of propyl alcohol; [11] by passing vapors of propyl alcohol over cadmium oxide at 325°; [12] by passing vapors of propionic acid and formic acid over titanium oxide at 250–300°; [13] by oxidizing propyl alcohol with a stream of air in the presence of copper bronze, nitrobenzene, and quinoline; [14] and by treating ethyl orthoformate with ethylmagnesium bromide.[15]

o-PROPIOPHENOL AND *p*-PROPIOPHENOL

(Propiophenone, *o*- and *p*-hydroxy-)

$$C_6H_5OCOC_2H_5 \xrightarrow[\text{(CS}_2\text{)}]{\text{(AlCl}_3\text{)}} o\text{- and } p\text{-}C_6H_4(OH)COC_2H_5$$

Submitted by Ellis Miller and Walter H. Hartung.
Checked by W. W. Hartman and L. J. Roll.

1. Procedure

In a 2-l. three-necked, round-bottomed flask, fitted with a reflux condenser, a mechanical stirrer (Note 1), and a 100-cc. dropping funnel, are placed 374 g. (2.8 moles) of anhydrous aluminum chloride and 400 cc. of carbon bisulfide (Note 2). To the stirred suspension is added slowly through the dropping funnel 375 g. (2.5 moles) of phenyl propionate (Note 3). Reaction sets in almost at once with the evolution of hydrogen chloride (Note 4), and the carbon bisulfide begins to reflux from the heat of reaction (Note 5). When all the propionate is in (one and one-half hours is required), the mixture is further heated to gentle refluxing on a steam bath until the evolution of hydrogen chloride has practically ceased (about two hours). The reflux condenser is then turned downward, and the carbon bisulfide is distilled. The steam bath is

[8] Linneman, ibid. **161**, 21 (1872).

[9] Bayer and Company, Ger. pat. 157,573 [Frdl. **8**, 156 (1905–07)].

[10] Sabatier and Senderens, Ann. chim. phys. (8) **4**, 398 (1905); (8) **16**, 72 (1909); Skita, Ber. **45**, 3316 (1912).

[11] Reitlinger, Z. Elektrochem. **20**, 261 (1914); Feyer, ibid. **25**, 142 (1919).

[12] Sabatier and Mailhe, Ann. chim. phys. (8) **20**, 303 (1910).

[13] Sabatier and Mailhe, Compt. rend. **154**, 563 (1912).

[14] Rosenmund and Zetzsche, Ber. **54**, 2034 (1921).

[15] Wood and Comley, J. Soc. Chem. Ind. **42**, 431T (1923).

replaced by an oil bath which is heated to 140° and maintained at 140–150° for three hours (Note 6). During this period a fresh evolution of hydrogen chloride takes place. The mixture thickens and finally congeals to a brown resinous mass. Stirring is continued as long as possible (Note 7).

The solid is then allowed to cool, and the aluminum complex is decomposed by slowly adding first a mixture of 300 cc. of concentrated hydrochloric acid with 300 cc. of water and then 500 cc. of water (Note 8), whereupon a black oil collects at the surface. After standing overnight in the ice box, a large portion of this layer solidifies and can be separated by filtration. This solid (p-propiophenol) is recrystallized from 400 cc. of methyl alcohol. The yield is 129–148 g. (34–39 per cent of the theoretical amount) of a light yellow product melting at 145–147°. A second recrystallization raises the melting point to 147–148°.

The oily filtrate combined with the concentrated mother liquors of the above recrystallization is dissolved in 500 cc. of 10 per cent sodium hydroxide and extracted with two 100-cc. portions of ether to remove non-phenolic products. The alkaline solution is acidified with hydrochloric acid and the oily layer is separated, dried over anhydrous magnesium sulfate, and distilled. The o-propiophenol boils at 110–115°/6 mm. The yield is 120–132 g. (32–35 per cent of the theoretical amount). About 40 g. of p-propiophenol boiling at 135–150°/11 mm. is obtained. The total yield of crude p-propiophenol is thus 169–188 g. (45–50 per cent of the theoretical amount) (Note 9).

2. Notes

1. The stirrer should be made of sturdy glass rod, bent preferably in the shape of a golf-club head, to provide for heavy duty when the reaction mass becomes thick.

2. When nitrobenzene was used as a solvent the products were always tarry.

3. Crude, undistilled phenyl propionate, made by gently heating a mixture of equivalent quantities of phenol and propionyl chloride until hydrogen chloride evolution ceases, is satisfactory for rearrangement. The ester can also be made conveniently by slowly adding one mole of thionyl chloride to a mixture of 1.05 moles each of phenol and propionic acid, driving off all the hydrogen chloride and sulfur dioxide, and distilling.

4. Because of the large volume of hydrogen chloride evolved, it is desirable to work under a good hood. If this is not available, the gas-trap described on p. 4 may be used.

5. The addition of the phenyl propionate is regulated so as to maintain gentle refluxing.

6. The temperature of the reaction mixture is about ten degrees lower than the bath temperature, or 130–135°. The reaction temperature determines the yield ratio of the two isomers. In general, higher temperature (above 160–170°) favors the formation of the ortho isomer.

7. It is necessary to stir the thickening mass continuously to permit the escape of hydrogen chloride formed during the heating; otherwise it may swell rapidly and choke the outlets of the flask.

8. This decomposition is strongly exothermic, and the dilute acid should be added slowly.

9. The above procedure has been found satisfactory for the preparation of a number of the homologs of *o*- and *p*-propiophenol, such as aceto-, butyro-, and caprophenol.

3. Methods of Preparation

Ortho and para propiophenols have been previously prepared by condensing propionic acid and phenol in the presence of zinc chloride,[1] and from phenol and propionyl chloride.[2] The procedure described above is an adaptation of one described by Cox[3] and by Hartung, Munch, Miller, and Crossley.[4] In an alternative procedure phenyl propionate is added directly to aluminum chloride; no solvent is used.[5]

l-PROPYLENE GLYCOL

(1,2-Propanediol, *l*-)

$$CH_3COCH_2OH + 2[H] \xrightarrow{\text{(Reductase of yeast)}} CH_3CHOHCH_2OH$$

Submitted by P. A. Levene and A. Walti.
Checked by Frank C. Whitmore and J. Pauline Hollingshead.

1. Procedure

A solution of 1 kg. of sucrose in 9 l. of water is placed in a 20-l. bottle provided with a gas trap. A paste of baker's yeast (Note 1) is

[1] Goldzweig and Kaiser, J. prakt. Chem. (2) **43**, 86 (1891).
[2] Perkin, J. Chem. Soc. **55**, 547 (1889).
[3] Cox, J. Am. Chem. Soc. **49**, 1028 (1927).
[4] Hartung, Munch, Miller, and Crossley, ibid. **53**, 4153 (1931).
[5] Farinholt, Harden, and Twiss, ibid. **55**, 3386 (1933).

made by breaking up 1 kg. of yeast and gradually stirring in 1 l. of water. This is then added to the sugar solution and the mixture is allowed to stand at room temperature until a lively evolution of gas starts (from one to three hours). To the vigorously fermenting solution 100 g. (1.35 moles) of freshly prepared acetol (p. 5) is added, and the mixture is allowed to stand at room temperature until the reaction subsides (Note 2). The bottle is then transferred to an incubator at 32°, when the fermentation recommences. At the end of three days the reaction is generally completed, and the solution when tested with Fehling's reagent gives a negligible test for reducing sugars.

At this point 20–30 g. of short glass fiber or asbestos is added and the yeast is filtered by suction. The filtrate is concentrated to a thick syrup under diminished pressure on a water bath, the temperature being kept below 40° (Note 3). The residue (about 200 cc.) is taken up in a mixture of 400 cc. of absolute alcohol and 100 cc. of dry ether (Note 4). The precipitate formed is removed by centrifuging, the supernatant liquid is decanted, and the residue is extracted with a mixture of 200 cc. of 98.5 per cent alcohol and 100 cc. of dry ether (Note 5). The combined alcohol-ether solutions are concentrated under diminished pressure at 35–40° to a thick syrup. The residue is again taken up in a mixture of 400 cc. of 98.5 per cent alcohol and 100 cc. of dry ether and centrifuged (Note 5). The supernatant liquid is concentrated under diminished pressure and distilled from a modified Claisen flask. The yield of the crude product boiling at 86–91°/12 mm. is approximately 100 g. The crude material is refractionated and collected at 88–90°/12 mm. or 187–189°/760 mm. The final product (Note 4) is a colorless liquid having a density 1.04 and specific rotation $[\alpha]_D^{20} = -15.0°$. The yield is 50–60 g. (49–58 per cent of the theoretical amount) (Note 6).

2. Notes

1. Fleischmann's yeast is satisfactory.

2. The addition of the acetol may cause the reaction to slacken for a time.

3. The evaporation must be carried out at as low a temperature as is practicable. A suitable device for this evaporation is given in Org. Syn. Coll. Vol. I, 1941, 427.

4. The product reacts slightly acid. If an entirely neutral l-propylene glycol is desired, the syrup first obtained should be made neutral with a solution of sodium methoxide in methyl alcohol, and again concentrated and extracted as indicated.

5. If a centrifuge is not available the same result may be obtained by

adding about 15 g. of short glass fiber or asbestos to the solution, stirring the solution mechanically or shaking it vigorously for five minutes, and filtering with suction.

6. The optically active glycols are convenient starting materials for the preparation of optically active carbinols, hydroxyacids, etc. The biological method of asymmetric reduction is perhaps the only convenient method for the preparation of these glycols. The steps in the preparation of other optically active glycols are identical with those in the preparation of *l*-propylene glycol from acetone. In the synthesis of certain glycols it is convenient to prepare the chloroketone by oxidizing the corresponding chlorohydrin, the succeeding steps being the same as those in the synthesis of *l*-propylene glycol.

3. Methods of Preparation

l-Propylene glycol has been obtained from the optically inactive glycol by the action of bacteria,[1] and by means of strychnine compounds.[2] The present method is based on that of Färber, Nord, and Neuberg.[3]

n-PROPYL SULFIDE

$$C_2H_5ONa + H_2S \rightarrow NaSH + C_2H_5OH$$

$$NaSH + C_2H_5ONa \rightarrow Na_2S + C_2H_5OH$$

$$Na_2S + 2CH_3CH_2CH_2Br \rightarrow (CH_3CH_2CH_2)_2S + 2NaBr$$

Submitted by R. W. Bost and M. W. Conn.
Checked by Reynold C. Fuson and Charles F. Woodward.

1. Procedure

In a 2-l. round-bottomed flask, fitted with a reflux condenser and a 6-mm. glass tube closed at the upper end with a rubber tube and pinchcock and reaching to the bottom of the flask, are placed 800 cc. of absolute alcohol and 50.6 g. (2.2 gram atoms) of clean sodium cut in small pieces. When the sodium is completely dissolved, one half of the solution is transferred to a 2-l. three-necked, round-bottomed flask fitted with a dropping funnel, a liquid-sealed mechanical stirrer, and a reflux

[1] LeBel, Jahresb. **1881**, 512.

[2] Grün, Ber. **52**, 260 (1919).

[3] Färber, Nord, and Neuberg, Biochem. Z. **112**, 313 (1920).

condenser closed by a calcium chloride tube. The flask containing the remainder of the solution is connected again to the reflux condenser, and hydrogen sulfide from a cylinder is introduced by means of the glass tube at the rate of about two bubbles per second until the solution is saturated (about six hours is required). This solution of sodium hydrogen sulfide is added to the sodium ethoxide solution in the three-necked flask, and the mixture is refluxed for one hour. After cooling to room temperature, sufficient absolute alcohol (about 200 cc.) is added to dissolve all the sodium sulfide (Note 1).

To this solution of sodium sulfide is added dropwise with stirring 246 g. (2 moles) of n-propyl bromide (Org. Syn. Coll. Vol. I, **1941, 37** and p. 359 above). After all the bromide has been added the flask is heated on a steam cone for eight hours (Note 2), during which time the mixture should not be stirred too vigorously. It is then cooled and added to 2 l. of 25 per cent aqueous sodium chloride solution contained in a separatory funnel. The mixture is shaken to ensure thorough mixing, allowed to stand until the layers have separated, and the upper oily layer of propyl sulfide removed and dried with anhydrous sodium sulfate. The lower layer is extracted with five 200-cc. portions of petroleum ether (b.p. 25–45°) (Note 3), the extract dried with 20 g. of sodium sulfate, and the petroleum ether distilled through a 60-cm. fractionating column until the temperature of the vapors passing over reaches 60°. The residue is added to the crude propyl sulfide previously separated and the combined portions distilled. The yield of product boiling at 140–143° is 80–100 g. (68–85 per cent of the theoretical amount) (Note 4).

2. Notes

1. It is best to have the sodium sulfide completely dissolved before it is added to the bromide; subsequent extraction is facilitated, however, if the solution is kept at a minimum volume.

2. In order to be certain that no propyl sulfide is being lost during the refluxing, the condenser should be connected to a trap containing an aqueous solution of mercuric chloride. Condenser water cooled to 5–10° is recommended.

3. To avoid contamination of the product the petroleum ether should be distilled before being used. Ethyl ether is not suitable for this extraction.

4. n-Butyl and sec.-amyl sulfides can be prepared by procedures essentially the same as that given here.

3. Method of Preparation

n-Propyl sulfide has always been prepared by the action of an alcoholic solution of an alkali sulfide on a *n*-propyl halide.[1]

PROTOCATECHUALDEHYDE

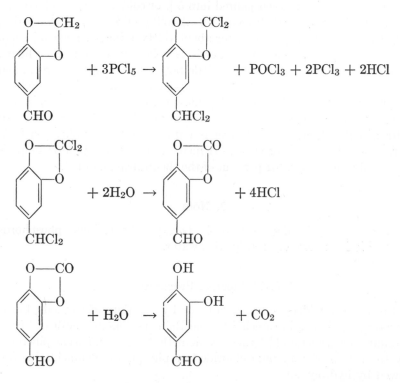

Submitted by JOHANNES S. BUCK and F. J. ZIMMERMANN.
Checked by REYNOLD C. FUSON and W. E. ROSS.

1. Procedure

To 108 g. (0.72 mole) of piperonal in a 3-l. round-bottomed flask is added, in portions of 20 to 30 g., 454 g. (2.18 moles) of fresh phosphorus pentachloride. The reaction is vigorous at first, and the flask is kept cold with ice; moisture must be excluded. After about half of the

[1] Cahours, Compt. rend. **76,** 133 (1873); Winssinger, Bull. soc. chim. (2) **48,** 109 (1887).

pentachloride has been added the reaction becomes sluggish and cooling is unnecessary. The entire addition requires about thirty minutes. The resulting green or blue liquid containing undissolved pentachloride is heated very gently over a flame for about sixty minutes to expel hydrogen chloride. From the turbid, light brown liquid thus formed, volatile material is removed on a steam bath under the reduced pressure of a water pump. This operation takes about thirty minutes. The contents of the flask are then poured into 5 l. of cold water contained in a 12-l. round-bottomed flask (Note 1). A milky oil is formed which rises and sinks in the water and, after about thirty minutes, becomes solid. After standing overnight, the mixture is boiled gently for three hours. The brown solution, containing a little tar, is cleared with charcoal and evaporated under reduced pressure to about 700 cc., when the aldehyde begins to separate. The solution is allowed to stand overnight at about 0°; a large crop of crystals separates, is collected on a filter, and washed with a little water. The product is purified by recrystallization from three times its weight of water. It melts at 153–154°, with decomposition, and weighs 61 g. (61 per cent of the theoretical amount).

2. Note

1. This must be done cautiously because the residual phosphorus pentachloride reacts vigorously with water.

3. Methods of Preparation

Protocatechualdehyde has been made by a variety of methods, but is usually prepared from catechol by the Reimer-Tiemann method;[1] from vanillin[2] or veratric aldehyde[3] by demethylation; and from piperonal by the action of aluminum chloride[4] or phosphorus pentachloride followed by hydrolysis.[5]

[1] Reimer and Tiemann, Ber. **9**, 1268 (1876); Tiemann and Koppe, ibid. **14**, 2015 (1881).

[2] Tiemann and Haarmann, ibid. **7**, 620 (1874).

[3] Dreyfus, Ger. pat. 193,958 [Frdl. **9**, 161 (1908–10)].

[4] Givaudan-Delawanna, Inc., U. S. pat. 2,027,148 [C. A. **30**, 1395 (1936)].

[5] Fittig and Remsen, Ann. **159**, 144 (1871); Pauly, Ber. **40**, 3096 (1907); Barger, J. Chem. Soc. **93**, 563 (1908); Hoering and Baum, Ber. **41**, 1914 (1908).

PYROMELLITIC ACID

Charcoal →

$$HO_2C \diagdown\diagup CO_2H$$
$$HO_2C \diagup\diagdown CO_2H$$

Submitted by E. PHILIPPI and R. THELEN.
Checked by H. T. CLARKE and S. GRAFF.

1. Procedure

IN a 5-l. round-bottomed flask, set on a wire gauze in a hood and provided with a thermometer, are placed 100 g. of finely powdered pine or spruce charcoal (Note 1), 650 cc. of 82–88 per cent sulfuric acid (sp. gr. 1.76–1.80) (Note 2), and a small drop of mercury (Note 3). The mixture is heated with a small flame so that the temperature is raised to 250° during four hours. During another half hour the temperature is raised to 290°, when the acid begins to volatilize and the reaction mixture froths and expands greatly (Note 4). During the next hour the temperature is raised to 300°, and during the following hour to 315°. The mixture thickens somewhat, then much bubbling and spurting take place and fine white needles of pyromellitic anhydride begin to collect in the neck of the flask (Note 5). Fifty cubic centimeters of sulfuric acid is now added—the acid being poured in so as to rinse the walls of the flask; the syrupy reaction mixture is again heated to 250° for a few minutes and is transferred while still hot to a 1-l. tubulated retort (Note 6). The flask is rinsed with the minimum amount of water and the washings are added to the retort.

The material in the retort is heated with a free flame until the water present has distilled, then 30 g. of acid potassium sulfate is added and the distillation is continued. At first almost colorless sulfuric acid distils. As soon as crystals of pyromellitic anhydride appear, the receiver is replaced by one containing 50 cc. of water and the distillation is continued until nothing more passes over. The retort is then rinsed with 100 cc. of water and the rinsings are filtered and evaporated on the steam bath to 25 cc. (Note 7). The last distillate is also evaporated to 25 cc. On cooling, pyromellitic acid crystallizes. The product from both portions is collected on a suction funnel, using hardened filter paper, washed with two 10-cc. portions of ice water (Note 8), and recrystallized from four parts of boiling water. After drying at 105° the product (containing 2 moles of water of crystallization) melts at 271° (262° uncorr.). The yield is 6–8 g. (Note 9).

2. Notes

1. The yield of pyromellitic acid depends considerably on the kind of charcoal used. An experiment with ordinary willow charcoal gave none of the desired product.

2. The specific gravity of the acid used should not vary from the limits indicated.

3. A series of experiments has proved the helpful action of mercuric sulfate.

4. If the time for the first stage (temperature below 250°) is shortened, the frothing is greater at this point, and the final yield is lowered.

5. The end of the reaction can be readily detected by the bubbling and spurting which take the place of the quiet boiling. If the mixture is not fairly thick, less sulfuric acid should be used in the next run. If it is solid, more acid should be used.

6. The transfer to the retort must be made before the mixture solidifies. The best method of closing the tubulure of the retort (unless this consists of Pyrex glass) is by means of a glass stopper too small for the hole and wrapped with asbestos paper moistened with sodium silicate solution.

7. If much pyromellitic anhydride collects in the neck of the retort, it should be rinsed into a clean dish rather than into the receiver for the receiver contains considerable sulfuric acid.

8. In recovering pyromellitic acid from the mother liquor it is advisable first to remove the sulfuric acid by adding a slight excess of barium hydroxide and acidifying to Congo red with hydrochloric acid.

9. The anhydride of pyromellitic acid may be obtained by boiling the dry anhydrous acid with acetic anhydride.

3. Method of Preparation

Pyromellitic acid may be obtained by the oxidation of benzene derivatives containing organic substituents in the 1-, 2-, 4-, and 5-positions: by the oxidation of durene with nitric acid;[1] by condensing benzene with diethylmalonyl chloride, reducing to the hydrocarbon, again condensing in the same way, and finally oxidizing the resulting tetraethylbenzodihydrindenedione;[2] by an analogous synthesis from m-xylene, involving condensation with acetyl chloride, reduction, and oxidation;[3] and by the oxidation of tetrahydro-5,6-benzindan-1-one.[4]

[1] Jacobsen, Ber. **17**, 2516 (1884).

[2] Freund, Fleischer, and Gofferjé, Ann. **414**, 26 (1918).

[3] Philippi, Seka, and Froeschl, ibid. **428**, 300 (1922).

[4] Darzens and Levy, Compt. rend. **201**, 904 (1935).

Mellitic acid can be decarboxylated to yield pyromellitic acid, either by the action of heat alone [5] or in the presence of sulfuric acid.[6]

The oxidation of wood charcoal by means of sulfuric acid leads to mellitic acid and its decarboxylation products;[7] nitric acid may also be employed.[8] Pyromellitic acid has also been obtained by the electrolytic oxidation of graphite in an alkaline medium.[9]

QUINONE *

$$3p\text{-}HOC_6H_4OH + NaClO_3 \xrightarrow{(V_2O_5)}$$

$$3O{=}C{\Big\langle}{\overset{\displaystyle CH{=}CH}{\underset{\displaystyle CH{=}CH}{}}}{\Big\rangle}C{=}O + NaCl + 3H_2O$$

Submitted by H. W. Underwood, Jr., and W. L. Walsh.
Checked by Louis F. Fieser and D. J. Potter.

1. Procedure

In a 2-l. round-bottomed flask equipped with a mechanical stirrer are placed 1 l. of 2 per cent sulfuric acid, 0.5 g. of vanadium pentoxide (Note 1), 110 g. (1 mole) of hydroquinone, and 60 g. (0.56 mole) of sodium chlorate. The mixture is vigorously stirred for about three hours, when the green quinhydrone first formed is converted to yellow quinone. The temperature of the reaction mixture rises to about 40° (Note 2). The reaction is usually complete in three and one-half to four hours. The flask is then disconnected and cooled under the tap, the mixture is filtered with suction, and the quinone is washed once with about 100 cc. of cold water. After drying in a desiccator over calcium chloride, the product weighs 86–90 g. and melts at 110–112°. This material is pure enough for most purposes. Extraction of the filtrate and washings with four 100-cc. portions of benzene yields a further 12–14 g. of quinone and brings the total amount to 99–104 g. (92–96 per cent of the theoretical yield) (Note 3).

[5] Erdmann, Ann. **80**, 281 (1851); Baeyer, Ann. Suppl. **7**, 36 (1870).

[6] Erdmann, Ann. **80**, 282 (1851); Silberrad, J. Chem. Soc. **89**, 1795 (1906).

[7] Verneuil, Compt. rend. **118**, 195 (1894); **132**, 1340 (1901); Philippi and Thelen, Ann. **428**, 296 (1922).

[8] Silberrad, Ger. pat. 214,252 [Frdl. **9**, 173 (1908–10)]; Philippi and Rie, Ann. **428**, 287 (1922).

[9] Bartoli and Papasogli, Gazz. chim. ital. **12**, 113 (1882); **13**, 37 (1883).

* Commercially available; see p. v.

Very pure quinone can be obtained either by vacuum sublimation or recrystallization from boiling ligroin (b.p. 90–120°). One hundred grams of the crude quinone requires about 1.2 l. of ligroin for recrystallization and yields 92–97 g. of bright yellow quinone, m.p. 111–113°.

2. Notes

1. Vanadium pentoxide may be obtained from the Vanadium Corporation of America or prepared from ammonium metavanadate as described in Note 2, p. 302.

2. If quantities larger than one mole of hydroquinone are used the temperature should not be allowed to rise above 40°.

3. By the use of a suitable organic solvent the same oxidizing agent can be employed for the preparation of anthraquinone. A mixture of 90 g. (0.51 mole) of finely powdered pure anthracene, 0.5 g. of vanadium pentoxide, 76 g. of sodium chlorate, 1 l. of glacial acetic acid, and 200 cc. of 2 per cent sulfuric acid is warmed under reflux until a vigorous reaction commences. The source of heat is removed, and the reaction allowed to proceed for about twenty minutes. The mixture is refluxed for one hour longer and then cooled in ice. The light yellow solid is filtered with suction, washed well with water, and dried at 110°. The yield is 92–96 g. (88–91 per cent of the theoretical amount) of a product melting at 273–275°.

The sodium chlorate-vanadium pentoxide mixture is not very powerful, and, although it attacks easily the particularly reactive anthracene, it is not suitable for the conversion of hydrocarbons of the naphthalene and phenanthrene series into the corresponding quinones or for the oxidation of acenaphthene or fluorene (observation of the checkers).

3. Methods of Preparation

The oxidation of hydroquinone by means of sodium dichromate in sulfuric acid solution and references to other methods of preparation are given in an earlier volume of this series.[1]

[1] Org. Syn. Coll. Vol. I, 1941, 482.

REINECKE SALT

$$NH_4[Cr(NH_3)_2(SCN)_4] \cdot H_2O$$

Submitted by H. D. DAKIN.
Checked by H. T. CLARKE and R. S. INGOLS.

1. Procedure

EIGHT HUNDRED grams (10.5 moles) of ammonium thiocyanate is gently heated, by means of three small flames (Note 1), in a white enameled cooking pot of about 4-l. capacity. The mass is stirred with a thermometer enclosed in a glass tube until the solid has partially melted and the temperature has reached 145–150°. At this point an intimate mixture of 170 g. (0.675 mole) of finely powdered ammonium dichromate and 200 g. (2.6 moles) of ammonium thiocyanate is added in portions of 10–12 g. with constant stirring. After about ten such portions have been added a fairly vigorous reaction takes place with evolution of ammonia and the temperature rises to 160°. The flames are extinguished, and the remainder of the mixture is added at such a rate that the heat of reaction maintains the temperature at 160° (Note 2). Stirring is continued while the mass cools, and any lumps of solid which form around the sides of the vessel are broken loose (Note 3).

The product, while still warm (Note 4), is finely powdered and stirred with 750 cc. of ice water in a large beaker. After fifteen minutes the insoluble portion is filtered by suction, freed as completely as possible from mother liquor without washing (Note 5), and stirred into 2.5 l. of water previously warmed to 65°. The temperature is then rapidly raised to 60° (Note 6), the solution is filtered at once through a hot-water funnel, and the filtrate is placed in a refrigerator overnight.

The resulting crystals are collected and the mother liquor employed for a second similar extraction of the residue at 60°. This yields a further crop of crystalline Reinecke salt. The mother liquor is finally concentrated to 250–300 cc. by evaporation at 40–50° under reduced pressure, when a small third crop (12–13 g.) is obtained. The total yield of air-dried crystals is 250–275 g. (52–57 per cent of the theoretical amount) (Note 7).

The undissolved residue from the second extraction consists chiefly of Morland salt (the guanidine salt of the Reinecke acid) and amounts to 130–135 g. (33–34 per cent of the theoretical amount) (Note 8).

2. Notes

1. Heat must be applied as uniformly as possible.
2. The addition of the mixture requires five to seven minutes.
3. The product is detached from the walls during cooling as it is difficult to remove when cold.
4. The material should be pulverized while warm before it has had an opportunity to attract moisture from the air.
5. The filtrate, which consists largely of unchanged ammonium thiocyanate and its decomposition products, contains too little Reinecke salt to repay further treatment.
6. Reinecke salt decomposes in aqueous solution with formation of a blue color and free hydrogen cyanide. At room temperature this decomposition occurs in about two weeks, and above 65° it takes place quite rapidly. A similar decomposition takes place in boiling alcohol.
7. Reinecke salt is of value as a precipitant for primary and secondary amines, proline and hydroxyproline, and certain amino acids.[1]
8. The Morland salt, which is soluble in acetone, contains a small proportion of a colorless sulfur compound insoluble in hot water. It can be partially converted into Reinecke salt by treatment in dilute ammonia solution with a large excess of ammonium chloride, but the amounts so obtainable are unprofitably small.

3. Method of Preparation

Reinecke salt has been prepared by adding either potassium dichromate[2] or ammonium dichromate[1,3] to fused ammonium thiocyanate.

[1] Kapfhammer and Eck, Z. physiol. Chem. **170**, 310 (1927); Grassmann and Lang, Biochem. Z. **269**, 223 (1934).

[2] Reinecke, Ann. **126**, 113 (1863); Christensen, J. prakt. Chem. (2) **45**, 213 (1892); Zeleny and Gortner, J. Biol. Chem. **90**, 430 (1931).

[3] Werner, Z. anorg. Chem. **15**, 260 (1897); Ann. **406**, 276 (1914).

β-RESORCYLIC ACID

$$m\text{-}C_6H_4(OH)_2 \xrightarrow{CO_2 + KHCO_3}$$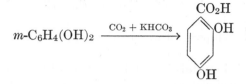

Submitted by M. Nierenstein and D. A. Clibbens.
Checked by Roger Adams and F. E. Kendall.

1. Procedure

In a 5-l. flask fitted with a reflux condenser, a solution containing 200 g. (1.8 moles) of resorcinol, 1 kg. (9.9 moles) of potassium acid carbonate (Note 1), and 2 l. of water (Note 2) is heated slowly on a steam bath for four hours. The flask is then placed over a flame and refluxed vigorously for thirty minutes; a rapid stream of carbon dioxide is passed through the solution during this heating process.

While still hot the solution is acidified by adding 900 cc. of concentrated hydrochloric acid (sp. gr. 1.19) from a separatory funnel with a tube-delivering the acid to the bottom of the flask; this prevents the formation of a layer of acid over the unneutralized solution. The flask is allowed to cool to room temperature and is then chilled in an ice bath. The resorcyclic acid crystallizes in prisms which are almost colorless but which, on exposure to air, turn pink owing to contamination with a small amount of resorcinol. The yield of crude acid is 225 g. By extracting the mother liquor with ether several times, 35 g. of resorcylic acid and some unchanged resorcinol can be recovered. The resorcylic acid is extracted from the ether by shaking with an aqueous solution of sodium bicarbonate. The aqueous solution is acidified with hydrochloric acid and again extracted with ether. The ether is then evaporated, leaving the resorcylic acid, which is usually highly colored and must be recrystallized several times from boiling water and charcoal to remove the color.

The combined yield of crude acid (260–270 g.) is dissolved in 1 l. of water, boiled with about 25 g. of Norite, filtered through a heated filter, and crystallized by placing in an ice-salt freezing mixture and stirring vigorously. A finely crystalline, colorless product is obtained in this way. If the acid is allowed to crystallize slowly the crystals are somewhat colored. The yield of pure resorcylic acid melting at 216–217° is 160–170 g. (57–60 per cent of the theoretical amount) (Note 3).

2. Notes

1. Instead of potassium acid carbonate, the sodium salt in corresponding quantity may be used.

2. If less than ten parts of water to one of resorcinol is used, the yield is diminished.

3. The air-dried crystals lose at 110° a quantity of water corresponding to a half mole of water of crystallization.

3. Method of Preparation

The method of preparing β-resorcylic acid described above is a modification [1] of the procedure given by Bistrzycki and Kostanecki.[2] A more rapid but less efficient variant of this procedure has been described.[3]

SODIUM 2-BROMOETHANESULFONATE

(Ethanesulfonic acid, 2-bromo-, sodium salt)

$$BrCH_2CH_2Br + Na_2SO_3 \rightarrow BrCH_2CH_2SO_3Na + NaBr$$

Submitted by C. S. MARVEL and M. S. SPARBERG.
Checked by FRANK C. WHITMORE and D. J. LODER.

1. Procedure

IN a 5-l. round-bottomed flask, fitted with a reflux condenser, a mechanical stirrer, and a separatory funnel, are placed 615 g. (3.3 moles) of ethylene dibromide (Note 1), 1250 cc. of 95 per cent alcohol, and 450 cc. of water (Note 2). The stirrer is started and the mixture heated to boiling. To the well-stirred boiling mixture a solution of 125 g. (1 mole) of anhydrous sodium sulfite in 450 cc. of water is added through the separatory funnel over a period of about two hours. The solution is boiled under a reflux condenser for two hours after all the sulfite solution has been added. The condenser is then set for distillation, and the alcohol and the ethylene bromide are distilled (Note 3). The remaining water solution is poured into a large evaporating dish and evaporated to dryness on the water bath. The sodium 2-bromoethanesulfonate is ex-

[1] Clibbens and Nierenstein, J. Chem. Soc. 107, 1494 (1915).
[2] Bistrzycki and Kostanecki, Ber. 18, 1984 (1885).
[3] Couturier, Ann. chim. (11) 10, 570 (1938).

tracted from the sodium bromide and unchanged sodium sulfite with 2 l. of boiling 95 per cent alcohol. On cooling the solution, most of the salt crystallizes; the mother liquor is used for a second extraction of the residue. The yield is 165–190 g. (78–90 per cent of the theoretical amount). The product (Note 4) may be further purified by recrystallizing from alcohol and drying in an oven at 110° (Note 5). The recovery on recrystallization is 75–80 per cent.

2. Notes

1. The large excess of ethylene dibromide is necessary to reduce the formation of the disulfonic acid.

2. The concentration of the alcohol seems to be important; poorer yields were obtained when it was changed in either direction.

3. By diluting the alcoholic distillate from the reaction mixture with 10 l. of water, it is possible to recover about 400 g. of ethylene dibromide.

4. This product may contain as much as 2 to 5 per cent of sodium bromide, but it is pure enough for the preparation of taurine (p. 563). A very pure product can be obtained by a second crystallization from alcohol.

5. The salt is slightly hygroscopic.

3. Methods of Preparation

The directions given in the procedure are based on those of Kohler.[1] Sodium 2-bromoethanesulfonate has also been prepared by treating ethylene oxide with sodium bisulfite and converting the isethionic acid thus obtained to the bromo acid with hydrobromic acid.[2]

[1] Kohler, Am. Chem. J. **20**, 692 (1898); Marvel, Bailey, and Sparberg, J. Am. Chem. Soc. **49**, 1835 (1927).

[2] Rumpf, Bull. soc. chim. (5) **5**, 879 (1938).

SUCCINIC ANHYDRIDE *

I

$$\begin{matrix} CH_2CO_2H \\ | \\ CH_2CO_2H \end{matrix} + CH_3COCl \rightarrow \begin{matrix} CH_2CO \\ | \qquad \diagdown \\ \qquad \qquad O \\ | \qquad \diagup \\ CH_2CO \end{matrix} + CH_3CO_2H + HCl$$

Submitted by Louis F. Fieser and E. L. Martin.
Checked by C. R. Noller.

1. Procedure

In a 1-l. round-bottomed flask, fitted with a reflux condenser attached to a gas trap, are placed 118 g. (1 mole) of succinic acid and 215 cc. (235 g., 3 moles) of acetyl chloride. The mixture is gently refluxed on the steam bath until all the acid dissolves; this requires one and a half to two hours. The solution is allowed to cool undisturbed and is finally chilled in an ice bath. The succinic anhydride, which separates in beautiful crystals, is collected on a Büchner funnel, washed with two 75-cc. portions of ether, and dried in a vacuum desiccator. The yield of material melting at 118–119° is 93–95 g. (93–95 per cent of the theoretical amount).

II

$$2 \begin{matrix} CH_2CO_2H \\ | \\ CH_2CO_2H \end{matrix} + POCl_3 \rightarrow 2 \begin{matrix} CH_2CO \\ | \qquad \diagdown \\ \qquad \qquad O \\ | \qquad \diagup \\ CH_2CO \end{matrix} + HPO_3 + 3HCl$$

Submitted by R. L. Shriner and H. C. Struck.
Checked by C. R. Noller, F. B. Hilmer, and J. D. Pickens.

1. Procedure

Two hundred thirty-six grams (2 moles) of succinic acid (Note 1) and 153.5 g. (1 mole) of phosphorus oxychloride are placed in a 1-l. Claisen flask. One neck of the flask is equipped with a reflux condenser connected to a gas trap; the other neck and the side arm of the flask are

* Commercially available; see p. v.

closed with corks. The mixture is heated (Note 2) with a free flame for about fifty minutes until no more hydrogen chloride is evolved. The condenser is removed and the flask is arranged for distillation, with a tube leading from the side arm of the receiver to the drain in order to carry off the vapors. A few cubic centimeters of distillate is collected below 255°, at which temperature the receiver is changed and succinic anhydride is collected from 255–260° (Note 3). The product melts at 118–120° and weighs 164–192 g. (82–96 per cent of the theoretical amount).

The succinic anhydride may be purified by dissolving 50 g. of the crude material in 35 cc. of acetic anhydride. The hot solution is cooled in an ice bath. The crystals are filtered with suction, washed with two 20-cc. portions of cold, absolute ether, and air-dried quickly at 40°. The yield of the pure anhydride, which melts at 119–120°, is 43.5 g.

2. Notes

1. Succinic acid having a melting point of 189–190° was used. Less pure material gives a lower-melting product.

2. The reaction foams considerably at the start, hence slow and careful heating is necessary. It is best to heat the flask directly with the flame, making certain that all parts of the mixture are heated about equally.

3. The tarry mass left in the distilling flask may be easily removed by warm dilute sodium hydroxide.

3. Methods of Preparation

Succinic anhydride has been prepared from succinic acid with phosphorus pentachloride,[1, 2] phosphorus oxychloride,[2, 3] phosphorus pentoxide,[4] thionyl chloride,[5] acetyl chloride,[3, 6, 7] or succinyl chloride;[8] from barium or sodium succinate with benzoyl chloride,[1] acetyl chloride,[9] acetic anhydride,[10] or benzophenone dichloride;[11] from succinyl chloride

[1] Gerhardt and Chiozza, Ann. **87**, 292 (1853).

[2] Volhard, ibid. **242**, 150 (1887).

[3] Verkade and Hartman, Rec. trav. chim. **52**, 947 (1933).

[4] d'Arcet, Ann. chim. phys. (2) **58**, 288 (1835).

[5] Meyer, Monatsh. **22**, 420 (1901).

[6] Anschütz, Ann. **226**, 8 (1884).

[7] Schulz, Ber. **18**, 2459 (1885).

[8] Anschütz, ibid. **10**, 1883 (1877).

[9] Heintz, Jahresb. **1859**, 279.

[10] Oddo and Manuelli, Gazz. chim. ital. **26** (II) 482 (1896).

[11] Evlampiev, J. Gen. Chem. (U.S.S.R.) **7**, 2934 (1937) [C. A. **32**, 5377 (1938)].

and oxalic acid [6] or sodium acetate; [9] and from ethyl succinate and benzoyl chloride.[12]

The procedure in Org. Syn. Coll. Vol. I, **1941,** 91, for preparing benzoic anhydride when applied to succinic acid and acetic anhydride gives a 72 per cent yield of succinic anhydride. Procedure I above has the advantage of convenience; Procedure II is more economical.

SUCCINIMIDE

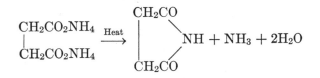

$$\begin{array}{l} CH_2CO_2NH_4 \\ | \\ CH_2CO_2NH_4 \end{array} \xrightarrow{\text{Heat}} \begin{array}{l} CH_2CO \\ \diagdown \\ \diagup \\ CH_2CO \end{array} NH + NH_3 + 2H_2O$$

Submitted by H. T. CLARKE and LETHA DAVIES BEHR.
Checked by W. H. CAROTHERS and W. L. McEWEN.

1. Procedure

IN a 1-l. distilling flask, fitted with a side arm 40 cm. in length and not less than 10 mm. in internal diameter (Note 1), is placed 236 g. (2 moles) of succinic acid. To this is added slowly, with cooling and shaking, 270 cc. (243 g., 4 moles) of 28 per cent aqueous ammonia (sp. gr. 0.90). Most of the acid dissolves, forming a clear solution. The flask is set for downward distillation, and a water-cooled 500-cc. distilling flask is attached to the side arm. Provision may be made for removal of ammonia from the side tube of the receiver. The mixture is heated gently over a free flame; solution takes place rapidly, and a small amount of uncombined ammonia passes over with the first portions of the distillate. The temperature of the vapor rises to 100° and remains at this point until about 200 cc. of water has distilled. The flame is then increased, and the ammonium succinate begins to decompose with evolution of ammonia; the temperature of the vapor falls to 97° during the distillation of the next 30 cc. When the vapor temperature has risen to 102°, the receiver is changed and an intermediate fraction collected from 102° to 275°. Succinimide then distils and is collected over the range 275–289°, largely at 285–289°. Decomposition takes place to a small extent with formation of a black tar; the distillation is stopped when the tarry residue begins to decompose with evolution of yellow fumes.

[12] Kraut, Ann. **137,** 254 (1866).

The crude succinimide, which solidifies completely, amounts to about 168 g. The intermediate fraction is redistilled from a smaller flask and furnishes about 10 g. more of crude succinimide boiling between 275° and 289°. The two portions of crude succinimide are combined and crystallized from 95 per cent ethyl alcohol, employing 1 cc. of solvent for every gram of product. If the mixture is chilled to 0° for some hours before filtration and about 25 cc. of cold alcohol is employed for washing the crystals, the first crop amounts to 163–164 g. (82–83 per cent of the theoretical amount). On concentrating the mother liquor to one-third of its volume, a second crop of 4–5 g. can be secured (Note 2). The product melts at 123–125° and contains no water of crystallization.

2. Notes

1. It is essential to employ a side arm of at least this diameter in order to avoid clogging by crystals of succinimide when this first passes over.

2. A further small quantity of a less pure product can be obtained by evaporating the dark mother liquor to dryness and recrystallizing the residue from fresh 95 per cent alcohol.

3. Methods of Preparation

Succinimide has usually been prepared by heating succinic acid in a current of ammonia [1,2] or by distilling ammonium succinate.[1,3]

TAURINE

I

$$BrCH_2CH_2SO_3Na + NH_3 \rightarrow NH_2CH_2CH_2SO_3H + NaBr$$

Submitted by C. S. MARVEL and C. F. BAILEY.
Checked by FRANK C. WHITMORE and D. J. LODER.

1. Procedure

A SOLUTION of 110 g. (0.52 mole) of sodium 2-bromoethanesulfonate (p. 558) in about 2 l. (28 moles) of concentrated aqueous ammonia (sp. gr. 0.9) is allowed to stand for five to seven days (Note 1) and is then

[1] Fehling, Ann. **49**, 198 (1844).

[2] Franchimont and Friedmann, Rec. trav. chim. **25**, 79 (1906).

[3] Bunge, Ann. Suppl. **7**, 118 (1870); Menschutkin, Ann. **162**, 166 (1872).

evaporated to dryness. The last of the water is removed by heating on a steam bath. The residue is dissolved in the minimum quantity of hot water (about 500 cc.) and, if necessary, treated with 5 g. of Norite. The colorless solution is concentrated to 65–70 cc., and 250 cc. of 95 per cent alcohol is added. In a short time taurine mixed with some sodium bromide separates. When crystallization is complete, the crude taurine is collected on a filter and recrystallized by dissolving in 100 cc. of hot water and then adding to the solution enough 95 per cent ethyl alcohol (about 500 cc.) to give a final concentration of 80 per cent of alcohol. The taurine which separates is usually free from bromides. However, occasional runs must be recrystallized four or five times to remove all the sodium bromide. The yield of pure taurine (Note 2) is 31–36 g. (48–55 per cent of the theoretical amount).

2. Notes

1. The reaction is about 25 per cent complete in five hours, 60 per cent complete in thirty hours, and 90 per cent complete in five days, as indicated by titration of the bromide ion.

2. The purity of the taurine prepared by this method was established by analysis.

II

$$BrCH_2CH_2NH_3Br + Na_2SO_3 \rightarrow NH_2CH_2CH_2SO_3H + 2NaBr$$

Submitted by FRANK CORTESE.
Checked by C. S. MARVEL and C. L. FLEMING.

1. Procedure

A SOLUTION of 615 g. (3 moles) of β-bromoethylamine hydrobromide (p. 91) and 416 g. (3.3 moles) of anhydrous u.s.p. sodium sulfite (Note 1) in 2.4 l. of water is concentrated on the steam bath to a minimum volume; thirty-six to forty-eight hours is required for this operation. After the mixture has cooled, the cold, moist cake is triturated with 1.5 l. of concentrated hydrochloric acid and collected on an asbestos mat in a Büchner funnel. The precipitate is washed ten times with 150-cc. portions of concentrated hydrochloric acid. The filtrate is mixed well, decanted from precipitated salts if necessary, and concentrated over a free flame to a volume of 600 cc.

Two and four-tenths liters of 95 per cent ethyl alcohol is added, with vigorous stirring, to the hot mixture. After fifteen minutes, the product is collected on a filter, washed with 95 per cent ethyl alcohol until it is colorless, and air-dried. The crude material is purified by dissolving it in four times its weight of hot water, adding Norite, filtering, and adding to the hot filtrate five volumes of 95 per cent ethyl alcohol.

This product is practically pure taurine; it decomposes at 305–310° (Maquenne block). The yield is 255–275 g. (68–73 per cent of the theoretical amount).

2. Note

1. An equivalent quantity (831 g.) of the more expensive crystalline sodium sulfite may be used.

3. Methods of Preparation

Taurine is generally prepared from ox bile [1] or the large muscle of the abalone.[2] It has been obtained from the oxidation of cystamine [3] and the decarboxylation of cysteic acid;[4] from ethyleneimine and sulfur dioxide;[5] from chloroethanesulfonic acid and ammonia;[6] from 2-mercaptothiazoline by oxidation with bromine water;[7] and from acetaldehyde by a complex set of reactions involving sulfonation, formation of the aldehyde ammonia and the imidosulfonic acid, and finally reduction.[8] Taurine is usually synthesized either from bromoethanesulfonic acid and ammonia [9]—Procedure I, or from β-bromoethylamine hydrobromide and sodium sulfite [10]—Procedure II.

[1] Hammarsten, Z. physiol. Chem. **32**, 456 (1901); Tauber, Beitr. chem. Physiol. Path. **4**, 324 (1904).

[2] Schmidt and Watson, J. Biol. Chem. **33**, 499 (1918).

[3] Schöberl, Z. physiol. Chem. **216**, 193 (1933).

[4] Friedmann, Beitr. chem. Physiol. Path. **3**, 1 (1903); White and Fishman, J. Biol. Chem. **116**, 457 (1936).

[5] Gabriel, Ber. **21**, 2667 (1888).

[6] Kolbe, Ann. **122**, 42 (1862); Anschütz, ibid. **415**, 97 (1918).

[7] Gabriel, Ber. **22**, 1154 (1889).

[8] Auzies, Rev. gén. chim. **14**, 278 (Chem. Zentr. **1911**, II, 1433).

[9] Marvel, Bailey, and Sparberg, J. Am. Chem. Soc. **49**, 1836 (1927).

[10] Reychler, Bull. soc. chim. Belg. **32**, 247 (1923); Cortese, J. Am. Chem. Soc. **58**, 191 (1936).

TETRAHYDROFURAN

(Furan, tetrahydro-)

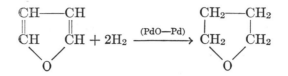

Submitted by DONALD STARR and R. M. HIXON.
Checked by JOHN R. JOHNSON and H. B. STEVENSON.

1. Procedure

(A) Preparation of Palladous Oxide.—In a 350-cc. casserole, 2.2 g. (0.02 gram atom) of palladium metal is dissolved in a small amount of aqua regia, and the solution (Note 1) is treated with 55 g. of c.p. sodium nitrate and enough distilled water to make a thick paste. The substances are thoroughly mixed and then heated gently to drive off the water. The heating is increased until the mixture melts (about 270–280°) and continued cautiously. Just above the melting point the mixture must be stirred and heated carefully as oxides of nitrogen are evolved and foaming occurs. After the evolution of gases is nearly complete (about five minutes) the full flame of a Bunsen burner is applied for about ten minutes. The entire time of heating should be about one-half hour. As the mass cools, the casserole is rotated to allow the melt to solidify on the sides of the dish. After digestion with about 200 cc. of distilled water until the sodium salts are completely dissolved, the dark brown precipitate of palladous oxide is filtered and washed thoroughly with 1 per cent sodium nitrate solution (Note 2). The oxide must not be washed with pure water since it shows a marked tendency to become colloidal. After drying in a vacuum desiccator the palladous oxide weighs 2.3–2.4 g. (91–95 per cent of the theoretical amount) (Note 3).

(B) Tetrahydrofuran.—In the pressure bottle of an apparatus for catalytic reduction (Note 4) are placed 10 g. of pure furan (Note 5) and 0.2 g. of palladous oxide. The bottle is swept out with hydrogen (Note 6), and an initial hydrogen pressure of about 7 atm. (100 lb.) is applied (Note 7). After a lag of about ten minutes the reduction proceeds smoothly, and in an hour the theoretical amount of hydrogen is absorbed; the reaction is noticeably exothermic. After the reaction has ceased 20 g. of furan and 0.2 g. of palladous oxide are added (Note 6), the bottle swept out with hydrogen, and the hydrogen pressure raised to 7 atm.

After this addition the lag is short and the reaction proceeds somewhat more rapidly than before; the temperature rises to 40–50°. When the reaction nears completion, 30 g. of furan and 0.2 g. of palladous oxide are added, and the reduction is continued. Successive portions of 30 g. of furan and 0.2 g. of palladous oxide are added in the same manner, until the bottle is about two-thirds filled. To ensure complete reduction, another portion of palladous oxide is added and the mixture shaken until no more hydrogen is taken up. The catalyst is allowed to settle (Note 8), and the tetrahydrofuran is decanted through a filter into a flask for distillation. The reduction product distils completely at 64–66°.

The reduction of 120 g. (128 cc., 1.76 moles) of furan requires about fifteen to twenty hours, depending upon the purity of the furan and the activity of the catalyst (Note 9). The yield of redistilled tetrahydrofuran is 114–118 g. (90–93 per cent of the theoretical amount). Since reduction is practically quantitative, the yield is determined largely by the care exercised in handling the volatile furan and tetrahydrofuran.

2. Notes

1. An aqueous solution of the equivalent quantity of Eimer and Amend's c.p. or Merck's REAGENT palladium chloride may be used.

2. The filtrates should be clear and colorless; if they show a yellow-orange opalescence, some of the oxide has become colloidal. The palladium may be recovered [1] as the oxide by evaporating the filtrates to dryness and re-fusing, or as palladium black by rendering the filtrates slightly alkaline with sodium carbonate and heating with formaldehyde.

3. A small amount of palladous oxide adheres to the casserole and cannot be removed by the ordinary means. The oxide is not dissolved readily by aqua regia but is easily removed by boiling with 48 per cent hydrobromic acid.

4. If initial pressures of 6–7 atm. are to be applied, the ordinary apparatus for catalytic reduction (Org. Syn. Coll. Vol. I, **1941,** 61) must be modified by using a flexible coil of copper tubing instead of rubber tubing for the connection between the hydrogen tank and the reduction bottle. To avoid dangerous accidents and loss of materials, it is advisable to cover the reduction bottle at all times with a screen of wire mesh and to test the bottle at the higher pressures before use. Brass fittings are used to hold the mouth of the bottle against a rubber gasket; before use the gasket is treated with alkali, washed thoroughly, and dried.

[1] Shriner and Adams, J. Am. Chem. Soc. **46,** 1684 (1924); Kern, Shriner, and Adams, ibid. **47,** 1147 (1925).

5. As in all catalytic reductions, the purity of the starting material is of great importance. Redistilled furan, b.p. 31–32°, prepared by the method of Wilson (Org. Syn. Coll. Vol. I, **1941**, 274) is quite satisfactory. It has been reported that furan prepared by the method of Gilman and Louisinian [2] should be dried over calcium chloride and fractionated carefully.[3] It is advisable to redistil the furan shortly before use and to avoid contact with rubber stoppers.

6. Owing to the high volatility of furan and tetrahydrofuran the bottle is not evacuated, as is customary, before the hydrogen pressure is applied; instead, the air is displaced by hydrogen. For the same reason, appreciable losses will occur if the reduction bottle is not cooled before the hydrogen pressure is released for the introduction of fresh portions of furan and catalyst. Effective and rapid cooling may be obtained by directing a jet of ether, from a wash bottle, over the surface of the reduction bottle while the shaking motor is in operation.

7. The hydrogenation of furan may be carried out with initial pressures of only 3 atm. (45 lb.), but the reduction is slower.

8. The catalyst may be removed, dried in a vacuum desiccator over sulfuric acid, and used again. The second reduction usually proceeds much more slowly, and it is advantageous to use one or two portions of fresh oxide along with the recovered material. Spent catalyst may be regenerated by conversion to the oxide as described in Part (A).

9. In the high-pressure apparatus of Adkins, 120 g. of furan can be hydrogenated with 10 g. of Raney nickel catalyst [4] in a single run. Using pressures of 100–150 atm. and a temperature range of 100–150° the reduction is extremely rapid and is strongly exothermic. Platinum oxide-platinum black is not a satisfactory catalyst for the reduction of furan.[3]

3. Methods of Preparation

The preparation of palladous oxide-palladium black and its use as a catalyst in the reduction of organic compounds have been studied by Shriner and Adams.[1] Palladium black and colloidal palladium have been widely used as hydrogenation catalysts.[5]

Tetrahydrofuran has been prepared by the reduction of furan in the vapor phase with a nickel catalyst at 170°,[6] in butyl alcohol at 50° with

[2] Gilman and Louisinian, Rec. trav. chim. **52**, 156 (1933).

[3] Starr and Hixon, J. Am. Chem. Soc. **56**, 1595 (1934).

[4] Covert and Adkins, ibid. **54**, 4116 (1932).

[5] Sabatier, "Catalysis in Organic Chemistry," transl. by E. Emmet Reid. New York, D. Van Nostrand Company, 1922.

[6] Bourguignon, Bull. soc. chim. Belg. **22**, 88 (1908); Shuĭkin and Bunina, J. Gen Chem. (U.S.S.R.) **8**, 669 (1938) [C. A. **33**, 1316 (1939)].

Raney nickel catalyst,[7] with palladous oxide-palladium black in the absence of a solvent,[3] and with an osmium-asbestos catalyst;[8] by the reduction of an alkyl succinate over a copper chromite catalyst;[9] and by the dehydration of 1,4-butyleneglycol.[10]

α-TETRALONE

[1(2)-Naphthalenone, 3,4-dihydro-]

$$C_6H_5(CH_2)_3CO_2H + SOCl_2 \rightarrow C_6H_5(CH_2)_3COCl + HCl + SO_2$$

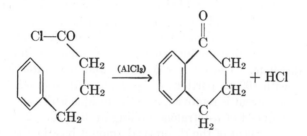

Submitted by E. L. Martin and Louis F. Fieser.
Checked by C. R. Noller.

1. Procedure

In a 500-cc. round-bottomed flask, fitted with a reflux condenser carrying at the top a tube leading to a gas absorption trap, are placed 32.8 g. (0.2 mole) of γ-phenylbutyric acid (p. 499) and 20 cc. (32 g., 0.27 mole) of thionyl chloride (Note 1). The mixture is carefully heated on a steam bath until the acid is melted, and then the reaction is allowed to proceed without the application of external heat. After twenty-five to thirty minutes hydrogen chloride is no longer evolved and the mixture is warmed on the steam bath for ten minutes. The flask is then connected to the water pump, evacuated, and heated for ten minutes on the steam bath and finally for two or three minutes over a small flame in order to

[7] Cloke and Ayers, J. Am. Chem. Soc. 56, 2144 (1934).

[8] Shuĭkin and Chilikina, J. Gen. Chem. (U.S.S.R.) 6, 279 (1936) [C. A. 30, 4855 (1936)].

[9] E. I. du Pont de Nemours and Company, U.S. pat. 2,130,501 [C. A. 32, 9101 (1938)].

[10] Strukov, Khim. Farm. Prom. 1935, No. 1, 35 [C. A. 30, 1769 (1936)]; General Aniline and Film Corporation, U. S. pat. 2,251,292 [C. A. 35, 6982 (1941)]; I. G. Farbenind. A.-G., Ger. pat. 700,036 [C. A. 35, 6982 (1941)].

remove the excess thionyl chloride. The acid chloride thus obtained is a nearly colorless liquid and needs no further purification. The flask is cooled, 175 cc. of carbon disulfide is added, and the solution cooled in an ice bath. Thirty grams (0.23 mole) of aluminum chloride is added rapidly in one lot, and the flask is immediately connected to the reflux condenser. After a few minutes, the rapid evolution of hydrogen chloride ceases and the mixture is slowly warmed to the boiling point on the steam bath. After heating and shaking the mixture for ten minutes the reaction is complete. The reaction mixture is cooled to 0°, and the aluminum chloride complex is decomposed by the careful addition, with shaking, of 100 g. of ice. Twenty-five cubic centimeters of concentrated hydrochloric acid is added and the mixture transferred to a 2-l. round-bottomed flask and steam-distilled (Note 2). The carbon disulfide distils first (Note 3), then there is a definite break in the distillation, after which the reaction product comes over completely in about 2 l. of the next distillate. The oil is separated, and the water is extracted three times with 100-cc. portions of benzene. The oil and extracts are combined, the solvent is removed, and the residue is distilled at reduced pressure. The yield of α-tetralone boiling at 105–107°/2 mm. is 21.5–26.5 g. (74–91 per cent of the theoretical amount based on the γ-phenylbutyric acid).

2. Notes

1. The thionyl chloride was purified by distilling 50 g. of commercial thionyl chloride from 10 cc. of quinoline and then from 20 cc. of boiled linseed oil in a system protected from moisture.

2. It is advisable to use an efficient condenser system, such as that described in Org. Syn. Coll. Vol. I, **1941**, 479, as α-tetralone is only moderately volatile with steam.

3. On recovery of the carbon disulfide there is not more than a trace of residue.

3. Methods of Preparation

α-Tetralone has been obtained from the catalytic hydrogenation of α-naphthol;[1] from γ-phenylbutyryl chloride and aluminum chloride;[2] from γ-phenylbutyric acid and concentrated sulfuric acid;[3] and by the oxidation of tetralin with chromic anhydride[4] or with atmospheric

[1] Schroeter, Ger. pat. 352,720 [C. A. **17**, 1245 (1923)].

[2] Kipping and Hill, J. Chem. Soc. **75**, 147 (1899); Amagat, Bull. soc. chim. (4) **41**, 940 (1927).

[3] Krollpfeiffer and Schäfer, Ber. **56**, 624 (1923); Horne and Shriner, J. Am. Chem. Soc. **55**, 4652 (1933); Cook and Hewett, J. Chem. Soc. **1934**, 373.

[4] Schroeter, Ger. pat. 346,948 [Frdl. **14**, 491 (1921–25)].

oxygen.[5] Detailed directions for preparing α-tetralone by this last-named method are given in Org. Syn. **20, 94.**

TETRAMETHYLENE CHLOROHYDRIN

(1-Butanol, 4-chloro-)

$$\begin{array}{cc} \text{CH}_2\!\!-\!\!\text{CH}_2 \\ | \quad\ \ | \\ \text{CH}_2 \quad \text{CH}_2 \\ \diagdown\ \diagup \\ \text{O} \end{array} + \text{HCl} \rightarrow \text{ClCH}_2\text{CH}_2\text{CH}_2\text{CH}_2\text{OH}$$

Submitted by Donald Starr and R. M. Hixon.
Checked by John R. Johnson and H. B. Stevenson.

1. Procedure

A 500-cc. three-necked flask containing 114 g. (1.58 moles) of tetrahydrofuran (p. 566) is fitted with a reflux condenser, a thermometer dipping into the liquid, and a bent glass tube arranged to introduce gaseous hydrogen chloride (Note 1) near the bottom of the flask. The upper end of the reflux condenser is connected to a 150-cc. distilling flask cooled in an ice-salt mixture to trap material entrained by the hydrogen chloride.

The tetrahydrofuran is heated to the boiling point (64–65°), and a slow stream of hydrogen chloride is bubbled into the liquid. As the reaction proceeds the temperature of the boiling liquid increases, slowly at first and then more rapidly, until it is above 100° (after about four hours' heating). At the end of about five hours the temperature remains practically constant in the range 103.5–105.5°, and the reaction is stopped. The light brown liquid is cooled, transferred to a 250-cc. Claisen flask having a 20-cm. fractionating side arm, and fractionated at reduced pressure, using a water aspirator. A large quantity of hydrogen chloride is evolved at the start of the distillation and a low pressure cannot be obtained until this has been removed. Throughout the fractionation a trap cooled to −15° in an ice-salt mixture is used to collect the recovered tetrahydrofuran.

After removal of a small amount of low-boiling material the main fraction distils in the range 80–90°/14 mm. or 65–75°/7 mm. (Note 2)

[5] I. G. Farbenind. A.-G., Ger. pat. 539,476 [C. A. **26,** 1614 (1932)]; Cook, J. Chem. Soc. **1938,** 1778; Hock and Susemihl, Ber. **66,** 61 (1933); Brown, Widiger, and Letang, J. Am. Chem. Soc. **61,** 2601 (1939).

and weighs 95–100 g. A small amount (5–10 g.) of high-boiling material remains. The crude product on refractionation yields 93–98 g. (54–57 per cent of the theoretical amount) of pure tetramethylene chlorohydrin boiling over a one-degree interval, 81–82°/14 mm. or 70–71°/7 mm. (Notes 3 and 4).

2. Notes

1. Hydrogen chloride prepared by dropping concentrated sulfuric acid into a mixture of sodium chloride and concentrated hydrochloric acid may be used directly without drying.

2. It has been reported [1] that tetramethylene chlorohydrin undergoes loss of hydrogen chloride when distilled at pressures appreciably above 15 mm. If an oil pump is used for the distillation of the main fraction, it should be protected from hydrogen chloride by means of soda-lime towers.

3. For recovery of tetrahydrofuran, the condensate from the cooling traps and the low-boiling material from the fractionations are combined, cooled in an ice bath, and treated carefully with 15–20 cc. of 40 per cent alkali. The upper layer is separated, dried with a little calcium chloride, and distilled. The recovered tetrahydrofuran, b.p. 64–67°, weighs 20–22 g. (17–19 per cent of the original material). The residue (12–14 g.) remaining after distillation of the tetrahydrofuran distils at 43–45°/10 mm. and is tetramethylene dichloride.

4. Tetramethylene chlorohydrin may be converted to the chlorobromide in excellent yields by the action of phosphorus tribromide.[2]

3. Methods of Preparation

Tetramethylene chlorohydrin has been prepared by the action of thionyl chloride on tetramethylene glycol in the presence of pyridine,[1] and by the method described above.[2]

[1] Kirner and Richter, J. Am. Chem. Soc. **51**, 2503 (1929).

[2] Starr and Hixon, ibid. **56**, 1595 (1934).

THIOBENZOPHENONE

(Benzophenone, thio-)

$$CH_3CH_2ONa + H_2S \rightarrow C_2H_5OH + NaSH$$
$$C_6H_5CCl_2C_6H_5 + 2NaSH \rightarrow C_6H_5CSC_6H_5 + 2NaCl + H_2S$$

Submitted by H. Staudinger and H. Freudenberger.
Checked by Roger Adams and E. H. Woodruff.

1. Procedure

An alcoholic solution of sodium hydrosulfide is prepared by dissolving 4.6 g. (0.2 gram atom) of sodium in 150 cc. of absolute alcohol and saturating the solution with dry hydrogen sulfide.

A 3-l. three-necked flask is equipped with a reflux condenser, a mechanical stirrer, a dropping funnel, and a tube for introducing dry carbon dioxide. Twenty-five grams (0.11 mole) of benzophenone dichloride (Note 1) is put into the flask, the air in the flask is displaced by a stream of carbon dioxide, and the sodium hydrosulfide solution is added slowly (Note 2). A vigorous reaction, which must be controlled by cooling, takes place, and the reaction mixture becomes deep blue in color.

After the reaction mixture has stood for one-half hour, water is added and the solution is extracted with ether. The ether solution is dried over calcium chloride and the ether is distilled. The residue is distilled under reduced pressure in an atmosphere of carbon dioxide (Note 3). Thiobenzophenone boils at 174°/14 mm.; it distils as a blue oil which, if pure and dry, forms beautiful blue crystals on cooling. The yield of crude product is 10–12.5 g. (50–63 per cent of the theoretical amount) (Note 4).

This product is approximately 75 per cent pure and is purified further by recrystallization from petroleum ether (b.p. 70–90°). The yield of purified product melting at 53–54° is 8.4–9.9 g. (42–50 per cent of the theoretical amount).

For preservation, the thiobenzophenone is sealed in a glass tube with dry carbon dioxide and placed in the dark.

2. Notes

1. The benzophenone dichloride was prepared by heating equivalent molecular quantities of benzophenone and phosphorus pentachloride to 145–150° for two hours and then fractionating the mixture under reduced

pressure. The product used for this preparation boiled at 201–202°/35 mm.

2. It is necessary to have an excess of the chloride present at all times in order to prevent the formation of dibenzohydryldisulfide, due to reduction of the thioketone with the hydrosulfide. If the chloride is added to the sodium hydrosulfide solution, a 70 per cent yield of the pure disulfide is obtained and no thioketone is formed.

3. All the operations in the purification should be done very quickly and out of contact with the air as much as possible.

4. Larger runs using 100 g. of benzophenone dichloride gave the same percentage yields of product.

3. Methods of Preparation

Thiobenzophenone has been prepared by the action of thiophosgene on benzene in the presence of aluminum chloride;[1] by the action of phosphorus pentasulfide[2] or ethyl thioacetoacetate[3] on benzophenone; and by treating benzophenone dichloride with alcoholic potassium sulfide,[4] alcoholic sodium hydrosulfide,[5] or thioacetic acid.[6]

p-THIOCYANODIMETHYLANILINE

(Aniline, N,N-dimethyl-p-thiocyano-)

$$C_6H_5N(CH_3)_2 + Br_2 + 2NH_4SCN$$
$$\rightarrow 2NH_4Br + p\text{-}NCSC_6H_4N(CH_3)_2 + HSCN$$

Submitted by R. Q. Brewster and Wesley Schroeder.
Checked by W. H. Carothers.

1. Procedure

A solution of 60.5 g. (0.5 mole) of dimethylaniline and 80 g. (1.05 moles) of ammonium thiocyanate in 250 cc. of glacial acetic acid, contained in a 1-l. beaker, is cooled to 10–20° in a bath of ice and water. The solution is stirred mechanically while a solution of 80 g. (25.7 cc., 0.5

[1] Bergreen, Ber. 21, 341 (1888).
[2] Gattermann, ibid. 28, 2877 (1895).
[3] Mitra, J. Indian Chem. Soc. 9, 637 (1932) [C. A. 27, 3922 (1933)].
[4] Gattermann and Schulze, Ber. 29, 2944 (1896).
[5] Staudinger and Freudenberger, ibid. 61, 1577 (1928).
[6] Schönberg, Schütz, and Nickel, ibid. 61, 1378 (1928).

mole) of bromine in 100 cc. of glacial acetic acid is added dropwise, over a period of twenty to thirty minutes, and the temperature is kept below 20° (Notes 1 and 2). After all the bromine has been added the reaction mixture is removed from the cooling bath and, after standing at room temperature for ten minutes, is poured into 5–6 l. of water. Most of the *p*-thiocyanodimethylaniline separates as a pale yellow solid (Note 3), which is collected on a suction filter and washed with water. After drying in air it weighs 50–55 g. and melts at 71–73°. An additional 10–15 g. of less pure product is obtained by making the filtrate alkaline to litmus; this requires about 1250 cc. of 20 per cent sodium hydroxide. The two fractions are combined, dissolved in about 1.2 l. of boiling ligroin (b.p. 90–100°), and filtered rapidly through a large fluted filter in a heated funnel. The product separates from the filtrate in the form of long yellow needles, and crystallization is completed by thorough chilling. The melting point of the purified product is 73–74°, and the total yield is 56–60 g. (63–67 per cent of the theoretical amount) (Note 4).

2. Notes

1. At higher temperatures a considerable quantity of a yellow thiocyanogen polymer is formed which contaminates the product.

2. Toward the end of the addition a heavy precipitate begins to accumulate on the walls of the beaker; this should be dislodged occasionally with a spatula.

3. *p*-Thiocyanodimethylaniline is a weak base, and its salts are easily hydrolyzed.

4. An additional 4–5 g. of low-melting product may be recovered by evaporation of the mother liquor.

3. Methods of Preparation

p-Thiocyanodimethylaniline has been prepared from dimethylaniline and thiocyanogen,[1] chlorothiocyanogen,[2] ammonium thiocyanate and bromine,[3,4] or lead thiocyanate and iodobenzene dichloride.[5] Thiocyanation of aromatic amines and phenols has also been accomplished electrolytically.[4,6]

[1] Söderbäck, Ann. **419**, 275 (1919).

[2] Lecher and Joseph, Ber. **59**, 2603 (1926).

[3] Likhosherstov and Petrov, J. Gen. Chem. (U.S.S.R.) **3**, 183 (1933) [C. A. **28**, 1677 (1934)].

[4] Helwig, U. S. pat. 1,816,848 [C. A. **25**, 5355 (1931)].

[5] Neu, Ber. **72**, 1505 (1939).

[6] Fichter and Schönmann, Helv. Chim. Acta **19**, 1411 (1936).

β-THIODIGLYCOL

(Ethanol, 2,2'-thiodi-)

$$2HOCH_2CH_2Cl + Na_2S \rightarrow HOCH_2CH_2SCH_2CH_2OH + 2NaCl$$

Submitted by E. M. FABER and G. E. MILLER.
Checked by W. L. McEWEN and W. H. CAROTHERS.

1. Procedure

IN a 3-l. round-bottomed flask equipped with a mechanical stirrer are placed 1.5 kg. (3.7 moles) of 20 per cent ethylene chlorohydrin solution (Note 1) and 750 g. of water. The flask is set in an empty pan of suitable size to serve as a bath in case cooling becomes necessary. With the stirrer in operation, 493 g. (2.05 moles) of crystalline sodium sulfide containing nine molecules of water of crystallization is added to the chlorohydrin solution at a rate which will maintain the temperature at 30–35°. This will require from forty to sixty minutes. After all the sodium sulfide has been added the solution is stirred for thirty minutes.

The stirrer is removed, and the flask is fitted with a reflux condenser and a thermometer which dips into the liquid. The flask is then heated on a steam bath until the temperature of the liquid is 90°, and for a period of forty-five minutes the temperature is held at 90–95°. The solution is then cooled to 25° and neutralized to turmeric paper by adding concentrated hydrochloric acid drop by drop (Note 2). After filtering, the solution is returned to the flask for concentration at reduced pressure.

The flask is equipped with a short column attached to a condenser set for distillation. A capillary is provided to prevent bumping. The water is then distilled at a pressure of 30–40 mm. by heating the flask in a water bath which is raised to the boiling point as rapidly as is consistent with smooth distillation. The residue in the flask, which consists of sodium chloride and thiodiglycol, is extracted twice with 500-cc. portions of hot absolute alcohol in order to dissolve the sulfide. After the second extraction, the salt is transferred to a Büchner funnel and is washed with a little hot alcohol (Note 3).

The extract and washings are returned to the distilling flask, and the alcohol is removed under reduced pressure. When practically all the alcohol has distilled, the temperature of the bath is raised to 100° and the residue is heated for three hours under 30-mm. pressure (Note 4).

The crude product, which is colorless or very pale yellow, weighs

200–215 g. It boils at 164–166°/20 mm. and may be purified by vacuum distillation (Note 5). The yield of pure material is 180–195 g. (79–86 per cent of the theoretical amount).

2. Notes

1. Aqueous solutions of ethylene chlorohydrin of 18–40 per cent are suitable for the preparation of thiodiglycol; a 20 per cent solution is convenient because the reaction proceeds very smoothly and is easy to control.

If the chlorohydrin is available in the form of a solution weaker than 20 per cent, it may be concentrated by distillation. Chlorohydrin and water form a constant-boiling-point mixture of 42.5 per cent chlorohydrin which boils at 95.8°/735 mm.

2. At the end of the reaction the liquid is alkaline and must be neutralized; otherwise considerable decomposition occurs during distillation. Care must be taken not to pass the neutral point, as a small amount of mustard gas may be formed. Furthermore, if much acid is present, the heat necessary for vacuum distillation causes resinification and the yield of distilled material falls to about 50 per cent. The use of litmus paper for the neutralization is not satisfactory.

3. If 95 per cent alcohol is used for this extraction, some salt is dissolved with the thiodiglycol. This salt may be filtered easily after the alcohol has been removed from the product.

4. The time of drying is dependent upon the pressure used. At 20 mm. the water and alcohol are removed in one hour. If the water has not been completely removed before the alcohol extraction, there may be a small amount of salt left in the material after the alcohol is removed. This may be removed by decanting the product or by pouring it through a glass-wool filter.

5. If chemicals of good quality are used in the preparation the crude product is practically water-white and is sufficiently pure for many purposes. A completely pure product can be obtained by vacuum distillation.

3. Methods of Preparation

The method described in the procedure is a modification of the one originally described by Meyer.[1] Irvine [2] showed that this general method could be adapted to works-scale production. β-Thiodiglycol can also

---

[1] Meyer, Ber. **19,** 3260 (1886); Clarke, J. Chem. Soc. **101,** 1583 (1912); Gomberg, J. Am. Chem. Soc. **41,** 1414 (1919).

[2] Irvine, British War Report.

be prepared from ethylene oxide and hydrogen sulfide, a process that works well only if some of the product is used as a solvent for the reactants.[3]

THIOPHENE

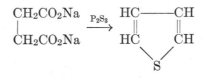

Submitted by Ross Phillips.
Checked by Roger Adams and H. D. Cogan.

1. Procedure

In a 3-l. round-bottomed flask is placed an intimate mixture of 486 g. (3 moles) of finely powdered anhydrous sodium succinate (Note 1) and 648 g. (4.1 moles) of finely ground phosphorus trisulfide (Note 2). The flask is fitted to a 100-cm. (40-in.) condenser set for distillation (Note 3), and a tube for introduction of carbon dioxide is extended through the stopper to the center of the flask. The condenser is connected to a 2-l. flask cooled in an ice-salt mixture. The uncondensed gases are bubbled through two 2-l. flasks connected in series with the receiving flask; each flask is cooled by an ice-salt mixture and contains 1 kg. of cracked ice and 200 cc. of 40 per cent sodium hydroxide solution.

The reaction flask is thoroughly swept out with carbon dioxide (Note 4) while the flask is rotated to remove any air that is trapped by the solid. The flask is then connected to the condenser, and a slow stream of carbon dioxide is passed through the system while the mixture is heated moderately for thirty minutes with a low Bunsen flame (Note 5), and then with the full force of the Bunsen burner until no more yellow vapors are produced (about thirty minutes). Carbon dioxide is passed through more rapidly during the latter heating period to ensure complete removal of the thiophene (Note 6). The contents of the receiver and the two absorption flasks are combined and steam-distilled from a 5-l. flask until no more oily drops are formed in the distillate. The thiophene layer in the distillate is separated, dried successively over solid sodium hydroxide and sodium, and fractionated. The yield of product boiling

[3] Chichibabin, Fr. pat. 769,216 [C. A. **29**, 481 (1935)]; Chichibabin and Bestuzhev, Compt. rend. **200**, 242 (1935); Nenitzescu and Scărlătescu, Ber. **68**, 587 (1935); Othmer and Kern, Ind. Eng. Chem. **32**, 160 (1940).

at 83–86° is 63–75 g. (25–30 per cent of the theoretical amount) (Notes 7 and 8).

2. Notes

1. The sodium succinate, prepared from succinic acid, was dried in shallow pans on a water bath for several days.

2. If phosphorus trisulfide is not obtainable, it may be prepared by the following method developed by A. M. Griswold: An intimate mixture of the calculated amounts of finely powdered sulfur and red phosphorus is placed in an earthenware flower pot, sealed at the bottom with a cork. The pot is imbedded in a bucket of sand, and a heavily weighted cover is held in readiness over the top. A lighted match is dropped into the mixture, the lid is quickly put in place, and the crevices are sealed with sand. The reaction should be carried on out-of-doors as it is extremely vigorous and often gives an excellent display of fireworks. When thoroughly cool, the flower pot is broken and the contents are removed and kept in a tightly stoppered bottle. It is desirable to keep this material for some time before grinding, as the freshly prepared product is not completely crystallized and is difficult to grind.

3. Because of the rapid evolution of a large volume of gas, the tube connecting the flask to the condenser should be about 2 cm. in diameter and all stoppers should fit tightly.

4. If the air is not completely removed before heating the flask, the reaction may take place with explosive violence.

5. The checkers found that continual heating was necessary during this part of the preparation.

6. The stopper should not be removed from the reaction flask until the flask is quite cool, in order to prevent ignition of the excess phosphorus trisulfide on contact with the air. The hard mass remaining may be softened by treatment with hot water, after which it may be broken up and shaken out.

7. The freshly prepared product often deposits a small amount of a brown precipitate on standing for some time. If this is removed by filtration, no further precipitate is formed.

8. Larger runs have been made successfully, but there is greater danger of a violent reaction.

3. Methods of Preparation

Thiophene is found in small amounts in coal gas and benzene.[1] It has been prepared by passing ethylene or acetylene into boiling sulfur;[2] by

[1] Meyer, Ber. **16,** 1471 (1883); **17,** 2642 (1884).
[2] Meyer and Sandmeyer, ibid. **16,** 2176 (1883).

passing ethyl sulfide through a hot tube; [3] by passing ethylene or illuminating gas over hot pyrites; [4] by heating succinic anhydride with phosphorus pentasulfide; [5] by treatment of erythritol with phosphorus pentasulfide; [6] by treatment of succinic aldehyde with phosphorus trisulfide; [7] by passing acetylene over pyrites at 300°; [8] by treatment of sodium succinate with phosphorus trisulfide; [9] by passing acetylene and hydrogen sulfide over bauxite at 320°, or over nickel hydroxide at 300°; [10] and by passing furan and hydrogen sulfide over heated alumina.[11]

THIOSALICYLIC ACID

(Benzoic acid, *o*-mercapto-)

$$\text{C}_6\text{H}_4(\text{CO}_2\text{H})(\text{NH}_2) \xrightarrow[\text{NaNO}_2]{\text{HCl}} o\text{-HO}_2\text{CC}_6\text{H}_4\text{N}_2\text{Cl}$$

$$2o\text{-HO}_2\text{CC}_6\text{H}_4\text{N}_2\text{Cl} \xrightarrow{\text{Na}_2\text{S}_2} o\text{-HO}_2\text{CC}_6\text{H}_4\text{SSC}_6\text{H}_4\text{CO}_2\text{H-}o'$$

$$o\text{-HO}_2\text{CC}_6\text{H}_4\text{SSC}_6\text{H}_4\text{CO}_2\text{H-}o' \xrightarrow{\text{(Zn + CH}_3\text{CO}_2\text{H)}} 2 \, \text{C}_6\text{H}_4(\text{CO}_2\text{H})(\text{SH})$$

Submitted by C. F. H. ALLEN and D. D. MACKAY.
Checked by ROGER ADAMS and A. E. KNAUF.

1. Procedure

In a 4-l. beaker, 290 cc. of water is heated to boiling, and 260 g. (1.1 moles) of crystallized sodium sulfide ($\text{Na}_2\text{S}\cdot9\text{H}_2\text{O}$) and 34 g. of powdered sulfur are dissolved by heating and stirring. A solution of 40 g. of sodium hydroxide in 100 cc. of water is then added and the

[3] Kekulé, ibid. **18**, 217 (1885).

[4] Nahnsen, ibid. **18**, 217 (1885.).

[5] Volhard and Erdmann, ibid. **18**, 454 (1885).

[6] Paal and Tafel, ibid. **18**, 689 (1885).

[7] Harris, ibid. **34**, 1496 (1901).

[8] Steinkopf and Kirchoff, Ann. **403**, 5 (1914); Ger. pat. 252,375 [C. A. **7**, 538 (1913)]; Steinkopf, Chem. Ztg. **35**, 1098 (1911).

[9] Volhard and Erdmann, Ber. **18**, 454 (1885); Friedburg, J. Am. Chem. Soc. **12**, 85 (1890).

[10] Stuer and Grob, U. S. pat. 1,421,743 [C. A. **16**, 3093 (1922)].

[11] Yur'ev, Ber. **69**, 440 (1936).

mixture cooled, first in cold water, and finally by a freezing mixture of ice and salt.

In a 2-l. beaker, set in a freezing mixture and provided with a stirrer and a thermometer for reading temperatures to 0°, are placed 500 cc. of water, 137 g. (1 mole) of anthranilic acid, and 200 cc. of concentrated hydrochloric acid; the stirrer is started and the mixture cooled to about 6°. Meanwhile 69 g. (1 mole) of sodium nitrite is dissolved in 280 cc. of hot water and the solution cooled in ice; portions are then placed in a separatory funnel of convenient size, supported in such a way that the lower end of the stem extends beneath the surface of the anthranilic acid solution. When the temperature has fallen to 5°, the nitrite solution is run in; about 500 g. of cracked ice is added at such a rate as to keep the temperature below 5°. This takes about ten minutes (Note 1). A drop of the solution should give an immediate blue color with starch-iodide paper.

The stirrer and thermometer are now transferred to the alkaline sulfide solution, the temperature of which must be below 5°. The diazo solution is added over a period of twenty to thirty minutes along with 950 g. of ice to prevent the temperature from rising above 5°. When addition is complete, the water bath is removed and the mixture allowed to warm up to room temperature; after two hours the evolution of nitrogen ceases (Note 2). About 180 cc. of concentrated hydrochloric acid is added until the solution is acid to Congo red paper, and the precipitate of dithiosalicylic acid is filtered and washed with water.

To remove the excess sulfur, the precipitate is dissolved by boiling with a solution of 60 g. of anhydrous sodium carbonate (soda ash) in 2 l. of water, and the mixture is filtered while hot. It is divided into five equal parts (Note 3), and the dithiosalicylic acid is reprecipitated as before with concentrated hydrochloric acid. The solid is filtered, the cake being sucked as dry as possible.

The moist cake is mixed with 27 g. of zinc dust and 300 cc. of glacial acetic acid in a 1-l. round-bottomed flask, and the mixture is refluxed vigorously for about four hours (Note 4). When the reduction is complete, the mixture is cooled and filtered with suction. The filter cake is washed once with water and then transferred to a 1-l. beaker. The cake is suspended in 200 cc. of water, and the suspension is heated to boiling. The hot solution is made strongly alkaline by the addition of about 40 cc. of 33 per cent aqueous sodium hydroxide solution. The alkaline solution is boiled for about twenty minutes to ensure complete extraction of the product from the filter cake, filtered from the insoluble material (Note 5), and the thiosalicylic acid is then precipitated by the addition of sufficient concentrated hydrochloric acid to make the solution acid to

Congo red paper. The product is filtered with suction, washed once with water, and dried in an oven at 100–110°. The yield of a product which melts at 162–163° is 110–130 g. (71–84 per cent of the theoretical amount based on the anthranilic acid).

This product is sufficiently pure for most purposes (Note 6).

For recrystallization 5 g. of this material is dissolved in 20 cc. of hot 95 per cent alcohol, and 40 cc. of water is added. The solution is boiled with a little decolorizing carbon, filtered hot, and then allowed to cool. The product crystallizes in yellow flakes. The yield of recrystallized material is 4.7 g.; the melting point of the material is 163–164°.

2. Notes

1. This method is much more rapid than when external cooling alone is used (Org. Syn. Coll. Vol. I, **1941,** 374). The total volume of the solution is not important since the insoluble dithiosalicylic acid is readily filtered.

2. Foaming sometimes becomes very serious during the evolution of nitrogen. The addition of a few cubic centimeters of ether from time to time helps to keep this foaming under control.

3. The dithiosalicylic acid may be precipitated all at once if desired and the entire amount reduced in one operation. If this is done, the reduction must be carried out in a 5-l. flask fitted with a good stirrer. The mixture needs to be refluxed about ten hours over a ring burner. In the laboratory, this is much less convenient than it is to divide the material and reduce in smaller amounts. The yield is not materially lowered by making the reduction in one portion.

4. The reduction does not always run smoothly. If the zinc lumps and becomes inactive more must be added. To determine whether reduction is complete, a sample is removed, cooled, and filtered. The precipitate is boiled with strong sodium hydroxide solution, filtered, and then acidified with hydrochloric acid. If the reduction is complete, the precipitated material will melt at 164° or lower. If the reduction is not complete, the precipitated material will melt above 164°. If the reduction is not complete, the refluxing of the main portion must be continued (and perhaps more zinc must be added) until a test portion shows that the reaction is complete.

In determining the melting point of the material, the capillary tube containing the test sample should be inserted in a bath previously heated to 163–164°.

5. When the reduction is carried out in five portions, one extraction with sodium hydroxide is usually sufficient for each portion. If the

reduction is carried out in one operation, several extractions are usually required. When the material is to be extracted more than once, it is best to boil the residue from the first alkaline treatment with hydrochloric acid, filter, and then treat again with the alkali.

6. Thiosalicylic acid is used for the preparation of oxythionaphthene and many thioindigoid dyes.

3. Methods of Preparation

Of the several methods described for the production of thiosalicylic acid, only the following are of preparative interest: heating *o*-halogenated benzoic acids with an alkaline hydrosulfide at 150–200° in the presence of copper or copper salts,[1, 2] or with sodium sulfide at 200°;[3] and reduction of dithiosalicylic acid with glucose,[4] or metals [5, 2] in alkaline solution. The dithiosalicylic acid is prepared by treating diazotized anthranilic acid with sodium disulfide in alkaline solution.[5]

p-TOLUALDEHYDE

$$CO + HCl \xrightarrow{(CuCl)} HCOCl$$

$$CH_3C_6H_5 + HCOCl \xrightarrow{(AlCl_3)} p\text{-}CH_3C_6H_4CHO + HCl$$

Submitted by G. H. COLEMAN and DAVID CRAIG.
Checked by HENRY GILMAN and J. B. DICKEY.

1. Procedure

The apparatus shown in Fig. 18 is set up in a hood. The narrow reaction bottle A, of about 500-cc. capacity and having a wide mouth, is provided with an efficient mercury-sealed mechanical stirrer, an inlet tube B for admitting the mixture of gases, and an outlet tube connected with the wash bottle E. Into A, contained in a water bath at 20°, is placed 200 g. (2.17 moles) of dry toluene; then 30 g. (0.3 mole CuCl) of cuprous chloride (Note 1) and 267 g. (2 moles) of finely

[1] (a) Cassella and Company, Ger. pat. 189,200 [C. A. **2**, 607 (1908)]; (b) Cain, "Intermediate Products for Dyes," p. 151.

[2] Chem. Age **21**, Dyestuffs Suppl. p. 11 (1929).

[3] Cassella and Company, Ger. pat. 193,290 [C. A. **2**, 1514 (1908)]; Ref. 1(b).

[4] Claasz, Ber. **45**, 2427 (1912).

[5] Kalle and Company, Ger. pat. 204,450 [C. A. **3**, 1695 (1909)]; Ref. 1(b).

powdered anhydrous aluminum chloride are added rapidly with active stirring.

A mixture of hydrogen chloride and carbon monoxide (Note 2) is led to the bottom of the reaction bottle through tube B at such rates that

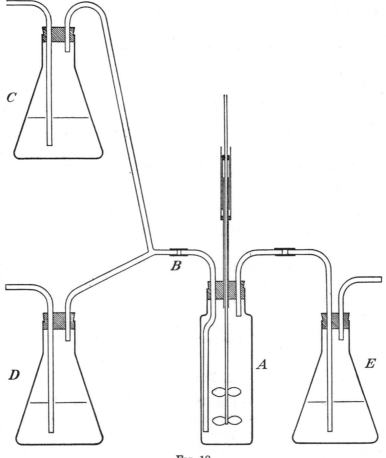

Fig. 18

the carbon monoxide is admitted uniformly during seven hours and the hydrogen chloride at about one-half this rate. The rates of flow of the two gases may be compared by the bubbling in wash bottles C and D. The extent of absorption can be estimated from the bubbling of the effluent gases in wash bottle E. The carbon monoxide is absorbed almost quantitatively at the beginning, and as the mixture thickens the absorption becomes less complete.

The mixture is hydrolyzed by adding it (Note 3) slowly and with shaking to 1.5 kg. of cracked ice in a 3-l. flask. The resulting mixture is then steam-distilled until all the aldehyde and unchanged toluene are driven over. Fifty cubic centimeters of ether is added to the distillate, the two layers are separated, and the aqueous layer is washed with 150 cc. of ether which is then added to the first non-aqueous layer (Note 4). After drying with calcium chloride, the aldehyde is obtained by distilling from a 500-cc. flask provided with a short fractionating column (Note 5). The yield of *p*-tolualdehyde distilling at 201–205° is 121–132 g. (46–51 per cent of the theoretical amount based on the toluene). Redistillation involves but a slight loss and gives an almost colorless product which distils at 203–205° (Note 6).

2. Notes

1. The cuprous chloride may be prepared according to Org. Syn. Coll. Vol. I, **1941**, 170. The precipitate is washed several times by decantation with sulfurous acid, collected on a large Büchner funnel, washed with glacial acetic acid, and dried in an evaporating dish until the odor of acetic acid has disappeared. The cuprous chloride obtained is pure white and should not be exposed unnecessarily to the air.

An alternative procedure for preparing cuprous chloride is given on p.131.

2. The carbon monoxide may be forced by water from a 75-l. container, after which it is dried by bubbling through two wash bottles containing concentrated sulfuric acid.

In place of this large container for carbon monoxide it is possible to generate the gas more conveniently by adding 170 g. (141 cc., 3.69 moles) of pure formic acid (sp. gr. 1.2) to 250 g. (135.8 cc.) of concentrated sulfuric acid contained in a 1-l. distilling flask heated on an oil bath to 70–80°. The side arm of the flask is connected to the sulfuric acid wash bottles. In order to ensure a smooth and constant flow of carbon monoxide, the formic acid is placed in a dropping funnel which extends through a two-holed rubber stopper in the neck of the distilling flask. The other hole of this stopper is connected by rubber tubing to a one-holed stopper in the dropping funnel. No frothing was evident under these conditions, but should frothing occur it is recommended that a small quantity of paraffin be added to the sulfuric acid in the generating flask.

3. The mixture frequently is too thick to be poured, and a spatula must be used to remove it.

4. The extraction of the aldehyde from the unchanged toluene with sodium bisulfite at this point in the preparation does not appreciably increase the purity of the product.

5. It is necessary to use a water-condenser until the temperature reaches 150°, after which an air-condenser is used.

6. The use of a few crystals of hydroquinone has been recommended for the storage of p-tolualdehyde.

3. Methods of Preparation

The method described is that of Gattermann and Koch.[1] p-Tolualdehyde has also been prepared from toluene, hydrogen cyanide, hydrogen chloride, and aluminum chloride;[2] from toluene, nickel carbonyl, and aluminum chloride;[3] from p-tolunitrile by reduction with stannous chloride followed by hydrolysis of the aldimine;[4] from p-xylene by oxidation, particularly with chromyl chloride;[5] and from p-tolylmagnesium bromide with formaldehyde phenylhydrazone,[6] ethyl orthoformate, methylformanilide, ethoxymethyleneaniline, or carbon disulfide. The syntheses involving the Grignard reagent, except the first mentioned, have been examined and evaluated by Smith and Bayliss[7] and Smith and Nichols:[7] the use of ethoxymethyleneaniline or of ethyl orthoformate is recommended.

o-TOLUAMIDE

$$2o\text{-}CH_3C_6H_4CN + 2H_2O_2 \xrightarrow{\text{(NaOH)}} 2o\text{-}CH_3C_6H_4CONH_2 + O_2$$

Submitted by C. R. NOLLER.
Checked by W. W. HARTMAN and L. A. SMITH.

1. Procedure

IN a 2-l. round-bottomed flask are placed 88 g. (0.75 mole) of o-tolunitrile (Org. Syn. Coll. Vol. I, **1941**, 514), 300 cc. (2.6 moles) of 30 per cent hydrogen peroxide, 400 cc. of 95 per cent alcohol, and 30 cc. of

[1] Gattermann and Koch, Ber. **30**, 1622 (1897).

[2] Bayer and Company, Ger. pat. 99,568 [Frdl. **5**, 98 (1897–1900)]; Hinkel, Ayling, and Morgan, J. Chem. Soc. **1932**, 2796; Hinkel, Ayling, and Beynon, ibid. **1935**, 677.

[3] Dewar and Jones, J. Chem. Soc. **85**, 216 (1904).

[4] Stephen, ibid. **127**, 1874 (1925); Williams, J. Am. Chem. Soc. **61**, 2248 (1939).

[5] Law and Perkin, J. Chem. Soc. **91**, 258 (1907).

[6] Grammaticakis, Compt. rend. **210**, 303 (1940).

[7] Smith and Bayliss, J. Org. Chem. **6**, 437 (1941); Smith and Nichols, ibid. **6**, 489 (1941).

6 N sodium hydroxide solution (Note 1). The mixture evolves oxygen and soon warms up owing to the heat of reaction; the temperature is kept at 40–50° by external cooling (Note 2). After about one hour, heat is no longer evolved; the temperature is then maintained at 50° by external heating for an additional three hours. The mixture, while still warm, is made exactly neutral to litmus with 5 per cent sulfuric acid and distilled with steam until 1 l. of distillate is collected. The residue, which has a volume of about 600 cc. (Note 3), is poured while hot into a 1-l. beaker and cooled to 20°. The crystals which form are filtered with suction. They are transferred to a mortar and ground to a paste with 100 cc. of cold water, filtered again, and then washed on the filter with an additional 100 cc. of cold water. The *o*-toluamide is obtained in the form of white crystals melting at 141–141.5°. The yield of air-dried product is 91–93 g. (90–92 per cent of the theoretical amount) (Note 4). The product may be recrystallized from water (10 g. per 100 cc.). The recovery is 92 per cent, and the melting point is not changed (Note 5).

2. Notes

1. This amount of alcohol is sufficient to provide a homogeneous solution.

2. If the temperature is allowed to rise much above 50° the evolution of oxygen will be sufficiently rapid to cause the mixture to foam out of the flask.

3. The volume of the solution in the flask is kept down by applying a small flame to the flask after most of the alcohol has been distilled.

4. An additional 3–4 g. of low-melting material may be obtained by concentrating the filtrate, but this is hardly worth while.

5. In general, amides may be prepared by this method from aliphatic nitriles in yields of 50–60 per cent and from aromatic nitriles in yields of 80–95 per cent. Slight variations in the above procedure may be necessary for carrying out the reaction and for isolating the amide, depending on the solubility of the nitriles and amides. Except for difficultly hydrolyzable nitriles such as the *o*-substituted aromatic nitriles, an equivalent amount of 6 to 12 per cent hydrogen peroxide gives better yields than the 30 per cent reagent.[1]

[1] McMaster and Noller, Wash. Univ. Studies **13**, 23 (1925); J. Indian Chem. Soc. **12**, 652 (1935) [C. A. **30**, 1736 (1936)].

3. Methods of Preparation

o-Toluamide has been prepared by the action of ammonia on *o*-toluyl chloride,[2] and by the action of alcoholic potassium hydroxide [3] or of an alkaline solution of hydrogen peroxide [4, 1] on *o*-tolunitrile.

o-TOLUIC ACID

$$+ 2H_2O + H_2SO_4 \rightarrow \qquad + NH_4HSO_4$$

Submitted by H. T. Clarke and E. R. Taylor.
Checked by C. S. Marvel and W. W. Moyer.

1. Procedure

In a 5-l. flask, equipped with a mechanical stirrer, a reflux condenser, and a separatory funnel, is placed 3 kg. of 75 per cent sulfuric acid (sp. gr. 1.67). The solution is heated to about 150°, the stirrer is started, and 1 kg. (8.54 moles) of *o*-tolunitrile (Org. Syn. Coll. Vol. I, **1941,** 514) is added during two hours. The temperature is maintained at 150–160° and the mixture is stirred for two hours after the addition of the nitrile is complete. The temperature is then raised to 190° and stirring is continued for another hour. Usually some crystalline material appears in the condenser at this stage. The reaction mixture is cooled, poured into ice water, and filtered. The crude material is dissolved in an excess of 10 per cent sodium hydroxide solution (Note 1), filtered hot, and the filtrate acidified with dilute sulfuric acid. The product is collected on a Büchner funnel, dried, and recrystallized from about 3 l. of benzene (Note 2). The yield is 930–1030 g. (80–89 per cent of the theoretical amount) of *o*-toluic acid which melts at 102–103° (Note 3).

[2] Remsen and Reid, Am. Chem. J. **21,** 289 (1899).

[3] Weith, Ber. **6,** 419 (1873).

[4] Kattwinkel and Wolffenstein, ibid. **37,** 3224 (1904); Dubsky, J. prakt. Chem (2) **93,** 137 (1916).

2. Notes

1. Any insoluble material which separates on conversion into the sodium salt is toluamide, which may be isolated. The appearance of this substance indicates too short a period of heating or too low a temperature.

The reaction can be stopped readily so that a considerable quantity of the amide is produced. The amide purified by recrystallization from water, melts at 139–140°.

2. An additional amount of pure *o*-toluic acid may be obtained by distilling the benzene mother liquor to a small volume and allowing to cool.

3. *p*-Toluic acid (m.p. 178°) may be obtained from *p*-tolunitrile by the same process and in the same yields. This acid is less soluble in benzene, and about 9 l. is needed for recrystallization.

3. Methods of Preparation

o-Toluic acid has been prepared by heating 1,3-naphthalenedisulfonic acid, 1,3-dihydroxynaphthalene, 1-naphthol-3-sulfonic acid, or 1-naphthylamine-3-sulfonic acid with sodium hydroxide;[1] by reduction of phthalide with hydriodic acid and phosphorus;[2] by electrolytic oxidation of *o*-xylene;[3] by oxidation of *o*-xylene with dilute nitric acid;[4] by catalytic hydrogenation of phthalic anhydride;[5] by hydrolysis of *o*-tolunitrile with 75 per cent sulfuric acid;[6] and by carbonating the ether solution of the reaction product from butyl lithium and *o*-bromotoluene.[7]

[1] Kalle and Company, Ger. pat. 79,028 [Frdl. **4**, 147 (1894–97)]; Friedlaender and Rüdt, Ber. **29**, 1611 (1896).

[2] Hessert, ibid. **11**, 238 (1878); Racine, Ann. **239**, 72 (1887).

[3] Fichter and Rinderspacher, Helv. Chim. Acta **10**, 41 (1927).

[4] Fittig and Bieber, Ann. **156**, 242 (1870).

[5] Willstätter and Jaquet, Ber. **51**, 771 (1918).

[6] Cahn, Ann. **240**, 280 (1887).

[7] Gilman, Langham, and Moore, J. Am. Chem. Soc. **62**, 2330 (1940).

p-TOLYL CARBINOL

(Benzyl alcohol, p-methyl-)

$$p\text{-}CH_3C_6H_4CHO + CH_2O + KOH \rightarrow p\text{-}CH_3C_6H_4CH_2OH + HCO_2K$$

Submitted by DAVID DAVIDSON and MARVIN WEISS.
Checked by REYNOLD C. FUSON and E. A. CLEVELAND.

1. Procedure

THE apparatus consists of a 3-l. three-necked flask fitted with a mercury-sealed mechanical stirrer, a reflux condenser, a dropping funnel, and a thermometer which reaches almost to the bottom of the flask. Five hundred grams of potassium hydroxide pellets (85 per cent potassium hydroxide) (7.6 moles) and 750 cc. of commercial absolute methyl alcohol (free from acetone) are placed in the flask, and stirring is begun. The bulk of the alkali dissolves in a few minutes, with the evolution of heat. The flask is now surrounded by an ample cold-water bath, and, when the internal temperature drops to 60°, addition of a mixture of 360 g. (353 cc., 3 moles) of p-tolualdehyde (Note 1), 300 cc. of formalin (3.9 moles) (Note 2), and 300 cc. of absolute methyl alcohol is begun at such a rate that the internal temperature remains at 60–70°. This addition requires about fifteen minutes. The internal temperature is then maintained at 60–70° for three hours, after which the reflux condenser is replaced by a downward condenser and the methyl alcohol distilled with the aid of a brine bath until the internal temperature reaches 101°. Nine hundred cubic centimeters of cold water is then added to the warm residue, and the mixture is cooled. The resulting two layers are separated at once (Note 3), and the aqueous layer is extracted with three 200-cc. portions of benzene. The combined oil and extracts are washed with five or six 50-cc. portions of water (Note 4), and the combined washings extracted with 50 cc. of benzene, the benzene layer being added to the washed extract. The benzene solution is cleared by shaking it with a few grams of anhydrous sodium sulfate and is then distilled under diminished pressure. After removal of the benzene, 331 g. (90 per cent of the theoretical amount) of p-tolyl carbinol (b.p. 116–118°/20 mm.) is obtained; the product solidifies in the receiver to a mass of oil-drenched crystals melting at 54–55°. Recrystallization from an equal weight of commercial heptane (b.p. 90–100°) gives an 80 per cent recovery of long needles which melt at 61°. A further 8 per cent is recoverable by concentration of the mother liquor (Notes 5 and 6).

2. Notes

1. A technical grade of *p*-tolualdehyde, obtained from Fritzsche Brothers, New York, New York, was found satisfactory. Directions for preparing *p*-tolualdehyde are given on p. 583.

2. The formaldehyde content of the solution is determined by analysis (p. 611).

3. The upper layer solidifies if allowed to stand.

4. This washing removes potassium *p*-toluate, which causes difficulty in the distillation of the product if allowed to remain.

5. The final residue from the mother liquor is an oil which does not solidify in a freezing mixture and which appears to be a mixture of *p*- and *m*-tolyl carbinols. Only a trace of phthalic acid (phenolphthalein test) was obtained by oxidizing this oil with permanganate; the portion of the oil which was more readily soluble in water yielded a phenylurethan which depressed the melting point of the phenylurethan of either *p*-tolyl carbinol or benzyl alcohol.

6. Under the same conditions, benzaldehyde yielded 80 per cent of benzyl alcohol and piperonal 86 per cent of piperonyl alcohol.

3. Methods of Preparation

p-Tolyl carbinol has been prepared from *p*-tolualdehyde by the action of alcoholic potassium hydroxide,[1] by electrolytic reduction,[2] and by the reducing action of the Grignard reagent,[3] as well as from *p*-toluic acid by electrolytic reduction.[4] The procedure described is an adaptation of a general method for reducing aromatic aldehydes to the corresponding alcohols.[5]

[1] Cannizzaro, Ann. **124**, 252 (1862).

[2] Law, J. Chem. Soc. **91**, 755 (1907).

[3] Oddo, Gazz. chim. ital. **41** (I) 285 (1911).

[4] Mettler, Ber. **39**, 2933 (1906).

[5] Davidson and Bogert, J. Am. Chem. Soc. **57**, 905 (1935).

sym.-TRIBROMOBENZENE

(Benzene, 1,3,5-tribromo-)

$$C_6H_5NH_2 + 3Br_2 \rightarrow C_6H_2Br_3NH_2 + 3HBr$$

$$C_6H_2Br_3NH_2 \cdot H_2SO_4 + HNO_2 \rightarrow C_6H_2Br_3N_2HSO_4 + 2H_2O$$

$$C_6H_2Br_3N_2HSO_4 + C_2H_5OH$$
$$\rightarrow 1,3,5\text{-}C_6H_3Br_3 + CH_3CHO + N_2 + H_2SO_4$$

Submitted by G. H. Coleman and William F. Talbot.
Checked by Reynold C. Fuson and Charles F. Woodward.

1. Procedure

The apparatus consists of a 12-l. round-bottomed flask and a 250-cc. suction flask provided with stoppers and glass tubes as shown in Fig. 19.

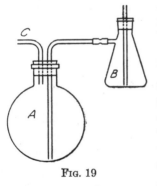

After the flasks have been charged, A is surrounded by an ice bath and B by a water bath heated to 40–50°. In flask A are placed 100 g. (1.1 moles) of aniline, 1 l. of water, and 100 cc. (1.2 moles) of concentrated hydrochloric acid; when the aniline has dissolved, sufficient water is added to bring the volume of the solution to 5 l. Flask B is charged with 577 g. (185 cc., 3.6 moles) of bromine; the baths are adjusted, and a rapid stream of air saturated with bromine vapor is drawn into A by applying suction at C (Notes 1 and 2).

Fig. 19

The introduction of bromine is continued until the solution assumes a distinctly yellow color; approximately three to four hours is required, and the reaction is then complete. The tribromoaniline is filtered on a Büchner funnel, thoroughly washed with water to remove hydrobromic acid, sucked as dry as possible, and taken without further drying to the next step.

The moist tribromoaniline, together with 2.1 l. of 95 per cent alcohol and 525 cc. of benzene, is placed in a 5-l. two-necked flask. One neck of the flask bears a reflux condenser; the other is closed with a stopper that can be removed momentarily for the addition of reagents. The tribromoaniline is brought into solution by heating the flask on a steam bath. To this solution is added 140 cc. of concentrated sulfuric acid, and then 140 g. (2.03 moles) of powdered sodium nitrite as rapidly as the violence of the reaction will permit. When the reaction has moderated, the solution is brought to boiling and maintained there as long as gas is evolved. It is then allowed to stand in a warm place for three hours more.

After the mixture has been cooled (preferably in an ice bath), the mother liquor is removed from the solid by decantation through a Büchner funnel; if necessary, a wire gauze is used to retain the solid in the flask. To the solid is then added a solution of 150 cc. of concentrated sulfuric acid in 1.5 l. of water. When the excess sodium nitrite has been decomposed, the solid is transferred to a Büchner funnel and washed first with water and then with a small amount of alcohol. The yield of crude, dry tribromobenzene is 250–260 g. (74–77 per cent of the theoretical amount). The product is reddish brown and melts at 112–116°.

For further purification, 100 g. of the crude product is dissolved in a boiling mixture of 1560 cc. of glacial acetic acid and 350 cc. of water; the solution is boiled for a few minutes with 25 g. of decolorizing carbon, filtered hot, and allowed to cool. The crystals are washed on a Büchner funnel with a small amount of chilled 95 per cent alcohol to remove the acetic acid. The whole of the crude product crystallized in this way furnishes 216–240 g. (64–71 per cent of the theoretical amount) of slightly colored tribromobenzene, melting at 121.5–122.5° (corr.).

The yield of crude product may be increased 50–60 g.—making a total of 300–320 g. (89–95 per cent of the theoretical amount)—by working up the mother liquors. The alcoholic liquors and washings are diluted to 6 l., and the aqueous layer is separated from the heavy oil at the bottom. The benzene is then distilled from the oil in a distilling flask, care being taken not to overheat the product after the benzene has been removed. The residual tribromobenzene may be recrystallized in the manner described above.

2. Notes

1. Considerable heat is generated by the reaction, and it is necessary to cool A to prevent the loss of bromine. Flask B must be kept at 40–50° to ensure a high concentration of bromine vapor in the air stream; the violent agitation produced by air saturated at lower temperatures whips the reaction mixture into a light foam that is likely to overflow the flask.

2. It is well to interpose a safety flask between the reaction flask and the aspirator. The safety flask is partially filled with water through which the air stream passes after the bromine vapor has been removed. This device has a double purpose; it enables one to gauge the rate of

aspiration by the rate at which the air current bubbles through the water, and it serves to detect the escape of bromine vapors from the reaction flask. A small amount of bromine will impart to the water a distinctly yellow color. In case bromine does escape into this trap a momentary release of the suction will cause the wash water to be drawn into the reaction flask, thus returning the bromine to the reaction mixture.

3. Methods of Preparation

Sym.-tribromobenzene has been prepared from 3,5-dibromoaniline by the replacement of the amino group by bromine;[1] from bromoacetylene by the action of light;[2] by the decomposition of 2,4,6-tribromophenylhydrazine;[3] by the reduction of 2,4,6-tribromobenzenediazonium sulfate;[4,5] and as a by-product in the preparation of 2,4,6-tribromobenzonitrile.[6]

TRICARBETHOXYMETHANE

(Methanetricarboxylic acid, triethyl ester)

$$CH_2(CO_2C_2H_5)_2 + Mg + C_2H_5OH \rightarrow$$
$$C_2H_5OMgCH(CO_2C_2H_5)_2 + H_2$$
$$C_2H_5OMgCH(CO_2C_2H_5)_2 + ClCO_2C_2H_5 \rightarrow$$
$$ClMgC(CO_2C_2H_5)_3 + C_2H_5OH$$
$$ClMgC(CO_2C_2H_5)_3 + CH_3CO_2H \rightarrow$$
$$CH(CO_2C_2H_5)_3 + CH_3CO_2MgCl$$

Submitted by Hakon Lund and Axel Voigt.
Checked by W. H. Carothers and W. L. McEwen.

1. Procedure

In a 1-l. round-bottomed flask, provided with an efficient and not too narrow reflux condenser, are placed 25 g. (1.03 gram atoms) of magnesium turnings (Grignard), 25 cc. of absolute alcohol (Note 1), 1 cc. of carbon tetrachloride (Note 2), and 30 cc. of a mixture of 160 g. (151 cc.,

[1] Körner, Gazz. chim. ital. **4,** 410 (1874).

[2] Sabanejew, J. Russ. Phys.-Chem. Soc. **17,** I, 176 (1885) [Ber. **18,** 374 (R) (1885)].

[3] Chattaway and Vonderwahl, J. Chem. Soc. **107,** 1508 (1915).

[4] Jackson and Moore, Am. Chem. J. **12,** 167 (1890).

[5] Jackson and Bentley, ibid. **14,** 335 (1892).

[6] Montagne, Rec. trav. chim. **27,** 347 (1908).

1 mole) of ethyl malonate and 80 cc. of absolute alcohol. Provision is made for cooling the flask when necessary in cold water, and the mixture is gently heated until hydrogen is evolved. The reaction may become so violent that external cooling is necessary. The ethyl malonate is gradually added through the condenser at such a rate that the reaction proceeds vigorously but not beyond control. When the reaction moderates the flask is cooled and 300 cc. of ether, dried twenty-four hours with calcium chloride, is added through the condenser. On gentle heating the crystals which have separated are dissolved, and hydrogen is again evolved for some time without further heating. The reaction is brought to completion on the steam bath. The flask is then removed from the steam bath, and a mixture of 100 cc. of ethyl chloroformate (1.05 moles) and 100 cc. of dry ether is added through the condenser from a dropping funnel at such a rate that vigorous boiling is maintained throughout the addition (Note 3). The reaction is complete after heating for fifteen minutes on the steam bath.

The viscous magnesium compound formed is cautiously decomposed with dilute acetic acid (75 cc. in 300 cc. of water), the flask being cooled under the tap. Two clear layers are formed, and, after separation, the aqueous layer is extracted with 100 cc. of ether, the combined ethereal solution is washed with water and dried with sodium sulfate, and the ether is distilled on the steam bath. The residue is distilled under reduced pressure. After a small fore-run the temperature rapidly rises to 130° at 10 mm. when the pure tricarbethoxymethane begins to distil. The yield of material collected over a five-degree interval is 204–215 g. (88–93 per cent of the theoretical amount). The product solidifies at 25°. The melting point of the pure substance is 28–29°.

2. Notes

1. Anhydrous alcohol is preferable, but a good grade of commercial absolute alcohol may be used without appreciably lowering the yield. The submitters used alcohol dehydrated by means of magnesium.[1]

2. Carbon tetrachloride (as well as a number of other halogen compounds) greatly accelerates the reaction between magnesium and alcohol. If anhydrous alcohol is used the reaction will start after some time without heating, whereas 99.5 per cent alcohol has to be heated nearly to boiling before the evolution of hydrogen becomes rapid.

3. Towards the end of the reaction the magnesium compound of ethyl methanetricarboxylate separates as a viscous mass which tends to prevent the remaining magnesium malonic ester and ethyl chloroformate

[1] Lund and Bjerrum, Ber. **64,** 210 (1931).

from reacting. Vigorous boiling keeps the material from forming a compact mass. If larger runs are made it may be necessary to stir during this part of the process.

3. Methods of Preparation

Tricarbethoxymethane has been prepared by the action of ethyl chloroformate upon sodiomalonic ester suspended in benzene,[2] toluene, or xylene;[3] by the distillation of ethyl ethoxalylmalonate;[4] from ethyl carbonate and malonic ester;[5] and by the procedure described above.[6]

According to Backer and Lolkelma,[3] the aromatic esters of methane-tricarboxylic acid and the ethyl ester of that acid are best prepared by the Lund procedure using magnesium; the other aliphatic esters of methanetricarboxylic acid are best prepared using sodium—compare the preparation of the trimethyl ester below—with toluene or xylene as the reaction medium.

TRICARBOMETHOXYMETHANE

(Methanetricarboxylic acid, trimethyl ester)

$$2CH_2(CO_2CH_3)_2 + 2Na \rightarrow 2NaCH(CO_2CH_3)_2 + H_2$$

$$NaCH(CO_2CH_3)_2 + ClCO_2CH_3 \rightarrow CH(CO_2CH_3)_3 + NaCl$$

Submitted by B. B. Corson and J. L. Sayre.
Checked by Roger Adams and A. E. Knauf.

1. Procedure

In a 2-l. three-necked flask, fitted with an upright condenser, a separatory funnel, and a mercury-sealed stirrer, are placed 400 cc. of dry xylene and 13 g. (0.56 gram atom) of sodium. The flask is heated in an oil bath until the sodium melts, and the mixture is stirred until the sodium is broken up into fine globules. Then 69 g. (0.57 mole) of methyl malonate is added over a period of five to ten minutes (Note 1).

The mixture is cooled while being well stirred, and, when the temperature reaches about 65°, 57 g. (0.6 mole) of methyl chloroformate is

[2] Conrad and Guthzeit, Ann. **214**, 32 (1882).

[3] Backer and Lolkelma, Rec. trav. chim. **57**, 1237 (1938).

[4] Bouveault, Bull. soc. chim. (3) **19**, 79 (1898); Scholl and Egerer, Ann. **397**, 353 (1913).

[5] Wallingford, Homeyer, and Jones, J. Am. Chem Soc. **63**, 2056 (1941).

[6] Lund, Ber. **67**, 938 (1934).

added during five to ten minutes. The mixture is then warmed slowly so that the boiling point is reached in about fifteen to twenty minutes. Boiling and stirring are maintained for five hours.

The mixture is cooled to room temperature, the flask is filled two-thirds full with water, and stirring is continued for five minutes. The xylene solution is separated, washed with water, dried over calcium chloride, filtered, and distilled under reduced pressure. After removal of the solvent, the tricarbomethoxymethane distils at 128–142°/18 mm. The yield of crude product is 50–51 g. (50–51 per cent of the theoretical amount). This material becomes semi-solid on cooling.

The crude product is purified by dissolving it in an equal volume of methyl alcohol and then cooling the solution in a freezing mixture until crystallization is complete. The crystals are filtered with suction, and the mother liquors are again cooled in a freezing mixture. If necessary this process is repeated a third time or until no further crop of crystals separates on chilling the mother liquors. The crystals are transferred from the funnel to a beaker, stirred with about 70 cc. of petroleum ether (b.p. 32–45°), filtered, pressed dry, and washed with a little petroleum ether. The yield of fine, snow-white crystals melting at 43–45° is about 40–42 g. (40–42 per cent of the theoretical amount) (Note 2).

2. Notes

1. A brisk evolution of hydrogen occurs, and sodium methyl malonate settles as a pasty mass. Stirring should be vigorous during the addition of the malonic ester and also during the subsequent heating, to avoid caking of the sodium salt.

2. Tricarbomethoxymethane is very soluble, and only by repeated cooling and filtration can a good recovery be obtained. Owing to this excessive solubility the apparent loss on crystallization is high. However, a considerable portion of this material can be recovered from the alcoholic mother liquors.

3. Methods of Preparation

The method described is very similar to one published by Adickes, Brunnert, and Lücker [1] for the preparation of the ethyl ester. In other procedures the sodium derivative of methyl malonate was isolated.[2] Compare the "Methods of Preparation" for tricarbethoxymethane on p. 596 above.

[1] Adickes, Brunnert, and Lücker, J. prakt. Chem. (2) **130**, 163 (1931).

[2] Scholl and Egerer, Ann. **397**, 355 (1913); Philippi, Hanusch, and von Wacek, Ber. **54**, 901 (1921); Backer and Lolkelma, Rec. trav. chim. **57**, 1237 (1938).

TRICHLOROETHYL ALCOHOL

(Ethanol, 2-trichloro-)

$$3CCl_3CHO + Al(OCH_2CH_3)_3 \rightarrow 3CH_3CHO + Al(OCH_2CCl_3)_3$$
$$2Al(OCH_2CCl_3)_3 + 3H_2SO_4 \rightarrow Al_2(SO_4)_3 + 6CCl_3CH_2OH$$

Submitted by WILLIAM CHALMERS.
Checked by REYNOLD C. FUSON and H. H. HULLY.

1. Procedure

In a 2-l. three-necked flask immersed in an oil bath are placed 250 g. (1.7 moles) of anhydrous chloral, 650 cc. of anhydrous alcohol (Note 1), and 75 g. (0.46 mole) of pure aluminum ethoxide (Note 2). An efficient fractionating column (Note 3) is attached to the flask through a cork in the central neck, and through one of the side necks is passed a tube leading to the bottom of the flask for the introduction of dry nitrogen (Note 4). The remaining neck is tightly stoppered. It is used to withdraw portions of liquid for testing.

To the outlet of the column is attached a U-tube and a Peligot tube each of about 100-cc. capacity. The U-tube is immersed in a freezing mixture, and the Peligot tube is filled with saturated sodium bisulfite solution. The oil bath is then heated to 135° and a slow current of gas is admitted. The mixture boils vigorously, but the alcohol is returned by the refluxing device whereas the acetaldehyde formed by the reaction is allowed to pass through and is caught in the freezing mixture and the Peligot tube.

The end of the reaction is easily determined by removing a few drops of the reaction mixture and treating with water in a test tube. After the aluminum hydroxide settles, the clear liquid is decanted and a few drops of yellow ammonium sulfide are added to it. So long as chloral is present, even in traces, a dark brown coloration will be produced on heating to incipient boiling. If the liquid collecting in the U-tube is removed every few hours, the completion of the reaction will be readily noticed by the diminishing quantity of acetaldehyde coming over. After twenty-three or twenty-four hours of heating (over a period of two or three days) the reaction is complete. The temperature of the bath is then allowed to fall to 120° and the alcohol is distilled through an ordinary condenser which replaces the fractionating column. When the residue of aluminum trichloroethoxide is nearly dry (Note 5), the flask is removed from the oil bath and the solid is treated with 250 cc. of 20 per cent sulfuric acid and stirred thoroughly to ensure complete decomposition.

The mixture is then subjected to steam distillation until no more trichloroethyl alcohol passes over. About 4 l. of distillate is obtained (Note 6). The oil is separated from the aqueous layer, which is saturated with sodium sulfate and extracted with three 200-cc. portions of ether. The ether solution is added to the main portion of the alcohol, and the whole is dried over anhydrous sodium sulfate.

The ether is removed by distillation and the product distilled under reduced pressure (Note 7). There is obtained 215 g. (84 per cent of the theoretical amount) of trichloroethyl alcohol boiling at 94–97°/125 mm. and melting at 16–17° (Note 8). A purer compound can be obtained by refractionating under reduced pressure and pressing the crystals on a cooled porous plate. Pure trichloroethyl alcohol has a melting point of 19° (Note 9).

2. Notes

1. The action of a small proportion of aluminum ethoxide on pure chloral or chloral diluted with benzene leads to the formation of trichloroethyl trichloroacetate in the same manner that ethyl acetate is formed from acetaldehyde by aluminum ethoxide, but the yields are small. A slightly better yield is obtained when molecular proportions are employed, but it is only in the presence of a large amount of an alcohol that good yields are obtained.

2. Aluminum ethoxide is now obtainable commercially. The submitter recommends the following method of preparation,[1] which has been checked:

In a flask fitted with a reflux condenser is placed 27 parts of aluminum filings or groats. This is treated with 276 parts of absolute alcohol, 0.1 to 0.25 part of mercuric chloride, and several crystals of iodine. After a few minutes a violent evolution of hydrogen takes place, and by heating on the water bath for several hours the ethoxide is produced in the form of a grayish powder.

Unchanged alcohol is removed by distillation from an oil bath until the residual material melts to a dark-colored liquid. It is then poured into a Claisen flask and distilled under reduced pressure, using a short air condenser and a Pyrex suction flask as a receiver. Since aluminum ethoxide tends to sublime, a glass-wool filter is inserted between the receiving flask and the vacuum line to prevent clogging. A free flame is necessary, and distillation should be rapid. While the distillate is still liquid it is poured into a Pyrex flask and allowed to cool. It forms a

[1] Meister, Lucius, and Brüning, Ger. pat. 286,596 [Frdl. 12, 29 (1914–16)].

tough, white mass which must be preserved in a well-stoppered flask to prevent adsorption of water vapor. A yield of 90 per cent of the theoretical amount may be obtained.

For the present reaction, where a trace of water does no harm, some of the ethoxide may be fused and poured into a mortar and covered with a watch glass until cool, when it can be powdered and weighed.

The submitter has found that it is possible to make a large quantity of the ethoxide by the above method with 95 per cent alcohol, but that success is more certain if the alcohol is refluxed and distilled twice over quicklime.

3. This column must be capable of separating acetaldehyde from ethyl alcohol with the greatest completeness. A very efficient type is that of Hahn,[2] in which the column is kept at a constant temperature by means of boiling methyl alcohol in a central tube, the vapors passing up an outer annular space. It is necessary that the lower part of the condenser employed be sufficiently broad (say 2 cm. in diameter) so that when in operation there will be no tendency for liquid to collect in the column.

A column involving the same principle can be made from two reflux condensers. The lower outlet of the outer jacket of the condenser connected to the flask is closed, and the jacket is filled to about two-thirds its height with methyl alcohol. A few small pieces of pumice are introduced to prevent bumping of the alcohol. The upper outlet of this jacket is joined to a second condenser to prevent loss of the methanol vapors. The inner tube of the fractionating column is filled with glass beads which are held in place by indentations made in the lower part of the tube. The outlet of this tube is connected to the U-tube immersed in the freezing mixture.

Care is essential to ensure the complete separation of aldehyde and alcohol, because any slight continual loss of solvent is multiplied into a serious loss in the long period of heating necessary. The boiling point of chloral is only twenty degrees above that of ethyl alcohol, and any loss of the latter would lead to some loss of the reagent.

4. The nitrogen is supplied from a cylinder and is dried by passing through a calcium chloride tube and a tube containing phosphorus pentoxide mixed with glass wool or beads. A tube filled with glass wool is placed between the phosphorus pentoxide tube and the flask to prevent the pentoxide from being blown into the reaction mixture if the nitrogen is turned on too rapidly.

5. The residue must not be heated too strongly as it becomes compact

[2] Hahn, Ber. **43,** 419 (1910). A modified form of this apparatus is described in Org. Syn. **20,** 27.

and some tarry material is formed which makes the action of the acid upon it very slow.

6. A small amount of trichloroaldol, $CCl_3CHOHCH_2CHO$, can be extracted from the residue of the steam distillation by means of ether.

7. The product can be distilled at ordinary pressure at a temperature of 151°, but the distillate has a brown color due to slight decomposition.

8. Even by simple refluxing of the mixture without separation of the aldehyde, a yield of 65 per cent may be obtained.[3] The reaction between the aluminum ethoxide and the chloral is in equilibrium with that between the aluminum trichloroethoxide and the acetaldehyde. A certain amount of the acetaldehyde is removed from the reaction by condensation to ethyl acetate under the catalytic influence of aluminum ethoxide, but the change takes place only slowly in this mixture. A great deal of the aldehyde is converted into paraldehyde and acetal, but these changes are reversible.

9. The use of aluminum ethoxide in the presence of alcohol offers a means of reducing many compounds which could not be reduced in the desired manner by the more customary methods. Meerwein and his students have used the method to convert bromal into tribromoethyl alcohol and have also prepared cinnamyl alcohol and various halogenated cinnamyl and crotyl alcohols from the corresponding aldehyes.[4,5] Excellent yields are obtained.

3. Methods of Preparation

The method given here is essentially that of Meerwein and his pupils Schmidt [5] and von Bock.[3] The theory of the reaction and applications are also discussed by Dworzak [4] and by Verley.[6] A number of patents have appeared covering this reaction, in some of which a secondary alcohol such as isopropyl alcohol [7] is used in place of ethyl alcohol. Trichloroethyl alcohol is one of the chlorination products of alcohol and is found in the high-boiling fractions in the production of chloral.[8] It was prepared by Garzarolli-Thurnlackh [9] and by Delacre [10] by the action of diethylzinc on chloral.

[3] v. Bock, Dissertation, Albertus-Universität, Königsberg i. Pr., Germany, 1926.

[4] Dworzak, Monatsh. **47**, 11 (1926).

[5] Meerwein and Schmidt, Ann. **444**, 233 (1925).

[6] Verley, Bull. soc. chim. (4) **37**, 537 (1925).

[7] Bayer and Company, Brit. pat. 235,584 [C. A. **20**, 917 (1926)]; I. G. Farbenind A.-G., Brit. pat. 286,797 [C. A. **23**, 395 (1929)]; Winthrop Chemical Company, U. S pat. 1,725,054 [C. A. **23**, 4709 (1929)].

[8] Altschul and Meyer, Ber. **26**, 2758 (1893).

[9] Garzarolli-Thurnlackh, Ann. **210**, 63 (1881).

[10] Delacre, Bull. soc. chim. (2) **48**, 784 (1887).

TRIETHYL CARBINOL

(3-Pentanol, 3-ethyl-)

$$C_2H_5Br + Mg \rightarrow C_2H_5MgBr$$

$$3C_2H_5MgBr + (C_2H_5)_2CO_3 \rightarrow (C_2H_5)_3COMgBr + 2C_2H_5OMgBr$$

$$(C_2H_5)_3COMgBr + H_2O \rightarrow (C_2H_5)_3COH + Mg(OH)Br$$

Submitted by W. W. Moyer and C. S. Marvel.
Checked by Frank C. Whitmore and D. J. Loder.

1. Procedure

In a 3-l. three-necked flask, fitted with a mechanical stirrer, a 500-cc. separatory funnel, and an efficient reflux condenser to which a calcium chloride tube is attached, are placed 107 g. (4.4 gram atoms) of magnesium turnings and 800 cc. of anhydrous ether. The reaction is started by adding 5 cc. (7 g., 0.06 mole) of ethyl bromide (Note 1) without stirring. The stirrer is started, and a solution of 480 g. (4.4 moles) of ethyl bromide in 1 l. of anhydrous ether is added as rapidly as the refluxing of the ether allows. The addition requires about two hours (Note 2). The reaction is practically complete when all the halide has been added, but stirring should be continued for fifteen minutes longer.

A solution of 156 g. (1.32 moles) of ethyl carbonate (Note 3) in 200 cc. of ether is added to the Grignard reagent, with rapid stirring, over a period of approximately three hours. The reaction is vigorous, and the ether refluxes continually. After all the diethyl carbonate has been added, the flask is heated on a water bath and stirring is continued for another hour.

The reaction mixture is hydrolyzed by pouring it, with frequent shaking, into a 5-l. round-bottomed flask containing 1.5 kg. of cracked ice and a solution of 300 g. of ammonium chloride in 600 cc. of water. The ether layer is separated in a large separatory funnel, and the aqueous residue is extracted with two 500-cc. portions of ether (Note 4).

The ether is distilled from the combined extracts, and the crude triethyl carbinol is dried with 10 g. of anhydrous potassium carbonate. The alcohol is then distilled at atmospheric pressure, and the portion (80–90 g.) boiling at 139–142° is collected. The low-boiling distillate is treated with 5 g. of anhydrous potassium carbonate, filtered, and redistilled, whereby another portion (about 25 g.) of triethyl carbinol boiling at 139–142° is obtained. The process is repeated once, or twice if necessary, and an additional 20 g. is collected (Note 5). The total yield is 125–135 g. (82–88 per cent of the theoretical amount) (Note 6). Triethyl carbinol is a viscous liquid with a penetrating, camphor-like odor.

2. Notes

1. The ethyl bromide used in this preparation was dried over calcium chloride and then distilled from phosphorus pentoxide. The fraction boiling at 38–39° was collected. Directions for preparing ethyl bromide are given in Org. Syn. Coll. Vol. I, **1941**, 29, 36.

2. The time of addition may be decreased by cooling the flask externally. A towel is folded in a narrow strip and wrapped about the flask above the ether line, and cracked ice is packed on top of the flask. This arrangement allows the ether vapor to be condensed without appreciable cooling of the reaction mixture.

3. The commercial "99 per cent" ester was used. It was purified according to the description given in Note 2 on p. 283 above.

4. The ether used for extraction may be obtained by distilling the ether from the triethyl carbinol solution.

5. In checking this preparation the first drying was allowed to continue for fifteen hours. The first fractionation of the carbinol yielded 122 g., the second 10 g., and the third yielded none.

6. The preparation of homologous trialkyl carbinols by use of the Grignard reagent and ethyl carbonate was found to be very satisfactory. The following compounds were prepared: tri-*n*-propyl carbinol (b.p. 89–92°/20 mm.) in 75 per cent yield; tri-*n*-butyl carbinol (b.p. 129–131/20 mm.) in 84 per cent yield; tri-*n*-amyl carbinol (b.p. 160–164°/19 mm.) in 75 per cent yield; and tri-*n*-heptyl carbinol (b.p. 195–200°/6 mm.) in 72 per cent yield.

3. Methods of Preparation

Triethyl carbinol has been prepared by the action of zinc on a mixture of ethyl iodide and diethyl ketone;[1] by the action of magnesium on ethyl bromide in diethyl ketone solution;[2] by the action of sodium and ethyl bromide on diethyl ketone or ethyl propionate;[3] as a by-product in the reaction between ethylmagnesium bromide and carbon oxy-

[1] Barataeff and Saytzeff, J. prakt. Chem. (2) **34**, 463 (1886).

[2] Davies and Kipping, J. Chem. Soc. **99**, 298 (1911).

[3] Morton and Stevens, J. Am. Chem. Soc. **53**, 2247 (1931).

sulfide;[4] and by the action of ethylmagnesium bromide on ethyl propionate,[5] ethyl chloroformate,[6] ethyl cyanoformate,[7] or ethyl carbonate.[8]

1,2,3-TRIIODO-5-NITROBENZENE

(Benzene, 1,2,3-triiodo-5-nitro-)

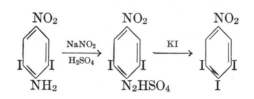

Submitted by R. B. Sandin and T. L. Cairns.
Checked by Frank C. Whitmore and L. H. Sutherland.

1. Procedure

Fifty grams (0.13 mole) of 2,6-diiodo-4-nitroaniline (p. 196) is dissolved in 200 cc. of concentrated sulfuric acid (sp. gr. 1.84) in a 1-l. two- or three-necked flask provided with a mechanical stirrer. The solution is cooled to 5° in an ice-salt mixture, and to it is added with stirring a mixture of 100 cc. of concentrated sulfuric acid and 12 g. (0.17 mole) of sodium nitrite, also cooled to 5° (Note 1). To liberate nitrous acid from the nitrosylsulfuric acid, there is then added slowly from a separatory funnel, with rapid stirring, 200 cc. of 85 per cent phosphoric acid (151 cc. of U.S.P. phosphoric acid diluted to 200 cc.). During the addition the temperature is kept below 10°. The mixture is removed from the ice-salt bath and stirred until diazotization is complete (one to two hours) (Note 2). At the end of this time the solution is poured, with stirring, into 2 l. of a mixture of cracked ice and water in a 4-l. beaker. To destroy excess nitrous acid about 15 g. of urea is added in small portions, with stirring, as long as gas is produced. The mixture is filtered if it is not clear and is then treated gradually with a solution of 30 g. (0.18 mole) of potassium iodide in 200 cc. of water. To complete the

[4] Weigert, Ber. 36, 1009 (1903).
[5] Schreiner, J. prakt. Chem. (2) 82, 295 (1910); Böeseken and Wildschut, Rec. trav. chim. 51. 169 (1932).
[6] Mazurewitsch, J. Russ. Phys.-Chem. Soc. 42, 1582 (1910) (Chem. Zentr. 1911, I, 1500).
[7] Bruylants, Bull. soc. chim. Belg. 33, 529 (1924).
[8] Whitmore and Badertscher, J. Am. Chem. Soc. 55, 1560 (1933).

reaction the mixture is heated until no more gas is evolved. Any free iodine is removed by sodium bisulfite, and the mixture is then filtered on a Büchner funnel, washed free from sulfuric acid and inorganic salts, pressed, and sucked as dry as possible. The product is then air dried to constant weight. The yield of light brown, crude material, m.p. 160–162°, is 60–61 g. (94–95 per cent of the theoretical amount). The pure product can be obtained by dissolving the crude material in 200 cc. of boiling benzene, filtering, and cooling the filtrate to 10°. The yield of yellow crystals melting at 161–162° is 40–42 g. (65–70 per cent recovery) (Note 3).

2. Notes

1. The sodium nitrite must be finely powdered and added slowly with vigorous stirring to the sulfuric acid, which is kept at 5° in an ice-salt mixture. The evolution of oxides of nitrogen during the addition of the nitrite must be avoided.

2. During this time the temperature of the mixture gradually rises to that of the room. When diazotization is complete a drop of the mixture will form a clear yellow solution if added to 10 cc. of cold water.

3. It has been reported that the procedure of Hodgson and Walker [1] is preferable to that described above, and that crude triiodonitrobenzene is best purified by crystallization from cellosolve. [CARL NIEMANN and C. E. REDEMANN, private communication, and J. Am. Chem. Soc. **63,** 1550 (1941).]

3. Methods of Preparation

1,2,3-Triiodo-5-nitrobenzene has been prepared by the diazotization of 2,6-diiodo-4-nitroaniline (without the use of phosphoric acid) and subsequent treatment with potassium iodide.[1] The present procedure is an example of a general method developed by Schoutissen [2] for diazotization of weakly basic amines, such as the 2,6-dihalogen derivatives of p-nitroaniline.

[1] Willgerodt and Arnold, Ber. **34,** 3343 (1901); Kalb, Schweizer, Zellner, and Berthold, ibid. **59,** 1866 (1926); Harington and Barger, Biochem. J. **21,** 175 (1927); Hodgson and Walker, J. Chem. Soc. **1933,** 1620.

[2] Schoutissen, J. Am. Chem. Soc. **55,** 4531 (1933).

TRIPHENYLETHYLENE

(Ethylene, triphenyl-)

$$C_6H_5CH_2MgCl + (C_6H_5)_2CO \rightarrow (C_6H_5)_2C(OMgCl)CH_2C_6H_5$$

$$(C_6H_5)_2C(OMgCl)CH_2C_6H_5 + H_2O \rightarrow$$
$$(C_6H_5)_2C(OH)CH_2C_6H_5 + Mg(OH)Cl$$

$$(C_6H_5)_2C(OH)CH_2C_6H_5 \xrightarrow{(H_2SO_4)} (C_6H_5)_2C{=}CHC_6H_5 + H_2O$$

Submitted by HOMER ADKINS and WALTER ZARTMAN.
Checked by C. R. NOLLER and F. M. McMILLAN.

1. Procedure

IN a 3-l. three-necked flask, fitted with a mechanical stirrer, a reflux condenser, and a separatory funnel, are placed 24.3 g. (1 gram atom) of magnesium turnings, 500 cc. of absolute ether, a crystal of iodine, and a 5- to 10-cc. portion of 126.5 g. (115 cc., 1 mole) of freshly distilled benzyl chloride (b.p. 177–179°). In a few minutes the reaction starts (Note 1) and is controlled if necessary by cooling with a wet towel. The stirrer is started, and the balance of the benzyl chloride is run in as fast as the refluxing will permit. The addition requires from one to two hours, and when completed the mixture is refluxed on the steam bath with stirring for three hours. With the stirrer still running, 182 g. (1 mole) of benzophenone (Org. Syn. Coll. Vol. I, **1941**, 95) dissolved in 500 cc. of absolute ether is added at such a rate that the mixture refluxes rapidly. This requires about twenty minutes, and then the reaction mixture is allowed to stand for two hours (Note 2).

The flask is placed in an ice bath, 700 g. of cracked ice is added, and the magnesium hydroxide is dissolved by adding 500 cc. of cold 20 per cent sulfuric acid. The ether layer is separated, and the water layer extracted with two 200-cc. portions of ether. The ether is distilled from the combined extracts, and the residual liquid is refluxed for two hours with 100 cc. of 20 per cent sulfuric acid to dehydrate the carbinol. The layers are separated and the product vacuum-distilled (Note 3). The fraction boiling at 215–225°/15 mm. weighs 160–170 g. and melts at 60–68° (Note 4). Crystallization from 900 cc. of hot 95 per cent alcohol and cooling to 0° gives 140–150 g. of the hydrocarbon melting at 68–69° (54–59 per cent of the theoretical amount). Concentration of the mother liquor to 150 cc. and cooling gives 5–10 g., m.p. 65–67.5° (Note 5).

2. Notes

1. If the reaction does not start within thirty minutes, the mixture is warmed on a water bath with the stirrer running.

2. Slightly better yields are obtained if the reaction mixture is allowed to stand overnight.

3. The product should not be washed with water as the presence of a trace of sulfuric acid during the distillation seems to be necessary to complete the dehydration.

4. Seeding is usually necessary to induce crystallization.

5. By the same procedure, stilbene may be prepared from benzaldehyde and benzylmagnesium chloride in 25–35 per cent yield.

3. Methods of Preparation

Triphenylethylene has been prepared by the reaction between phenylmagnesium bromide and desoxybenzoin or ethyl phenylacetate;[1] and by the reaction between diphenylketene-quinoline and benzaldehyde.[2] The procedure given above is an adaptation of that published by Hell and Wiegandt.[3] The intermediate diphenylbenzylcarbinol can be dehydrated very effectively by heating with potassium bisulfate or by boiling its solution in glacial acetic acid.[4]

TRIPHENYLMETHYLSODIUM

(Sodium, triphenylmethyl-)

$$(C_6H_5)_3CCl + 2NaHg \rightarrow Na[C(C_6H_5)_3] + NaCl + 2Hg$$

Submitted by W. B. Renfrow, Jr., and C. R. Hauser.
Checked by Lee Irvin Smith and E. C. Ballard.

1. Procedure

A solution of 63 g. (0.226 mole) of pure triphenylchloromethane (Note 1) in 1.5 l. of pure anhydrous ether is prepared in a 2-l. bottle provided with a tight ground-glass stopper, and 1150 g. of freshly pre-

[1] Klages and Heilmann, Ber. **37**, 1455 (1904).
[2] Staudinger and Kon, Ann. **384**, 89 (1911).
[3] Hell and Wiegandt, Ber. **37**, 1429 (1904).
[4] Van de Kamp and Sletzinger, J. Am. Chem. Soc. **63**, 1880 (1941).

pared 1 per cent sodium amalgam (0.5 gram atom of sodium) is added
(Note 2). The stopper is greased with a small amount of Lubriseal
and the bottle is stoppered tightly. The bottle is clamped firmly in a
mechanical shaker and shaken vigorously. The reaction is strongly exothermic, and the bottle should be cooled with wet towels during the
shaking operation. A persistent blood-red color develops rapidly,
generally within ten minutes. After shaking for three hours *the bottle is
cooled to room temperature*, removed from the shaker, and allowed to
stand undisturbed until the sodium chloride has settled (Note 3).

The ether solution of triphenylmethylsodium is separated from the
sodium chloride and amalgan by the following procedure. The stopper
of the reaction bottle is removed and replaced immediately by a closely
fitting two-holed cork carrying a short glass tube that protrudes about
1 cm. into the bottle, and a long glass tube bent into an inverted U-
shape. The short tube is connected through a drying train to a cylinder
of nitrogen. One arm of the U-tube reaches to within about 4 cm. of the
bottom of the bottle, and the other end extends just below the bottom of
a two-holed cork that is fitted tightly into a 2-l. Erlenmeyer flask. The
flask is filled with nitrogen and serves as receiver for the decanted solution. The cork of the Erlenmeyer flask is provided with a short-stemmed
separatory funnel which serves to release nitrogen during the decantation and may be used subsequently to introduce reactants (Note 4).
The corks are sealed by a coating of paraffin-wax. The stopcock of the
dropping funnel is opened slightly, and the ether solution of triphenylmethylsodium is forced slowly and steadily into the nitrogen-filled flask
by means of a small pressure of nitrogen from the cylinder. With
proper adjustment of the height of the glass tube above the surface of
the sodium chloride and amalgam, it is possible to remove all but 50 to
75 cc. of the ether solution.

If a good grade of triphenylchloromethane and freshly prepared
amalgam are used the yield of triphenylmethylsodium is almost quantitative. The solution may be analyzed approximately, as follows: A
50-cc. aliquot portion is allowed to flow into 25 cc. of water in a separatory funnel. The aqueous layer is drawn off and the ether solution
washed with three 25-cc. portions of water. The aqueous solutions are
combined, boiled to expel ether, and titrated with 0.2 N sulfuric acid,
methyl red being used as indicator.

2. Notes

1. A good grade of triphenylchloromethane, m.p. 112–113°, should be
used. The commercial product may be recrystallized conveniently from

a mixture of five parts of ligroin (b.p. 90–110°) and one part of acetyl chloride, using about 1.8 g. of solvent per gram of material. Directions for preparing triphenylchloromethane are given in Volume 23 of this series.

2. A 1 per cent sodium amalgam is prepared by cutting 11.5 g. (0.5 gram atom) of sodium into pieces about 5 mm. square and dissolving them, one at a time, in 1150 g. of purified mercury. A piece of sodium is speared on the sharpened end of a long glass rod and thrust quickly beneath the surface of the mercury, in a 500-cc. wide-mouthed Erlenmeyer flask. To avoid damage from flying pieces of sodium resulting from the vigorous reaction, the glass rod is inserted through a piece of heavy cardboard which serves to cover the mouth of the flask during the reaction.

3. If solid 3 per cent sodium amalgam is used, solutions containing almost 0.67 mole of triphenylmethylsodium per liter—about five times as concentrated as the solution described in the procedure above—can be prepared.

One and seven-tenths kilograms of 3 per cent sodium amalgam (2.2 gram atoms of sodium) is prepared by adding 1649 g. of mercury to 51 g. of molten sodium under mineral oil—compare Org. Syn. Coll. Vol. I, **1941**, 554, Note 1. The amalgam while still hot is poured into a large evaporating dish and left to cool. It is then transferred to a large iron mortar, broken into pieces about 1 mm. on a side, washed free from oil with ligroin, and placed in a 2-l. glass-stoppered bottle. One and one-half liters of dry ether and 278 g. (1 mole) of triphenylchloromethane are added, and the glass stopper, greased with Lubriseal, is inserted. The bottle is shaken in a mechanical shaker until no solid amalgam remains, and then for two hours longer. If the shaker has a 4- to a 5-in. stroke and makes three to four strokes per second, the total time on the shaker is from five to eight hours. With less effective shaking, more time is required. The reaction is sufficiently slow so that cooling with towels is unnecessary. The rest of the procedure is the same as that on p. 608. (Private communication from CHARLES R. HAUSER and BOYD E. HUDSON, JR. Checked by LEE IRVIN SMITH and R. W. LIGGETT.)

4. Triphenylmethylsodium is a useful reagent for the preparation of the sodium derivatives of very weak acids (aliphatic esters, acid anhydrides, etc.). An example of this procedure, using ethyl isobutyrate, is given on p. 268.

3. Method of Preparation

The procedure described above is a modification of the method of Schlenk and Ochs.[1]

---

[1] Schlenk and Ochs, Ber. **49**, 608 (1916).

sym.-TRITHIANE

$$3CH_2O + 3H_2S \xrightarrow{\text{(HCl)}} S\underset{CH_2-S}{\overset{CH_2-S}{<}}\!\!\!>CH_2 + 3H_2O$$

Submitted by R. W. Bost and E. W. Constable.
Checked by Reynold C. Fuson and C. F. Woodward.

1. Procedure

A mixture of 326 g. (3.9 moles) of a 36 per cent formaldehyde solution (Note 1) and 700 cc. of concentrated hydrochloric acid (sp. gr. 1.18) is placed in a tall glass cylinder (Note 2), and hydrogen sulfide is passed through the solution until no more precipitate is formed. In order to facilitate the process, the accumulated mass of crystals is removed from time to time by filtration. The time required for completion of the reaction varies from twelve to twenty-four hours. A crude yield of 176 g. (98 per cent of the theoretical amount) of fine, nearly colorless needles melting at 210–213° is obtained.

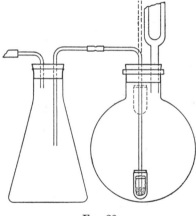

Fig. 20

The product is purified by the inverted filtration method. The apparatus used is shown in Fig. 20. A 2-l. round-bottomed flask is equipped with a reflux condenser and a bent glass tube, 8–10 mm. in diameter (Note 3). To the lower end of this tube is attached, by means of a cork, a 25-mm. tip prepared from a paper Soxhlet thimble and packed with glass wool. A 2-l. conical flask serves as a receiver for the hot filtrate.

The crude product is placed in the round-bottomed flask, 1 l. of benzene is added, and heat is applied until the solvent boils vigorously. After a few minutes the source of heat is withdrawn and the mixture is allowed to become quiet. The filtering thimble, which up to this point is kept at the top of the flask (see Fig. 20), is now lowered to its normal position and the conical flask is attached. Gentle suction is applied, and the liquid is drawn over into the conical flask. The hot solution is removed, allowed to cool, and filtered. In the meantime, the extraction process is repeated with a second 1-l. portion of benzene. The two portions of benzene are used alternately over and over in the manner described until all the crude product has been recrystallized. This requires about ten separate extractions, using each 1-l. portion of benzene five times.

The yield of pure product, melting at 214–215°, is 165–169 g. (92–94 per cent of the theoretical amount).

2. Notes

1. The formaldehyde content of the solution is determined by analysis, for which the iodimetric method of Borgstrom and Horsch [1] is recommended. The yield is calculated upon the basis of the amount of formaldehyde actually present, as shown by analysis. Since commercial solutions of formaldehyde contain methyl alcohol, the formaldehyde present cannot be estimated accurately by reference to specific-gravity tables—compare Org. Syn. Coll. Vol. I, **1941**, 378, Note 1.

2. A tall cylindrical vessel ensures good contact between the solution and the gas bubbling through it. Hydrogen sulfide from a commercial cylinder was used.

3. Successful use of the inverted filtration method requires attention to these details: (*a*) use of a sufficiently wide transfer tube; (*b*) minimum exposure of tubing between the flasks; (*c*) rapid transfer of the hot solution; (*d*) avoidance of too strong an application of suction.

The inverted filtration device is convenient for simple recrystallizations as well as repeated extractions. It is of particular advantage for the manipulation of volatile, inflammable solvents and of lachrymatory solutions.

3. Methods of Preparation

sym.-Trithiane has been prepared by treating carbon bisulfide,[2] ethyl isothiocyanate,[3] or potassium thiocyanate [3] with zinc and hydrochloric acid; by heating methylene iodide with alcoholic sodium hydrosulfide; [4] and by treating aqueous formaldehyde with hydrogen sulfide and concentrated hydrochloric acid.[5]

[1] Borgstrom and Horsch, J. Am. Chem. Soc. **45**, 1493 (1923).
[2] Girard, Compt. rend. **43**, 396 (1856); Ann. **100**, 306 (1856).
[3] Hofmann, Ber. **1**, 176, 179 (1868).
[4] Husemann, Ann. **126**, 293 (1863).
[5] Hofmann, ibid. **145**, 360 (1868); Baumann, Ber. **23**, 67 (1890).

l-TRYPTOPHANE

(By-product, l-Tyrosine)

Digestion of casein →

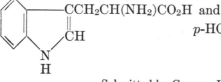

$-CCH_2CH(NH_2)CO_2H$ and

$p\text{-}HOC_6H_4CH_2CH(NH_2)CO_2H$

Submitted by Gerald J. Cox and Harriette King.
Checked by H. T. Clarke and Jessica P. Leland.

1. Procedure

(A) *Tryptophane.*—In an 8-l. (2-gal.) bottle is placed 600 g. of commercial casein (coarse powder), which is then covered with about 3.2 l. of tap water at 37° (Note 1). The bottle is shaken until all the casein is moistened. A solution of 60 g. of anhydrous sodium carbonate (Note 2) and 6 g. of sodium fluoride (Note 3) in 1 l. of water at 37° is added. A thin paste of 20 g. of commercial pancreatin in 100 cc. of water (37°) is poured in. The mixture is covered with a layer of toluene (80 cc.), diluted to 6 l., stoppered, shaken thoroughly, and placed in a warm room or bath at 37°.

After four or five days, with daily shakings, most of the casein is in solution and chalky masses of tyrosine begin to separate. After five days, a second 20-g. portion of pancreatin in 100 cc. of water is added. After twelve days, the bottle is cooled in an icebox overnight and the undissolved material is filtered (Note 4) and reserved for the preparation of tyrosine.

The filtrate (6.9–7 l.) is measured into a 16-l. (4-gal.) stone jar, and for every liter there is added 163 cc. of dilute sulfuric acid (one volume of 95 per cent sulfuric acid and one volume of water, cooled to room temperature). The first part of the acid must be added cautiously on account of the liberation of carbon dioxide.

The tryptophane is precipitated by adding a solution of 200 g. of mercuric sulfate (Note 5) in a mixture of 1860 cc. of water and 140 cc. of 95 per cent sulfuric acid. After standing for twenty-four to forty-eight hours, the clear liquid is siphoned out and the yellow precipitate is filtered and washed (Note 6) with a solution of 100 cc. of concentrated sulfuric acid in 1.9 l. of distilled water containing 20 g. of mercuric sulfate, until the filtrate is colorless and Millon's test is atypical (Note 7); about 1.5 l. is necessary. The precipitate is washed with three successive 500-cc. portions of distilled water to remove most of the sulfuric acid.

The moist precipitate (120–130 g.) is suspended with mechanical stirring in 1.2–1.3 l. of distilled water, and a hot, 20 per cent aqueous solution of barium hydroxide is added until the mixture is permanently alkaline to phenolphthalein (about 120 cc. is required). A rapid stream of hydrogen sulfide is passed in with stirring until the mercury is completely precipitated (Note 8). The precipitate is filtered and washed with water until a sample of the washings gives a negative test for tryptophane with bromine water (Note 9). The barium is removed from the combined filtrate and washings by adding the exact amount of dilute sulfuric acid (Note 10) and filtering. The filtrate is concentrated under reduced pressure to about 80 cc.

The tryptophane is extracted from the aqueous solution by repeated shaking in a separatory funnel with 25-cc. quantities of *n*-butyl alcohol; water is added from time to time to keep the volume approximately constant (Note 11). The butyl alcohol extract is distilled under reduced pressure. After the water present has distilled, the tryptophane precipitates in the distilling flask and may cause bumping. When all the water has been removed, as is indicated by non-formation of drops on the side of the condenser, the distillation is stopped and, after cooling, the tryptophane is filtered and washed with a little fresh butyl alcohol. Such extractions and distillations are continued until the quantities of tryptophane obtained are negligibly small (Note 11).

The tryptophane so produced (7–8 g.) varies somewhat in quality in different runs. It is purified by recrystallization from 60 cc. of dilute alcohol (two volumes of 95 per cent alcohol to one volume of water), filtering from the hot solution an appreciable quantity of insoluble matter, and subjecting this to a second extraction with an additional 10 cc. of aqueous alcohol. The solution is decolorized by the addition of 1 g. of Norite and allowed to stand in the icebox; the silvery leaflets of tryptophane are filtered and washed successively with cold 70 per cent, 80 per cent, 95 per cent alcohol, and, finally, with a little ether. Less than half the tryptophane is obtained in each crystallization (Note 12). The yield of pure (Note 13) tryptophane is 4.0–4.1 g., together with under 0.1 g. of less pure product.

(*B*) *Tyrosine.*—The insoluble material (160–170 g.) obtained on filtering the digestion mixture (p. 612) is suspended in 320 cc. of water and 80 cc. of 36 per cent hydrochloric acid, and the mixture is boiled gently for thirty minutes (Note 14). After straining through cheesecloth, decolorizing with 6 g. of Norite (Note 15), and filtering hot (Note 16), the warm (60–70°) solution is shaken with three 20-cc. portions of benzene (Note 17) and heated to boiling (Note 18). A slight excess (120–150 cc.) of 28 per cent ammonia is cautiously added, and the mixture is allowed to stand overnight in the icebox. The crystalline product is then filtered and washed with three 40-cc. portions of ice-water. After drying, it weighs about 23 g. The mother liquor and washings are evaporated to about 200 cc., when a second crop is obtained, weighing slightly under 1 g.

The combined product is suspended in 400 cc. of water and dissolved by adding 8 g. of sodium hydroxide in 20–30 cc. of water (Note 19); 2 g. of Norite is added, and the solution filtered. The residue is washed on the funnel with 20–30 cc. of hot distilled water. The filtrate is heated to boiling (Note 18) and treated with 13 cc. of hydrochloric acid (Note 20), when crystallization usually begins. The mixture is then acidified to litmus with acetic acid (Note 21) and allowed to stand overnight in the refrigerator. The resulting tyrosine is filtered and washed with ice-cold distilled water (130–150 cc. is necessary) until the washings are free of chloride. The product is dried in air or in a vacuum oven. The yield is 17.0–18.2 g. of pure white, silky needles of tyrosine. A second crop (about 0.5 g.) of a slightly less pure product may be obtained on concentrating the mother liquor to about 120 cc.

2. Notes

1. Tryptic action is more rapid if all water used is at 37°. Distilled water is not necessary at this stage.

2. This is a considerable excess of sodium carbonate. Smaller quantities might be satisfactory.

3. The sodium fluoride probably inhibits the action of the oxidases.

4. This filtration may be slow. Büchner funnels of 20-cm. diameter are best used; the material from a single filling is allowed to suck dry and the filter paper then changed.

5. Approximately this quantity of mercuric sulfate is necessary to precipitate the tryptophane completely, as judged by the Hopkins-Cole glyoxylic acid test.

6. This washing is to remove tyrosine, which is precipitated as a mercury compound somewhat more soluble than the tryptophane precipitate. The mercuric sulfate addition tends to reduce the tryptophane solubility.

7. A persistent red color is always obtained in the filtrates, but the final color is distinctly different from that due to tyrosine.

8. Excess hydrogen sulfide must remain in the solution after standing.

A sample of the filtrate, after acidifying with acetic acid, should give a copious black preciptate with lead acetate.

9. The bromine water test is somewhat more satisfactory for pure tryptophane than the glyoxylic acid test. Hydrogen sulfide may interfere (owing to sulfur formation) and must be boiled out first. The solution to be tested must be acid with acetic acid.

10. This amount is best determined in a 20-cc. aliquot sample, employing 2 per cent sulfuric acid in a buret.

11. In checking, it was found satisfactory to extract in a continuous apparatus (Fig. 21). Extraction is continued until the liquid in the

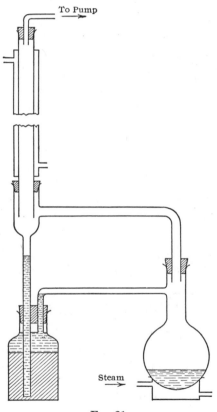

To Pump

Steam

FIG. 21

flask begins to bump on account of the separation of solid; a new charge of butyl alcohol is then employed, about five charges being necessary. This process is repeated until the residue after a three-hour period of extraction fails to give the red color, characteristic of tryptophane, with bromine water. The time necessary, of course, depends upon the rate of boiling; in checking, it was found to be twenty-eight to thirty hours.

12. The recrystallization of the crude tryptophane is an extremely troublesome process. Not only must a less soluble by-product be removed, but the mother liquors contain a more soluble, gummy impurity in considerable proportion. After collecting each crop, the mother liquor must be evaporated to a small volume on the steam bath and treated with a double volume of alcohol. This process is repeated until no further crystals are obtained, but only a gum.

13. The purity of the tryptophane has been checked by the optical rotation ($[\alpha]_D = -28$ to $-33°$) and by analysis for amino nitrogen (6.8–6.9 per cent) by Van Slyke's method.

14. The boiling acid solution hydrolyzes protein material that otherwise greatly retards filtration.

15. The Norite used in this preparation is insufficient to decolorize the solutions completely but gives a white final product.

16. All the filtrations in the purification of tyrosine, except possibly the last, are best done on a 20-cm. Büchner funnel. Whenever charcoal is used, kieselguhr may be employed to obtain a clear filtrate.

17. Benzene extracts traces of substances, probably fatty acid, that retard filtration and greatly alter the quality of the final product.

18. The tyrosine crystallizes in long, silky needles, easy to filter, if the solution is neutralized at the boiling point.

19. A small amount of flocculent impurity remains undissolved.

20. The hydrochloric acid is added to provide chloride ion as an index of complete washing.

21. Tyrosine is very slightly soluble in all concentrations of acetic acid. Therefore any excess of acetic acid does not redissolve the tyrosine.

3. Methods of Preparation

The above procedure for preparing tryptophane is an adaptation from the methods of Hopkins and Cole,[1] Dakin,[2] and Onslow.[3]

Tyrosine, as a primary product, may be readily prepared by hydrolyzing silk with hydrochloric acid, neutralizing the acid with sodium hydroxide, and finally acidifying with acetic acid.

[1] Hopkins and Cole, J. Physiol. **27,** 418 (1902).

[2] Dakin, Biochem. J. **12,** 302 (1918).

[3] Onslow, ibid. **15,** 392 (1921).

URAMIL

NH—CO
| |
CO   CHNO$_2$·3H$_2$O + 6HCl + 3Sn →
| |
NH—CO

NH—CO
| |
CO   CHNH$_2$ + 5H$_2$O + 3SnCl$_2$
| |
NH—CO

Submitted by W. W. Hartman and O. E. Sheppard.
Checked by C. S. Marvel and B. H. Wojcik.

1. Procedure

In a 5-l. flask are placed 100 g. (0.44 mole) of nitrobarbituric acid (p. 440) and 600 cc. of concentrated hydrochloric acid, and the mixture is heated on a boiling water bath. To the hot mixture are added 250 g. (2.1 gram atoms) of mossy tin and 400 cc. of hydrochloric acid over a period of about thirty minutes; heating is continued until there is no yellow color in the liquid (Note 1). The solution is treated with about 3 l. more of concentrated hydrochloric acid and heated until all the white solid is in solution. Norite is added, and the hot mixture is filtered through a sintered-glass funnel (Note 2). The filtrate is allowed to stand in an icebox overnight, and then the precipitate (Note 3) of uramil is collected on a filter and washed with liberal quantities of dilute hydrochloric acid and finally with water (Note 4). The filtrate is concentrated under reduced pressure to about 1 l. and cooled overnight. The additional uramil thus obtained is collected on a Büchner funnel and added to the first product (Note 5). The uramil is dried in a desiccator over concentrated sulfuric acid, and finally over 40 per cent sodium hydroxide to remove hydrochloric acid (Note 6).

Uramil is a fine, white powder which becomes pink to red on standing. The yield of a product which does not melt below 400° is 40–46 g. (63–73 per cent of the theoretical amount).

2. Notes

1. Nitrobarbituric acid forms a yellow aqueous solution, but, as the color is weak in concentrated hydrochloric acid solution, no trace of it should show at the end of the reaction.

2. If a sintered-glass funnel is not available, the solution may be filtered through a half-inch layer of decolorizing carbon on a double filter paper. After the uramil is once dissolved in the concentrated hydrochloric acid it comes out of solution very slowly, and, if filtered promptly, the solution may be cooled to 60–80° with little loss of product.

3. If, as happens occasionally, the uramil does not crystallize, the solution must be concentrated and cooled again.

4. Unless the product is washed thoroughly it will contain tin salts.

5. To test for uramil an ammoniacal solution is boiled in the air. A positive test is the appearance of a pink color which gradually grows deeper. The reaction proceeds more rapidly in the presence of mercuric oxide.

6. If the material is dried in the air or in an oven, a pink product is almost always obtained. The pink color is produced more rapidly if ammonia or amines are present in the air.

3. Methods of Preparation

Uramil has been obtained by boiling alloxantin with ammonium chloride; [1] by reduction of nitrosobarbituric acid or nitrobarbituric acid with hydrogen iodide; [2] and by reduction of alloxan phenylhydrazone with tin and hydrochloric acid. [3]

[1] Wöhler and Liebig, Ann. **26**, 309 (1838).
[2] Baeyer, ibid. **127**, 223 (1863).
[3] Kühling, Ber. **31**, 1973 (1898).

VERATRALDEHYDE

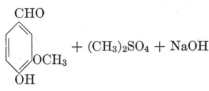

+ (CH₃)₂SO₄ + NaOH

→ $\rightarrow$ + CH₃SO₄Na + H₂O

Submitted by JOHANNES S. BUCK.
Checked by JOHN R. JOHNSON and H. R. SNYDER.

1. Procedure

IN a 3-l. three-necked flask a mixture of 182 g. (1.2 moles) of vanillin (Note 1) and 450 cc. of boiling water is heated on a steam bath. A solution of sodium hydroxide is prepared by dissolving 150 g. of U.S.P. sodium hydroxide in 200–300 cc. of water and diluting to 750 cc. A 360-cc. portion of this solution is heated to about 100° and added in one lot to the hot mixture of vanillin and water (Note 2). The flask is now fitted with a reflux condenser, mechanical stirrer, and 250-cc. separatory funnel.

Heating is continued on a steam bath, and 189 g. (142 cc.) of methyl sulfate (Note 3) is added through the separatory funnel just rapidly enough to maintain the gentle ebullition which starts after the addition of the first 10–15 cc. (Note 4). After the addition of all the methyl sulfate, which requires about one hour, the reaction mixture is heated for forty-five minutes longer and an additional portion of 39 g. (30 cc.) of methyl sulfate is added at the same rate as the first portion. At the end of this addition the reaction mixture should show an acid reaction to litmus (Note 5). After ten minutes' heating the reaction mixture is rendered slightly alkaline by the addition of about 60 cc. of the sodium hydroxide solution (prepared above), and another 39-g. portion of methyl sulfate is added. The alternate addition of sodium hydroxide solution and of methyl sulfate (39-g. portions) is repeated twice more, so that a total of 345 g. (2.7 moles) of methyl sulfate is added. The reaction mixture is then made strongly alkaline by the addition of 150 cc. of the sodium hydroxide solution and is heated for twenty minutes after the last addition of methyl sulfate. The reaction mixture is cooled rapidly to 25° (Note 6) with continued stirring, and the veratraldehyde is extracted with three 300-cc. portions of ether.

The combined ether extracts are dried over anhydrous magnesium sulfate, and the ether is distilled. There remains a slightly yellow oil which soon solidifies. The yield is 164–173 g. (82–87 per cent of the theoretical amount) of veratraldehyde melting at 43–44.5°. This product is sufficiently pure for many purposes, but it can be purified further with slight loss by distillation under diminished pressure. From 164 g. of the above product there is obtained 156 g. of pure veratraldehyde, boiling at 153°/8 mm. and melting at 46° (Note 7). Since this aldehyde is easily oxidized in the air it should be stored in a tightly corked or sealed container.

2. Notes

1. A good grade of vanillin (m.p. 81–82°) should be used.

2. The reactants are mixed while hot in order to avoid precipitation of the sodium salt of vanillin.

3. A practical grade of methyl sulfate gives satisfactory results. Methyl sulfate is very toxic; the vapor must not be inhaled or the liquid spilled. Ammonia is a specific antidote. The preparation is preferably carried out under a good hood.

4. The yield is usually lowered if this ebullition does not occur.

5. In order to obtain a good yield the reaction mixture must be allowed to become acid at several times. For this reason the last 156 g. of methyl sulfate is added in four small portions, and the reaction mixture is allowed to become acid after each portion is added. It is convenient to test the reaction of the solution by lowering through the condenser tube a glass tube with a piece of litmus paper attached at the end.

6. If the reaction mixture is cooled too slowly or below 25°, the veratraldehyde may solidify at this stage. If this occurs it is advisable to remelt the product before extracting with ether.

7. Veratraldehyde of sufficient purity for most synthetic purposes may be prepared rapidly and conveniently by the method of Barger and Silberschmidt, J. Chem. Soc. 133, 2924 (1928). In a 1-l. three-necked, round-bottomed flask (or a wide-mouthed bottle), fitted with a mechanical stirrer, a reflux condenser, and two separatory funnels, 152 g. (1 mole) of vanillin is melted by warming on a steam bath. With vigorous stirring, a solution of 92 g. (1.5 moles) of 90 per cent potassium hydroxide in 150 cc. of water is run in at the rate of two or three drops a second; twenty seconds after this is started, the addition of 160 g.

(120 cc., 1.25 moles) of methyl sulfate is begun at about the same rate. (Just before use, the methyl sulfate is washed with an equal volume of ice-water, followed by one-third its volume of cold, saturated sodium bicarbonate solution.) The external heating is stopped after a few minutes, and the mixture continues to reflux from the heat of reaction. A turbidity soon develops, and separation into two layers occurs after about one-half of the methyl sulfate has been added. The addition of both reagents should be completed in about twenty minutes.

The color of the reaction mixture, which is purplish brown at the beginning, changes abruptly to yellow towards the end. A temporary greenish yellow during the earlier part of the reaction indicates that the solution has become acid, and this condition should be corrected by increasing the rate of addition of the alkali. The final yellow color is permanent, and at the end the reaction mixture is alkaline to litmus.

The reaction mixture is transferred at once to a large beaker covered with a watch glass and allowed to cool without disturbance, preferably overnight. The hard, crystalline mass of veratraldehyde is removed, ground in a mortar with 300 cc. of ice-water, filtered with suction, and dried in a vacuum desiccator. The yield is 152–158 g. (92–95 per cent of the theoretical amount) of a product melting at 42.5–43.5°. This material gives a 90 per cent yield of the oxime (m.p. 89–90°) and an overall yield of 68–70 per cent of veratronitrile (m.p. 65–66°)—compare p. 622, below. The nitrile thus produced is satisfactory for the preparation of aminoveratrole (p. 44). (JOHN R. JOHNSON and H. B. STEVENSON, private communication.)

3. Methods of Preparation

Veratraldehyde has been prepared by treatment of veratrole with hydrogen cyanide in the presence of aluminum chloride;[1] by condensing veratrole with formylpiperidine and hydrolyzing the product;[2] and by methylating vanillin with methyl iodide,[3] methyl sulfate,[4] methyl p-toluenesulfonate,[5] or trimethylphenylammonium hydroxide.[6]

[1] Gattermann, Ann. **357,** 367 (1907).

[2] Akabori and Senoh, Bull. Chem. Soc. Japan **14,** 166 (1939) [C. A. **33,** 6270 (1939)].

[3] Tiemann, Ber. **8,** 1135 (1875); **11,** 663 (1878); Juliusberg, ibid. **40,** 119 (1907).

[4] v. Kostanecki and Tambor, ibid. **39,** 4022 (1906).

[5] Kanevska, Arch. Pharm. **271,** 462 (1933).

[6] Rodionow, Bull. soc. chim. (4) **45,** 116 (1929).

VERATRONITRILE

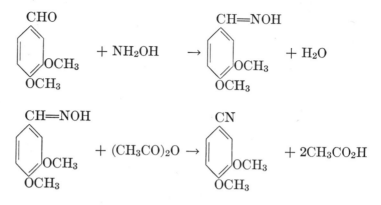

Submitted by J. S. Buck and W. S. Ide.
Checked by John R. Johnson and E. Amstutz.

1. Procedure

In a 1-l. round-bottomed flask, 83 g. (0.5 mole) of veratraldehyde (p. 619) is dissolved in 200 cc. of warm 95 per cent alcohol, and a warm solution of 42 g. (0.6 mole) of hydroxylamine hydrochloride (Org. Syn. Coll. Vol. I, **1941**, 318) in 50 cc. of water is added. The two solutions are mixed thoroughly, and a solution of 30 g. (0.75 mole) of sodium hydroxide in 40 cc. of water is introduced. After the mixture has stood for two and one-half hours at room temperature, 250 g. of crushed ice is added and the solution is saturated with carbon dioxide. This causes the separation of the aldoxime as an oil which solidifies on standing overnight in an ice chest (Note 1). The crystalline oxime is filtered with suction, washed thoroughly with water, and allowed to dry in the air. The yield of oxime is 88–89 g. (97–98 per cent of the theoretical amount).

The veratraldoxime is placed with 100 g. of acetic anhydride (94–96 per cent) in a 300-cc. round-bottomed flask provided with a ground-glass air condenser (Note 2), and heated cautiously. A vigorous reaction takes place, and when this occurs the flame is removed. After the reaction has subsided the solution is boiled gently for twenty minutes and then poured carefully, with stirring, into 300 cc. of cold water. The stirring is continued, and on cooling the nitrile separates in small, almost colorless crystals, which are filtered and dried in the air. The veratronitrile thus obtained is quite pure; it weighs 57–62 g. (70–76 per cent of the theoretical amount, based upon the veratraldehyde), and melts at 66–67°.

2. Notes

1. Occasionally the oxime does not solidify after standing overnight. If this happens it is advisable to separate the oily layer, treat with crushed ice, and induce crystallization by scratching. If a seed crystal of the oxime is available no difficulty is experienced.

2. If a cork or rubber stopper is used the product is likely to be colored.

3. Methods of Preparation

Veratronitrile has been obtained from 4-aminoveratrole by diazotization and treatment with cuprous cyanide,[1] and by heating veratryl-glyoxylic acid with hydroxylamine.[2] The general method employed above has been used for the preparation of a variety of substituted aromatic nitriles.[3]

[1] Moureu, Bull. soc. chim. (3) **15,** 650 (1896).

[2] Garelli, Gazz. chim. ital. **20, 700** (1890).

[3] Marcus, Ber. **24,** 3650 (1891); Pschorr, Ann. **391,** 33 (1912).

TYPE OF REACTION INDEX

This index lists most of the preparations contained in this volume in accordance with some general type of reaction. Only those preparations are included which can be classified under the selected headings with some definiteness. The arrangement of types and of preparations is alphabetical.

CONDENSATION (see, also, *Addition; Alkylation; Dehydration; Diazotization; Friedel-Crafts Reaction; Grignard Reaction; Rearrangements; Reduction*). The term "condensation" is used here in a restricted sense and applies to those reactions, intramolecular or intermolecular, which result in a carbon-carbon linkage, generally by the elimination of a simple inorganic molecule.

The subheadings illustrate the types of compounds formed. The reagents used are included in brackets before the names of the preparations.

(A) HETEROCYCLIC COMPOUNDS

(B) KETO-ALCOHOLS

(C) KETO-ESTERS

(D) MISCELLANEOUS

CONDENSATION—*Continued.*

DECARBOXYLATION

DEHYDRATION.
The reagent used is included in brackets before the names of the preparations.

DEHYDROGENATION (see, also, *Oxidation*).
The reagent used is included in brackets before the names of the preparations.

HYDROLYSIS. The subheadings illustrate the types of compounds hydrolyzed. The reagent used to effect hydrolysis is included in brackets before the name of the preparation.

TYPE OF COMPOUND INDEX

Preparations are listed, where possible, according to the group introduced. For example, *m*-bromonitrobenzene, prepared by the bromination of nitrobenzene, would be listed under halogen derivatives and not under nitro compounds. Unless otherwise stated, the ethylenic linkage is not considered a substituent. Salts are included with the corresponding acids and bases.

FORMULA INDEX

All preparations are listed in this index. The system of indexing is that used by *Chemical Abstracts*. The essential principles involved are as follows: (1) The arrangement of symbols in formulas is alphabetical except that in carbon compounds C always comes first, followed immediately by H if hydrogen is also present. (2) The arrangement of formulas is also alphabetical except that the number of atoms of any specific kind influences the order of compounds: *e.g.*, all formulas with one carbon atom precede those with two carbon atoms, thus: CH_2I_2, CH_3NO_2, CH_5N, C_2H_2O. (3) The arrangement of entries under any heading is strictly alphabetical according to the names of the isomers. (4) Inorganic salts of organic acids and inorganic addition compounds of organic compounds are listed under the formulas of the compounds from which they are derived. (5) Water of crystallization is not included in the formulas indexed.

640

FORMULA
INDEX

ILLUSTRATION INDEX

GENERAL INDEX

The name of a compound in SMALL CAPITAL LETTERS together with a number in **bold-faced type** indicates complete preparative directions for the substance named. The name of a compound in ordinary type together with a number in **bold-faced type** indicates directions, usually adequate but not in full detail, for preparing the substance named. A name in ordinary type together with a number in light-faced type indicates a compound or an item mentioned in connection with a preparation.